Towards Future Smart Power Systems with High Penetration of Renewables

Towards Future Smart Power Systems with High Penetration of Renewables

Emerging Technologies, new Tools, and Case Studies

Edited by

Marcos Tostado-Véliz

Department of Electrical Engineering, University of Jaén, Jaén, Spain

Ahmad Rezaee Jordehi

Department of Electrical Engineering, Rasht Branch, Islamic Azad University, Rasht, Iran

Seyed Amir Mansouri

Institute for Research in Technology, Comillas Pontifical University, Madrid, Spain

Andrés Ramos Galán

Institute for Research in Technology, Comillas Pontifical University, Madrid, Spain

Francisco Jurado Melguizo

Department of Electrical Engineering, University of Jaén, Jaén, Spain

ELSEVIER

ACADEMIC PRESS

An imprint of Elsevier

Academic Press is an imprint of Elsevier
125 London Wall, London EC2Y 5AS, United Kingdom
525 B Street, Suite 1650, San Diego, CA 92101, United States
50 Hampshire Street, 5th Floor, Cambridge, MA 02139, United States

ISBN: 978-0-443-29871-4

For Information on all Academic Press publications
visit our website at https://www.elsevier.com/books-and-journals

Publisher: Megan Ball
Acquisitions Editor: Edward Payne
Editorial Project Manager: Aleksandra Packowska
Production Project Manager: Surya Narayanan Jayachandran
Cover Designer: Mark Rogers

Typeset by MPS Limited, Chennai, India

Contents

Part 4 Optimization and control for power systems with renewables

List of contributors

José A. Aguado Department of Electrical Engineering, University of Málaga, Malaga, Spain

Roque Aguado Department of Electrical Engineering, University of Jaén, Jaén, Spain

Esteban Albornoz Faculty of Engineering, Department of Electrical, Electronics and Telecommunications Engineering (DEET), Universidad de Cuenca, Balzay Campus, Cuenca, Azuay, Ecuador

Mohamad Al-Widyan Amman Arab University, Department College of Engineering, Mubis, Amman, Jordan

Paul Arévalo Department of Electrical Engineering, University of Jaén, Jaén, Spain; Faculty of Engineering, Department of Electrical, Electronics and Telecommunications Engineering (DEET), Universidad de Cuenca, Balzay Campus, Cuenca, Azuay, Ecuador

Mohammad Bayat Department of Electrical Engineering, Institute of Advanced Technology, Arak University, Arak, Iran; Department of Electrical Engineering, Faculty of Engineering, Arak University, Arak, Iran

Darío Benavides Department of Electrical Engineering, University of Jaén, Jaén, Spain; Department of Electrical Engineering, University of Málaga, Malaga, Spain

José Pablo Chaves-Ávila Institute for Research in Technology (IIT), ICAI School of Engineering, Comillas Pontifical University, Madrid, Spain

Adrián Criollo Faculty of Engineering, Department of Electrical, Electronics and Telecommunications Engineering (DEET), Universidad de Cuenca, Balzay Campus, Cuenca, Azuay, Ecuador

Carlos Cruz Department of Electronics, University of Alcalá, Alcalá de Henares, Madrid, Spain

Antonio Escámez Department of Electrical Engineering, University of Jaén, Jaén, Spain

Ali Esmaeel Nezhad Department of Electrical Engineering, School of Energy Systems, LUT University, Lappeenranta, South Karelia, Finland

Juan Leonardo Espinoza Faculty of Engineering, Department of Electrical, Electronics and Telecommunications Engineering (DEET), Universidad de Cuenca, Balzay Campus, Cuenca, Azuay, Ecuador

Danny Espín-Sarzosa Department of Electrical Engineering, Universidad Técnica Federico Santa María, Valparaíso, Chile

Ali Asghar Ghadimi Department of Electrical Engineering, Institute of Advanced Technology, Arak University, Arak, Iran; Department of Electrical Engineering, Faculty of Engineering, Arak University, Arak, Iran

Farideh Ghanavati Department of Industrial Engineering and Management, University of Aveiro, Aveiro, Portugal

Guillermo Gutiérrez-Alcaraz Department of Electrical Engineering, Tecnológico Nacional de México/I.T. Morelia, Morelia, Michoacán, Mexico

Pedro H.J. Nardelli Department of Electrical Engineering, School of Energy Systems, LUT University, Lappeenranta, South Karelia, Finland

Víctor Hinojosa Department of Electrical Engineering, Universidad Técnica Federico Santa María, Valparaíso, Chile

Mohammad Sadegh Javadi Institute for Systems and Computer Engineering, Technology and Science (INESC TEC), Porto, Portugal

Saeid Jowkar Department of Electrical, Electronic, and Information Engineering, University of Bologna, Bologna, Italy

Francisco Jurado Melguizo Department of Electrical Engineering, University of Jaén, Jaén, Spain

João Mello Institute for Systems and Computer Engineering, Technology and Science, Porto, Portugal; Faculty of Engineering, University of Porto, Porto, Portugal

Patricio Mendoza-Araya Department of Electrical Engineering, Universidad de Chile, Santiago, Chile

Andrés Montenegro Department of Electrical and Electronic Engineering, Universidad Nacional de Colombia, Bogotá, Colombia

Sahar Moslemi Department of Electrical Engineering, Science and Research Branch, Islamic Azad University, Tehran, Iran

Ra'ed Nahar Myyas Department of Electrical Engineering, University of Jaén, Linares, Spain; Jordan Design & Development Bureau (JODDB), R&D Department, Amman, Jordan; Amman Arab University, Department College of Engineering, Mubis, Amman, Jordan

Morsy Nour Institute for Research in Technology (IIT), ICAI School of Engineering, Comillas Pontifical University, Madrid, Spain; Department of Electrical Engineering, Faculty of Energy Engineering, Aswan University, Aswan, Egypt

Danny Ochoa-Correa Faculty of Engineering, Department of Electrical, Electronics and Telecommunications Engineering (DEET), Universidad de Cuenca, Balzay Campus, Cuenca, Azuay, Ecuador

Abouzar Samimi Department of Electrical Engineering, Arak University of Technology, Arak, Iran

Daniel Sánchez-Lozano Department of Electrical Engineering, University of Jaén, Jaén, Spain

Toktam Tavakkoli Sabour Department of Electrical and Computer Engineering, Texas Tech University, Lubbock, TX, United States

Matteo Troncia Institute for Research in Technology (IIT), ICAI School of Engineering, Comillas Pontifical University, Madrid, Spain

Felipe Valencia-Arroyave Electricity and Electronics Insitute, Faculty of Engineering Sciences, Universidad Austral de Chile, Valdivia, Los Ríos, Chile

David Vera Department of Electrical Engineering, University of Jaén, Jaén, Spain

Edisson Villa-Ávila Department of Electrical Engineering, University of Jaén, Jaén, Spain; Faculty of Engineering, Department of Electrical, Electronics and Telecommunications Engineering (DEET), Universidad de Cuenca, Balzay Campus, Cuenca, Azuay, Ecuador

Michael Villa-Ávila Faculty of Engineering, Department of Electrical, Electronics and Telecommunications Engineering (DEET), Universidad de Cuenca, Balzay Campus, Cuenca, Azuay, Ecuador

Jose Villar Institute for Systems and Computer Engineering, Technology and Science, Porto, Portugal

Alaa Zetawi Department of Electrical Engineering, University of Jaén, Linares, Spain

Preface

Aligned with the Green Deal objectives, the decarbonization of the energy sector becomes a critical issue. Existing infrastructures will dramatically evolve to accommodate renewable sources massively. This paradigm constitutes a new framework for existing power and energy systems, which calls up for upgrading existing infrastructures, enabling new communication channels, integrating new generation and storage technologies, providing more flexibility from consumers and developing new markets and businesses models. These new concepts suppose a formidable and exciting challenge. This book aims at collecting some of the recent developments toward the massive integration of renewable technologies in energy systems, especially focusing on new technologies and management models. Some real cases are reviewed, focusing on those projects where the integration of renewable sources supposes a major challenge either by social, economic, or legislative barriers. This book also studies different markets and businesses models as well as shedding light into the different techniques and tools necessary to manage and control the energy system under the massive integration of renewable sources.

To ease the readability of the book, it is divided into four parts, each one devoted on a specific topic.

Part 1 reviews emerging technologies for generation and storage, based on renewable sources, as well as new energy vectors or coupling technologies among carriers. Chapter 1 dedicates to hydrogen energy, more particularly, its for providing storage capability is explained. Chapter 2 studies the gasification process of biomass, as a way to provide clean energy in isolated and rural areas. Chapter 3 reviews the different processes and technologies that allow to obtain energy from waste products. Chapter 4 studies an emerging nexus between water and electricity carriers, as well as the new technologies and infrastructures involved in the process. This part closes with Chapter 5, whereby heating production from biomass is analyzed and discussed.

Part 2 focuses on upcoming markets and businesses structures, as a way to enabling the massive integration of clean technologies. Chapter 6 includes a review of the concept of energy community, which empowers final users and allows best leveraging of own resources. Chapter 7 reviews the concept of microgrids and introduces the definition of microgrid clusters, as an aggregation of microgrids pursuing common objectives. Chapter 8 focuses on peer-to-peer energy transactions as well as blockchain technologies and protocols, which are essential for energy trading among users. Finally, Chapter 9 studies new markets models for enabling the active participation of storage assets, resulting essential to provide energy arbitrage and integrate renewable sources economically.

Part 3 collects some key real-life cases and projects and provide insights in their more significant features. Chapter 10 focuses on the case of Ecuador. More specifically, this chapter reviews the measures and projects launched in this country to electrify some isolated areas like Galapagos Islands or Amazonas, through renewable sources. Chapter 11 provides a state-of-art of different power smoothing techniques and their application on real cases, as well as their importance to ensure network stability in the presence of multiple renewable generation technologies. Chapter 12 extends some previous chapters by reviewing different real cases and energy community projects worldwide. Lastly, Chapter 13 focuses in the case of Jordan and reviews the barriers and countermeasures put on practise in this country to increase the renewable penetration into its existing infrastructure.

Part 4 provides insight into the problem of controlling and ensuring reliability of the energy system under the massive integration of renewable sources. Specifically, Chapter 14 reviews the existing tools and models for energy management in energy communities, establishing their particular features and difficulties for implementation. Chapter 15 revisits the power flow analysis for microgrids, which presents some particularities with respect to the conventional tool widely employed in transmission and distribution networks. Chapter 16 introduces some classical and emerging techniques for forecasting in power systems, highlighting their crucial role for ensuring the reliability of existing networks in the presence of renewable technologies. Finally, Chapter 17 focuses on the energy management problem involving storage technologies, including electric vehicles as a way of enabling energy arbitrage through on-board storage systems.

Acknowledgment

The editors thank the authors for all their valuable contributions. Without them, this book would not have been possible. This is your book. Thanks to Elsevier for the opportunity to collaborate with this prestigious editorial and to all the editorial team for their help and support during the entire process. Special thanks to Aleksandra Packowska and her invaluable help with the submission process.

Part 1

Emerging technologies and energy vectors

Hydrogen for storage and renewable integration

1

Darío Benavides[1,2], *Edisson Villa-Ávila*[1,3], *Adrián Criollo*[3] *and José A. Aguado*[2]

[1]Department of Electrical Engineering, University of Jaén, Jaén, Spain, [2]Department of Electrical Engineering, University of Málaga, Malaga, Spain, [3]Faculty of Engineering, Department of Electrical, Electronics and Telecommunications Engineering (DEET), Universidad de Cuenca, Balzay Campus, Cuenca, Azuay, Ecuador

Introduction

Background

Currently, hydrogen (H_2) has versatile applications and can significantly contribute to a clean and secure energy future. The production of hydrogen consists of the process of electrolysis with water, and it decomposes into hydrogen and oxygen. When the electrical energy used for electrolysis comes from renewable energy sources (RES) like wind or solar, the resulting hydrogen is termed "green." However, there is a considerable amount of photovoltaic and wind energy that is not usable in terms of constant demand. Therefore, the use of hydrogen has been proposed as a potential in energy storage systems. This technology is promising for clean energy conversion with an efficiency of 60% of hydrogen energy into electricity (Hassan et al., 2023). Likewise, hybrid photovoltaic/wind systems offer reliability, cost savings, and environmental benefits (Arsad et al., 2022). Researchers continue to explore innovative approaches to optimize their design and maximize their impact (Arsad et al., 2023; Lin et al., 2020; Stetson & Wieliczko, 2020; Tahan, 2022). The influence of different parameters on the performance of the electrolyzer is required to know and propose efficient operation of the system (Falcão & Pinto, 2020). In this chapter, the benefits of hydrogen production through the electrolysis process are analyzed. Modeling and simulation are considered in a case study that integrates a microgrid with storage systems. Finally, some resource optimization criteria are presented to improve the efficiency of hydrogen production.

Motivation

The production of hydrogen through the electrolysis process is currently considered a promising technology in the generation of green fuel. However, it is important to take into account the high cost of its production. Consequently, the use of RES has been studied to improve their viability in the energy sector. This integration into microgrids offers several benefits for green hydrogen production (GH_2P). The storage of energy converted into hydrogen in different seasons is also considered to

Towards Future Smart Power Systems with High Penetration of Renewables. DOI: https://doi.org/10.1016/B978-0-443-29871-4.00001-4

have a promising future in the energy sector due to its nondegradative characteristics compared to other types of batteries, where long-term production costs can be optimized. The current points of interest are highlighted below.

Sustainability: GH_2P supports decarbonization efforts by leveraging renewable energy.

Versatility: Hydrogen can be used in various applications, including energy storage, transportation, and industrial processes.

Electrification: It facilitates the broader adoption of RES by providing a stable and scalable energy storage solution.

Challenges

Despite its advantages, GH_2P faces significant challenges:

High energy consumption: Electrolysis requires substantial electrical energy, making efficiency a critical factor.

Safety: Handling and storing hydrogen safely remains a key concern.

Cost: The current high costs of electrolyzers and renewable energy infrastructure pose economic barriers.

Simulation needs: Developing high-fidelity models to accurately simulate hydrogen production and integration within microgrids is essential.

This technology offers significant advantages such as sustainability and versatility for electrification. However, it faces challenges such as high energy consumption, safety issues in handling hydrogen, and substantial costs. A major challenge affecting the cost per unit of energy in hydrogen is low energy efficiency (Egeland-Eriksen et al., 2021). Therefore, in this study, some issues related to the models for hydrogen generation and adjustment of experimental parameters in the case of study will be analyzed.

Review literature

Hydrogen generation is crucial for emerging domestic sectors, which consider both conventional and renewable hydrogen essential for intermediate and end uses. The viability of the H2@Scale initiative hinges on hydrogen demand from existing markets, such as petroleum refining and ammonia production, as well as the expansion of new markets. If currently imported ammonia (NH_3) were instead produced domestically, production could increase by approximately 30% without the corresponding rise in domestic NH_3 demand. The Energy Information Administration (EIA) projects this trend to continue, with distillation capacity increasing by only 9% from 2016 to 2050 (Elgowainy et al., 2020).

A study examines the potential role of hydrogen and fuel cell (FC) technologies in the future energy system, focusing on reducing CO_2 emissions and transitioning to RES (Norouzi, 2021). It highlights the challenges and opportunities in the hydrogen economy, including the need for increased renewable distribution funding and infrastructure development. The study also addresses the growth in global energy consumption, the importance of reducing greenhouse gas emissions, and the potential for hydrogen to play a significant role in transportation and energy storage. Overall, the study emphasizes the necessity of transitioning to a more sustainable energy system to combat climate change and ensure energy security.

Kaniz Farhana et al. reported that optimization can reduce total annualized costs by 14.9%–15.1%, CO_2 emissions by 4.4%–5.2%, and the cost of hydrogen production to \$2.4–\$2.9 kg^{-1} (Farhana et al., 2024). On the other hand, the results of a study in Zhangbei (China) show that the proposed system has the potential for the production of 931.39 kg of hydrogen per year with an energy efficiency of around 16.03% (Liu et al., 2023). Methane decomposition becomes economically viable compared to steam methane reforming (SMR) when carbon production costs reach €280–€310 per ton. Electrolysis incurs higher costs due to electricity use, whereas using methane in SMR lowers production costs. Consequently, the cost of hydrogen production significantly increases by 20%–55% when including storage and transportation in the overall plant costs. Similarly, Yang et al. (2019) reported on a hierarchical self-regulation control for the economic operation of an AC/DC hybrid microgrid with a hydrogen energy storage system. This method minimizes energy storage system operation costs, reducing them by 27.71% compared to droop control. In addition, it decreases direct current (DC) bus voltage and alternating current (AC) frequency fluctuations. The control method improves system stability and inertia and determines critical power for charging/discharging based on energy costs.

Niknam et al. (2015) studied the impact of hydrogen production and thermal energy recovery of proton exchange membrane FC power plant (PEMFCPP) on optimal management of renewable microgrids. The interaction between a PEMFCPP and a utility is explored in a stochastic environment, considering thermal energy recovery and hydrogen production. Three different operation statuses are analyzed, showing that utilizing these capabilities reduces total costs. The study investigates the effects of hydrogen production and thermal energy recovery on optimal energy management of a microgrid with various RES. A metaheuristic method based on the bat algorithm is used to find the optimal solution to the energy management problem. The utilization of renewable microgrids and RES in competitive markets is highlighted as beneficial from both economic and environmental perspectives. Previous works in the field have been criticized for neglecting uncertainties associated with problem parameters.

Additionally, Akbari-Dibavar et al. (2020) proposed a net zero-emission model for the coordinated operation of integrated carbon free energy systems in a micro power grid, including hydrogen energy systems, wind turbines, and PV panels for clean energy production. The study focuses on optimal energy management in a deregulated environment, using robust optimization to handle wind power uncertainties. The goal is to minimize energy costs and reduce harmful emissions, with a focus on FC and hydrogen storage systems for net-zero emissions. Energy storage systems are proposed to address electricity supply volatility, with hydrogen playing a key role in producing clean electrical energy.

In a related project, Sproul et al. successfully demonstrated the seamless integration of hydrogen production and renewable electricity generation, highlighting the broad array of services a behind-the-meter microgrid can deliver (Sproul et al., 2021). This was achieved through the combination of a grid-forming battery energy storage system and a real-time microgrid control system, supporting an industrial facility in achieving lower energy costs, lower greenhouse emissions, and improved reliability.

Likewise, a novel approach using a capacitive energy buffer connected to the inverter DC bus to manage the intermittency of renewable energy microgrids was reported by Ganeshan et al. (2019); an experimental hydrogen-based energy storage system with a ripple energy buffer significantly reduced the required DC bus capacitance value and maintained a smooth sinusoidal output voltage. The goal is to balance the double-frequency components and reduce the DC link capacitor size. Additionally, the ripple buffer phase leg current controller uses a proportional-resonant strategy to track the ripple current component.

Furthermore, Fu et al. discussed the development of distributed generations in microgrids, the integration of energy storage devices, and the implementation of a model predictive control controller for optimal scheduling in a power to heat and hydrogen microgrid system (Fu et al., 2017). The study shows the potential for improved system efficiency and emphasizes the importance of balancing heat and hydrogen production.

A sizing method was developed using a genetic algorithm and rule-based strategy to optimize component sizes for a 100% renewable energy power station (Li, 2020). Simulation results showed the importance of hydrogen tanks in storing energy, and a rule-based operation strategy was presented for the hydrogen-centric storage system.

A novel unit commitment strategy for an islanded microgrid was used using model predictive control with wavelet neural network forecasts. This strategy focuses on reducing unnecessary FC use and increasing electrolyzer use to minimize solar power curtailment, thereby improving overall microgrid durability and efficiency (Morin et al., 2019).

A two-stage coordinated operational strategy for an emission-free microgrid with hybrid hydrogen-battery energy storage is proposed in Fan et al. (2022). This strategy optimizes power exchange with the main grid, FC output, and electrolysis device input under uncertain conditions, showing optimized operational costs and robustness.

On the other hand, various energy management systems based on hysteresis band control and adaptive neuro-fuzzy controllers for electric-hydrogen hybrid systems have been studied (Pei et al., 2022). They emphasize the need for efficient power distribution, energy management, and coordination in renewable energy systems, highlighting the challenges and strategies for optimizing the efficiency and performance of hydrogen production units.

A multienergy microgrid was optimized with a waste processing capacity for electricity-hydrogen charging services, proposing an electricity-hydrogen and microgrid control system based on renewable energy supply, waste disposal, and power grid regulation (Yan et al., 2022). This system includes pyrolysis furnaces, FCs, and energy storage devices to meet energy demands and provide regulation capacity, showing economic benefits and improved waste utilization. Also, a study presents daily seasonal green hydrogen storage and hydrogen refueling station with promising results that reduces costs, increases profits, and efficiently manages fluctuations of renewable energy (Hemmati et al., 2024).

A hydrogen FC electrolyzer energy storage system that can act as a dynamic load compensator, improving the operation of the FC and storing maximum PV power in the storage tank, despite incurring a 30% energy loss during operation (Ganeshan et al., 2017). Islam (2020) discussed the use of hydrogen energy and methylcyclohexane for power generation and energy storage in Bangladesh. The proposed microgrid system utilizes wind energy for hydrogen production, with methylcyclohexane used for long-term energy storage, aiming to reduce fossil fuel consumption and ensure a stable energy supply.

Fundamental concept and structure of green hydrogen production

What is a hydrogen electrolyzer?

A hydrogen electrolyzer is an electrochemical apparatus that uses electrical energy to separate water into hydrogen and oxygen. These devices are integral to hydrogen production and form a component of the green energy production–storage distribution system when integrated with RES, hydrogen storage units, and FC technologies, such as those in FC electric vehicles (Hydrogen Electrolyzer, 2024).

Currently, there are three primary types of hydrogen electrolyzers: alkaline, polymer electrolyte membrane (PEM), and solid oxide, each distinguished by the type of electrolyte material used (Arsad et al., 2023). The PEM hydrogen electrolyzer operates by breaking down water with a semipermeable membrane that permits proton movement while inhibiting electron flow, which is why it is also termed a proton exchange membrane electrolyzer.

In operation, water is introduced into the electrolyzer at the anode, as seen in Fig. 1.1; DC across the electrodes causes the oxygen in the water to release electrons, producing protons, electrons, and oxygen at the anode (+). These protons traverse through the membrane to the cathode (−), where these fuse with electrons to generate hydrogen.

The principle of chemical reaction Eq. (1.1) at the anode and the cathode during the electrolysis of water is to produce hydrogen and global reaction, respectively (Falcão & Pinto, 2020; Hydrogen Electrolyzer, 2024):

$$H_2O \rightarrow 2H^+ + \frac{1}{2}O_2 + 2e^- \text{Anode } 2H^+ + 2e^- \rightarrow H_2 \text{ Cathode} \cdots \quad H_2O \rightarrow H_2 + \frac{1}{2}O_2 \tag{1.1}$$

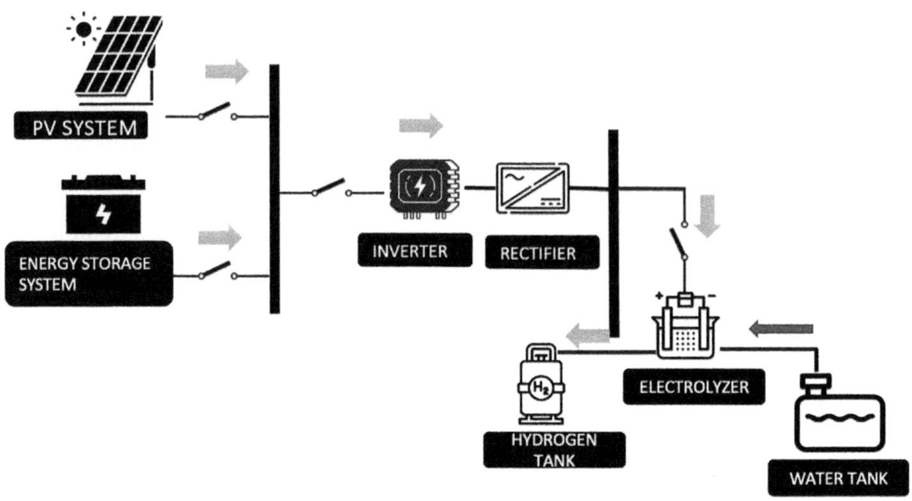

Figure 1.1 Representation of green hydrogen production.

In modeling, an electrolysis stack comprises several series-connected individual electrolyzer cells, where it can specify the potential of hydrogen (pH) parameter assumption in two ways:

pH-dependent state: In this approach, the pH of the electrolyte is treated as a variable that changes based on the electrochemical reactions and other process conditions. This is more realistic for detailed simulations where pH variations affect the performance and safety of the electrolysis process.

Ideal state with constant pH: Here, the pH is assumed to be constant throughout the operation. This simplification is useful for more straightforward models where pH impacts are considered negligible or when the focus is on other aspects of the stack's performance.

Equation–electrolyzer model

The electrolyzer block calculates the electrical power Eq. (1.2) by using Schalenbach et al. (2018):

$$P_E = \frac{v^2}{R_T} \tag{1.2}$$

where v is the voltage and R_T is the total resistance Eq. (1.3). The R_T, is defined by:

$$\frac{1}{R_T} = \frac{1}{R_{el}} + \frac{1}{R} \quad R_{el} = \frac{\rho(pH) \times x}{A} \tag{1.3}$$

Where R is the external resistance, A value is a resistive term obtained from the cross-sectional transport area, x distance anode–cathode, $\rho(pH)$ the electrical resistivity of the solution. The energy per mole stored over the electrical load facilitates water electrolysis and can be computed using Eq. (1.4) (Schalenbach et al., 2018):

$$G_E = \frac{P_E}{\frac{1\text{mol}}{s}} \tag{1.4}$$

To perform electrolysis, the electrolyzer needs a minimum level of energy Eq. (1.5):

$$G_{\min} = \Delta H - \frac{T}{T_o} T_{\Delta s} \tag{1.5}$$

where ΔH is the water enthalpy, $T_{\Delta s}$ is the reaction entropy, T is the water temperature, and To is the reference temperature.

If there is not enough electrical energy to split the water molecules, the electrolyzer will not generate hydrogen. On the other hand, the electrolyzer block determines the rates of electrons, moles, and hydrogen production, as shown in Eqs. (1.6) and (1.7).

$$\text{mol}_e = \begin{cases} \dfrac{\mu(G_E - G_{\min})}{e_V N_A} & G_E > G_{\min} \\ 0 & G_E \leq G_{\min} \end{cases} \tag{1.6}$$

$$\text{mol}_{\text{H2}} = \begin{cases} \dfrac{N_c mol_e}{2} & G_E > G_{\min} \\ 0 & G_E \leq G_{\min} \end{cases} \tag{1.7}$$

where N_A is the Avogadro constant, e_V is the energy per electron, N_c is the number of cells, and μ is the temperature-dependent efficiency of the electrolysis.

The current passing through the electrolyzer tank is then determined as follows, where F is the Faraday constant, as shown in Eq. (1.8).

$$i = \text{mol}_e \times F \tag{1.8}$$

The electrolyzer block computes the mass rates of water consumption and hydrogen production using the following Eqs. (1.9) and (1.10), respectively:

$$\dot{m}_{H20} = mol_{H2} M_{H20} \tag{1.9}$$

$$\dot{m}_{H2} = mol_{H2} M_{H2} \tag{1.10}$$

where M_{H20} is the molar mass of water and M_{H2} molar mass of hydrogen.

Case study modeling

This section presents the comparison of the electrolyzer model from the University of Cuenca's Microgrid Laboratory with the simulation-based model using MATLAB® Simulink. The data and parameters are presented in detail in Table 1.1. The mathematical modeling and the parameters of GH_2P were based on the characteristics of the renewable generation equipment (Espinoza et al., 2018). The microgrid features a 35 kWp on-grid photovoltaic solar plant connected to AC bus. It includes an ELS500 electrolyzer producing 500 L of hydrogen per hour, with a 30-L water tank, and four hydrogen storage tanks with a total capacity of 200 L at 35 bar pressure. Additionally, it has a hydrogen FC connected to a battery bank (48 V, with an external power source for the FC and voltage reference) and a 3 kW/220 V single-phase/60 Hz power converter. Likewise, the FC technical specifications are presented in Table 1.2.

An experimental stage of the electrolyzer of case study Fig. 1.2 is considered. The laboratory mainly consists of a photovoltaic generation system (35 kW), a mini

Table 1.1 System and control parameter values.

Description	Symbol	Parameter value	Unit
Electrolyzer power	$PELy$	3	kW
Stack cell number	NC	72	
Active cell area	A	30	cm^2
Starting hydrogen pressure at the cathode	$p^o_{H2} = pc$	1.013×10^5	Pa
Starting anode pressure	ρ_A	1.013×10^5	Pa
Final hydrogen tank pressure	ρ^1_{H2}	2×10^6	Pa
Water viscosity	μ	0.011	g/cm.s
Dry membrane thickness	ϕ	1.27×10^{-2}	cm
Membrane humidification degree	λ	12.7	
Ohmic resistance	R_{ele}	3.5×10^{-5}	Ω
Anode thickness	ϕ_A	$4.15. \times 10^{-2}$	cm
Cathode thickness	ϕ_C	4.15×10^{-2}	cm
Cathode charge transfer coefficient	α_C	$0.4 - 0.5$	
Anode charge transfer coefficient	α_A	$0.3 - 0.6$	
Cathode exchange current density	$i_{C,o}$	$10^{-4} - 10^{-2}$	A/cm^2
Anode exchange current density	$i_{A,o}$	$10^{-11} - 3^{-2}$	A/cm^2
Starting operating temperature	T_0	313	K
Final operating temperature	T_f	343	K
Stoichiometric factor Sh	Sh	10	
Hydrogen tank volume	V_{H_2}	0.5	m^3
Oxygen tank volume	V_{H_2}	0.5	m^3
Simulation time	V_{H_2}	0.5	m^3

Table 1.2 PEM fuel cell technical specifications (TW Horizon Fuel Cell Technologies, 2024).

Description	Parameter value	Unit
Type of fuel cell	1	PEM
Number of cells	72	
Rated power	3000	W
Performance	43.2 V@70 A	
H2 supply valve voltage	12	V
Purgin valve voltage	12	V
Blower voltage	12	V
Reactants	Hydrogen and air	
External temperature	5–30	°C
Max stack temperature	65	°C
H_2 pressure	0.45–0.55	bar
Hydrogen purity	> 99.995%	dry H_2
Humidification	Self-humidified	
Cooling	Air (integrated cooling fan)	
Weight	15	kg
Controller	2.5	kg
Dimension	41.8 × 35 × 18.3	cm
Flow rate at max output	39	L/min
Start-up time	<30 (at temperature ambient)	s
Efficiency of stack	40% @ 43.2	V
Low voltage shut down	36	V
Over current shut down	90	A
Over temperature shut down	65	°C
External power supply	13–5	V-A

PEM, Polymer electrolyte membrane.

wind farm (15 kW), storage systems with valve-regulated lead-acid gel battery bank (15 kWh), and a programmable electrical load that allows for the creation of step scenarios every 5 kW (Espinoza et al., 2018). The microgrid systems are connected at a point common connection on the bus in island mode as described below.

Solar power generation: The photovoltaic generation system comprises three units with capacities of 15, 15, and 5 kWp, respectively, as illustrated in Fig. 1.2. Collectively, these units provide a nominal power output of 35 kWp. Each PV system is independently connected to inverters equipped with maximum power point tracking control. The bus configuration permits operation in either grid-connected mode or island mode (Espinoza et al., 2018).

Energy storage system: A lead battery bank composed of 84 cells placed in series, A600/Solar (1130 Ah), each connected to a double flow inverter, allows for the charging and discharging of energy with the electrical grid (Criollo et al., 2024; Espinoza et al., 2018).

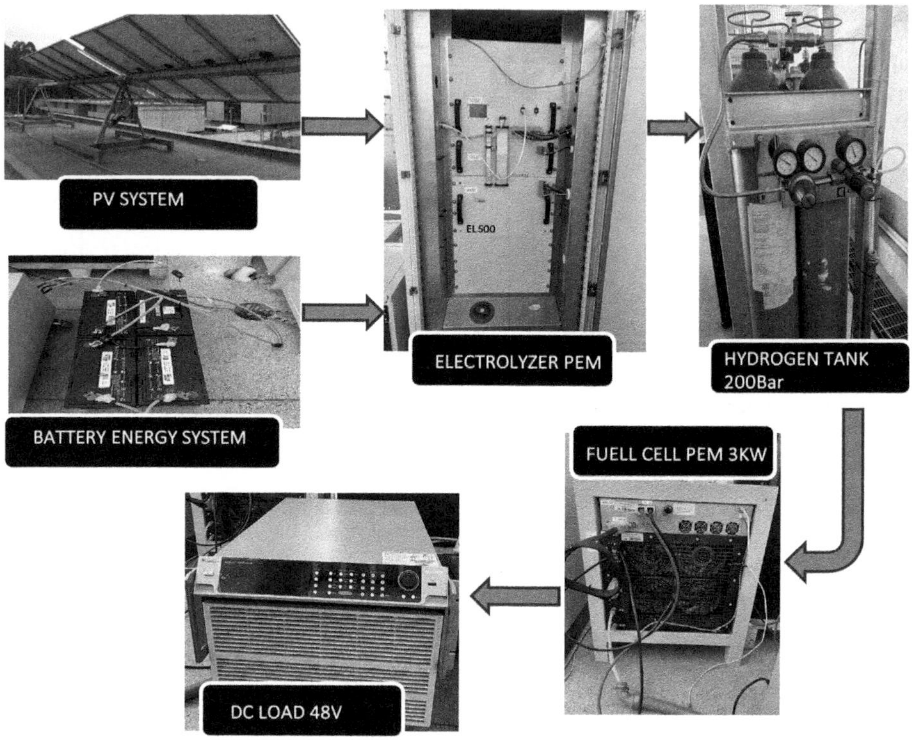

Figure 1.2 Experimental stage with Microgrid Laboratory Equipment of the Centro Científico Tecnológico y de Investigación Balzay (CCTI-B).

Simulation results and discussion

Fig. 1.3 represents a scenario with generation solar and energy storage (batteries). This model was developed to demonstrate that an electrolyzer can be integrated into a DC bus microgrid GH_2P system with a solar array and an energy storage system.

Fig. 1.4 illustrates a mapping of the variation in electrical resistance from 100% (representing 0.25 ohms) to 40% (representing 0.14 ohms). The graph shows that as the ohmic value decreases, the voltage in the DC microgrid also decreases. However, below 14% of the ohmic value, the equations do not converge and the simulation stops.

In the scenario of current consumption in the DC microgrid, the representation of current consumption can be analyzed in Fig. 1.5. Essentially, when the ohmic value of the resistance is lower (40% of the value in Eq. [1.3]), there are current peaks around 230 A with an average of 49 A. This will allow us to forecast protection calculations and short circuit studies for future research.

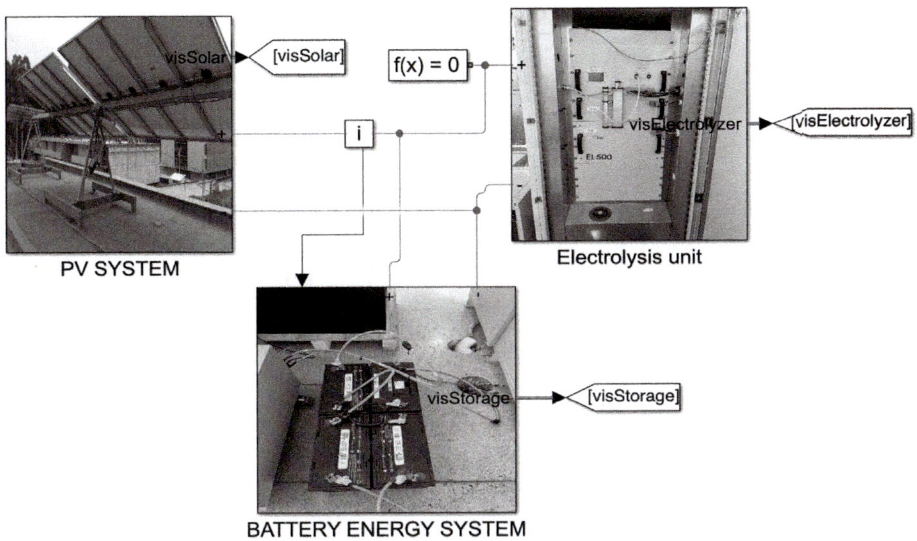

Figure 1.3 Simulink model electrolyzer operating in the DC microgrid.

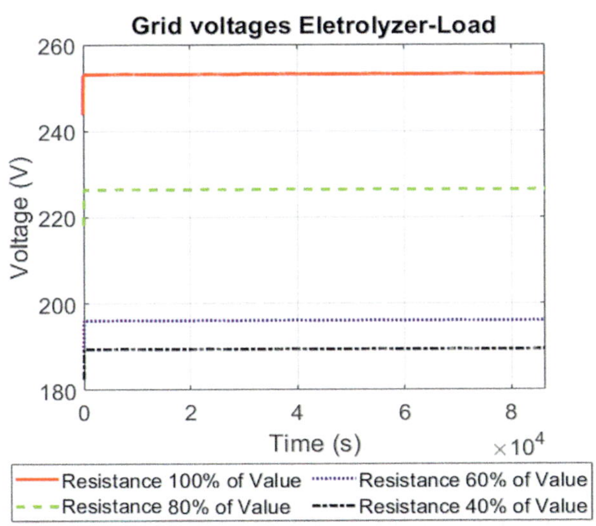

Figure 1.4 Grid voltage in DC microgrid with an electrolyzer as load.

The state of charge of the storage system (lead-acid batteries) is a crucial variable in determining the condition of the batteries. In the study, a condition is maintained in which the SOC should never fall below 30% to guarantee the longevity of the batteries. According to Fig. 1.6, when the resistance is at 100% and the

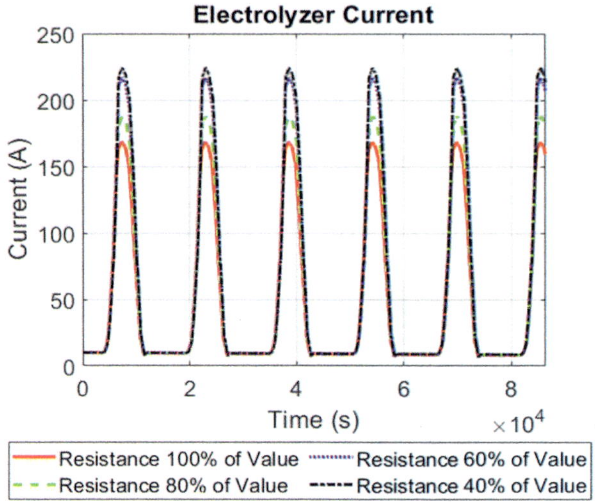

Figure 1.5 Electrolyzer current as a function of varying resistance.

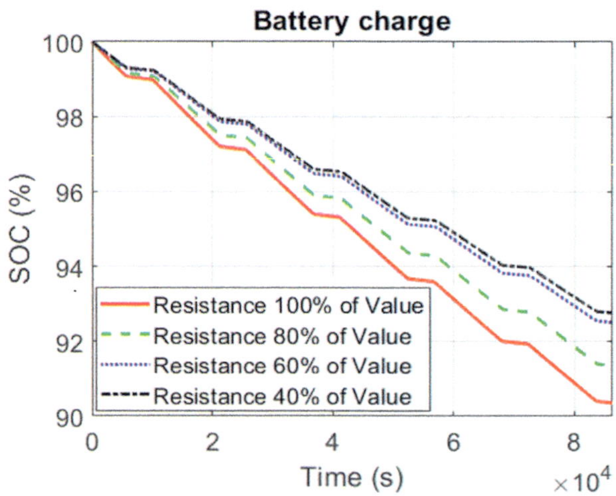

Figure 1.6 SOC battery process.

electrolyzer has been running for 1 hour, the SOC is at 99.41% of its nominal capacity. In the scenario where the ohmic value is reduced to 40%, the SOC is 99.55%. The figure also allows us to project SOC values when the electrolyzer has been operating continuously for one day, with SOC results of 90.35% and 92.76%, respectively.

Regarding hydrogen production, Fig. 1.7 shows that as the ohmic resistance value decreases to 40%, hydrogen production increases. Analyzing daily generation,

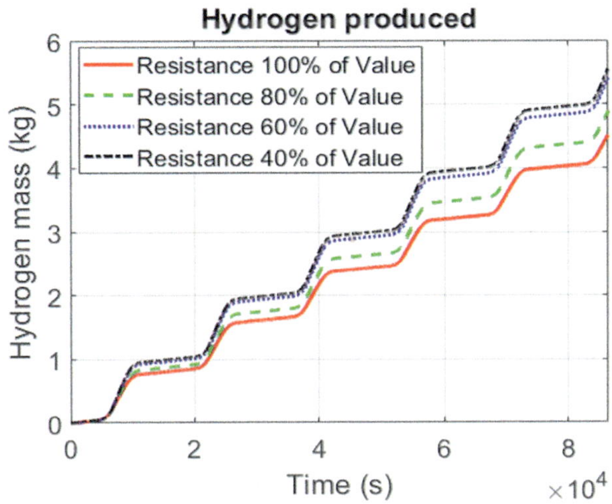

Figure 1.7 Hydrogen production per day.

we observe that at 40%, there is a production of 5.50 kg/day, whereas at 100% resistance, the production is 4.48 kg/day.

One of the challenges is to reduce energy consumption for generating green hydrogen from renewable sources. In this study, within the DC microgrid, Fig. 1.8 shows that when the ohmic value is reduced to 40%, the energy consumption is 49 kWh/kg, whereas at 100% resistance, the energy consumption increases to 60 kWh/kg.

Optimization of parameters to reduce energy consumption

Below is a mapping of the most important resources that will allow us to reduce kWh energy consumption. To reduce power consumption in the production of hydrogen through electrolysis, the analysis should focus on optimizing the following parameters:

- Resistance: Lowering the resistance in the system can significantly reduce power consumption. This can be achieved by improving the conductivity of the electrolyte or the materials used in the electrodes.
- Distance anode–cathode: Reducing the distance between the anode and cathode can decrease the ohmic resistance, thereby reducing power consumption.
- Number of electrode plates: Increasing the number of electrode plates can enhance the surface area for the reaction, improving efficiency and potentially reducing power consumption. However, this should be balanced with the overall design and cost considerations.
- Number of cells: Optimizing the number of cells in series can help in distributing the voltage more efficiently, leading to lower power consumption.
- Heating resistor: Minimizing or optimizing the heating resistor can help in reducing unnecessary power losses due to heating.

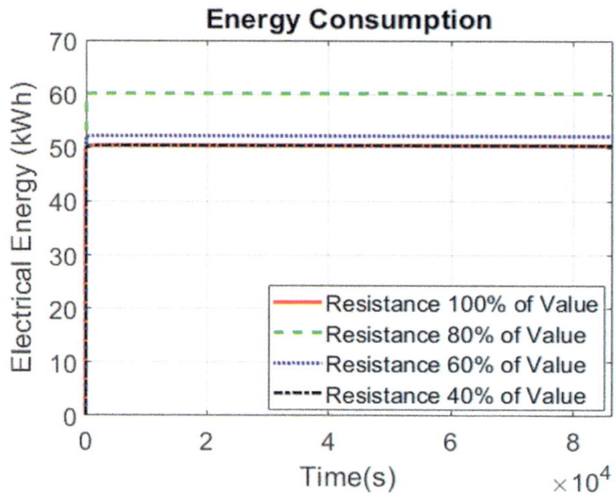

Figure 1.8 Energy consumption kWh/kg per day.

- Temperature vector: Operating at an optimal temperature can improve the efficiency of the electrolysis process. Higher temperatures generally reduce the required voltage for electrolysis, but this must be balanced with material limitations and safety considerations.

Reduce electric resistance: Ensure good conductivity of the electrolyte. Use highly conductive materials for electrodes. Maintain cleanliness and avoid scale buildup on electrodes.

Minimize anode–cathode distance: Design the cell to have a minimal gap between anode and cathode without causing short circuits or operational issues.

Optimize the number of electrode plates and cells: Find a balance between the number of electrodes and cells to maximize surface area while keeping the system cost-effective and efficient.

Optimize the heating resistor: Use efficient heating elements or consider alternative methods to maintain optimal temperatures.

Optimize operating temperature: Choose a temperature that minimizes the required voltage for electrolysis while considering the limitations of the materials used.

Estimation of hydrogen production per year and cost of deionized water consumption

Fig. 1.9 displays the energy consumption by year for hydrogen production, measured in kWh/kg. As shown in the figure, it is evident that optimizing resistance leads to variations in electrical energy consumption over time. The graph indicates that as resistance is optimized, energy consumption fluctuates, suggesting periodic changes in efficiency. Hence, in a year, the consumption will be 30 kWh/kg.

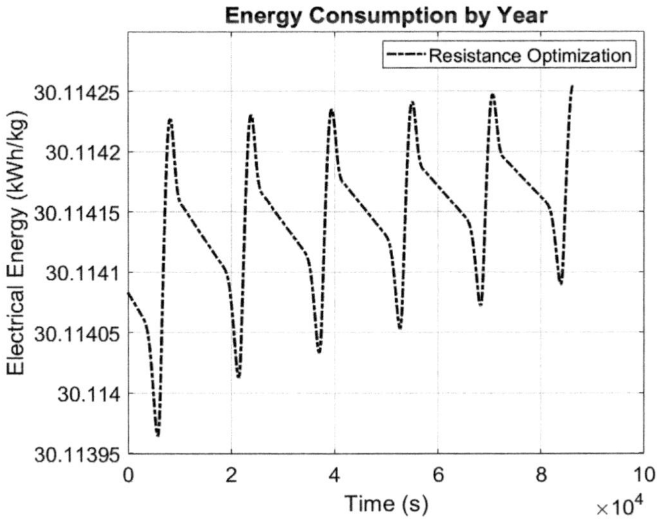

Figure 1.9 Energy consumption kWh/kg by year.

Conclusions

Green hydrogen holds great promise as a clean energy carrier and supports the transition to sustainable alternatives. This chapter has presented an approach to hydrogen production for renewable storage and integration in a case study applied to a microgrid. Simulation-based R&D is seen as a way to reduce these risks and steer projects toward success. Starting with a solar-based microgrid, key components include photovoltaic panels, energy storage units such as batteries, and an electrolyzer that converts electrical energy into hydrogen. For solar system-based setups, importing on-site irradiation data into simulation models is crucial for accurate projections.

Simulation fidelity is key; different levels are required depending on the project's focus. For detailed component analysis, high fidelity is necessary, but for broader system performance or feasibility studies, lower fidelity may suffice. Therefore, a series of data and adjustment parameters have been integrated that allow the electrolyzer models to be analyzed and characterized. Furthermore, the importance of simulation extends to system-level analysis as well, where different fidelity levels coexist along the development cycle to optimize both the physical unit and system-level operations. For instance, examining the impacts of grid frequency changes on an electrolyzer demonstrates the capability of simulation tools like Simscape Electrical to perform detailed analyses and optimize component configurations.

During hydrogen production in the experimental phase of the electrolyzer, the residual protons in the solution inside the electrolyzer naturally raise the pH and conductivity of the water in the tank. Therefore, this greater conductivity can cause a leak phenomenon. To avoid this, it is necessary to purge the tank by draining its volume and filling it with fresh water.

On the other hand, the price of hydrogen varies significantly depending on the production method and location in different countries. Green hydrogen, which is produced using RES, generally costs between \$4.5 and \$6.5 kg^{-1} in the United States. In the present case of study, the price is approximated as \$16.5 kg^{-1}.

At the system level, challenges include selecting the optimal concept for green hydrogen production and understanding the energy balances. Algorithmic strategies need to consider energy storage contributions and manage assets efficiently under varying conditions. Finally, simulation not only helps to understand and refine hydrogen production systems, but also to optimize the performance and sustainability of the entire system in the long term.

Acknowledgments

The authors wish to thank to the Spanish Ministry of Science and Innovation and European Union NextGenerationEU/PRTR, under the research Project "Desarrollo de herramientas computacionales para microrredes multienergía con vectores de electricidad, hidrógeno y gas" (TED2021−129631B-C31).

The authors thank Universidad de Cuenca for easing access to the facilities of the Microgrid Laboratory of the Centro Científico Tecnológico y de Investigación Balzay (CCTI-B), for allowing the use of its equipment, and for authorizing its staff the provision of technical support necessary to carry out the experiments described in this article.

The icons used in this document were developed by Freepik, monkik, Smashicons, and Pixel perfect, from http://www.flaticon.com [access 21 May 2024].

References

Akbari-Dibavar, A., Daneshvar, M., Mohammadi-ivatloo, B.., Zare, K., & Anvari-Moghaddam, A. (2020). Optimal robust energy management of microgrid with fuel cells, hydrogen energy storage units and responsive loads. In *2020 International conference on smart energy systems and technologies (SEST)* (pp. 1−6). Available from https://doi.org/10.1109/SEST48500.2020.9203215. Unpublished content.

Arsad, A. Z., Hannan, M. A., Al-Shetwi, A. Q., Hossain, M. J., Begum, R. A., Ker, P. J., Salehi, F., & Muttaqi, K. M. (2023). Hydrogen electrolyser for sustainable energy production: A bibliometric analysis and future directions. *International Journal of Hydrogen Energy*, *48* (13). Available from https://doi.org/10.1016/j.ijhydene.2022.11.023, http://www.journals.elsevier.com/international-journal-of-hydrogen-energy/.

Arsad, A. Z., Hannan, M. A., Al-Shetwi, A. Q., Mansur, M., Muttaqi, K. M., Dong, Z. Y., & Blaabjerg, F. (2022). Hydrogen energy storage integrated hybrid renewable energy systems: A review analysis for future research directions. *International Journal of Hydrogen Energy*, *47*(39). Available from https://doi.org/10.1016/j.ijhydene.2022.03.208, http://www.journals.elsevier.com/international-journal-of-hydrogen-energy/.

Criollo, A., Minchala-Avila, L.I., Benavides, D., Arévalo, P., Tostado-Véliz, M., Sánchez-Lozano, D., & Jurado, F. (2024). Enhancing virtual inertia control in microgrids: A novel frequency response model based on storage systems. *Batteries*, 10(1), 18. Available from https://doi.org/10.3390/batteries10010018.

Egeland-Eriksen, T., Hajizadeh, A., & Sartori, S. (2021). Hydrogen-based systems for integration of renewable energy in power systems: Achievements and perspectives. *International Journal of Hydrogen Energy*, *46*(63). Available from https://doi.org/10.1016/j.ijhydene.2021.06.218, http://www.journals.elsevier.com/international-journal-of-hydrogen-energy/.

Elgowainy, A., Mintz, M., Lee, U., Stephens, T., Sun, P., Reddi, K., Zhou, Y., Zang, G., Ruth, M., Jadun, P., Connelly, E., & Boardman, R. (2020). Assessment of potential future demands for hydrogen in the United States. https://www.osti.gov/biblio/1710201.

Espinoza, J.L., González, L.G., & Sempértegui, R. (2018). Micro grid laboratory as a tool for research on non-conventional energy sources in Ecuador. In *IEEE international autumn meeting on power, electronics and computing (ROPEC)*. https://doi.org/10.1109/ROPEC.2017.8261615. Unpublished content.

Falcão, D. S., & Pinto, A. M. F. R. (2020). A review on PEM electrolyzer modelling: Guidelines for beginners. *Journal of Cleaner Production*, *261*. Available from https://doi.org/10.1016/j.jclepro.2020.121184. https://www.journals.elsevier.com/journal-of-cleaner-production.

Fan, F., Zhang, R., Xu, Y., & Ren, S. (2022). Robustly coordinated operation of an emission-free microgrid with hybrid hydrogen-battery energy storage. *CSEE Journal of Power and Energy Systems*, *8*(2). Available from https://doi.org/10.17775/CSEEJPES.2021.04200.

Farhana, K., Shadate Faisal Mahamude, A., & Kadirgama, K. (2024). Comparing hydrogen fuel cost of production from various sources—A competitive analysis. *Energy Conversion and Management*, *302*. Available from https://doi.org/10.1016/j.enconman.2024.118088, https://www.sciencedirect.com/science/journal/01968904.

Fu, C., Lin, J., Song, Y., Zhou, Y., & Mu, S. (2017). Model predictive control of an integrated energy microgrid combining power to heat and hydrogen. In: *2017 IEEE conference on energy internet and energy system integration (EI2)*. https://doi.org/10.1109/EI2.2017.8245756. Unpublished content

Ganeshan, A., Holmes, D.G., Meegahapola, L., & McGrath, B.P. (2017). Enhanced control of a hydrogen energy storage system in a microgrid. In *2017 Australasian universities power engineering conference (AUPEC)*. https://doi.org/10.1109/AUPEC.2017.8282434. Unpublished content.

Ganeshan, A., Holmes, D.G., Meegahapola, L., & McGrath, B.P. (2019). Ripple energy buffer for microgrid connected hydrogen energy storage system. In *2019 IEEE 10th international symposium on power electronics for distributed generation systems (PEDG)* (pp. 634–641). https://doi.org/10.1109/PEDG.2019.8807461. Unpublished content.

Hassan, Q., Sameen, A. Z., Olapade, O., Alghoul, M., Salman, H. M., & Jaszczur, M. (2023). Hydrogen fuel as an important element of the energy storage needs for future smart cities. *International Journal of Hydrogen Energy*, *48*(78). Available from https://doi.org/10.1016/j.ijhydene.2023.03.413, http://www.journals.elsevier.com/international-journal-of-hydrogen-energy/.

Hemmati, R., Bornapour, S. M., & Saboori, H. (2024). Standalone hybrid power-hydrogen system incorporating daily-seasonal green hydrogen storage and hydrogen refueling station. *Energy*, *295*. Available from https://doi.org/10.1016/j.energy.2024.131122, https://www.sciencedirect.com/science/journal/03605442.

Hydrogen electrolyzer (2024). https://la.mathworks.com/discovery/hydrogen-electrolyzer.html.

Islam, M.R. (2020). Operation planning of a microgrid incorporating hydrogen energy and RE technology. In *International conference on system, computation, automation and networking (ICSCAN)* (pp. 1–6). Available from https://doi.org/10.1109/ICSCAN49426.2020.9262348. Unpublished content.

Li, B. (2020). Build 100% renewable energy based power station and microgrid clusters through hydrogen-centric storage systems. In *2020 4th international conference on HVDC (HVDC)*. Available from https://doi.org/10.1109/HVDC50696.2020.9292836. Unpublished content.

Lin, R. H., Zhao, Y. Y., & Wu, B. D. (2020). Toward a hydrogen society: Hydrogen and smart grid integration. *International Journal of Hydrogen Energy*, *45*(39). Available from https://doi.org/10.1016/j.ijhydene.2020.01.047, http://www.journals.elsevier.com/international-journal-of-hydrogen-energy/.

Liu, L., Zhai, R., & Hu, Y. (2023). Performance evaluation of wind-solar-hydrogen system for renewable energy generation and green hydrogen generation and storage: Energy, exergy, economic, and enviroeconomic. *Energy*, *276*. Available from https://doi.org/10.1016/j.energy.2023.127386, https://www.journals.elsevier.com/energy.

Morin, D., Stevenin, Y., & Brault, P. (2019). Energy management of isolated DC microgrids with hybrid batteries-hydrogen storage system using model predictive control and wavelet neural networks based forecasts. In *2019 21st European conference on power electronics and applications (EPE '19 ECCE Europe)*. Available from https://doi.org/10.23919/EPE.2019.8915136. Unpublished content.

Niknam, T., Kavousi-Fard, A., & Ostadi, A. (2015). Impact of hydrogen production and thermal energy recovery of PEMFCPPs on optimal management of renewable microgrids. *IEEE Transactions on Industrial Informatics*, *11*(5). Available from https://doi.org/10.1109/TII.2015.2475715.

Norouzi, N. (2021). An overview on the renewable hydrogen market. *International Journal of Energy Studies*, *6*(1). Available from https://dergipark.org.tr/en/pub/ijes/issue/63190/934518.

Pei, W., Zhang, X., Deng, W., Tang, C., & Yao, L. (2022). Review of operational control strategy for DC microgrids with electric-hydrogen hybrid storage systems. *CSEE Journal of Power and Energy Systems*, *8*(2). Available from https://doi.org/10.17775/CSEEJPES.2021.06960.

Schalenbach, M., Zeradjanin, A. R., Kasian, O., Cherevko, S., & Mayrhofer, K. J. J. (2018). A perspective on low-temperature water electrolysis—Challenges in alkaline and acidic technology. *International Journal of Electrochemical Science*, *13*(2). Available from https://doi.org/10.20964/2018.02.26. http://www.electrochemsci.org/papers/vol13/130201173.pdf.

Sproul, S., Cherevatskiy, S., & Mohan, S.L. (2021). Grid forming energy storage with microgrid controls provides green hydrogen, enhanced reliability, reduced site costs and lower emissions—ATCO clean energy innovation hub (a case study). In *CIRED 2021—the 26th international conference and exhibition on electricity distribution*. Available from https://doi.org/10.1049/icp.2021.2161. Unpublished content.

Stetson, N., & Wieliczko, M. (2020). Hydrogen technologies for energy storage: A perspective. *MRS Energy & Sustainability*, *7*(1), 41. Available from https://doi.org/10.1557/mre.2020.43. https://doi.org/10.1557/mre.2020.43.

Tahan, M. R. (2022). Recent advances in hydrogen compressors for use in large-scale renewable energy integration. *International Journal of Hydrogen Energy*, *47*(83). Available from https://doi.org/10.1016/j.ijhydene.2022.08.128, http://www.journals.elsevier.com/international-journal-of-hydrogen-energy/.

TW horizon fuel cell technologies. (2024). https://www.horizonfuelcell.com/.

Yan, J., Teng, Y., & Chen, Z. (2022). Optimization of multi-energy microgrids with waste process capacity for electricity-hydrogen charging services. *CSEE Journal of Power and Energy Systems*, *8*(2). Available from https://doi.org/10.17775/CSEEJPES.2021.05060.

Yang, H., Li, Q., Zhao, S., Chen, W., & Liu, H. (2019). A hierarchical self-regulation control for economic operation of AC/DC hybrid microgrid with hydrogen energy storage system. *IEEE Access*, *7*. Available from https://doi.org/10.1109/ACCESS.2019.2923794.

Gasification for electrification of rural areas

2

Roque Aguado, Daniel Sánchez-Lozano, Antonio Escámez, David Vera and Francisco Jurado Melguizo
Department of Electrical Engineering, University of Jaén, Jaén, Spain

Introduction

Rural areas across the globe face diverse challenges that significantly impact their development and overall quality of life (Beltramo et al., 2024; Gkartzios, Gallent, & Scott, 2022). These regions are often characterized by geographical isolation, lower population densities, and an economy predominantly based on agriculture and related activities (Taubenböck et al., 2023). Despite their crucial role in global food supply and raw material production, rural areas frequently lag behind urban centers in terms of infrastructure and access to essential services. This disparity is particularly evident in developing countries, where the gap between urban and rural development is stark, leading to persistent poverty and limited opportunities for economic advancement (Sakketa, 2022).

One of the most pressing issues in rural areas is the lack of access to stable and reliable electricity. Electricity is a fundamental driver of modern development, enabling advancements in healthcare, education, and economic activities. However, many rural regions remain without access to this critical resource (Aceituno Dardon & Farzaneh, 2024). The extension of national electricity grids to remote areas is often economically unfeasible due to the high costs and logistical challenges involved (Feyissa et al., 2024; Ayodele et al., 2021). Rugged terrain, sparse populations, and limited economic activities further complicate the expansion of grid infrastructure (Rajbongshi, Borgohain, & Mahapatra, 2017). As a result, rural communities frequently rely on traditional energy sources such as kerosene, firewood, and diesel generators (Pandiyan et al., 2022), which are not only expensive but also present serious health and environmental hazards.

Addressing the electrification needs of rural areas requires innovative and sustainable solutions that can provide stable electricity without the need for extensive grid infrastructure. Renewable energy technologies, particularly those that leverage local resources, offer a compelling path forward. Among these technologies, gasification stands out as a particularly promising solution for rural electrification (Tostado-Veliz et al., 2024). Gasification is a process that converts organic materials, such as biomass residues, into a combustible gas mixture through partial oxidation at high temperatures (Basu, 2018). This gas, known as producer gas, can be used to generate electricity using gas engines or turbines (Aguado et al., 2022; Basu, 2018; Vera et al., 2014).

Towards Future Smart Power Systems with High Penetration of Renewables. DOI: https://doi.org/10.1016/B978-0-443-29871-4.00002-6

Rural economies, heavily dependent on agriculture, produce substantial amounts of biomass residues. These residues, including crop stalks, husks, leaves, and animal manure, are often considered waste products. Traditionally, biomass residues are either burned in open fields (Bako, Mamai, & Istifanus, 2020), leading to air pollution, or left to decompose, releasing methane, a potent greenhouse gas. This traditional handling of biomass residues represents a significant missed opportunity for rural communities. If properly managed and utilized, biomass residues can become a valuable resource for electricity generation. Converting these residues into energy provides a dual benefit: it helps manage agricultural waste and provides a more sustainable source of energy than traditional fossil fuel-based energy sources.

Gasification technology offers several advantages for rural electrification. First, it leverages locally available biomass residues, reducing dependency on external energy sources and enhancing energy security. Second, by converting biomass residues into energy, gasification reduces the need for open burning of agricultural waste and decreases greenhouse gas emissions (Awopone, 2021; Fernández-Lobato et al., 2022). Third, local energy production can boost economic activities, create jobs, and improve living standards in rural communities. Finally, gasification systems are scalable and flexible, allowing them to be designed to match the energy needs and capacities of different rural areas (Sharma et al., 2021). This decentralized approach aligns well with the dispersed population and economic activities of rural regions, offering a practical and sustainable solution for electrification.

The potential of gasification technology to transform agricultural waste into valuable energy can be used to address the energy needs of rural communities, fostering sustainable development, and improving quality of life (Arranz-Piera et al., 2018; Tostado-Veliz et al., 2024). This chapter aims to explore the feasibility, benefits, and challenges of the biomass gasification technology for rural electrification. It delves into the technical aspects of gasification, assesses its economic and environmental benefits, and discusses the main challenges and barriers to its implementation. By examining case studies and pilot projects, the chapter highlights best practices and provides insights into how gasification can be effectively deployed to enhance energy access in rural regions. The ultimate goal is to demonstrate that gasification technology offers a potentially viable and sustainable solution to the electrification challenges faced by rural areas, contributing to their socioeconomic development and environmental sustainability.

Challenge of access to stable electricity supply

Stable and reliable electricity is a cornerstone of modern development, enabling advancements in healthcare, education, and economic activities. However, many rural areas remain without access to this critical resource (The World Bank, 2023). The disparity between urban and rural areas is stark, leading to a cycle of poverty and underdevelopment that is difficult to break. The primary challenge in electrifying these rural regions lies in the high cost and logistical difficulties associated with extending national electricity grids to remote locations (United Nations Development Programme, 2016). The rugged terrain, sparse population, and limited economic activities make it economically unfeasible for utility companies to invest in grid expansion (United Nations Development Programme, 2016). Currently, access to electricity as a percentage of the world's population (see Fig. 2.1) stands at 91.4% on average globally. Despite this, there are more rural areas, especially in Central African regions, where access to energy is below 20%.

In the absence of grid electricity, rural communities often depend on traditional energy sources such as kerosene, firewood, and diesel generators. These alternatives are not only costly, but also pose significant health and environmental risks (*World Health Organization*, 2023; Muyanja et al., 2017). Kerosene lamps and diesel generators contribute to indoor air pollution, which can lead to respiratory illnesses, while the reliance on firewood exacerbates deforestation and environmental degradation (*World Health Organization*, 2023). The intermittent and unreliable nature of these energy sources also hinders the use of modern appliances and machinery, limiting economic productivity and quality of life.

To overcome these challenges, there is a growing need for decentralized and sustainable energy solutions that can provide stable electricity to rural areas without the need for extensive grid infrastructure (Agency, 2023). Renewable energy technologies, particularly those that leverage local resources, offer a promising path forward (Agency, 2023).

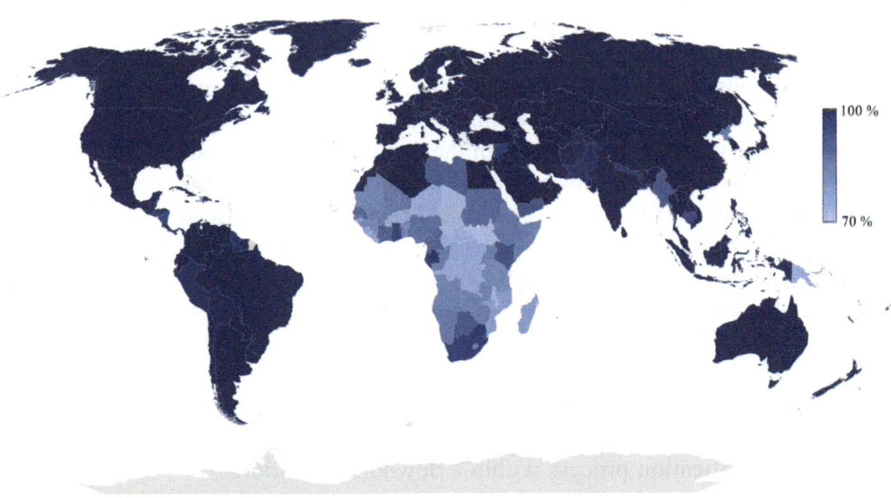

Figure 2.1 Access to electricity by country (% of the population) (The World Bank, 2023). Map lines delineate study areas and do not necessarily depict accepted national boundaries.

Rural economy and biomass residues

The economies of rural areas are heavily dependent on agriculture, which produces a substantial amount of biomass residues. These residues, including crop stalks, husks, leaves, and animal manure, are often considered waste products (Masters et al., 2013). In many rural regions, biomass residues are either burned in open fields, leading to air pollution, or left to decompose, releasing methane, a potent greenhouse gas (Bako, Mamai, & Istifanus, 2020). This traditional handling of biomass residues represents a significant missed opportunity for rural communities.

Biomass residues, if properly managed and utilized, can become a valuable resource for energy production (Aberilla et al., 2019). Converting these residues into energy can provide a dual benefit: it helps manage agricultural waste and offers a sustainable source of energy (Aberilla, Gallego-Schmid, & Azapagic, 2019). The use of biomass for energy production is not a novel concept; however, recent advancements in technology have made it more efficient and viable, particularly for rural electrification (Odoi-Yorke et al., 2022; Tostado-Veliz et al., 2024; Ansong et al., 2017).

In rural regions of developing countries, internal combustion engines are widely used for stationary applications, including power generation, water pumping, and milling. Consequently, there is a keen interest in technologies such as small-scale gasification. These areas, often tropical, are also characterized by abundant feedstock availability due to favorable conditions for biomass production.

The biomass gasification process

Gasification is a thermochemical process that holds significant promise in addressing the energy needs of rural areas (Arranz-Piera et al., 2018; Chambon et al., 2020). It offers a versatile solution for converting organic or carbonaceous materials, such as biomass, into a valuable fuel source. In the context of rural electrification, the gasification process typically occurs in the presence of air as gasifying agent at temperatures higher than 700°C (Tezer et al., 2022), leading to a lean fuel gas mixture, commonly referred to as producer gas. This gas mixture primarily contains combustible gases such as carbon monoxide (CO), hydrogen (H_2), and methane (CH_4), as well as inert gases without energy content such as carbon dioxide (CO_2) and nitrogen (N_2) (Aguado et al., 2023; Basu, 2018; Jha et al., 2022; Tezer et al., 2022).

The gasification process can be broken down into four primary stages: drying/dehydration, pyrolysis/devolatilization, combustion/oxidation, and gasification/reduction (Basu, 2018; Tezer et al., 2022). Fig. 2.2 presents the various stages of the biomass gasification process within a downdraft gasifier.

In the drying zone, the biomass is heated to evaporate the moisture content. The moisture content in fresh biomass typically ranges from 30% to 60% and must be reduced to below 15% for optimal gasification (Tezer et al., 2022). Drying occurs at relatively low temperatures, around 100°C–200°C, where the moisture is converted to steam (Tezer et al., 2022).

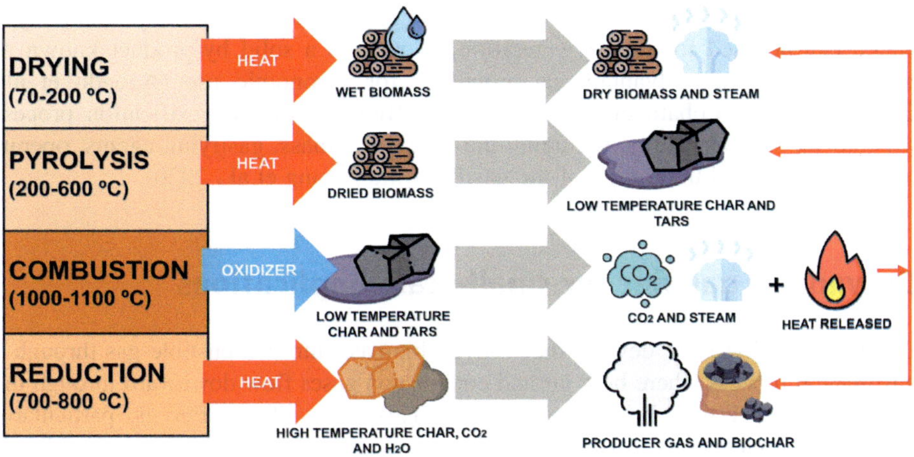

Figure 2.2 Stages of the biomass gasification process in a downdraft gasifier.

During pyrolysis, the biomass undergoes thermochemical decomposition in an oxygen-free environment. This stage breaks down large molecular groups into smaller hydrocarbons, producing biochar, liquid products (tar), and gaseous molecules. Pyrolysis temperatures range from 125°C to 500°C, affecting the product distribution. Hemicellulose decomposes between 150°C and 350°C, cellulose between 275°C and 350°C, and lignin between 250°C and 500°C (Molino, Chianese, & Musmarra, 2016). The reactions in this zone are endothermic, requiring heat from the combustion stage (Molino, Chianese, & Musmarra, 2016).

The combustion zone involves exothermic reactions where part of the biomass is oxidized to generate the necessary heat for the endothermic processes of drying and pyrolysis. This stage occurs at high temperatures, typically between 1000°C and 1100°C, producing CO, CO_2, H_2, and H_2O (Basu, 2018; Mishra & Upadhyay, 2021). The major oxidation reactions are:

Complete oxidation of carbon: $C + O_2 \leftrightarrow CO_2$ ($\Delta H = -394$ kJ/mol).

Oxidation of carbon monoxide: $CO + ½\, O_2 \leftrightarrow CO_2$ ($\Delta H = -284$ kJ/mol)

Complete oxidation of methane: $CH_4 + 2\, O_2 \leftrightarrow CO_2 + 2\, H_2O$ ($\Delta H = -803$ kJ/mol)

Oxidation of hydrogen: $H_2 + ½\, O_2 \leftrightarrow H_2O$ ($\Delta H = -242$ kJ/mol)

In the reduction zone, the products from the pyrolysis and combustion stages react to form a combustible gas mixture. This stage involves several key reactions (Basu, 2018; Mishra & Upadhyay, 2021), which include:

Water–gas or primary steam gasification reaction: $C + H_2O \leftrightarrow CO + H_2$ ($\Delta H = 131$ kJ/mol).

Boudouard reaction: $C + CO_2 \leftrightarrow 2\, CO$ ($\Delta H = 172$ kJ/mol)

Water–gas shift reaction: $CO + H_2O \leftrightarrow CO_2 + H_2$ ($\Delta H = -41$ kJ/mol)

Hydrogasification reaction: $C + 2\, H_2 \leftrightarrow CH_4$ ($\Delta H = -75$ kJ/mol)

These reactions are endothermic and occur at high temperatures around 800°C (Basu, 2018). The reduction stage is critical for minimizing tar content and optimizing the producer gas composition (Tezer et al., 2022).

Overall, gasification transforms solid biomass into a versatile producer gas, which can be used for power generation, as well as a solid by-product known as biochar, which has advantageous properties for agronomic use (Aguado et al., 2023; Fernández-Lobato et al., 2022). The efficiency of the gasification process depends on several factors, including the type of biomass, gasifying agents, operating conditions, and the design of the gasifier (Situmorang et al., 2020).

Downdraft gasifier for small-scale applications

Downdraft gasifiers are devices that convert biomass into combustible gas through a gasification process where both air and combustion gases flow downward through the reactor (Basu, 2018; Bridgwater, 1995). This type of gasifier is particularly suitable for applications in rural areas due to its operational characteristics and specific benefits (Bridgwater, 1995). In a downdraft gasifier, biomass is fed into the top of the reactor and descends as it sequentially undergoes drying, pyrolysis, combustion, and gasification. Fig. 2.3 illustrates a schematic diagram of the reaction zones and temperature profile within a downdraft gasifier.

A key advantage of downdraft gasifiers is their low tar production, typically less than 1 g/Nm^3 of gas, facilitating their use in internal combustion engines and reducing maintenance and pollution issues (Basu, 2018). Rural areas often have easy access to biomass resources such as agricultural residues, wood, and other organic waste. These materials, available in large quantities, can be used as feedstock for gasifiers, providing a local and renewable energy source (Awopone, 2021; Odoi-Yorke et al., 2022; Ansong, Mensah, & Adaramola, 2017). Downdraft gasifiers can

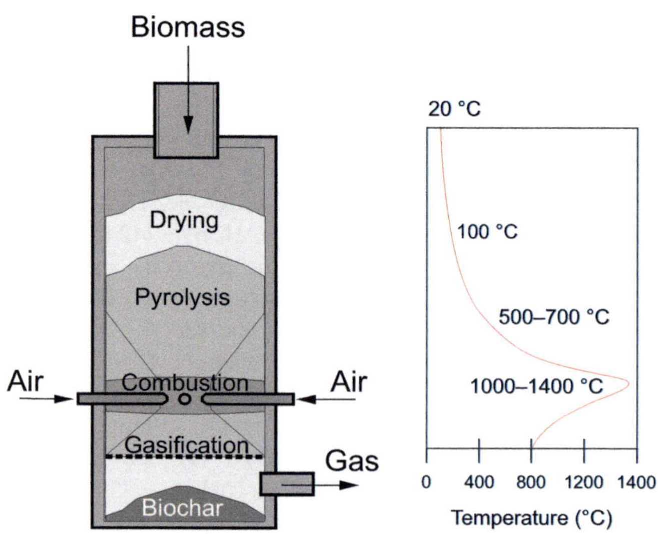

Figure 2.3 Schematic diagram of the reaction zones and temperature profile in a downdraft gasifier.

be installed in rural communities to provide decentralized energy, essential in areas with limited or no access to the electric power grid (Tostado-Veliz et al., 2024). The capacity of downdraft gasifiers can vary widely, with small models generating between 10 and 20 kW, suitable for domestic or small business applications, and larger models capable of producing up to 1 MW (Basu, 2018; Sharma et al., 2021), supplying power to entire communities or local industries. The specific fuel consumption of downdraft gasifier systems equipped with internal combustion engines varies between 1.1 and 1.5 kg/kWh when using wood as fuel and between 1.8 and 3.6 kg/kWh when fueled by rice husk (LTS International Ltd, 2021). A broader range of specific fuel consumption values between 1.3 and 3.8 kg/kWh is reported for gasification of olive pomace pellets in a downdraft gasifier (Aguado et al., 2023). However, it is noteworthy that the specific fuel consumption typically increases when the gasification equipment operates at partial load (Aguado et al., 2023). Downdraft gasifiers require regular maintenance to ensure efficiency, such as cleaning ashes and combustion residues, typically performed on a weekly or bi-weekly basis, depending on usage and biomass quality (Palit, Malhotra, & Kumar, 2011). Additionally, the relatively low tar production compared to other gasifier types reduces the need for frequent cleaning of the gasification system and associated equipment. The energy conversion efficiency of downdraft gasifiers can reach up to a higher efficiency in cogeneration mode, making them an interesting option for sustainable combined heat and power (CHP) generation in rural areas (Basu, 2018; Demirbas, 2006).

Challenges and barriers of biomass gasification in rural areas

Biomass gasification plants promote sustainability and strengthen the local economy by creating jobs and adding value to agricultural residues. Moreover, they reduce dependence on fossil fuels and decrease carbon footprint (Awopone, 2021; Odoi-Yorke et al., 2022; Ansong, Mensah, & Adaramola, 2017; Fernández-Lobato et al., 2022). However, despite these benefits, several challenges and barriers limit the widespread adoption of biomass gasification for rural electrification.

Gasification projects face a significant challenge in ensuring a consistent and affordable supply of feedstock that meets the quality standards essential for reactor system functionality. Quantity, quality, and cost are intricately intertwined, as gasifiers cannot tolerate wide quality variations, thereby constraining the variety of potential feedstocks viable for a particular unit. Consequently, ensuring both price stability and a reliable supply of the preferred feedstock becomes essential for gasification projects compared to other bioenergy technologies. Gasifiers demand feedstock with specific characteristics, such as low moisture content, high heating value, and uniform particle size, typically requiring clean, homogeneous material with low ash content and a moisture content below 13% (Owen & Ripken, 2017). Moreover, most gasifiers are intolerant of mixed feedstocks, thereby hindering operators from blending various biomass streams based on local availability.

Biomass gasifiers used for electrification of rural areas often exhibit a high sensitivity to operational parameters, including feedstock quality, consistency, feed rate, and air control. Deviations from precise manufacturer specifications can lead to suboptimal temperatures within the reactor chamber, resulting in relatively high production of char, tar, and particulate matter. These deviations can subsequently trigger a cascade of issues with cyclones, scrubbers, filters, precipitators, and engines, exacerbating the challenge of disposing of serious toxic byproducts (Owen & Ripken, 2017).

Moreover, small-scale gasification lacks the same degree of power dispatchability as other off-grid energy technologies, introducing difficulties in synchronizing power output with demand due to potential mismatches in timing and scale (Owen & Ripken, 2017). Accordingly, deployment of a gas storage system, a battery bank, or supplementary power sources such as photovoltaic or backup diesel generators is typically required to effectively manage periods of downtime (Tostado-Veliz et al., 2024). Fig. 2.4 illustrates an example of this hybrid small-scale power generation approach for rural electrification, a strategy explored in various studies (Arranz-Piera et al., 2018; Chambon et al., 2020; Rajbongshi et al., 2017; Ramchandran et al., 2016; Ribo-Perez et al., 2021; Tostado-Veliz et al., 2024).

Last but not least, it is noteworthy that gasification is a complex process that operates within stringent technical parameters. The feedstock requirements and operating conditions in the gasifier must be well understood and carefully managed

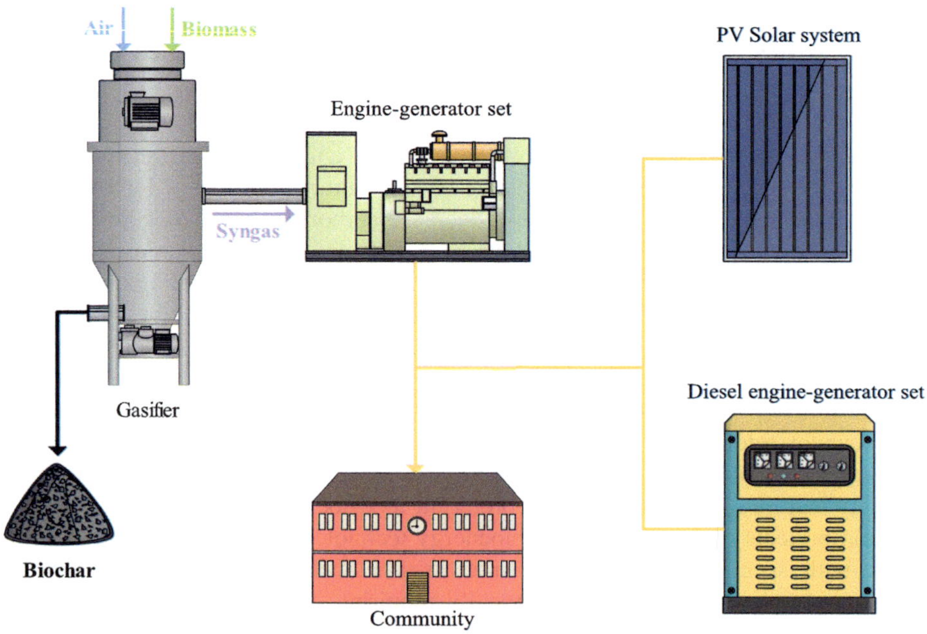

Figure 2.4 Schematic diagram of a hybrid biomass gasification plant with PV and diesel generators as supplementary power sources.

to minimize unwanted production of char, tar and particulate matter in favor of producer gas. Failure to do so can result in environmental pollution, especially from tar, and compromise engine performance, leading to wear and premature breakdowns. Securing skilled operators for gasification systems, particularly in rural and often remote areas poses another significant challenge (Owen & Ripken, 2017). Beyond technical expertise, operators and engineers must possess strong motivation to overcome the numerous barriers inherent in operating a gasifier in such settings. Unfortunately, the scarcity of qualified workers exacerbates concerns regarding the reliability of the biomass gasification technology, tarnishing the sector's reputation.

Review of case studies and technical reports

In recent years, extensive research has been conducted globally on the potential and challenges of biomass gasification for sustainable energy solutions, particularly in developing countries. Studies underscore the viability and socioeconomic benefits of utilizing local biomass resources for gasification-based power generation.

Table 2.1 compiles various case studies from 2007 onward in which gasification has been applied in rural electrification across various geographical regions, which have been carefully selected through the Scopus scientific database. These studies highlight key aspects and findings essential for understanding the implementation and impact of biomass gasification in addressing energy access challenges, promoting rural electrification, and fostering sustainable development in various regions of the world. They highlight not only the technical feasibility of gasification technology, but also its socioeconomic and environmental implications, emphasizing the need for further research, investment, and policy support to realize its full potential.

As shown in Figure Fig. 2.5, the selection of case studies covers a wide geographical scope, including Asia (e.g., Cambodia, India, and Sri Lanka), Africa (e.g., South Africa, Kenya, and Ghana), and other countries in America and Oceania such as Honduras and Vanuatu. These regions, characterized by their rural and often remote populations, exhibit significant potential for biomass gasification due to abundant biomass resources. Among these countries, India has been at the forefront of implementing biomass gasification projects to address rural electrification needs.

One of the earliest studies by Abe et al. (2007) examined the potential of biomass gasification for rural electrification in Cambodia. They found that a significant number of villages without access to electricity had ample land for tree planting to generate biomass for electricity, suggesting biomass gasification as a potentially viable solution for mini-grid installations. Similarly, studies in South Africa (Mamphweli & Meyer, 2009) and Sri Lanka (Blunck & Dimpl, 2010) highlighted the need for low-cost electricity in rural areas and evaluated the economic potential of biomass gasification for electrification.

Gasification offers a sustainable pathway for rural electrification, leveraging locally available biomass resources to generate electricity. For instance, studies like those by Abe et al. (2007) in Cambodia and Buragohain, Mahanta, and Moholkar

Table 2.1 Summary of studies on biomass gasification for rural electrification in the last 20 years across various geographical regions.

References	Geographical scope	Key findings
Abe et al. (2007)	Cambodia	Examined the potential of biomass gasification for rural electrification in Cambodia. Found that 77% of villages without electricity had sufficient land for tree planting for electricity generation. Identified 6418 villages as appropriate candidates for minigrid installation by biomass gasification. Demonstrated that biomass gasification becomes economically competitive compared to diesel generation when the plant capacity factor exceeds 13%.
Mamphweli and Meyer (2009)	South Africa	Identified a need for low-cost electricity for small businesses. Addressed air pollution and greenhouse gas emissions from traditional biomass waste burning. Successfully installed a 300 Nm^3/h (150 kVA) System Johansson Biomass Gasifier (SJBG).
Blunck and Dimpl (2010)	Sri Lanka	Evaluated the economic potential of biomass gasification for rural electrification with small power plants ranging from 10 to over 100 kW, supporting household and small machinery use. Demonstrated the ability to produce 1 kWh of electricity from 1.1 to 1.5 kg of wood, 0.7–1.3 kg of charcoal, or 1.8–3.6 kg of rice husks. Highlighted the overall system efficiencies of 16%–19% for wood gasification systems and 7%–14% for rice husk gasification systems, with potential efficiency reaching up to 80% when integrated into combined heat and power (CHP) systems. Found that the technology, while promising CO_2 neutrality and renewable energy benefits, faced significant obstacles such as high costs and environmental hazards. Concluded that current gasification technology has significant challenges, including high costs, environmental threats from carcinogenic waste, and the need for specialized maintenance and technical expertise.

(*Continued*)

Table 2.1 (Continued)

References	Geographical scope	Key findings
Buragohain, Mahanta, and Moholkar (2010)	India	Noted an annual biomass production of 200 million tons, with potential for further increase. Identified a significant potential of 20 GW power generation from biomass within India's overall renewable energy capacity of 85 GW. Proposed biomass gasification as a viable solution for electrifying remote and hilly areas, reducing dependency on state electricity boards. Estimated a reduction of 1 kg of CO_2 for every kilowatt-hour of coal-based electricity replaced by biomass gasification. Demonstrated advantages of biomass gasification over other decentralized generation options, citing proven technology, flexibility of installation and operation, as well as social and environmental benefits. Recommended actions for effective utilization, including large-scale production of gasifiers, enhancing capacity utilization, encouraging community involvement, and subsidizing capital costs.
Dasappa (2011)	Africa and India	Explored the biomass-based power generation potential of Africa, estimating a power generation potential of about 5000 and 10,000 MW using 30% of agricultural residues and 10% of forest residues, respectively. Identified potential to generate 100 terawatt-hours (TWh) of electricity, equivalent to approximately 15% of current generation in sub-Saharan Africa. Summarized experiences in biomass gasification technology for power generation in Africa and India, highlighting its potential to address energy access challenges in sub-Saharan Africa. Emphasized the importance of overcoming critical barriers such as resource mapping, capacity building, and technology demonstration to promote biomass energy adoption in sub-Saharan Africa.

(*Continued*)

Table 2.1 (Continued)

References	Geographical scope	Key findings
Fischer and Pigneri (2011)	Vanuatu	Identified up to 13 geographic matches between areas of rural electricity demand and potential areas of gasifier feedstock cultivation, currently occupied by obsolete *Cordia* plantations, in Vanuatu. Technoeconomic analysis showed that the levelized cost of electricity (LCOE) from gasification is competitive with diesel systems at capacity factors of 15%–40%, corresponding to USD 1.60 kWh^{-1} or less. Demonstrated that a 30 kW gasifier operating at a 15%–20% capacity factor is cost-competitive with a 30 kW diesel generator due to economies of scale and spreading out of initial capital costs. Recommended further research into *Cordia* plantation stands' properties and socioeconomic barriers to biomass gasifier deployment, including the lack of technical expertise and challenges in sustaining cooperative energy projects.
Palit, Malhotra, and Kumar (2011)	India	Designed a sustainable model for the financial viability of biomass gasifier power projects to enhance electricity access in India and other developing countries. Estimated that biomass gasification plants should operate at more than 30% capacity utilization factor for financial sustainability, with an average load of 70%–80% of the rated capacity for 8–10 h/day and about 350 days per year. Found that larger capacity systems, such as a 25 kW system, are more economically favorable, with a minimum delivered price of about INR 6.62 kWh^{-1}, compared to INR 9.93 kWh^{-1} for a 10 kW system. Advocated for developing a regulatory mechanism to extend tariff fixation and provide cross-subsidies to ensure long-term sustainability of off-grid projects.

(*Continued*)

Table 2.1 (Continued)

References	Geographical scope	Key findings
Sou et al. (2010)	Cambodia	Designed a hybrid renewable energy system for electrification of a rural village in Cambodia, comprising a 1.27 kWp solar photovoltaic generator, a 15 kW biomass gasification system, and a 7.28 kWh battery backup. Achieved an electrical generation capacity of 34,376 kWh per year, with 96% generated by the biomass gasification system and 4% by the photovoltaic system. Met 90% of the annual load demand directly from the biomass gasification system, 3% from the photovoltaic system, and 7% from the backup battery, ensuring 24-h power supply throughout the year. Estimated an annual energy surplus of 6980 kWh, allowing for the charging of external batteries, thus ensuring self-sufficiency in electricity supply for the village independent of imported conventional fuel.
Sanchez, Torres, and Kalid (2015)	Brazil	Evaluated the performance and cost-effectiveness of renewable energy technologies in isolated Amazon communities. Revealed that some renewable energy options outperform diesel generation in terms of cost and sustainability after a 10-year evaluation period. Highlighted the significance of mixed solar and wind systems with occasional diesel support for reliability and reduced operational costs. Emphasized the potential of biomass and biofuels to add value to local production schemes and improve energy access in impoverished rural communities.
Flores, Furubayashi, and Nakata (2016)	Honduras	Developed a model that selected electricity generation technologies, focusing on solar PV, wind turbines, biomass combustion, and biomass gasification, aiming to minimize LCOE. Conducted a case study in the Honduran rural residential sector, revealing solar PV for off-grid, and wind turbines for minigrid and grid-connected cases as optimal technologies, with LCOE ranging from USD 0.27 to USD $0.81\ kWh^{-1}$.

(*Continued*)

Table 2.1 (Continued)

References	Geographical scope	Key findings
		Identified lead-acid batteries as a cost-effective backup option compared to ICE, with costs ranging from USD 0.16–0.43 kWh^{-1}-cycle and USD 0.25–0.57 kWh^{-1}, respectively, across the three cases. Found that the off-grid scheme could only provide electricity to 37% of households in the study area, due to limitations in solar and wind potential, emphasizing the necessity of minigrids.
Naqvi et al. (2016)	Pakistan	Investigated waste gasification for off-grid electrification in rural Pakistan, using mixed biomass composts. Estimated off-grid electricity prices were consistently higher than government tariffs across all scenarios. Demonstrated that variable tariffs, charging different rates based on electricity usage, were economically beneficial compared to flat tariffs. Analyzed the impact of biomass mix on LCOE, revealing that a 100% agricultural residue mix resulted in the lowest cost (USD 0.28 kWh^{-1}), while 100% manure compost led to a higher cost (USD 0.34 kWh^{-1}).
Njenga et al. (2016)	Kenya	Determined that the gasifier stove saved 27%–40% of fuel, reduced cooking time by 19%–23%, and decreased emissions by 40%–90% compared to traditional and improved stoves. Found that women preferred the gasifier stove due to its perceived benefits, including rapid cooking and reduced indoor air pollution, making it appropriate for rural areas. Suggested the gasifier stove has the potential to alleviate energy and time poverty among small-scale farmers while improving kitchen air quality, with further development needed for stability and safety.
Ramamurthi et al. (2016)	Ghana	Investigated the role of rice residues in addressing electrification challenges in developing countries, particularly focusing on Ghana.

(Continued)

Table 2.1 (Continued)

References	Geographical scope	Key findings
		Found rice straw combustion plants to be economically viable in Ghana with LCOE between USD 0.116 and USD 0.130 kWh^{-1}, supported by a bioenergy feed-in-tariff of USD 0.295 kWh^{-1}. Determined husk gasification minigrids as a suitable and cost-effective electrification solution for rural populations, with LCOE ranging from USD 0.05–0.53 kWh^{-1}, making them comparable or cheaper than solar mini grids. Concluded that agroresidue-based bioenergy solutions could be crucial for increasing renewable energy capacity and rural electrification, especially in scattered rural populations of developing countries like Ghana.
Ramchandran, Pai, and Parihar (2016)	India	Tested the feasibility of off-grid electrification in Uttar Pradesh (India), including primary surveys assessing socio-economic characteristics, renewable energy availability, and energy demand. Conducted cash flow analysis, revealing that the solar PV system with subsidy was the most robust model, while biomass standalone and solar PV–biomass hybrid systems showed potential for healthy returns. Examined the A-B-C model's viability for solar PV, biomass gasification, and solar-biomass hybrid systems. Identified challenges including dependence on biomass vendors and competition from private companies.
Singh (2017)	India	Evaluated state-wise power potential of agricultural residues in India. Estimated highest biomass power potential in Uttar Pradesh (6045 MW), followed by Punjab (3585 MW), Maharashtra (3139 MW), and Haryana (2300 MW). Identified major barriers to biomass exploitation including logistics of biomass collection and moisture contents. Highlighted the importance of selecting appropriate technology (combustion or gasification) and implementing favorable policies for biomass utilization.

(Continued)

Table 2.1 (Continued)

References	Geographical scope	Key findings
Arranz-Piera et al. (2018)	Ghana	Investigated the technical and financial feasibility of decentralized electrification based on agricultural waste gasification in five Ghanaian communities. Found that when a customer tariff equal to the current expenditure on electricity equivalent uses in the communities was applied, along with a subsidy of about 35% on initial investment, it could enable a private entrepreneur an internal rate of return of 15%. Concluded that planning rural electrification projects in developing countries, particularly utilizing agricultural residues for biomass gasification, required careful consideration of socioeconomic, technical, organizational, and financial aspects. Applied the national electricity uniform tariff, which required a 65% of the investment subsidies to enable a 15% profitability, with the remaining 35% coming from private cofunding.
Piapdje Tamo et al. (2018)	Cameroon	Utilized a gasification power generation system with a capacity of 20 kW, achieving higher electric power generation with close-top operation compared to open-top operation. Monitored the higher temperatures obtained during close-top operation at reduction and oxidation layers. Found that the average concentration of CO gas was 22.64% for 40 kg biomass consumed, generating 15.4 kWh power at 2800 rpm. Noted that close-top configuration yielded higher power output than the open-top one, suggesting biomass gasification as a promising renewable energy technology for rural areas in Cameroon.
Getahun et al. (2019)	Ethiopia	Designed and fabricated laboratory-scale updraft natural and forced draft gasifier stoves, with thermal efficiencies of 22.7% and 25%, respectively.

(Continued)

Table 2.1 (Continued)

References	Geographical scope	Key findings
		Evaluated stove performance, showing gasifier stove efficiencies of 84% and 72% for natural and forced draft, respectively, surpassing traditional open fire stoves. Achieved fuel saving efficiencies of 84% and 72% for natural and forced draft gasifier stoves, respectively, exceeding conventional stove efficiencies which typically range between 25% and 50%. Recorded burning times of 65 and 40 min for natural and forced draft gasifier stoves, respectively, sufficient for complete cooking processes in Ethiopian households.
Aberilla, Gallego-Schmid, and Azapagic (2019)	Africa and Asia-Pacific	Evaluated gasification, which had up to 12 times lower impacts per kWh than combustion, except for resource depletion. Compared combustion and gasification to diesel generators, finding that they had 85% to 2 times lower impacts, except for eutrophication, ecotoxicity, and human toxicity. Observed anaerobic digestion, with its global warming potential ranging from 170% lower to 41% higher than diesel generators, depending on the feedstock. Found anaerobic digestion to be the most environmentally sustainable option, outperforming others in 14 out of 18 impact categories.
Chambon et al. (2020)	Ghana	Conducted surveys to collect energy demand data in a rural village in Uttar Pradesh, India, revealing seasonal load profiles and enabling simulation of single-source and hybrid mini-grids. Found that hybrid PV-biomass or PV-diesel systems offered the highest reliability for off-grid power at the lowest cost, with renewable options being around half the cost of diesel generation. Compared solar PV, biomass gasification, and diesel generation for minigrids, determining that standalone PV systems had the lowest LCOE, while biomass gasification showed significant cost advantages for high reliability systems.

(*Continued*)

Table 2.1 (Continued)

References	Geographical scope	Key findings
		Concluded that hybrid solar and biomass gasification-based minigrids theoretically offer electricity at a cheaper cost, emphasizing the need for careful consideration of consumer demand, service reliability, and operational feasibility.
Littlejohns et al. (2020)	Canada	Investigated feedstock characteristics for stable gasifier operation with Canadian softwood, finding that reducing moisture increased system efficiency from 8.2% to 12%. Utilized equilibrium simulations to assess deviation from ideal conversion, with actual producer gas deviating by 9% on average. Concluded that while there was deviation from ideal conversion, stable operation was achieved, with the system compensating for reduced gas heating value. Determined that produced biochar met Canadian standards for soil amendment, enhancing the economic viability of the units for rural electrification.
Ejiofor et al. (2020)	Nigeria	Conducted detailed load assessment and evaluated rice husk availability for power generation, resulting in the design of a gasification system. Estimated total load demand of the study area to be 850.054 kW, with daily, monthly, and yearly energy demands of 8991, 269,736.42, and 3,236,837.04 kWh, respectively. Identified annual availability of rice husk at 3636 tons/year, with potential for 1.52 MW power generation, equating to daily, monthly, and yearly energy generation of 36.48, 1094.4, and 13,132.8 MWh. Determined an LCOE of the designed power system at 3.6 Naira/kWh, which is lower than the 30.93 Naira/kWh tariff from the national power grid extension, enabling economical electricity generation and contributing to sustainable development.

(Continued)

Table 2.1 (Continued)

References	Geographical scope	Key findings
Kaoma and Gheewala (2021)	Zambia	Identified economically viable options for supplying cooking/heating energy services in nonelectrified rural growth centers of Zambia: pellets and briquettes. Found gasification and combustion technologies to be more financially attractive for electricity production compared to diesel generators, despite not being cost-competitive with grid-electricity due to subsidized tariffs. Analyzed technoeconomically, revealing that on a comparative basis, 10 and 40 kW combustion plants outperformed their gasification counterparts, while the 100 kW gasification plant performed significantly better than the combustion plant of the same size and feedstock type. Recommended establishing a decision support system framework for biomass exploitation to guide policymakers in identifying optimal and sustainable means of using biomass for modern energy services in rural areas, considering technology-feedstock heterogeneity and associated environmental and socio-economic impacts.
Ribo-Perez et al. (2021)	Honduras and Zambia	Presented a procedure to model biomass gasification in HOMER, with parameters tailored for downdraft gasifiers. Demonstrated the viability through case studies in Honduras and Zambia, showcasing lower LCOE compared to extending the electric grid, with LCOE as low as USD 0.06 kWh^{-1} in El Santuario. Highlighted the operational benefits, such as prolonging battery life by covering demand during low state of charge, with minimal unmet load (0.002% in El Santuario). Addressed concerns regarding environmental impact, social implications, and resource management, emphasizing the need for community involvement and consideration of local data for economic balance and sustainability.

(*Continued*)

Table 2.1 (Continued)

References	Geographical scope	Key findings
Salisu, Gao, and Quan (2021)	Nigeria	Modeled a steam cogasification system for rice husk and sachet water plastic waste using Aspen Plus, achieving an overall CHP efficiency of 74% at optimum conditions. Demonstrated the economic feasibility of the system with a NPV of $1.68 million over 15 years, a payback period of 6 years, and an IRR of 24%, making it an attractive alternative to traditional energy sources in Nigeria. Evaluated the LCOE at USD 0.07–0.11 kWh^{-1}, indicating cost-attractiveness compared to diesel generators, with potential annual emissions savings of 3182 tCO_2/year and 12,867 tCO_2/year compared to diesel and wood combustion, respectively. Identified the potential for Clean Development Mechanism adoption, allowing for carbon credits trading as earnings, aligning with UN Sustainable Development Goals for affordable and clean energy.
Odoi-Yorke et al. (2022)	Ghana	Estimated that about 29 million tons of surplus crop residues is available for bioenergy production in Ghana, generating approximately 401 PJ/year. Found that the LCOE from gasification and combustion plants ranged between USD 0.29 kWh^{-1} and USD 0.34 kWh^{-1}, higher than residential electricity tariffs in Ghana. Concluded that with state support and infrastructural development, including subsidies and tax credits, investment in bioenergy could become economically feasible for rural electrification in Ghana.
Sanchez-Lozano et al. (2023)	Morocco	Explored the feasibility of a biomass gasification CHP plant for off-grid olive oil mills in Morocco, achieving net electrical efficiencies and CHP efficiencies around 18% and 35%, respectively. Demonstrated that biomass gasification CHP plants fueled with olive cake pellets exhibit better performance compared to olive pruning chips.

(*Continued*)

Table 2.1 (Continued)

References	Geographical scope	Key findings
		Highlighted the transformative impact of biomass gasification CHP plants on olive oil mills' self-sufficiency by converting biomass waste into valuable byproducts, reducing dependence on external energy sources and generating potentially profitable byproducts like biochar. Conducted a profitability assessment revealing a 68.5% accumulated profit based on capital investment with a 30% subsidy from the Moroccan government, suggesting biomass gasification CHP plants as economically viable and sustainable solutions for electrifying off-grid areas in Morocco and similar regions.
Tostado-Veliz et al. (2024)	Ghana	Developed an innovative optimal sizing tool for hybrid power plants integrating biomass gasifiers, tailored for isolated areas with local biomass resources. Utilized a methodology based on representative weeks combined with multicut Benders' decomposition for efficient data management and problem tractability. Applied the method to a benchmark off-grid community in Ghana, showcasing significant cost reduction (90%) due to gasifiers, driven by lower biomass costs and reduced CO_2 emissions (83%) compared to diesel generators. Demonstrated biomass gasification's effectiveness as base load power generation, providing reliable and inexpensive local electricity.
Qudrat-Ullah (2024)	Uganda	Identified bamboo gasification as a promising social innovation for sustainable energy and rural development in Uganda. Demonstrated bamboo's potential as a fast-growing and versatile biomass source for reliable gasification, capable of generating electricity, heat, and valuable byproducts. Highlighted the significant contribution of bamboo gasification to Sustainable Development Goals, including poverty reduction, climate change mitigation, forest conservation, and job creation, despite facing challenges such as lack of awareness, research, investment, and support in Uganda.

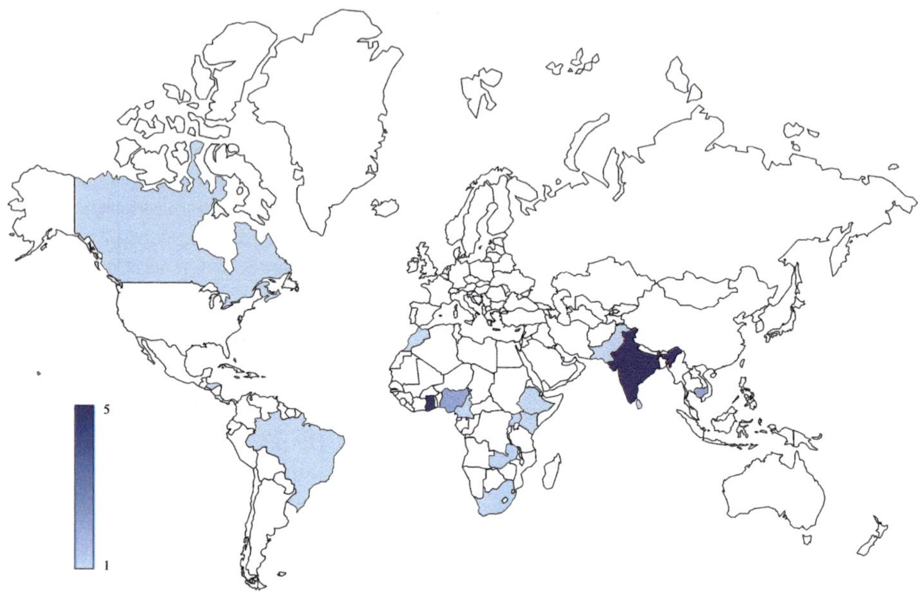

Figure 2.5 Geographical distribution of case studies included in the literature review, covering the past 20 years of research from the Scopus scientific database. Map lines delineate study areas and do not necessarily depict accepted national boundaries.

(2010) in India emphasize the potential of biomass gasification in addressing energy poverty by providing access to reliable electricity. Furthermore, research by Singh (2017) in India and Ramamurthi et al. (2016) in Ghana underscores the vast biomass power generation potential in countries with abundant agricultural residues, positioning gasification as a viable solution for rural electrification.

Economic viability and technological feasibility are crucial factors in the adoption of gasification for rural electrification. Economic analysis across multiple studies consistently shows that biomass gasification can be cost-competitive with or even cheaper than diesel generation, especially at higher capacity factors. For example, in Vanuatu, a 30 kW gasifier system becomes economically viable compared to diesel at capacity factors of 15%–40% (Fischer & Pigneri, 2011). Similarly, in India, large-scale biomass gasification is suggested as a means to achieve significant reductions in CO_2 emissions by replacing coal-based electricity (Buragohain, Mahanta, & Moholkar, 2010). Studies like Blunck and Dimpl (2010) in Sri Lanka and Ramchandran, Pai, and Parihar (2016) in India evaluated the economic potential of biomass gasification systems, indicating aspects such as system efficiency, capacity utilization, and levelized cost of electricity (LCOE).

Technical advancements and optimization strategies have also been explored to enhance gasification efficiency and performance. Studies in countries like Cameroon (Piapdje Tamo et al., 2018) and Canada (Littlejohns et al., 2020) investigated operational parameters and feedstock characteristics to improve gasifier design and operation, while others utilized simulation tools like HOMER (Ribo-Perez et al., 2021) and Aspen Plus (Salisu, Gao, & Quan, 2021) for system optimization and economic analysis.

Community engagement and policy support are essential in the successful deployment of gasification technologies. Many studies underscore the importance of community involvement and capacity building to ensure the successful implementation of biomass gasification projects. In particular, studies such as those by Sou, Siemers, and Exell (2010) in Cambodia and Arranz-Piera et al. (2018) in Ghana emphasize the importance of community involvement in project planning and implementation, ensuring local acceptance and ownership. Moreover, research by Palit et al. (2011) in India and Kaoma and Gheewala (2021) in Zambia advocate for supportive policies and regulatory mechanisms such as tariff fixation and cross-subsidies to incentivize investment in rural electrification projects, enhancing financial sustainability and scalability.

Environmental sustainability is another key consideration in gasification-based electrification initiatives. Studies such as Mamphweli and Meyer (2009) and Njenga et al. (2016) underline the reduction in air pollution and improved indoor air quality. Aberilla, Gallego-Schmid, and Azapagic (2019) and Salisu, Gao, and Quan (2021) assessed the environmental impacts and carbon emissions associated with gasification technologies, highlighting their potential to mitigate greenhouse gas emissions compared to conventional energy sources like diesel generators. Additionally, research by Sanchez-Lozano et al. (2023) in Morocco underscored the transformative impact of biomass gasification on waste management and resource utilization, contributing to circular economy principles. While biomass gasification offers environmental benefits, such as CO_2 neutrality and reduced reliance on fossil fuels, concerns about carcinogenic waste and indoor air pollution from gasifier stoves need to be addressed through improved technology and safety measures (Blunck & Dimpl, 2010).

A relatively recent technical report by LTS International Ltd. (2021) provides a detailed overview of various biomass gasification projects in the region of Sub-Saharan Africa, including countries such as Ethiopia, Ghana, Kenya, Mozambique, Nigeria, South Africa, Tanzania, Uganda, and Zambia. A total of 36 biomass gasification projects were identified in this region, all based on biomass gasification coupled to reciprocating engines. Most of these projects used downdraft, fixed-bed gasification technology, with only a few featuring bubbling, fluid-bed gasifiers. The gasification projects aimed at the electrification of rural areas in Sub-Saharan Africa have targeted off-grid villages and peri-urban areas. Each project faced unique challenges such as technical problems, economic feasibility issues, and grid connection delays, contributing to their varied statuses ranging from experimental phases to being nonoperational. These projects are typically developed by various organizations such as nonprofit institutions, government agencies, NGOs, for-profit consulting firms, technology developers, or social enterprises. Specific details about each project, including locations, ownership, manufacturers, project developers, and funding sources, can be consulted in this report.

Brazil, Cuba, and Honduras are some examples in Latin America where biomass gasification has been explored. In Brazil, several biomass gasification projects have been implemented, driven by the country's vast agricultural resources and government support for renewable energy initiatives. These projects typically focus on utilizing agricultural residues such as sugarcane bagasse, a byproduct of the country's large sugarcane industry, which provides a consistent and abundant feedstock. Despite various efforts and substantial government support, biomass gasification projects in Latin America have faced similar challenges as those in Sub-Saharan Africa, such as feedstock supply issues, technical problems, and insufficient operator skills (LTS International Ltd, 2021).

In Asia, countries such as China, India, Indonesia, Thailand, and the Philippines have had significant involvement in biomass gasification projects. Notably, China hosts numerous research and development organizations, with several spin-off companies emerging from these efforts. However, India stands out as the leading nation in the development and deployment of biomass gasifiers (Dasappa, 2011; LTS International Ltd, 2021). The Ministry of New and Renewable Energy in India has supported extensive research and development in this field, resulting in the manufacture and installation of thousands of gasifiers. Indian manufacturers such as Ankur Scientific and Husk Power Systems have exported their technology to various countries, including those in Sub-Saharan Africa, Latin America, and Southeast Asia, and most of their current business is overseas (LTS International Ltd, 2021). However, scant information is available in the public domain regarding the technical, economic, and socioenvironmental performance of these systems. Most performance data comes directly from manufacturers, but tends to be overestimated, often leading to perceptual discrepancies between the technology provider and the end-user (Dasappa, 2011).

Gasification holds promise for electrifying rural areas in Insulindia and Oceania, with initiatives dating back to the early 1980s in Indonesia and the Philippines, and further deployments across the Pacific region, notably in Vanuatu (LTS International Ltd, 2021). Insulindia has seen significant activity in biomass gasification, driven by its extensive agricultural sector, to address the energy needs of rural communities, particularly those that are off-grid. Biomass gasification projects in Insulindia often focused on using rice husk, coconut shells, palm oil residues, and other agricultural byproducts to produce electricity. While Insulindia initially witnessed various biomass gasification projects, interest waned after grid connectivity became widespread. By contrast, in the Pacific region, a remarkable biomass gasification project was financed by the European Commission under the LOME II programme. This initiative involved the installation of several gasifiers, with the first being set up at a school in Onesua, Vanuatu.

Conclusions and recommendations

Biomass gasification has demonstrated significant potential as a renewable energy solution for rural electrification across various geographical regions. This technology can play a crucial role in socioeconomic development by providing low-cost and reliable energy access, fostering economic activities, and improving the quality of life in rural communities. Key findings from the reviewed studies highlight that biomass gasification often becomes economically competitive with diesel generation, particularly when plants operate above certain capacity utilization factors. In some rural regions, it can be cost-effective at lower capacity factors due to the wide availability of feedstock and lower operational costs. Many studies emphasize the environmental advantages of biomass gasification, including reductions in CO_2 emissions compared to traditional fossil fuel-based generation. Technological advancements have improved the efficiency of biomass gasifiers, with the potential for achieving higher overall efficiencies when integrated with CHP systems. However, challenges such as high initial costs, maintenance requirements, and the need for technical expertise persist.

To harness the full potential of biomass gasification for rural electrification, several recommendations emerge from the analysis. Firstly, policies and incentives should be tailored to support the initial investment and ongoing operational costs of biomass gasification projects, especially in low-income regions. Secondly, there should be a concerted effort to develop local technical expertise through training programs and partnerships with educational institutions to ensure the sustainability and efficiency of biomass gasification systems. Thirdly, comprehensive feasibility studies should be conducted to identify the optimal locations and feedstock sources for biomass gasification plants, taking into account the local economic, environmental, and social contexts. Fourthly, integrating biomass gasification with other renewable energy sources and energy storage solutions can enhance grid stability and reliability. Lastly, ongoing research and development should be supported to further improve the efficiency and reduce the costs of biomass gasification technologies, ensuring they remain competitive with other renewable and nonrenewable energy sources. Implementing these recommendations can significantly advance the deployment of biomass gasification, thereby contributing to sustainable rural electrification and broader renewable energy goals.

Acknowledgments

This research work was supported by the project entitled: "Renewables energies for Africa: Effective valorization of agri-food wastes (REFFECT AFRICA)", funded by the European Commission under the Horizon 2020 European Framework Programme (Grant agreement ID: 101036900).

References

Abe, H., Katayama, A., Sah, B. P., Toriu, T., Samy, S., Pheach, P., Adams, M. A., & Grierson, P. F. (2007). Potential for rural electrification based on biomass gasification in Cambodia. *Biomass and Bioenergy*, *31*(9). Available from https://doi.org/10.1016/j.biombioe.2007.06.023.

Aberilla, J. M., Gallego-Schmid, A., & Azapagic, A. (2019). Environmental sustainability of small-scale biomass power technologies for agricultural communities in developing countries. *Renewable Energy, 141*. Available from https://doi.org/10.1016/j.renene.2019.04.036, http://www.journals.elsevier.com/renewable-and-sustainable-energy-reviews/.

Aceituno Dardon, J. D., & Farzaneh, H. (2024). Techno-economic analysis of a hybrid photovoltaic-wind-biomass-battery system for off-grid power in rural Guatemala. *Utilities Policy, 88*, 101762. Available from https://doi.org/10.1016/j.jup.2024.101762.

Agency I. R. E. (2023). *Off-grid renewable energy solutions to expand electricity access*. IREA.

Aguado, R., Escámez, A., Jurado, F., & Vera, D. (2023). Experimental assessment of a pilot-scale gasification plant fueled with olive pomace pellets for combined power, heat and biochar production. *Fuel, 344*, 128127. Available from https://doi.org/10.1016/j.fuel.2023.128127.

Aguado, R., Vera, D., Jurado, F., & Beltrán, G. (2022). An integrated gasification plant for electric power generation from wet biomass: Toward a sustainable production in the olive oil industry. *Biomass Conversion and Biorefinery*. Available from https://doi.org/10.1007/s13399-021-02231-0.

Ansong, M., Mensah, L. D. M. S. (2017). Techno-economic analysis of a hybrid system to power a mine in an off-grid. Sustainable Energy Technologies and Assessments. Available from http://doi.org/10.1016/j.seta.2017.09.001

Arranz-Piera, P., Kemausuor, F., Darkwah, L., Edjekumhene, I., Cortes, J., & Velo, E. (2018). Mini-grid electricity service based on local agricultural residues. *Energy, 153*. Available from https://doi.org/10.1016/j.energy.2018.04.058.

Awopone, A. K. (2021). Feasibility analysis of off-grid hybrid energy system for rural electrification in Northern Ghana. *Cogent Engineering, 8*(1). Available from https://doi.org/10.1080/23311916.2021.1981523, https://www.cogentoa.com/journal/engineering.

Ayodele, T. R., Mosetlhe, T. C., Yusuff, A. A., & Ogunjuyigbe, A. S. O. (2021). Off-grid hybrid renewable energy system with hydrogen storage for South African rural community health clinic. *International Journal of Hydrogen Energy, 46*(38). Available from https://doi.org/10.1016/j.ijhydene.2021.03.140.

Bako, T., Mamai, E. A., & Istifanus, A. B. (2020). Effects of groundnut shells on soil properties, growth and yield of maize. *Net Journal of Agricultural Science, 8*(4).

Basu, P. (2018). *Biomass gasification, pyrolysis and torrefaction* (3). Academic Press. Available from http://doi.org/10.1016/C2016-0-04056-12018.

Beltramo, R., Peira, G., Pasino, G., & Bonadonna, A. (2024). Quality of life in rural areas: A set of indicators for improving wellbeing. *Sustainability, 16*(5), 1804. Available from https://doi.org/10.3390/su16051804.

Blunck, M., & Dimpl, E. (2010). Biomass gasification—Is it a practicable option for rural electrification? *Appropriate Technology, 37*(2). Available from http://www.researchinformation.co.uk/aptearch/2010-2-Apr-Jun/pageflip.html.

Bridgwater, A. V. (1995). The technical and economic feasibility of biomass gasification for power generation. *Fuel, 74*(5). Available from https://doi.org/10.1016/0016-2361(95)00001-l.

Buragohain, B., Mahanta, P., & Moholkar, V. S. (2010). Biomass gasification for decentralized power generation: The Indian perspective. *Renewable and Sustainable Energy Reviews, 14*(1). Available from https://doi.org/10.1016/j.rser.2009.07.034.

Chambon, C. L., Karia, T., Sandwell, P., & Hallett, J. P. (2020). Techno-economic assessment of biomass gasification-based mini-grids for productive energy applications: The case of rural India. *Renewable Energy, 154*. Available from https://doi.org/10.1016/j.renene.2020.03.002.

Dasappa, S. (2011). Potential of biomass energy for electricity generation in subSaharan Africa. *Energy for Sustainable Development*, *15*(3). Available from https://doi.org/10.1016/j.esd.2011.07.006.

Demirbas, A. (2006). Biomass gasification for power generation in Turkey. *Energy Sources, Part A: Recovery, Utilization, and Environmental Effects*, *28*(5). Available from https://doi.org/10.1080/009083190913584.

Ejiofor, O. S., Okoro, P. A., Ogbuefi, U. C., Nnabuike, C. V., & Okedu, K. E. (2020). Off-grid electricity generation in Nigeria based on rice husk gasification. *Cleaner Engineering and Technology*, *1*. Available from https://doi.org/10.1016/j.clet.2020.100009.

Fernández-Lobato, L., Aguado, R., Jurado, F., & Vera, D. (2022). Biomass gasification as a key technology to reduce the environmental impact of virgin olive oil production: A life cycle assessment approach. *Biomass and Bioenergy*, *165*, 106585. Available from https://doi.org/10.1016/j.biombioe.2022.106585.

Feyissa, E. A., Tibba, G. S., Binchebo, T. L., Bekele, E. A., & Kole, A. T. (2024). Energy potential assessment and techno−economic analysis of micro hydro−photovoltaic hybrid system in Goda Warke village, Ethiopia. *Clean Energy*, *8*(1). Available from https://doi.org/10.1093/ce/zkad080, https://academic.oup.com/ce/pages/About.

Fischer, B., & Pigneri, A. (2011). Potential for electrification from biomass gasification in Vanuatu. *Energy*, *36*(3). Available from https://doi.org/10.1016/j.energy.2010.12.066.

Flores, H. F. V., Furubayashi, T., & Nakata, T. (2016). Decentralised electricity generation system based on local renewable. *Clean Technologies and Environmental Policy*, *18*(3). Available from https://doi.org/10.1007/s10098-015-1067-x.

Getahun, E., Tessema, D., & Gabbiye, N. (2019). Design and development of household gasifier cooking stoves: natural versus forced draft. In F. A. Zimale, T. Enku Nigussie, & S. W. Fanta (Eds.), *Advances of science and technology* (298, p. 314). Springer International. Available from http://doi.org/10.1007/978-3-030-15357-1_25.

Gkartzios, M., Gallent, N., & Scott, M. (2022). A capitals framework for rural areas: 'Place-planning' the global countryside. *Habitat International*, *127*, 102625. Available from https://doi.org/10.1016/j.habitatint.2022.102625.

Jha, S., Nanda, S., Acharya, B., & Dalai, A. K. (2022). A review of thermochemical conversion of waste biomass to biofuels. *Energies*, *15*(17). Available from https://doi.org/10.3390/en15176352, http://www.mdpi.com/journal/energies/.

Kaoma, M., & Gheewala, S. H. (2021). Techno-economic assessment of bioenergy options using crop and forest residues for nonelectrified rural growth centres in Zambia. *Biomass and Bioenergy*, *145*, 105944. Available from https://doi.org/10.1016/j.biombioe.2020.105944.

Littlejohns, J. V., Butler, J., Luque, L., Kannangara, M., & Totolo, S. (2020). Analysis of the performance of an integrated small-scale biomass gasification system in a Canadian context. *Biomass Conversion and Biorefinery*, *10*(2). Available from https://doi.org/10.1007/s13399-019-00442-0.

Mamphweli, N. S., & Meyer, E. L. (2009). Implementation of the biomass gasification project for community empowerment at Melani village, Eastern Cape, South Africa. *Renewable Energy*, *34*(12). Available from https://doi.org/10.1016/j.renene.2009.06.011.

Masters, W. A., Ghosh, S., Daniels, J. A., Sarpong, D. B. (2013). Comprehensive assessment of the peanut value chain for nutrition improvement in Ghana. https://sites.tufts.edu/willmasters/files/2013/11/TuftsReportForGAIN_PeanutValueChainInGhana_PublicVersion_Sept2013.pdf.

Mishra, S., & Upadhyay, R. K. (2021). Review on biomass gasification: Gasifiers, gasifying mediums, and operational parameters. *Materials Science for Energy Technologies*, *4*. Available from https://doi.org/10.1016/j.mset.2021.08.009, http://www.keaipublishing.com/en/journals/materials-science-for-energy-technologies/.

Molino, A., Chianese, S., & Musmarra, D. (2016). Biomass gasification technology: The state of the art overview. *Journal of Energy Chemistry*, *25*(1). Available from https://doi.org/10.1016/j.jechem.2015.11.005.

Muyanja, D., Allen, J. G., Vallarino, J., Valeri, L., Kakuhikire, B., Bangsberg, D. R., Christiani, D. C., Tsai, A. C., & Lai, P. S. (2017). Kerosene lighting contributes to household air pollution in rural Uganda. *Indoor Air*, *27*(5). Available from https://doi.org/10.1111/ina.12377.

Naqvi, M., Yan, J., Dahlquist, E., & Naqvi, S. R. (2016). Waste Biomass gasification based off-grid electricity generation: A case. *Energy Procedia*, *103*. Available from https://doi.org/10.1016/j.egypro.2016.11.307, http://doi.org/10.1016/j.egypro.2016.11.307.

NIRAS-LTS, E4tech, AIGUASOL and Aston University. (2021). *Bioenergy for sustainable local energy services and energy access in Africa*. LTS International Ltd. https://tea.carbontrust.com/wp-content/uploads/2021/09/BSEAA2_Prospects-for-commercial-biomass-gasification-in-Sub-Saharan-Africa.pdf.

Njenga, M., Iiyama, M., Jamnadass, R., Helander, H., Larsson, L., de Leeuw, J., Neufeldt, H., Röing de Nowina, K., & Sundberg, C. (2016). Gasifier as a cleaner cooking system in rural Kenya. *Journal of Cleaner Production*, *121*. Available from https://doi.org/10.1016/j.jclepro.2016.01.039.

Odoi-Yorke, F., Abaase, S., Zebilila, M., Atepor, L., & Ai, Q. (2022). Feasibility analysis of solar PV/biogas hybrid energy system for rural electrification in Ghana. *Cogent Engineering*, *9*(1). Available from https://doi.org/10.1080/23311916.2022.2034376.

Owen, M., & Ripken, R. (2017). LTS International Limited, The University of Edinburgh and E4tec.

Palit, D., Malhotra, R., & Kumar, A. (2011). Sustainable model for financial viability of decentralized biomass gasifier based power projects. *Energy Policy*, *39*(9). Available from https://doi.org/10.1016/j.enpol.2011.06.026.

Pandiyan, P., Sitharthan, R., Saravanan, S., Prabaharan, N., Ramji Tiwari, M., Chinnadurai, T., Yuvaraj, T., & Devabalaji, K. R. (2022). A comprehensive review of the prospects for rural electrification using stand-alone and hybrid energy technologies. *Sustainable Energy Technologies and Assessments*, *52*, 102155. Available from https://doi.org/10.1016/j.seta.2022.102155.

Piapdje Tamo, M. U., Branco, J. P., Iddi, H., & Dei, T. (2018). Introduction of biomass gasification for rural electrification in Cameroon. *IOP Conference Series: Earth and Environmental Science*, *188*(1), 012024. Available from https://doi.org/10.1088/1755-1315/188/1/012024.

Qudrat-Ullah, H. (2004). Bamboo gasification for sustainable energy and rural development in Uganda. In *Exploring the dynamics of renewable energy and sustainable development in Africa*. Cham: Springer Nature Switzerland. http://doi.org/10.1007/978-3-031-48528-2_4. doi:10.1007/978-3-031-48528-2_4.

Rajbongshi, R., Borgohain, D., & Mahapatra, S. (2017). Optimization of PV-biomass-diesel and grid base hybrid energy systems for rural electrification by using HOMER. *Energy*, *126*. Available from https://doi.org/10.1016/j.energy.2017.03.056, https://www.sciencedirect.com/science/article/pii/S0360544217304310.

Ramamurthi, P. V., Fernandes, M. C., Nielsen, P. S., & Nunes, C. P. (2016). Utilisation of rice residues for decentralised electricity generation in Ghana: An economic analysis. *Energy*, *111*. Available from https://doi.org/10.1016/j.energy.2016.05.116, http://www.elsevier.com/inca/publications/store/4/8/3/.

Ramchandran, N., Pai, R., & Parihar, A. K. S. (2016). Feasibility assessment of anchor-business-community model for off-grid rural electrification in India. *Renewable Energy*,

97. Available from https://doi.org/10.1016/j.renene.2016.05.036, http://www.journals.elsevier.com/renewable-and-sustainable-energy-reviews/.

Ribo-Perez, D., Herraiz-Cañete, A., Alfonso-Solar, D., Vargas-Salgado, C., & Gomez-Navarro, T. (2021). Modelling biomass gasifiers in hybridrenewable energy microgrids; A complete procedure for enabling gasifiers simulation. *Renewable Energy, 174*, 501–512. Available from https://doi.org/10.1016/j.renene.2021.04.083.

Sakketa, T. G. (2022). Urbanisation and rural development in developing countries: A review of pathways and impacts. *Research in Globalization, 6*. Available from https://doi.org/10.23661/dp5.20222022.

Salisu, J., Gao, N., & Quan, C. (2021). Techno-economic assessment of co-gasification of rice husk and plastic waste as an off-grid power source for small scale rice milling—an aspen plus model. *Journal of Analytical and Applied Pyrolysis, 158*, 105157. Available from https://doi.org/10.1016/j.jaap.2021.105157.

Sanchez, A. S., Torres, E. A., & Kalid, R. A. (2015). Renewable energy generation for the rural electrification of isolated. *Renewable and Sustainable Energy Reviews, 49*. Available from https://doi.org/10.1016/j.rser.2015.04.075, http://doi.org/10.1016/j.rser.2015.04.075.

Sanchez-Lozano, D., Escamez, A., Aguado, R., Oulbi, S., Hadria, R., & Vera, D. (2023). Techno-economic assessment of an off-grid biomass gasification CHP plant. *Applied Sciences, 13*(10). Available from https://doi.org/10.3390/app13105965, http://doi.org/10.3390/app13105965.

Sharma, P., Gupta, B., Pandey, M., Bisen, K. S., & Baredar, P. (2021). Downdraft biomass gasification: A review on concepts, designs analysis, modelling and recent advances. *Materials Today: Proceedings, 46*. Available from https://doi.org/10.1016/j.matpr.2020.08.789.

Singh, J. (2017). Management of the agricultural biomass on decentralized basis for producing sustainable power in India. *Journal of Cleaner Production, 142*, 3985–4000. Available from https://doi.org/10.1016/j.jclepro.2016.10.056.

Situmorang, Y. A., Zhao, Z., Yoshida, A., Abudula, A., & Guan, G. (2020). Small-scale biomass gasification systems for power generation (<200 kW class): A review. *Renewable and Sustainable Energy Reviews, 117*, 109486. Available from https://doi.org/10.1016/j.rser.2019.109486.

Sou, S., Siemers, W., & Exell, R. H. B. (2010). The design of a photovoltaic/biomass hybrid electrical energy system for a rural village in Cambodia. *International Journal of Ambient Energy, 31*(1), 3–12. Available from https://doi.org/10.1080/01430750.2010.9675803.

Taubenböck, H., Otto, C., Gülzau, F., & Mau, S. (2023). Border regions across the globe: Analyzing border typologies, economic and political disparities, and development dynamics. *Applied Geography, 151*, 102866. Available from https://doi.org/10.1016/j.apgeog.2022.102866.

Tezer, Ö., Karabağ, N., Öngen, A., Çolpan, C. Ö., & Ayol, A. (2022). Biomass gasification for sustainable energy production: A review. *International Journal of Hydrogen Energy, 47*(34). Available from https://doi.org/10.1016/j.ijhydene.2022.02.158, http://www.journals.elsevier.com/international-journal-of-hydrogen-energy/.

The World Bank. (2023). Access to electricity (% of the population). https://datos.bancomundial.org/indicador/EG.ELC.ACCS.ZS.

Tostado-Veliz, M., Escamez, A., Aguado, R., Sanchez-Lozano, D., Jurado, F., & Vera, D. (2024). Optimal sizing of hybrid PV–diesel–biomass gasification plants for electrification of off-grid communities: An efficient approach based on Benders' decomposition. *Journal of Cleaner Production, 458*, 2024.

United Nations Development Programme. (2016). *Delivering sustainable energy in a changing climate*. https://www.un-energy.org/wp-content/uploads/2017/01/UNDP-Energy-Strategy-2017-2021.pdf.

Vera, D., Jurado, F., Margaritis, N. K., & Grammelis, P. (2014). Experimental and economic study of a gasification plant fuelled with olive industry wastes. *Energy for Sustainable Development*, *23*. Available from https://doi.org/10.1016/j.esd.2014.09.011, http://www.elsevier.com.

World Health Organization (2023). https://www.who.int/news-room/fact-sheets/detail/household-air-pollution-and-health.

Energy from waste

3

Antonio Escámez, Daniel Sánchez-Lozano, Roque Aguado, David Vera and Francisco Jurado Melguizo
Department of Electrical Engineering, University of Jaén, Jaén, Spain

Introduction

The management and disposal of waste have become increasingly significant environmental and societal challenges globally (Association, I.S.W, 2024). In recent years, there has been a growing emphasis on finding sustainable and innovative solutions to address these challenges, including the utilization of waste as a valuable resource for energy production. This paradigm shift has led to the emergence of waste-to-energy (WtE) or energy-from-waste technologies, which aim to convert various types of waste into usable forms of energy such as electricity and/or heat, thereby reducing dependence on finite fossil fuels, mitigating greenhouse gas emissions, and promoting a circular economy approach.

Biogenic waste

One of the key categories of waste targeted for energy recovery is biogenic waste, which consists of organic materials derived from plants or animals. Biogenic waste encompasses a diverse range of sources, including agricultural waste, food waste, municipal organic waste, and wastewater sludge. These materials possess inherent energy content that can be harnessed through various WtE processing technologies, contributing to both renewable energy generation and waste management objectives.

Biogenic waste can be classified into primary and secondary categories based on their origin and processing. Primary wastes, directly generated from forestry and agricultural activities, retain original characteristics without significant processing. Examples include logs, branches, leaves, and agricultural residues. These materials serve as vital sources of organic matter and nutrients for composting, biomass product manufacturing, and energy generation. Secondary wastes, resulting from primary waste processing and industrial activities, include wood shavings, sawdust, and agricultural byproducts like sugarcane bagasse and nut shells (Gupta et al., 2022). These secondary wastes find applications in various industries such as fiberboard production, paper manufacturing, renewable energy generation, and biofuel production.

Towards Future Smart Power Systems with High Penetration of Renewables. DOI: https://doi.org/10.1016/B978-0-443-29871-4.00003-8

Primary waste

According to the classification of waste discussed above, this section focuses on primary waste, which is waste that comes directly from agricultural, forestry, livestock, and fishing activities; see Fig. 3.1. These wastes retain their original characteristics and have not undergone notable alterations. They include numerous types of waste such as organic matter, plant remains, animal waste, and fishery products, all generated in the primary production of food and natural resources.

Agricultural and forestry practices are abundant sources of derived residues that pose significant challenges in their management. Annually, it is estimated that around 140 Gt of biomass residues is generated worldwide, highlighting the importance of

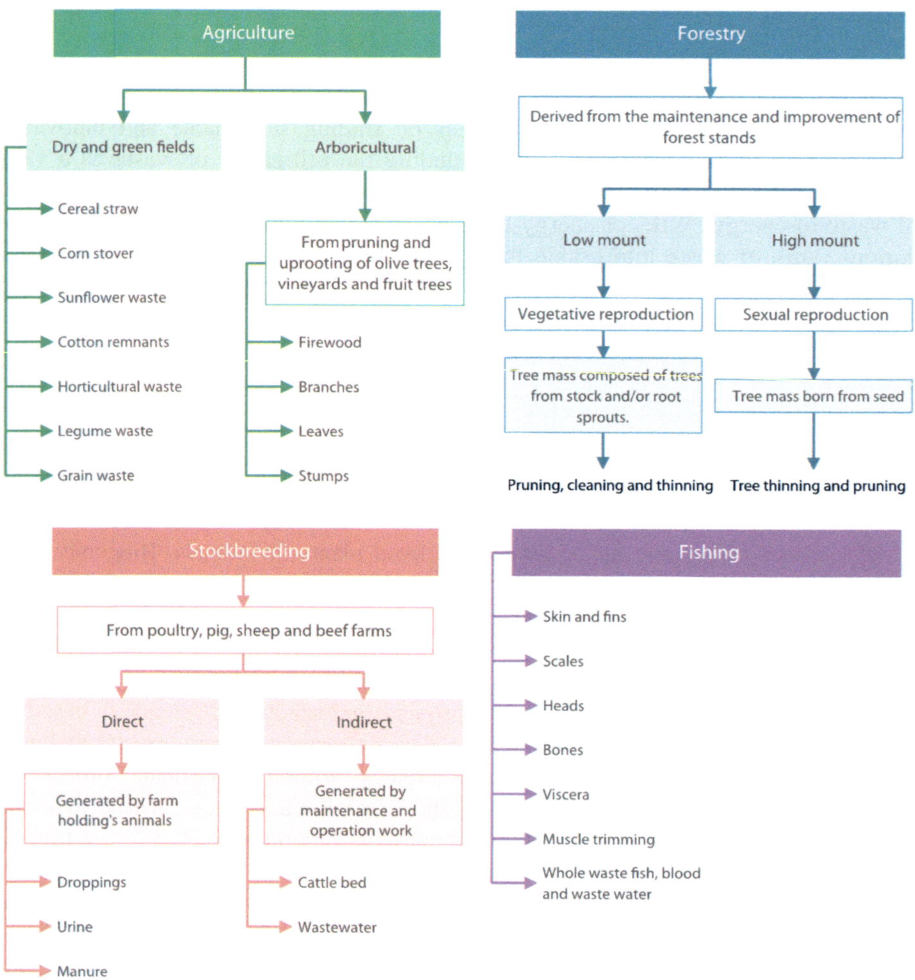

Figure 3.1 Classification of primary waste.

effectively addressing this issue to avoid adverse environmental impacts (Tripathi et al., 2019). Agriculture stands out as the sector with the highest biomass production (Duque-Acevedo et al., 2020). Primary agricultural residues encompass a wide range, from crop residues such as cereals, oilseeds, and vegetables, to waste from fruit trees and vineyards. Often, these residues are discarded or burned on-site; however, their reuse represents a valuable source of raw material with multiple applications (Parfitt et al., 2021; Tripathi et al., 2019). Similarly, this occurs in the case of primary forest residues, which include remnants from forest management activities such as maintenance and thinning of forests. The accumulation of these forest residues poses additional challenges, as they can not only contribute to the spread of wildfires but also have a negative impact on ecosystem biodiversity (Tripathi et al., 2019).

In livestock farms, the generation of large amounts of waste can translate into environmental and social deterioration in many rural areas (Balaman, 2019). Livestock waste is produced as a result of intensive or extensive breeding of livestock in any of its typologies (Komova et al., 2023). Animal waste has been used as an organic fertilizer in traditional agriculture (Liu & Chen, 2014); however, nowadays, animal waste can be utilized as sources for biomass-based conversion processes, especially in the production of bioenergy and biofertilizers (Balaman, 2019). It is important to note that there is no precise classification of the primary waste generated by the livestock industry, as these differ for each country (Komova et al., 2023). In this case, waste has been classified into direct waste, which originates directly from and is generated by the animal, such as slurry or urine and indirect waste, which is generated from the use and maintenance of the livestock farm. Only organic waste was considered in this classification, while zoosanitary or hazardous waste was not taken into account due to the lack of energy recovery potential

Fishery-related waste from the primary sector amounts to 17.9–39.5 million tons/year worldwide (Tan et al., 2024). This fishery waste is composed of the putrescible nature of the product and the presence of inedible fractions such as bones, skin, scales, and so forth. Of the total biogenic waste from fisheries (Coppola et al., 2021; Tan et al., 2024), the majority consists of skin and fins (3%), scales (5%), heads (12%), bones (15%), viscera (18%), muscle-trimming (20%), and miscellaneous waste such as whole waste fish, blood, waste water, and so forth (27%) (Tan et al., 2024).

Fig. 3.2 shows the data on the amount of primary waste generated by each sector organized by continent.

Secondary waste

The production of secondary waste arises from the conversion of raw materials obtained from the primary sector into final products (Naik et al., 2023), generating a substantial amount of untreated organic waste and byproducts (Bibi et al., 2023). These wastes can cause environmental harm and impact both humans and animals. Most of these wastes are not handled correctly and are disposed of improperly, leading to increased greenhouse gas emissions and exacerbating issues related to climate change (Bibi et al., 2022, 2023).

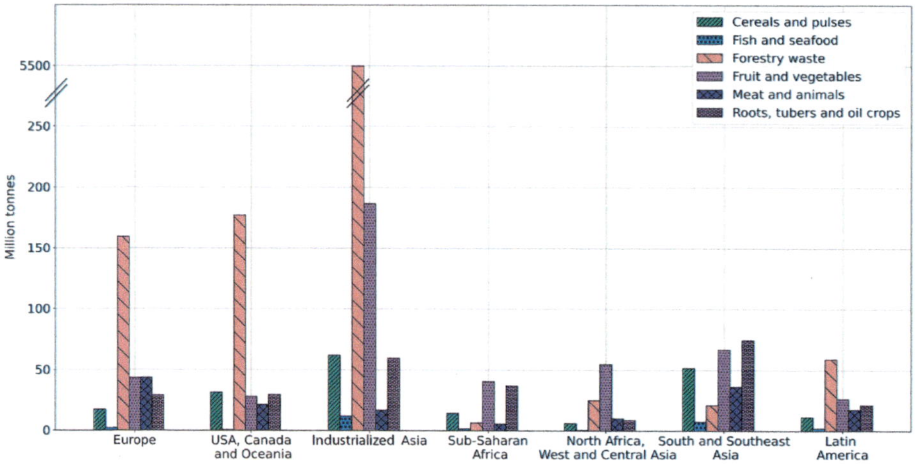

Figure 3.2 Amount of primary waste generated by continent (Parfitt et al., 2021; Tripathi et al., 2019).

These wastes are generated in various industries in different sectors; therefore, the waste classification shown in Fig. 3.3 has been followed. For instance, within agriculture-related sectors like olive oil mills, byproducts such as olive pomace and olive stones emerge during the extraction process. Livestock industries typically yield meat waste from product cleaning and bones. Wood processing industries are known for generating wood shavings and sawdust throughout their operations. Similarly, in the fishing industry, waste results from the cleaning of raw materials, mirroring the pattern seen in the livestock sector.

Municipal solid waste

Municipal solid waste (MSW) constitutes a diverse array of materials generated within residential, commercial, and industrial settings (Funk et al., 2020). These materials range from garden waste and office paper to printed newspapers, fruit and vegetable scraps, food leftovers, and yard trimmings, with food waste comprising the bulk of the organic fraction. Additionally, MSW also contains nonbiodegradable items such as metal and glass (Funk et al., 2020; Ghosh et al., 2020). It is noteworthy that hazardous industrial waste is not part of MSW (Rao et al., 2017).

The classification of MSW, as depicted in Fig. 3.4, categorizes them into two main groups: putrescible and nonputrescible. Putrescible waste consists of organic materials prone to decomposition, like food scraps and garden waste, which may emit unpleasant odors if not handled properly. In contrast, nonputrescible waste includes inorganic materials such as plastics, metals, and glass, which do not readily decompose but can pose environmental risks if improperly managed (Dixon & Langer, 2006; Zhou et al., 2014).

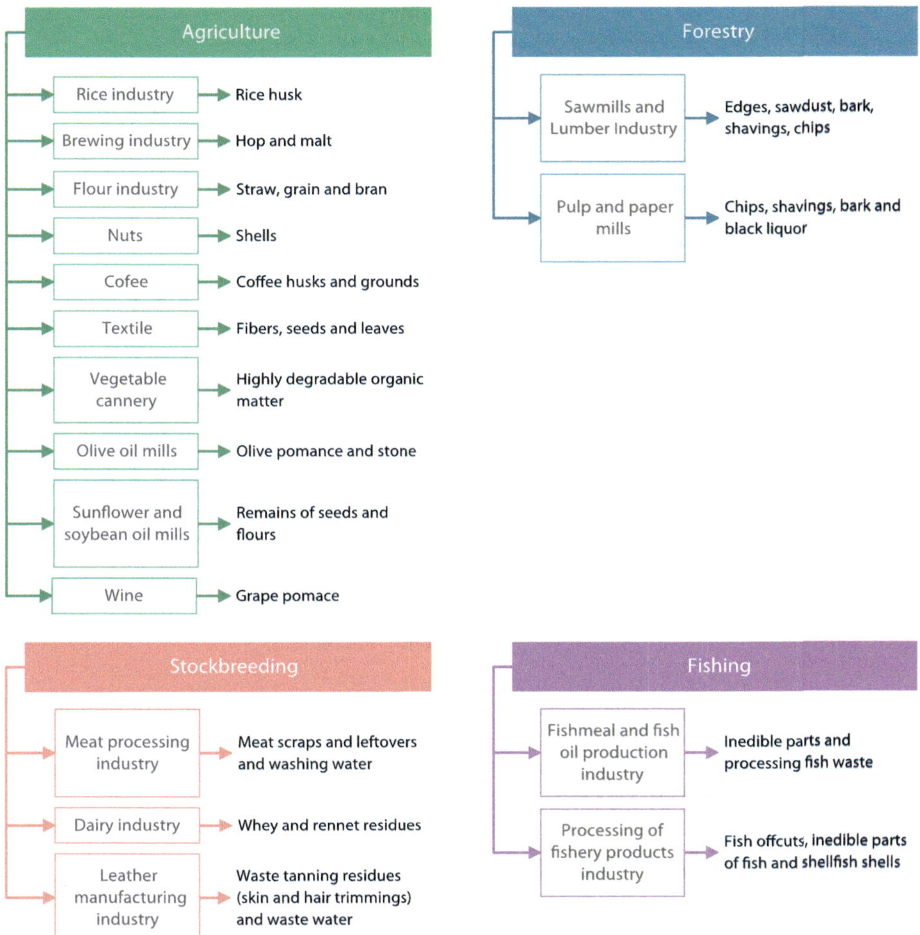

Figure 3.3 Classification of secondary waste.

The generation of MSW is heavily influenced by various factors including economic conditions and urban development within each nation (Najafi et al., 2024). Recent data suggests that global MSW generation reached an estimated 2010 million tons in 2016 (Reconstruction & Development, 2018). When examining waste generation by regions, it is evident that countries in the East Asia and Pacific as well as Europe and Central Asia regions collectively account for 43% of global waste production. Conversely, the Middle East and North Africa, along with Sub-Saharan Africa regions, contribute the least, comprising only 15% of the total. Specifically, the East Asia and Pacific region stands out as the largest waste generator, producing approximately 468 million tons in 2016, whereas the Middle East and North Africa region generates the least, totaling 129 million tons (Reconstruction & Development, 2018).

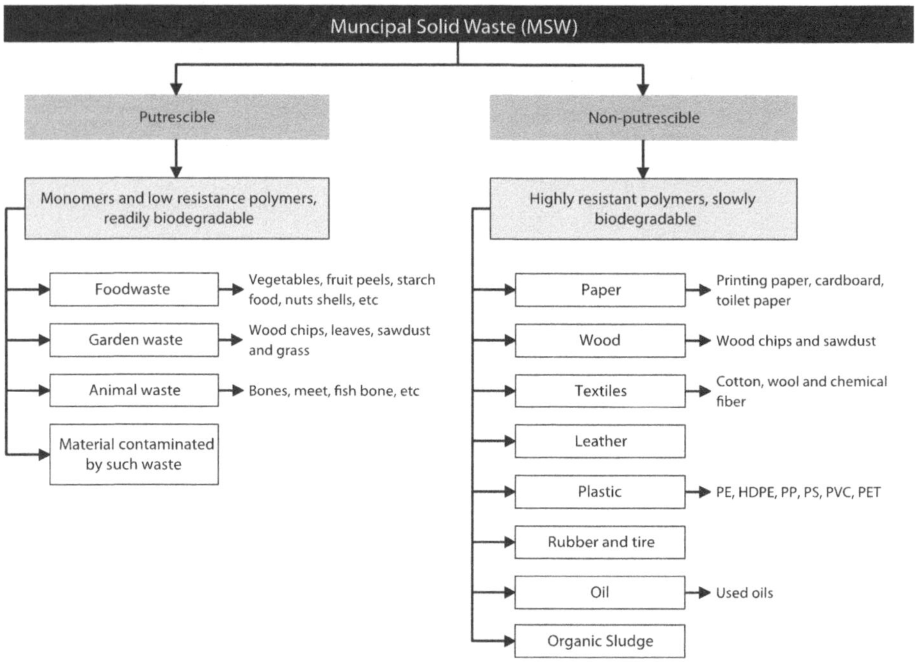

Figure 3.4 Classification of municipal solid waste (MSW) (Dixon & Langer, 2006; Zhou et al., 2014).

Waste processing technologies

Thermochemical processing

Three existing thermochemical methods are available for energy production from waste, namely, combustion, gasification, and pyrolysis in the descending order of their popularity (Vukovic & Makogon, 2022).

Combustion

Combustion represents a fundamental chemical process integral to numerous industrial applications and energy generation methods. It involves the rapid oxidation of a fuel in the presence of oxygen, resulting in the release of heat, light, and the formation of combustion products such as water vapor, carbon dioxide, and other byproducts contingent upon the fuel composition (Oriakhi, 2009). This exothermic reaction, governed by the combustion triangle encompassing fuel, oxygen, and heat, necessitates precise proportions of these components for initiation and sustenance.

The combustion process typically proceeds through several stages, commencing with ignition, where an external heat source elevates the fuel and oxidizer mixture's temperature to its ignition point. Subsequently, the flame propagates through the fuel-air mixture, accompanied by rapid heat liberation as chemical bonds within the fuel molecules break and recombine with oxygen molecules (Wang et al., 2012). This progression, marked by the formation of a flame front, consumes fuel as it advances. As combustion advances, the released heat further raises the temperature of the surrounding gases, intensifying reactions and accelerating the process.

Combustion processes facilitate the conversion of the chemical energy inherent in waste into thermal energy through the utilization of surplus air, typically at temperatures ranging from 850°C to 1200°C (Hupa et al., 2017; Richards, 2019; Wang et al., 2012). The resultant heat can be utilized for various purposes, including power generation and propulsion. Different combustion modes exist, including premixed and diffusion combustion, each characterized by distinct fuel-air mixing and reaction kinetics. Spontaneous combustion, occurring without external heat sources, and heterogeneous combustion, involving complex fuel compositions like solid fuels, further illustrate combustion's versatility.

Often used interchangeably with incineration, particularly when the primary aim is the complete destruction of materials through burning, combustion stands as one of the most established and prevalent WtE techniques. The efficiency of harnessing useful work from heat energy is constrained by the Carnot cycle's limitations. To qualify as a viable fuel for combustion, a waste must possess the characteristic of releasing thermal energy upon vigorous oxidation. Various waste streams lend themselves to WtE via combustion, including agricultural and forestry residues, paper waste, MSW, and sewage sludge. Depending on the waste type, pretreatment processes may be necessary to adjust moisture levels, particle sizes, or densities prior to combustion. Furthermore, the chemical composition of the waste influences the quality of emissions produced during combustion.

Despite its widespread application, combustion emits greenhouse gases and pollutants (Breeze, 2018; Richards, 2019), necessitating cleaner and more efficient combustion technologies to mitigate environmental impacts while meeting energy demands. Researchers and engineers continue to explore innovations such as catalytic converters and renewable energy sources to address these challenges (Richards, 2019).

Gasification

Gasification consists of the production of a gaseous fuel, referred to as syngas, from carbonaceous material using a gasifying agent that partially oxidizes the feedstock (Aguado et al., 2023; Sanchez-Lozano et al., 2023). Syngas, also known as synthesis gas, is a mixture of combustible gases (CH_4, H_2, and CO) along with small amounts of noncombustible gases (H_2O, CO_2, or N_2), as well as some undesirable compounds (Aguado et al., 2023; Basu, 2018; Sanchez-Lozano et al., 2023). The final composition of the gas is defined by both the raw material used and the gasifying agent used, the most common being air, oxygen, and steam (Basu, 2018). The product gas can be used for power generation, thermal uses, or value-added chemical products (Basu, 2018).

In addition, a highly valued byproduct called biochar is obtained from this process (Aguado et al., 2023; Basu, 2018). This product can be used in different and varied applications, from being used to incorporate it into the soil of plantations, thereby improving the productivity (Van Nguyen et al., 2022), to adding to concrete to improve its mechanical properties (Sirico et al., 2022).

Fig. 3.5 illustrates the gasification process schematically. The waste used is mixed with the oxidizing agent in the gasification reactor where a series of chemical reactions take place. These result in the main product, syngas, and byproducts such as biochar.

Certainly, gasification entails several distinct stages (Basu, 2018). Initially, in the drying phase, the feedstock is heated to eliminate moisture, optimizing subsequent processes. Then, during pyrolysis, the dried feedstock undergoes thermal decomposition in the absence of oxygen, yielding volatile gases, tar, and char. Subsequently, combustion occurs where char reacts with a gasifying agent to generate heat, sustaining the gasification reactions. Finally, in the reduction phase, the gasifying agent interacts with the char, producing syngas (Basu, 2018)

Various waste materials can serve as feedstocks for gasification, including pomace (Aguado et al., 2023; Vera et al., 2013), chips, chunks, or pellets from different varieties of trees (Soares & Oliveira, 2020), sunflower seeds (Simone et al., 2012), rice husks (Yoon et al., 2012), sugar cane bagasse (Awais et al., 2022), nut shells (Mazhkoo et al., 2021), and so forth. Pretreatment measures may be necessary to manage moisture levels and adjust particle sizes for optimal gasification efficiency.

Pyrolysis

Pyrolysis is a thermochemical decomposition process that converts organic materials into solid biochar, biooil, and pyrolytic gas through the application of heat under anaerobic conditions (Basu, 2018; Collard et al., 2016; Grande et al., 2021; Singh et al., 2016). A general system of the pyrolysis process can be seen in Fig. 3.6. In this process, the waste enters the pyrolysis reactor to extract the three aforementioned fractions. Additionally, a portion of the generated gas can be recirculated to maintain the process temperature if necessary.

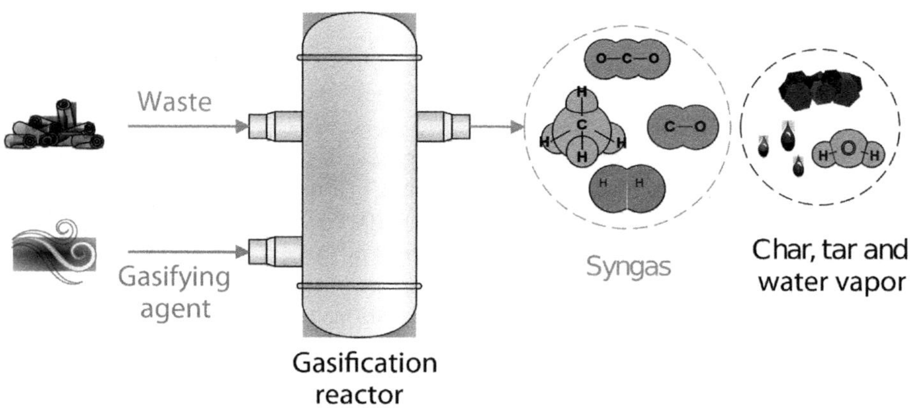

Figure 3.5 Generic gasification process.

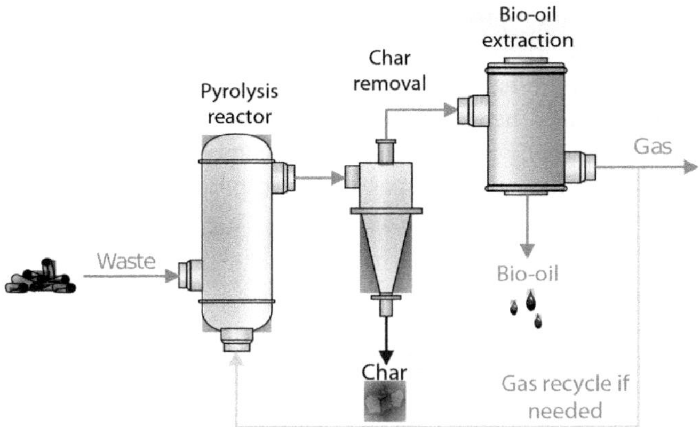

Figure 3.6 Generic pyrolysis process.

- **Biochar**, a solid residue rich in carbon, retains considerable energy content and finds applications in soil amendment and carbon sequestration (Basu, 2018).
- **Biooil**, a liquid mixture of organic compounds, serves as a potential precursor for biofuels and chemical feedstocks (Basu, 2018; Grande et al., 2021)
- **Gas**, a mixture of carbon monoxide, hydrogen, and methane, offers versatile energy applications such as heat, electricity generation, and synthesis of fuels and chemicals (Basu, 2018).

The quantity of each fraction obtained from the pyrolysis process will depend on factors such as temperature, heating rate, residence time, particle size, pressure, type of biomass, moisture content, and the pretreatment method of the biomass (Collard et al., 2016; Grande et al., 2021; Singh et al., 2016).

Slow pyrolysis, renowned for its traditional approach and the production of charcoal, can be divided into two main subtypes: carbonization and conventional pyrolysis. In carbonization, biomass undergoes a prolonged heating process at moderate temperatures (around 400°C) over several days, yielding charcoal as the primary product. Conversely, conventional pyrolysis involves heating at higher temperatures (approximately 600°C) for shorter periods, resulting in a broader range of pyrolysis products, from charcoal to gases and liquids (Singh et al., 2016). In contrast, intermediate pyrolysis is distinguished by its ability to control heating rates and prevent tar formation, allowing for greater versatility in biomass feedstock selection. This modality is characterized by longer residence times and more precise manipulation of reaction conditions (Basu, 2018; Singh et al., 2016). On the other hand, fast pyrolysis is designed to maximize biooil production, employing extremely high heating rates and very short material residence times (Basu, 2018; Singh et al., 2016). This approach, involving temperatures in the range of 425°C–600°C and rapid cooling of the gas product, is essential for achieving optimal biooil yields from biomass. Additionally, flash pyrolysis stands out for its speed, with even faster heating rates and significantly reduced material residence times. Waste is supplied in the powdered form, facilitating rapid decomposition and the production of a mixture of gases and liquids. This modality is known for its high biooil yield, which can exceed 70% of the total product, and a corresponding reduction in charcoal production (Basu, 2018; Carneiro et al., 2016; Singh et al., 2016). Finally, ultrafast pyrolysis involves a combination of high-speed biomass mixing with a solid heat carrier, followed by rapid heat transfer and cooling of the primary product. This approach is characterized by precise control of material residence time and specific operating temperatures to maximize gas or liquid production depending on process requirements (Basu, 2018; Singh et al., 2016).

Liquefaction

The thermochemical liquefaction of biomass (Fig. 3.7) has attracted significant attention due to its ability to generate high-energy density liquids swiftly, while also accommodating a wide array of materials (Gollakota et al., 2018; Grande et al., 2021; Shahbeik et al., 2024). This process is capable of handling diverse feedstocks such as lignocellulosic biomass, algae, plastics, and organic residues (Gollakota et al., 2018; Grande et al., 2021). Liquefaction operates under elevated temperatures (250°C–450°C) and pressures (100–300 bar) (Aysu & Durak, 2016; Grande et al., 2021; Usman et al., 2024). This process relies on water's unique properties to enable depolymerization reactions like hydrolysis, dehydration, and decarboxylation (Aysu & Durak, 2016). The products of liquefaction are biocrude oil or biocarbon, alongside gases such as CO_2, CO, H_2, or CH_4, as well as soluble organic substances (Aysu & Durak, 2016; Gollakota et al., 2018; Grande et al., 2021).

Various liquefaction techniques are available, including hydrothermal liquefaction (HTL), which uses water as a solvent and operates under moderate temperatures and high pressures (Aysu & Durak, 2016; Grande et al., 2021; Shahbeik et al., 2024). Unlike pyrolysis, which often produces liquid products with high oxygen

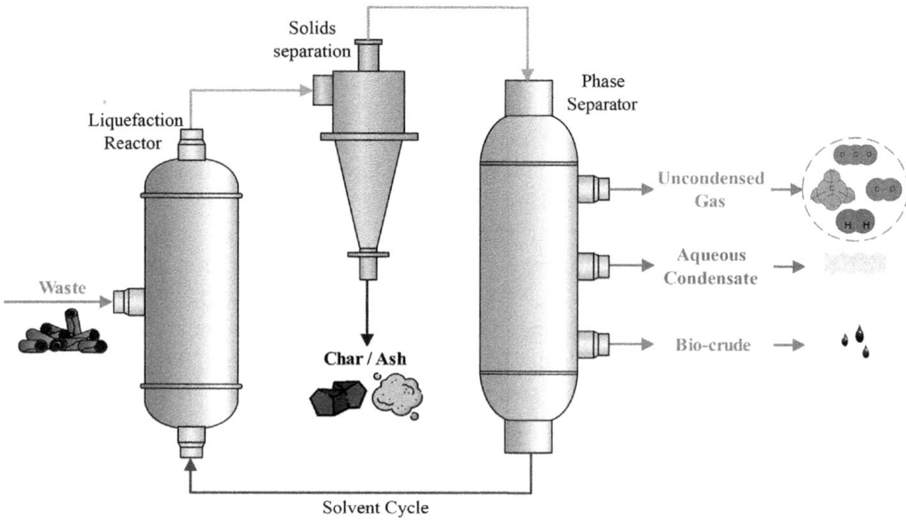

Figure 3.7 Generic liquefaction process.

content, making them unsuitable as fuels, HTL offers the prospect of generating liquids with lower oxygen content. Notably, HTL allows processing both wet and dry biomass without lipid content restrictions, yielding a renewable oil substitute known as biooil (Grande et al., 2021; Shahbeik et al., 2024). This liquid product represents a renewable alternative to fossil fuels for both electricity generation and transportation. Moreover, it can undergo refinement into various valuable products, including biofuels, biochemicals, and chemicals (Usman et al., 2024).

Biochemical processing

Anaerobic digestion

Anaerobic digestion is a biological process where microorganisms break down organic matter in the absence of oxygen to produce biogas and stabilized residues. The general process can be seen in Fig. 3.8. This process finds widespread use in wastewater treatment plants and biogas facilities to manage sludge generated during wastewater treatment and other organic waste. Anaerobic digestion takes place within an enclosed tank or vessel, commonly referred to as a digester, to create an environment devoid of oxygen (Uddin & Wright, 2023).

During anaerobic digestion, microorganisms, mainly bacteria, decompose the organic compounds present in the sludge within an oxygen-free environment. This process unfolds in several distinct stages. Initially, hydrolysis occurs, where large organic molecules are broken down into smaller molecules, such as fatty acids and sugars, by extracellular enzymes produced by bacteria. The next stage is acidogenesis, during which the hydrolysis products are transformed into short-chain organic acids such as acetic acid, propionic acid, and butyric acid by acidogenic bacteria. Following acidogenesis, acetogenesis takes place, where organic acids are primarily converted into acetate, hydrogen, and carbon dioxide by acetogenic bacteria. Finally, in the methanogenesis phase, the products of acetogenesis are further metabolized into methane (CH_4) and carbon dioxide (CO_2) by methanogenic bacteria, which thrive in oxygen-deprived environments (Uddin & Wright, 2023).

The biogas produced during anaerobic digestion primarily comprises methane (50%–70%) and carbon dioxide, with trace amounts of other gases such as hydrogen sulfide, nitrogen, and oxygen (Achinas et al., 2017; Lisbona et al., 2023). This biogas can be used to generate electricity and heat, or it can be refined and used as

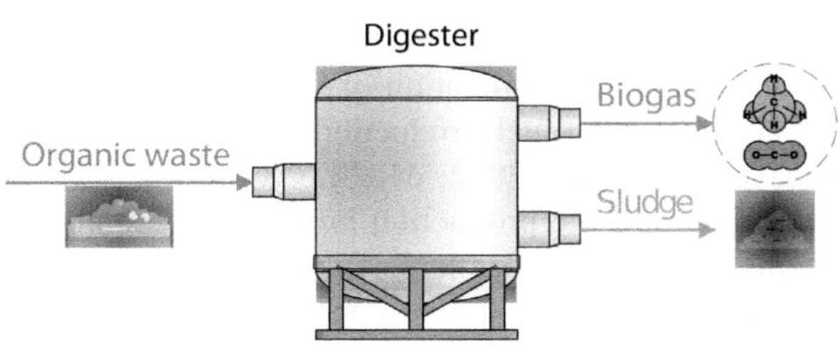

Figure 3.8 Anaerobic digestion schematic diagram.

a biofuel. In addition to biogas production, anaerobic digestion reduces the volume of sludge and eliminates pathogens (Jiang et al., 2020; Momayez et al., 2019; Rocamora et al., 2020), thus facilitating its safe disposal or utilization as fertilizer in agriculture.

Fermentation

Fermentation is a biological process that has been utilized by humans for thousands of years, with applications ranging from food and beverage production to biofuel generation.

Before fermentation can occur, it is often necessary to undergo a hydrolysis process to extract sugars from the raw material or waste substrate (Mohd Thani et al., 2020; Zhang et al., 2020). Hydrolysis involves breaking down complex carbohydrates, such as cellulose or starch, into their constituent sugar molecules, such as glucose or maltose. This process typically requires the use of enzymes, such as cellulases or amylases, which catalyze the cleavage of glycosidic bonds within the polysaccharide structure. By hydrolyzing the raw material, sugars become more readily available for subsequent fermentation by microorganisms (Cruz-Casas et al., 2021). Hydrolysis is a crucial step in various biotechnological processes, including the production of biofuels, biochemicals, and other value-added products from renewable biomass sources.

Biochemical pathways involved in sugar fermentation encompass several interconnected processes. Glycolysis, the initial step, occurs in the cytoplasm, breaking down glucose into pyruvate while generating ATP and NADH. Subsequently, under anaerobic conditions, pyruvate can undergo fermentation. Various microorganisms utilize distinct pathways: alcoholic fermentation, conducted by yeast like Saccharomyces cerevisiae, converts pyruvate into ethanol and carbon dioxide, pivotal in the production of alcoholic beverages. Lactic acid fermentation, executed by lactic acid bacteria such as Lactobacillus and Streptococcus species, transforms pyruvate into lactic acid (Song et al., 2022), crucial in dairy product fermentation like yogurt and cheese. Additionally, alternative fermentation pathways employed by certain bacteria and fungi yield diverse organic acids, alcohols, and gases, contributing to the broader spectrum of fermentation processes (Perez-Alva et al., 2022).

As depicted in Fig. 3.9, the emphasis is placed on the utilization of lignocellulosic residues such as prunings, nut shells, agricultural residues, microalgae, and so forth. This emphasis arises from the fact that these residues yield advanced biofuels, also known as second-generation biofuels, produced from raw materials not in direct competition with food and feed crops (Limayem & Ricke, 2012; Mizik & Gyarmati, 2022). In contrast, the utilization of sugary and starchy biomasses, typically used in first-generation biofuel production, would indeed compete with food consumption (Limayem & Ricke, 2012; Mizik & Gyarmati, 2022). Therefore, the focus will be directed toward the production process of second-generation bioethanol, which commonly employs lignocellulosic residues.

The conversion of lignocellulosic biomass to ethanol involves several stages; see Fig. 3.9. In the pretreatment, lignocellulose undergoes physical, chemical, or biological processes to break down its complex structure and release cellulose, hemicellulose, and

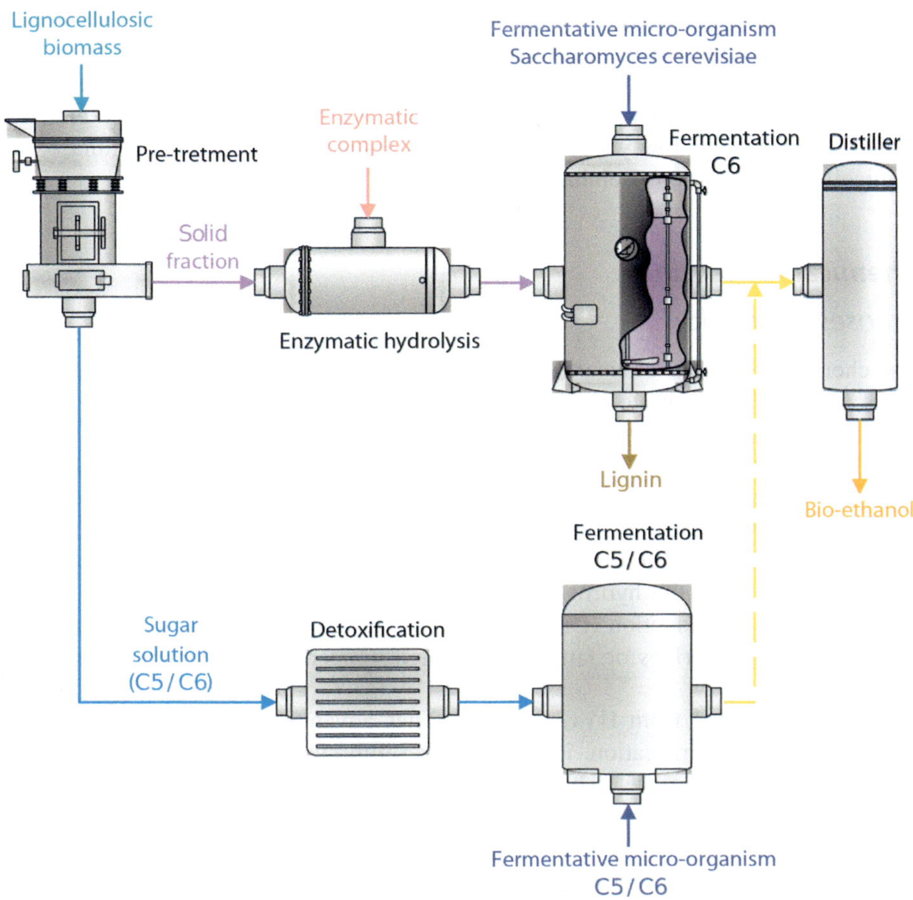

Figure 3.9 Fermentation of lignocellulosic biomass process diagram.

lignin, resulting in a solid fraction (composed of cellulose and lignin) and a liquid fraction (composed of hemicellulosic sugars) (Zhao et al., 2012). As for the solid fraction, a hydrolysis stage is required, where cellulose (C6) is broken down into simple sugars through enzymatic or acidic processes, providing the sugars necessary for fermentation (Limayem & Ricke, 2012; Zhao et al., 2012). During fermentation, microorganisms such as yeasts or bacteria convert these sugars into ethanol and other byproducts. Typically, Saccharomyces cerevisiae is used in this stage (Beluhan et al., 2023; Limayem & Ricke, 2012; Zhao et al., 2012). On the other hand, the liquid fraction undergoes a detoxification process where part of the inhibitors present in the prehydrolysate are removed. Inhibitors are compounds that prevent enzymatic hydrolysis from occurring or reduce the reaction rate. Subsequently, the fermentation phase begins, where microorganisms capable of fermenting C5 and C6 sugars such as Pichia stipitis, Pachysolen tannophilus, or Candida shehatae are used (Zhao et al., 2012). Finally, in the distillation stage, ethanol is concentrated to increase its purity.

Despite second-generation bioethanol production having significant advantages in terms of utilizing waste materials, as it does not compete with food-grade products, it presents numerous challenges that require further investigation. These include high production costs compared to the use of starchy or sugary biomass, higher enzyme costs, low yields, and inhibition issues with microbial, cellulase, and hemicellulase inhibition (Beluhan et al., 2023).

Chemical processing

Transesterification

The chemical reactions of esterification and transesterification offer promising pathways for converting waste materials into biofuels. Transesterification, also called alcoholysis, is the primary method for producing biodiesel from oil-based waste (Galadima & Muraza, 2020; Garg et al., 2024), using low molecular weight alcohols like methanol or ethanol to react with triglycerides found in oil sources. This process yields biodiesel, such as fatty acid methyl ester (Yadav & Sharma, 2018) or fatty acid ethyl ester, alongside glycerol as a byproduct (Ahmed & Huddersman, 2022). Additionally, the hydrogenation of vegetable oils (HVO) shows promise in renewable diesel production (Di Blasio et al., 2022), involving the catalytic hydrotreatment of triglycerides or fatty acids at high temperatures to obtain oxygen-free hydrocarbons.

Renewable diesel from HVO production has advantageous properties over biodiesel from transesterification, including lower emissions, high calorific value, and suitability for aviation fuel (Hunicz et al., 2021). Furthermore, existing fossil-fuel refineries can be repurposed for renewable biodiesel production from HVO technology, utilizing similar refinery elements and offering cost-effective processing of low-cost oil-based feedstocks.

Biodiesel production is categorized based on the feedstock used, from edible vegetable oils to waste oils and animal fats (Pikula et al., 2020). While first-generation biodiesel primarily utilizes edible vegetable oils, second-generation biodiesel from nonedible vegetable oils faces challenges due to competition with food production and high alcohol requirements for transesterification. Third-generation feedstocks, such as algae oils and waste oils, offer cost-effective and environmentally sustainable alternatives for biodiesel production (Pydimalla et al., 2023). Algae cultivation, in particular, can yield lipid-rich biomass for biodiesel production via transesterification (Singh et al., 2020), with ongoing research focusing on process optimization for increased productivity and cost reduction.

Waste oils are a low-cost and readily available feedstock for biodiesel production, offering significant cost reductions and addressing waste disposal issues. Transesterification of waste oil-based materials can occur with or without catalysis (Abd Rabu et al., 2013), offering flexibility in the production process. The bulk of waste oils originates from the cooking or frying processes, with waste cooking oils (WCOs) being collected from households, restaurants, and food processing industries. Typically, WCO contains high concentrations of free fatty acids, water, and solid impurities, which pose challenges to biodiesel production due to their interference with the transesterification process (Suzihaque et al., 2022). Recent literature has emphasized the development of various catalysts for waste oil transesterification, including enzymatic catalysts, homogeneous catalysts with acid-based esterification, and heterogeneous catalysts derived from waste or readily available natural materials (Orege et al., 2022). Enzymatic biocatalysts show promising results, achieving biodiesel yields ranging from 93.8% to 98.76% under favorable conditions (Lisbona et al., 2023).

The process of transforming waste oil into biodiesel, as shown in Fig. 3.10, involves the selection and pretreatment of the oil, followed by the transesterification reaction where the oil reacts with an alcohol in the presence of a catalyst to produce biodiesel and glycerin as a byproduct (Bello et al., 2012; MEHER et al., 2006).

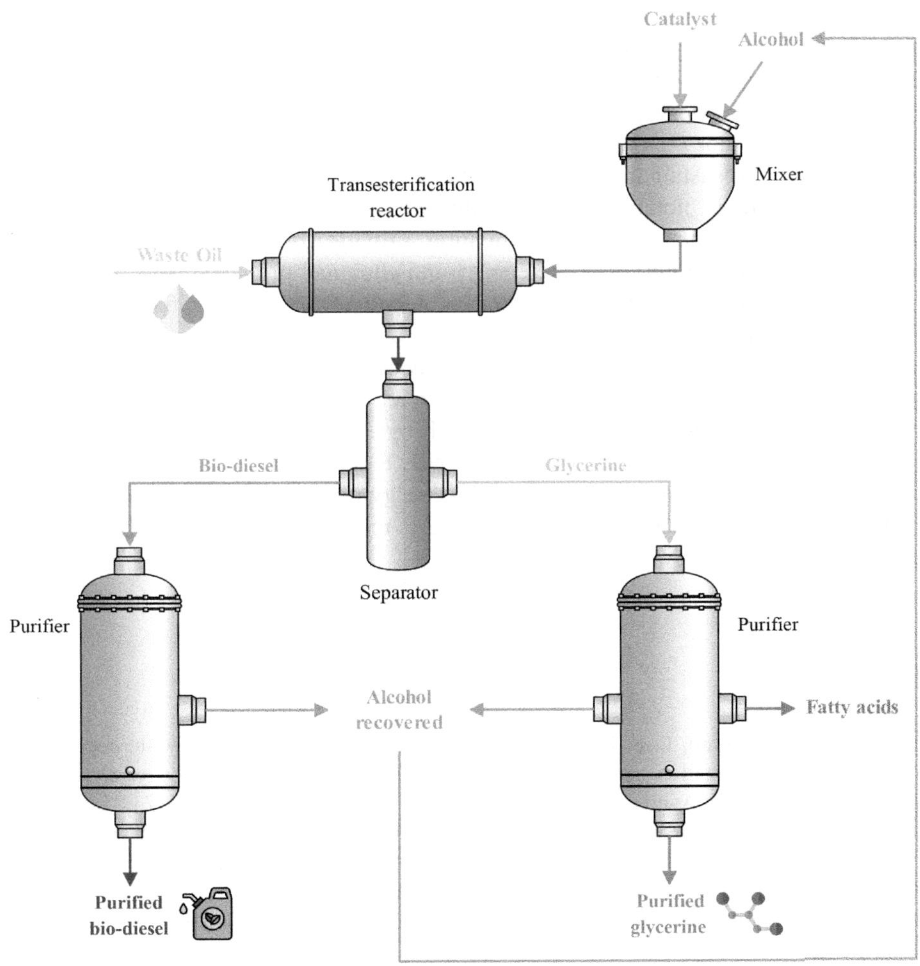

Figure 3.10 Transesterification process diagram.

Then, the biodiesel and glycerin phases are separated, which are subsequently purified (Bello et al., 2012). Part of the alcohol is recovered from this process and can be reused (Bello et al., 2012).

Continued research and development efforts are essential for maximizing the potential of waste oils and algae as promising feedstocks for sustainable biofuel production, contributing to long-term environmental sustainability and energy security.

Conclusions

WtE technologies, including both thermochemical and biochemical processes, present significant potential for addressing the global challenges of waste management and renewable energy production. Thermochemical processing methods such as combustion, gasification, and pyrolysis offer efficient means of converting various types of waste into energy products such as biogas, syngas, biochar, and biooil. These processes not only mitigate the adverse environmental impact of waste but also contribute to sustainable energy solutions and resource utilization. Biochemical processing methods, notably anaerobic digestion and fermentation, as well as other chemical methods such as transesterification, provide additional pathways for converting organic waste into valuable energy and chemical products. Anaerobic digestion is particularly effective in generating biogas from organic materials, while fermentation and transesterification processes hold promise for producing biofuels such as bioethanol and biodiesel from lignocellulosic biomass and waste oils, respectively. These processes encompass efficient utilization of waste streams and contribute to the circular economy model by repurposing waste into valuable resources. By embracing these innovative WtE technologies, the twin goals of sustainable waste management and renewable energy production can be achieved, ultimately contributing to the transition toward a more environmentally friendly and resource-efficient future.

References

Abd Rabu, R., Janajreh, I., & Honnery, D. (2013). Transesterification of waste cooking oil: Process optimization and conversion rate evaluation. *Energy Conversion and Management*, *65*. Available from https://doi.org/10.1016/j.enconman.2012.02.031.

Achinas, S., Achinas, V., & Euverink, G. J. W. (2017). A technological overview of biogas production from biowaste. *Engineering*, *3*(3). Available from https://doi.org/10.1016/J.ENG.2017.03.002, http://www.journals.elsevier.com/engineering/.

Aguado, R., Escámez, A., Jurado, F., & Vera, D. (2023). Experimental assessment of a pilot-scale gasification plant fueled with olive pomace pellets for combined power, heat and biochar production. *Fuel*, *344*, 128127. Available from https://doi.org/10.1016/j.fuel.2023.128127.

Ahmed, R., & Huddersman, K. (2022). Review of biodiesel production by the esterification of wastewater containing fats oils and grease (FOGs). *Journal of Industrial and Engineering Chemistry*, *110*. Available from https://doi.org/10.1016/j.jiec.2022.02.045.

Association, I.S.W (2024). *Global waste management outlook 2024: Beyond an age of waste: Turning rubbish into a resource* https://www.unep.org/resources/global-waste-management-outlook-2024.

Awais, M., Omar, M. M., Munir, A., li, W., Ajmal, M., Hussain, S., Ahmad, S. A., & Ali, A. (2022). Co-gasification of different biomass feedstock in a pilot-scale (24 kWe) downdraft gasifier: An experimental approach. *Energy*, *238*. Available from https://doi.org/10.1016/j.energy.2021.121821, https://www.journals.elsevier.com/energy.

Aysu, T., & Durak, H. (2016). Bio-oil production via catalytic supercritical liquefaction of Syrian mesquite (Prosopis farcta). *The Journal of Supercritical Fluids*, *109*. Available from https://doi.org/10.1016/j.supflu.2015.11.002.

Balaman, Ş. Y. (2019). *Introduction to biomass—Resources, production, harvesting, collection, and storage*. Elsevier BV. Available from http://doi.org/10.1016/b978-0-12-814278-3.00001-7.

Basu, P. (2018). *Gasification theory. Biomass gasification, pyrolysis and torrefaction: Practical design and theory* (p. 211) Canada: Elsevier. Available from http://www.sciencedirect.com/science/book/9780128129920, http://doi.org/10.1016/B978-0-12-812992-0.00007-8.

Bello, B. Z., Nwokoagbara, E., & Wang, M. (2012). Comparative techno-economic analysis of biodiesel production from microalgae via transesterification methods. *Computer Aided Chemical Engineering*, *30*. Available from https://doi.org/10.1016/B978-0-444-59519-5.50027-7, http://www.elsevier.com/wps/find/bookdescription.cws_home/BS_CCE/description#description.

Beluhan, S., Mihajlovski, K., Šantek, B., & Šantek, M. I. (2023). The production of bioethanol from lignocellulosic biomass: Pretreatment methods, fermentation, and downstream processing. *Energies*, *16*(19), 7003. Available from https://doi.org/10.3390/en16197003.

Bibi, F., Ilyas, N., Arshad, M., Khalid, A., Saeed, M., Ansar, S., & Batley, J. (2022). Formulation and efficacy testing of bio-organic fertilizer produced through solid-state fermentation of agro-waste by Burkholderia cenocepacia. *Chemosphere*, *291*, 132762. Available from https://doi.org/10.1016/j.chemosphere.2021.132762.

Bibi, F., Ilyas, N., Saeed, M., Shabir, S., Shati, A. A., Alfaifi, M. Y., Amesho, K. T. T., Chowdhury, S., & Sayyed, R. Z. (2023). Innovative production of value-added products using agro-industrial wastes via solid-state fermentation. *Environmental Science and Pollution Research*, *30*(60). Available from https://doi.org/10.1007/s11356-023-28765-6, https://www.springer.com/journal/11356.

Breeze, P. (2018). *Traditional waste combustion technologies*. Elsevier BV. Available from http://doi.org/10.1016/b978-0-08-101042-6.00006-6.

Carneiro, T. F., Timko, M., Prado, J. M., & Berni, M. (2016). *Biomass pretreatment with carbon dioxide. Biomass fractionation technologies for a lignocellulosic feedstock based biorefinery*. Brazil: Elsevier Inc. Available from http://www.sciencedirect.com/science/book/9780128023235, http://doi.org/10.1016/B978-0-12-802323-5.00017-7.

Collard, F. X., Carrier, M., & Gorgens, J. (2016). Chapter 4—Fractionation of lignocellulosic material with pyrolysis processing. *Biomass Fractionation Technologies for a Lignocellulosic Feedstock Based Biorefinery* (pp. 81−101). Amsterdam: Elsevier. Available from https://doi.org/10.1016/B978-0-12-802323-5.00004-9.

Coppola, D., Lauritano, C., Palma Esposito, F., Riccio, G., Rizzo, C., & de Pascale, D. (2021). , Fish waste: From problem to valuable resource. *Marine Drugs*, *19*, 116.

Cruz-Casas, D. E., Aguilar, C. N., Ascacio-Valdés, J. A., Rodríguez-Herrera, R., Chávez-González, M. L., & Flores-Gallegos, A. C. (2021). Enzymatic hydrolysis and microbial fermentation: The most favorable biotechnological methods for the release of bioactive peptides. *Food Chemistry: Molecular Sciences*, *3*. Available from https://doi.org/10.1016/j.fochms.2021.100047, http://www.journals.elsevier.com/food-chemistry-molecular-sciences.

Di Blasio, G., Ianniello, R., & Beatrice, C. (2022). Hydrotreated vegetable oil as enabler for high-efficient and ultra-low emission vehicles in the view of 2030 targets. *Fuel*, *310*, 122206. Available from https://doi.org/10.1016/j.fuel.2021.122206.

Dixon, N., & Langer, U. (2006). Development of a MSW classification system for the evaluation of mechanical properties. *Waste Management*, *26*(3). Available from https://doi.org/10.1016/j.wasman.2005.02.018.

Duque-Acevedo, M., Belmonte-Ureña, L. J., Cortés-García, F. J., & Camacho-Ferre, F. (2020). Agricultural waste: Review of the evolution, approaches and perspectives on alternative uses. *Global Ecology and Conservation*, *22*, e00902. Available from https://doi.org/10.1016/j.gecco.2020.e00902.

Funk, K., Milford, J., & Simpkins, T. (2020). *Waste not, want not: Analyzing the economic and environmental viability of waste-to-energy technology for site-specific optimization of renewable energy options. Bioenergy: Biomass to biofuels and waste to energy*. Elsevier. Available from https://www.elsevier.com/books/bioenergy/dahiya/978-0-12-815497-7, http://doi.org/10.1016/B978-0-12-815497-7.00019-1.

Galadima, A., & Muraza, O. (2020). Waste materials for production of biodiesel catalysts: Technological status and prospects. *Journal of Cleaner Production*, *263*, 121358. Available from https://doi.org/10.1016/j.jclepro.2020.121358.

Garg, A., Basu, S., Shetti, N. P., Bhattu, M., Alodhayb, A. N., & Pandiaraj, S. (2024). Biowaste to bioenergy nexus: Fostering sustainability and circular economy. *Environmental Research*, *250*, 118503. Available from https://doi.org/10.1016/j.envres.2024.118503, https://www.sciencedirect.com/science/journal/00139351.

Ghosh, P., Shah, G., Sahota, S., Singh, L., & Vijay, V. K. (2020). *Biogas production from waste: Technical overview, progress, and challenges. Bioreactors: Sustainable design and industrial applications in mitigation of GHG emissions*. India: Elsevier. Available from https://www.elsevier.com/books/bioreactors/singh/978-0-12-821264-6, http://doi.org/10.1016/B978-0-12-821264-6.00007-3.

Gollakota, A. R. K., Kishore, N., & Gu, S. (2018). A review on hydrothermal liquefaction of biomass. *Renewable and Sustainable Energy Reviews*, *81*. Available from https://doi.org/10.1016/j.rser.2017.05.178.

Grande, L., Pedroarena, I., Korili, S. A., & Gil, A. (2021). Hydrothermal liquefaction of biomass as one of the most promising alternatives for the synthesis of advanced liquid biofuels: A review. *Materials*, *14*(18). Available from https://doi.org/10.3390/ma14185286, https://www.mdpi.com/1996-1944/14/18/5286/pdf.

Gupta, J., Kumari, M., Mishra, A., Swati., Akram, M., & Thakur, I. S. (2022). Agro-forestry waste management—A review. *Chemosphere*, *287*, 132321. Available from https://doi.org/10.1016/j.chemosphere.2021.132321.

Hunicz, J., Krzaczek, P., Gęca, M., Rybak, A., & Mikulski, M. (2021). Comparative study of combustion and emissions of diesel engine fuelled with FAME and HVO. *Combustion Engines*, *184*(1). Available from https://doi.org/10.19206/ce-135066.

Hupa, M., Karlström, O., & Vainio, E. (2017). Biomass combustion technology development—It is all about chemical details. *Proceedings of the Combustion Institute*, *36*(1). Available from https://doi.org/10.1016/j.proci.2016.06.152, http://www.sciencedirect.com/science/journal/15407489.

Jiang, Y., Xie, S. H., Dennehy, C., Lawlor, P. G., Hu, Z. H., Wu, G. X., Zhan, X. M., & Gardiner, G. E. (2020). Inactivation of pathogens in anaerobic digestion systems for converting biowastes to bioenergy: A review. *Renewable and Sustainable Energy Reviews, 120*, 109654. Available from https://doi.org/10.1016/j.rser.2019.109654.

Komova, E. L., Palacios Hidalgo, A., & Viamontes, C. M. S. (2023). La economía circular en la gestion de residuos ganaderos. *Economia y Desarrollo, 167*, 2023.

Limayem, A., & Ricke, S. C. (2012). Lignocellulosic biomass for bioethanol production: Current perspectives, potential issues and future prospects. *Progress in Energy and Combustion Science, 38*(4). Available from https://doi.org/10.1016/j.pecs.2012.03.002.

Lisbona, P., Pascual, S., & Pérez, V. (2023). Waste to energy: Trends and perspectives. *Chemical Engineering Journal Advances, 14*, 100494. Available from https://doi.org/10.1016/j.ceja.2023.100494.

Liu, Y., & Chen, J. (2014). *Phosphorus cycle*. Elsevier BV. Available from http://doi.org/10.1016/b978-0-12-409548-9.09043-6.

Mazhkoo, S., Dadfar, H., HajiHashemi, M. S., & Pourali, O. (2021). A comprehensive experimental and modeling investigation of walnut shell gasification process in a pilot-scale downdraft gasifier integrated with an internal combustion engine. *Energy Conversion and Management, 231*, 113836. Available from https://doi.org/10.1016/j.enconman.2021.113836.

MEHER, L., VIDYASAGAR, D., & NAIK, S. (2006). Technical aspects of biodiesel production by transesterification—a review. *Renewable and Sustainable Energy Reviews, 10* (3). Available from https://doi.org/10.1016/j.rser.2004.09.002.

Mizik, T., & Gyarmati, G. (2022). Three pillars of advanced biofuels' sustainability. *Fuels, 3* (4). Available from https://doi.org/10.3390/fuels3040037.

Mohd Thani, N., Mustapa Kamal, S. M., Taip, F. S., Sulaiman, A., Omar, R., & Siajam, S. I. (2020). Hydrolysis and characterization of sugar recovery from bakery waste under optimized subcritical water conditions. *Journal of Food Science and Technology, 57*(8). Available from https://doi.org/10.1007/s13197-020-04345-1.

Momayez, F., Karimi, K., & Taherzadeh, M. J. (2019). Energy recovery from industrial crop wastes by dry anaerobic digestion: A review. *Industrial Crops and Products, 129*. Available from https://doi.org/10.1016/j.indcrop.2018.12.051, http://www.elsevier.com/inca/publications/store/5/2/2/8/2/5.

Naik, B., Kumar, V., Rizwanuddin, S., Chauhan, M., Gupta, A. K., Rustagi, S., Kumar, V., & Gupta, S. (2023). Agro-industrial waste: A cost-effective and eco-friendly substrate to produce amylase. *Food Production, Processing and Nutrition, 5*.

Najafi, S., Khosravani, F., Yousefi, M., Jandaghi, J., Valizadeh, B., & Torkashvand, J. (2024). Development of municipal solid waste management guidelines in biological crisis based on international experiences and considering local techno-economic characteristics. *Results in Engineering, 22*, 102052. Available from https://doi.org/10.1016/j.rineng.2024.102052.

Orege, J. I., Oderinde, O., Kifle, G. A., Ibikunle, A. A., Raheem, S. A., Ejeromedoghene, O., Okeke, E. S., Olukowi, O. M., Orege, O. B., Fagbohun, E. O., Ogundipe, T. O., Avor, E. P., Ajayi, O. O., & Daramola, M. O. (2022). Recent advances in heterogeneous catalysis for green biodiesel production by transesterification. *Energy Conversion and Management, 258*. Available from https://doi.org/10.1016/j.enconman.2022.115406, https://www.journals.elsevier.com/energy-conversion-and-management.

Oriakhi, C. O. (2009). *Oxidation and reduction reactions*. Oxford University Press (OUP). Available from http://doi.org/10.1093/oso/9780195367997.003.0026.

Parfitt, J., Croker, T., & Brockhaus, A. (2021). Global food loss and waste in primary production: A reassessment of its scale and significance. *Sustainability*, *13*(21), 12087. Available from https://doi.org/10.3390/su132112087.

Perez-Alva, A., MacIntosh, A. J., Baigts-Allende, D. K., García-Torres, R., & Ramirez-Rodrigues, M. M. (2022). Fermentation of algae to enhance their bioactive activity: A review. *Algal Research*, *64*, 102684. Available from https://doi.org/10.1016/j.algal.2022.102684.

Pikula, K., Zakharenko, A., Stratidakis, A., Razgonova, M., Nosyrev, A., Mezhuev, Y., Tsatsakis, A., & Golokhvast, K. (2020). The advances and limitations in biodiesel production: Feedstocks, oil extraction methods, production, and environmental life cycle assessment. *Green Chemistry Letters and Reviews*, *13*(4). Available from https://doi.org/10.1080/17518253.2020.1829099.

Pydimalla, M., Husaini, S., Kadire, A., & Verma, R. K. (2023). Sustainable biodiesel: A comprehensive review on feedstock, production methods, applications, challenges and opportunities. *Materials Today: Proceedings*, *92*. Available from https://doi.org/10.1016/j.matpr.2023.03.593.

Rao, M. N., Sultana, R., & Kota, S. H. (2017). *Municipal solid waste*. Elsevier BV. Available from http://doi.org/10.1016/b978-0-12-809734-2.00002-x.

Reconstruction and Development, (2018). *What a waste 2.0 a global snapshot of solid waste management to 2050*. Reconstruction and Development.

Richards, T. (2019). *Combustion of waste in combined heat and power plants. Sustainable resource recovery and zero waste approaches*. Sweden: Elsevier. Available from http://www.sciencedirect.com/science/book/9780444642004, http://doi.org/10.1016/B978-0-444-64200-4.00013-X.

Rocamora, I., Wagland, S. T., Villa, R., Simpson, E. W., Fernandez, O., & Bajón-Fernández., Y. (2020). Dry anaerobic digestion of organic waste: A review of operational parameters and their impact on process performance. *Bioresource Technology*, *299*. Available from https://doi.org/10.1016/j.biortech2020.

Sanchez-Lozano, D., Escamez, A., Aguado, R., Oulbi, S., Hadria, R., & Vera, D. (2023). Techno-Economic assessment of an off-grid biomass gasification CHP plant for an olive oil mill in the region of Marrakech-Safi, Morocco. *Applied Sciences*, *13*(10), 5965. Available from https://doi.org/10.3390/app13105965.

Shahbeik, H., Panahi, H. K. S., Dehhaghi, M., Guillemin, G. J., Fallahi, A., Hosseinzadeh, H., et al. (2024). Biomass to biofuels using hydrothermal liquefaction: A comprehensive review. *Renewable and Sustainable Energy Reviews*, *189*(Part B), 113976. Available from https://doi.org/10.1016/j.rser.2023.113976.

Simone, M., Barontini, F., Nicolella, C., & Tognotti, L. (2012). Gasification of pelletized biomass in a pilot scale downdraft gasifier. *Bioresource Technology*, *116*. Available from https://doi.org/10.1016/j.biortech.2012.03.119.

Singh, D., Sharma, D., Soni, S. L., Sharma, S., Sharma, P. K., & Jhalani, A. (2020). A review on feedstocks, production processes, and yield for different generations of biodiesel. *Fuel*, *262*, 116553. Available from https://doi.org/10.1016/j.fuel.2019.116553.

Singh, R., Krishna, B. B., Mishra, G., Kumar, J., & Bhaskar, T. (2016). Strategies for selection of thermo-chemical processes for the valorisation of biomass. *Renewable Energy*, *98*. Available from https://doi.org/10.1016/j.renene.2016.03.023, http://www.journals.elsevier.com/renewable-and-sustainable-energy-reviews/.

Sirico, A., Belletti, B., Bernardi, P., Malcevschi, A., Pagliari, F., Fornoni, P., & Moretti, E. (2022). Effects of biochar addition on long-term behavior of concrete. *Theoretical and Applied Fracture Mechanics*, *122*, 103626. Available from https://doi.org/10.1016/j.tafmec.2022.103626.

Soares, J., & Oliveira, A. C. (2020). Experimental assessment of pine wood chips gasification at steady and part-load performance. *Biomass and Bioenergy*, *139*, 105625. Available from https://doi.org/10.1016/j.biombioe.2020.105625.

Song, L., Yang, D., Liu, R., Liu, S., Dai, L., & Dai, X. (2022). Microbial production of lactic acid from food waste: Latest advances, limits, and perspectives. *Bioresource Technology*, *345*, 126052. Available from https://doi.org/10.1016/j.biortech.2021.126052.

Suzihaque, M. U. H., Alwi, H., Ibrahim, U. K., Abdullah, S., & Haron, N. (2022). Biodiesel production from waste cooking oil: A brief review. *Materials Today: Proceedings*, *63*, S490. Available from https://doi.org/10.1016/j.matpr.2022.04.527.

Tan, A. S. T., Uthayakumar, H., Yeo, L. S., Kong, K. G. H., Lo, S. L. Y., Andiappan, V., Loy, A. C. M., Teng, S. Y., & How, B. S. (2024). Shapley-shubik agents within superstructure-based recycling model: Circular economy approaches for fish waste eco-industrial park. *Process Integration and Optimization for Sustainability*. Available from https://doi.org/10.1007/s41660-024-00391-w, https://www.springer.com/journal/41660.

Tripathi, N., Hills, C. D., Singh, R. S., & Atkinson, C. J. (2019). Biomass waste utilisation in low-carbon products: Harnessing a major potential resource. *npj Climate and Atmospheric Science*, *2*(1). Available from https://doi.org/10.1038/s41612-019-0093-5, https://www.nature.com/npjclimatsci/.

Uddin, M. M., & Wright, M. M. (2023). Anaerobic digestion fundamentals, challenges, and technological advances. *Physical Sciences Reviews*, *8*(9). Available from https://doi.org/10.1515/psr-2021-0068, https://www.degruyter.com/journal/key/psr/html.

Usman, M., Cheng, S., Boonyubol, S., & Cross, J. S. (2024). From biomass to biocrude: Innovations in hydrothermal liquefaction and upgrading. *Energy Conversion and Management*, *302*, 118093. Available from https://doi.org/10.1016/j.enconman.2024.118093.

Van Nguyen, T. T., Phan, A. N., Nguyen, T. A., Nguyen, T. K., Nguyen, S. T., Pugazhendhi, A., & Phuong, H. H. K. (2022). Valorization of agriculture waste biomass as biochar: As first-rate biosorbent for remediation of contaminated soil. *Chemosphere*, *307*. Available from https://doi.org/10.1016/j.chemosphere.2022.135834, http://www.elsevier.com/locate/chemosphere.

Vera, D., de Mena, B., Jurado, F., & Schories, G. (2013). Study of a downdraft gasifier and gas engine fueled with olive oil industry wastes. *Applied Thermal Engineering*, *51*(1-2). Available from https://doi.org/10.1016/j.applthermaleng.2012.09.012.

Vukovic, N., & Makogon, E. (2022). Waste-to-energy generation: Complex efficiency analysis of modern technologies. *Sustainability*, *14*(21), 13814. Available from https://doi.org/10.3390/su142113814.

Wang, Q., Zhao, W., Liu, H., Jia, C., & Xu, H. (2012). Reactivity and kinetic analysis of biomass during combustion. *Energy Procedia*, *17*. Available from https://doi.org/10.1016/j.egypro.2012.02.181.

Yadav, M., & Sharma, Y. C. (2018). Process optimization and catalyst poisoning study of biodiesel production from kusum oil using potassium aluminum oxide as efficient and reusable heterogeneous catalyst. *Journal of Cleaner Production*, *199*. Available from https://doi.org/10.1016/j.jclepro.2018.07.052, https://www.journals.elsevier.com/journal-of-cleaner-production.

Yoon, S. J., Son, Y.-I., Kim, Y.-K., & Lee, J.-G. (2012). Gasification and power generation characteristics of rice husk and rice husk pellet using a downdraft fixed-bed gasifier. *Renewable Energy*, *42*. Available from https://doi.org/10.1016/j.renene.2011.08.028.

Zhang, C., Kang, X., Wang, F., Tian, Y., Liu, T., Su, Y., Qian, T., & Zhang, Y. (2020). Valorization of food waste for cost-effective reducing sugar recovery in a two-stage enzymatic hydrolysis platform. *Energy*, *208*, 118379. Available from https://doi.org/10.1016/j.energy.2020.118379.

Zhou, H., Meng, A., Long, Y., Li, Q., & Zhang, Y. (2014). Classification and comparison of municipal solid waste based on thermochemical characteristics. *Journal of the Air and Waste Management Association*, *64*(5). Available from https://doi.org/10.1080/10962247.2013.873094, http://www.tandfonline.com/loi/uawm20?.

Zhao, X. Q., Zi, L. H., Bai, F. W., Lin, H. L., Hao, X. M., Yue, G. J., & Ho, N. W. Y. (2012). Bioethanol from Lignocellulosic Biomass. In F. W. Bai, C. G. Liu, H. Huang, & G. Tsao (Eds.), *Biotechnology in China III: Biofuels and Bioenergy. Advances in Biochemical Engineering Biotechnology* (128, pp. 25–51). Berlin, Heidelberg: Springer. Available from https://doi.org/10.1007/10_2011_129.

4 Water–energy nexus

Ali Esmaeel Nezhad[1], *Toktam Tavakkoli Sabour*[2], *Mohammad Sadegh Javadi*[3], *Pedro H.J. Nardelli*[1], *Saeid Jowkar*[4] *and Farideh Ghanavati*[5]

[1]Department of Electrical Engineering, School of Energy Systems, LUT University, Lappeenranta, South Karelia, Finland, [2]Department of Electrical and Computer Engineering, Texas Tech University, Lubbock, TX, United States, [3]Institute for Systems and Computer Engineering, Technology and Science (INESC TEC), Porto, Portugal, [4]Department of Electrical, Electronic, and Information Engineering, University of Bologna, Bologna, Italy, [5]Department of Industrial Engineering and Management, University of Aveiro, Aveiro, Portugal

Nomenclature

Indices/Sets

t	Index for the time intervals of scheduling
s	Index for the scenario
N_T	Scheduling period
N_s	Total number of scenarios

Variables

$PG_{s,t}^{G2H}$	Power flowing from the grid to the hub in scenario s and time t (kW)
$P_s(g,t)PG_{s,t}^{H2G}$	Power flowing from the hub to the grid in scenario s and time t (kW)
$F_{s,t}^{\mathrm{CHP}}$	Cost function of the CHP unit in scenario s and time t ($)
$\overline{P_s}(g,t){:}F_{s,t}^{\mathrm{Boiler}}$	Cost function of the boiler in scenario s and time t ($)
$P_h(j,t)PG_{s,t}^{\mathrm{CHP}}$	Electrical power generation of the CHP unit (kW)
$PH_{s,t}^{\mathrm{CHP}}$	Heating power generation of the CHP unit in scenario s and time t (kWt)
$PH_{s,t}^{\mathrm{Boiler}}$	Heat generated by boiler in scenario s and time t (kWt)
$PH_{s,t}^{\mathrm{EHP,Heating}}$	Heat generated by EHP in scenario s and time t (kWt)
$PC_{s,t}^{\mathrm{Chiller}}$	Cooling power generated by the chiller in scenario s and time t (kWt)
$PC_{s,t}^{\mathrm{EHP}}$	Cooling power generated by the EHP in scenario s and time t (kWt)
$PG_{s,t}^{\mathrm{EES,Ch}}$	Charging power of the EES system in scenario s and time t (kW)
$PG_{s,t}^{\mathrm{EES,Dis}}$	Discharging power of the EES system in scenario s and time t (kW)
$W_{s,t}^{\mathrm{Des}}$	Water demand (m^3)
$W_{s,t}^{\mathrm{Tank}}$	Water available in the tank in scenario s and time t (m^3)

Towards Future Smart Power Systems with High Penetration of Renewables. DOI: https://doi.org/10.1016/B978-0-443-29871-4.00004-X

$W_{s,t}^{L}$	Water discharged from the tank in scenario s and time t (m^3)
$PG_{s,t}^{\mathrm{Des}}$	Power consumed by the desalination system (kW)
$P_s(g,t)I_{s,t}^{\mathrm{CHP}}$	Binary variable showing the commitment status of the CHP unit
$P_s(g,t)I_{s,t}^{\mathrm{Boiler}}$	Binary variable showing the commitment status of the boiler
$P_s(g,t)I_{s,t}^{\mathrm{Chiller}}$	Binary variable showing the commitment status of the chiller
$P_s(g,t)I_{s,t}^{\mathrm{EHP,Cooling}}$	Binary variable showing the cooling mode of the EHP
$P_s(g,t)I_{s,t}^{\mathrm{EHP,Heating}}$	Binary variable showing the heating mode of the EHP
$I_{s,t}^{\mathrm{Heater}}$	Binary variable showing the commitment status of the heater
$I_{s,t}^{\mathrm{EES,Ch}}$	Binary variable showing the charging mode of the EES system
$I_{s,t}^{\mathrm{EES,Dis}}$	Binary variable showing the discharging mode of the EES system
$I_{s,t}^{G2H}$	Binary variable showing power flow from grid to hub
$I_{s,t}^{H2G}$	Binary variable showing power flow from hub to grid
$\mathrm{Eng}_{s,t}^{\mathrm{EES}}$	Energy available in the EES system (kWh)

Parameters

$a_s(g), b_s(g), c_s(g), d_s(g), e_s(g){:}\lambda_t^{\mathrm{Buy}}$	Energy buying price at time t
$\lambda_t^{\mathrm{Sell}}$	Energy purchasing price at time t
λ_t^{NG}	Hourly natural gas price at time t
η_E^{CHP}	Efficiency of the CHP unit in the electrical power generation mode
$\alpha_s(g), \beta_s(g), \gamma_s(g), \omega_s(g), \delta_s(g){:}\eta_H^{\mathrm{CHP}}$	Efficiency of the CHP unit in the heating power generation mode
$\alpha_s(g), \beta_s(g), \gamma_s(g), \omega_s(g), \delta_s(g){:}\eta_H^{\mathrm{Boiler}}$	Efficiency of the boiler
$\eta^{\mathrm{EES,Ch}}$	Efficiency of the EES system in the charging mode
$\eta^{\mathrm{EES,Dis.}}$	Efficiency of the EES system in the discharging mode
$C_1(j), C_2(j), C_3(j), C_4(j), C_5(j), C_6(j)PS_{s,t}^{\mathrm{CHP,min}}$	Minimum capacity of the CHP unit in scenario s and time t (kW)
$PS_{s,t}^{\mathrm{CHP,max}}$	Maximum capacity of the CHP unit in scenario s and time t (kW)
$\eta_{\mathrm{g}}(\mathrm{i}){:}\eta_{\mathrm{p}}(\mathrm{i}){:}PG_{s,t}^{\mathrm{CHP,min}}$	Minimum electrical power generation of the CHP unit in scenario s and time t (kW)
$PG_{s,t}^{\mathrm{CHP,max}}$	Maximum electrical power generation of the CHP unit in scenario s and time t (kW)
$\mathrm{P}_{\mathrm{P-S}}^{\max}(i){:}PH_{s,t}^{\mathrm{CHP,min}}$	Minimum heating power generation of the CHP unit in scenario s and time t (kWt)
$\mathrm{P}_{\mathrm{P-S}}^{\min}(i){:}PH_{s,t}^{\mathrm{CHP,max}}$	Maximum heating power generation of the CHP unit in scenario s and time t (kWt)
$\mathrm{P}_{\mathrm{P-S}}^{\max}(i){:}PH_{s,t}^{\mathrm{Boiler,min}}$	Minimum heat generation of the boiler in scenario s and time t (kWt)
$P_s(g,t)PH_{s,t}^{\mathrm{Boiler,max}}$	Maximum heat generation of the boiler in scenario s and time t (kWt)
$P_s(g,t)PC^{\mathrm{Chiller,min}}$	Minimum cooling power of the chiller (kWt)
$P_s(g,t)PC^{\mathrm{Chiller,max}}$	Maximum cooling power of the chiller (kWt)

$PG_{s,t}^{EES,Ch,max}$	Maximum charging power of the EES system in scenario s and time t (kW)
$PG_{s,t}^{EES,Dis,max}$	Maximum discharging power of the EES system in scenario s and time t (kW)
$PG_{s,t}^{PV}$	Power generation of the PV system (kW) in scenario s and time t (kW)
PL^{max}	The capacity of the feeder connecting the hub to the electrical grid (kW)
W^{min}	Minimum water demand of the desalination system (m^3)
W^{max}	Maximum water demand of the desalination system (m^3)
W_{min}^{Tank}	Minimum capacity of the desalination tank (m^3)
W_{max}^{Tank}	Maximum capacity of the desalination tank (m^3)
RD^{Des}	Ramp-down limit of the water demand of the desalination system (m^3)
RU^{Des}	Ramp-up limit of the water demand of the desalination system (m^3)
$P_s(g,t)COP^{Chiller}$	Energy conversion factor of the chiller
$COP^{EHP,Cooling}$	Energy conversion factor of the EHP in the cooling mode
$COP^{EHP,Heating}$	Energy conversion factor of the EHP in the heating mode
COP^{Heater}	Energy conversion factor of the heater
$P_s(g,t)PC^{EHP,min}$	Minimum cooling power of the EHP (kWt)
$P_s(g,t)PC^{EHP,max}$	Maximum cooling power of the EHP (kWt)
$P_s(g,t)PH^{EHP,min}$	Minimum heating power of the EHP (kWt)
$P_s(g,t)PH^{EHP,max}$	Maximum heating power of the EHP (kWt)
Δt	Duration of each time slot (hour)
K^{Des}	Performance coefficient of the water desalination system (m^3/kW)

Introduction

Motivation

The technological issues related to energy hubs and their integration in the energy–water nexus is complicated, comprising a wide range of subjects and methods aimed at enhancing the performance of the system, resource utilization maximization, and promoting sustainability. This chapter focuses on efficaciously operating various energy sources, storage technologies, and consumption devices in energy hubs to achieve optimal performance in the face of changing energy supply and demand patterns.

The literature mainly addresses the design and optimization of energy hubs. Researchers deploy mathematical models, optimization methods, and simulation platforms to give out efficient plans for integrating and operating renewable energy sources (RESs), energy storage systems, and thermal devices in a cohesive system. The target of these attempts is to optimize the system efficiency, mitigate operating costs, and attain performance targets, considering different aspects like the availability of resources, energy consumption profiles, and system constraints.

Techniques like mixed-integer linear programming (MILP), dynamic programming, and metaheuristic methods are often employed to tackle the complicated, multiobjective characteristics of EH management problems.

Recently, the ways to efficiently operate RESs, characterizing the uncertainty of renewable power generation, and optimally utilizing such intermittent energy sources in energy hubs have changed. Methods such as model predictive control, forecasting, and advanced control techniques are employed to enhance the incorporation of volatile RESs comprising solar PV, wind, and hydropower. These studies are mostly aimed at ensuring the stability and reliability of the power system while optimally using such sources.

Energy storage technologies are also crucial in energy hubs as they enable using the surplus renewable power generation over off-peak periods and also stabilize the RESs' output during abrupt changes in weather conditions. The research in this area addressed the modeling, analysis, and improvement of energy storage technologies, such as batteries, thermal storage systems, and pumped hydro storage systems. Furthermore, the potential synergies between energy storage systems and demand-side management programs have been discussed so far. The most common objective functions also include operating cost minimization, emissions minimization, system stability enhancement, and so forth.

Thermal devices such as heat pumps, heaters, boilers, and combined heat and power (CHP) units play a vital role in energy hubs, particularly in meeting the load demand for heating and cooling. In this respect, modeling such devices, integration aspects, and scheduling issues within energy hubs are the most addressed topics in this area. In addition, there are attempts made to address the potential interaction between CHP units and RESs, waste heat recovery, and district heating and cooling systems for optimizing energy efficiency and minimizing environmental pollutant emissions.

Furthermore, incorporating water treatment and desalination technologies into energy hubs is an emerging topic focusing on solving water shortage and quality issues. The studies carried out thus far in this area address desalination procedures and their impacts on the energy system, assess the energy efficiency of various desalination methods, and investigate the potential merits of joint operation of renewable power generation and water desalination systems in the presence of energy storage systems and demand response programs.

Scientometric analysis

Recently, there has been an increasing trend in focusing on the evaluation of bibliometric networks, involving coauthorship, bibliographic coupling, and cocitation networks, in scientific texts (Perianes-Rodriguez et al., 2016). This section includes information on the number of papers published in the field of water and energy nexus, utilizing the Scopus database. A total of 5243 documents have been used in the analysis. These statistics give researchers data on the publication rates, publishers, and journals for finding related documents, the type of published documents, focus areas, top authors in the field, and the most common keywords in the field of water and energy nexus. Researchers will be able to utilize the knowledge as a reference for their forthcoming studies. In this respect, the number of published papers each year in the interval of 1985–2023 is given in Fig. 4.1. The increasing number of published documents shows the increasing importance of this topic.

Besides, Fig. 4.2 illustrates the number of papers published by each author in this focus area. By using these data, it will be possible to find the researchers who

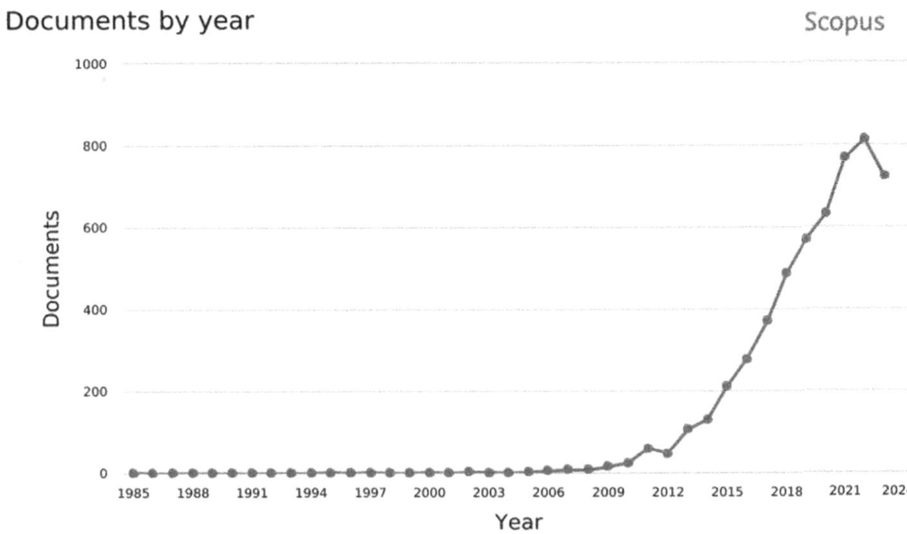

Figure 4.1 Number of published papers by year.

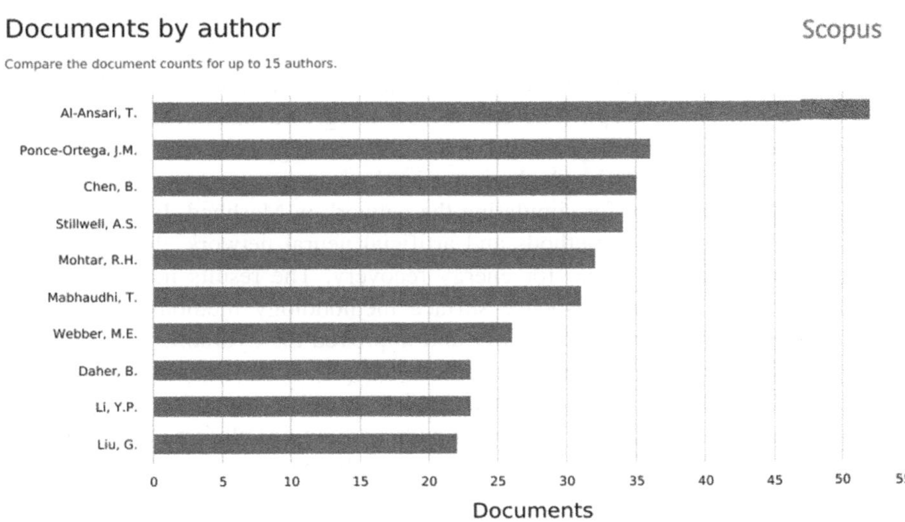

Figure 4.2 Number of published papers by each author.

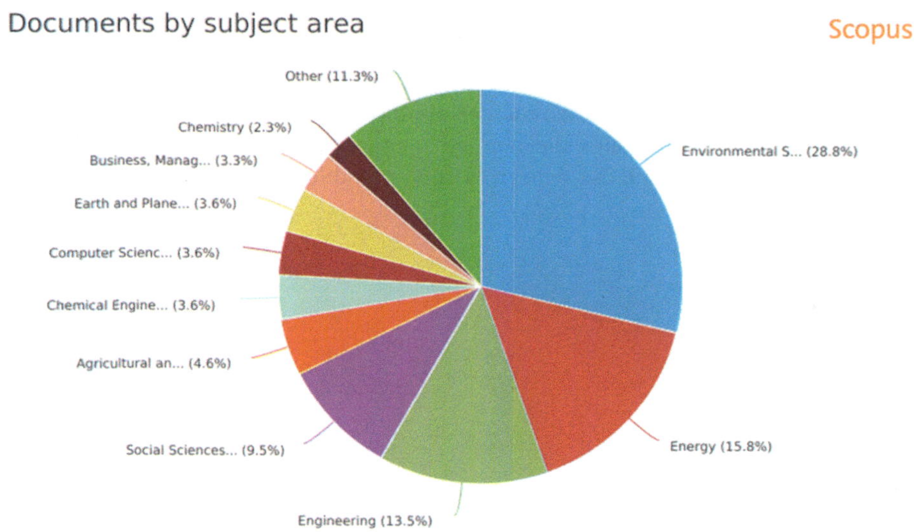

Figure 4.3 Subject areas of the published documents.

are actively doing research in the field of water and energy nexus. Fig. 4.3 also depicts the subject areas of the published papers. As expected, environmental science, energy, and engineering have the highest contributions, 28.8%, 15.8%, and 13.5%, of the total published documents, respectively.

Literature review

There are many research works focusing on the water–energy nexus and the related issues. In this regard, Nakhaei et al. (2022) developed an innovative technique for evaluating the technical performance of water distribution networks employing the water–energy nexus concept. The framework integrated statistical optimization, modeling, and artificial intelligence methods for the energy recovery potential maximization in such networks by deploying microhydropower systems. The research utilized EPANET software for simulating the network in Mashhad, Iran. Afterward, the design of experimental methods and artificial neural network algorithms were used to maximize the potential for energy recovery. The results indicate that the integration of Taguchi and response surface methodology methods efficaciously maximized the energy recovery capacity and improved the hydraulic characteristics of the studied network. The research also derived optimal locations for microhydropower systems deployment and indicated that it is possible to precisely approximate energy harvesting and hydraulic responses with a strong correlation coefficient. The study emphasized the significance of considering the water-energy connection and using microhydropower systems for energy retrieval in the mentioned network. A multiple input, multiple output model predictive control method has been proposed by Saskia et al. (2023) to optimally manage the demand of an integrated water distribution network, fed by an islanded microgrid. The presented framework aimed to promote the efficiency of the water and energy systems' synchronization. The developed closed-loop energy management system considered the dynamics of assets, water and energy system coupling, and the capability to recover fast from disruptions. The approach deployed a state-space model and sequential quadratic programming to tackle the problem. The reliability of the integrated approach to optimize water and energy demand was verified by simulating a real-world case study. Rezaei et al. (2023) developed a new framework for virtual power plants (VPPs), known as water-energy VPPs. The proposed VPP comprises different energy and water sources, such as wind turbines, compressed air energy storage (CAES), gas and electrical boilers, absorption chiller, ice storage, water storage, as well as a water well. This integration enables supplying water and energy demand, such as cooling, electricity, and heat. In this regard, a mixed-integer nonlinear programming (MINLP) model was deployed to investigate the stochastic self-scheduling of the water-energy VPP. Their methodology to manage the intermittent output of wind turbines includes utilizing scenarios and data clustering methods to minimize the number of scenarios. The numerical results depicted that the expected daily profit of the VPP goes up once the CAES system is employed. The paper also highlighted the interrelationship significance between water and electricity networks within the VPP optimization problem. Xie et al. (2023) presented an optimization framework for various sources of energy in a typical iron and steel mill while taking into consideration the interconnections between water, energy, and carbon. In this respect, a complicated mathematical model named "multiobjective MINLP" was introduced to enhance the system efficiency. This model addressed the interconnection between gas, steam, and electricity networks. The objective functions are cost, CO_2 emissions, water production, as well as global energy efficiency. The paper proposed a curve accurately representing the relationship between the energy efficiency of the equipment and other factors. It indicated that the iron and steel factory has the capability to mitigate economic costs and emissions while raising water production and enhancing global energy efficiency. The model was indeed aimed at assisting the facility in achieving its carbon peak and carbon neutrality objectives.

Mirzaei et al. (2023) introduced a framework for simultaneously optimizing energy and water sources within a microgrid, aimed at promoting system efficiency and economic performance. In this regard, a look-ahead optimization model was utilized for an energy–water microgrid including different energy conversion units and multienergy demand response programs. The mentioned model deployed a hybrid method based on the info-gap decision theory and robust optimization to characterize uncertainties due to solar power and market prices. The obtained results showed that employing energy–water storage systems and a proactive planning method results in alleviating the total costs. The paper also addresses the importance of efficacious energy management systems, especially for electric and hydrogen vehicles and the interdependence of energy and water networks. Yu et al. (2023) evaluated the performance of an efficient method to control the interconnection between energy, water, and carbon by deploying carbon capture and storage devices. The main objective is to mitigate the total operating costs and CO_2 emissions while ensuring an environmentally friendly energy and water supply. The paper utilized a risk-constrained framework to handle uncertainties due to renewable power generation, electrical power, and water demands. The problem has been presented in a mixed-integer programming (MIP) framework, and the obtained results recommend employing carbon capture and storage systems to handle carbon-related concerns. The results also showed that cap regulations have a tangible effect on enhancing carbon capture system performance. The performance of different techniques for optimizing thermal energy usage in surface water-source heat pumps was assessed by Shin et al. (2024) while operating at low water temperatures via the implementation of heat source adjustment methods. The paper addressed the difficulties of running the mentioned system at extremely low temperatures and recommended a compensation-based water-source heat pump and a hybrid variation as solutions to enhance thermal energy usage in cold weather. Hybrid systems which include boilers show the minimum value of emissions, which leads to significantly higher costs. This paper shows the economic and environmental advantages of the new techniques in mitigating energy use and emissions. Akbar et al. (2023) took into consideration climate as a key factor in analyzing the water–food–energy–land nexus. The water–food–energy–land–climate nexus index has been employed to evaluate the wheat crop production system in Pakistan. The value of the nexus index has been evaluated by normalizing and aggregating 12 different indexes. The obtained results showed remarkable alteration in the sustainable production of wheat crops and land suitability throughout Punjab, Pakistan. A comprehensive scheduling model has been developed by Sun et al. (2023) for the water–energy–food nexus, while emphasizing the Yulin City case. The model deployed a bilevel programming and uncertainty characterization framework, which was based on the chance-constrained programming and fuzzy credibility constrained programming. In this respect, the interrelationships between water, energy, and food resources as vital sources to the society were assessed. This intricate relationship was optimized while aimed at maximizing the allocation of resources in the presence of uncertainties such as unforeseen oscillations in water availability and subjective ambiguity in water demand forecasts. The proposed framework not only provides a strategic tool for managing resources efficiently, but also adds to the wider discussion on sustainable development by addressing the significant interconnection between water, energy, and food security. The review paper by Gui et al. (2023) proposed a framework for assessing of solar and water energy usage in a hybrid solar evaporation system dedicated to the concurrent water and electricity generation. The analysis investigated the energy transfer within the system and gave a brief performance of several techniques for harnessing energy, comprising photovoltaic, thermoelectric, piezoelectric, triboelectric, reverse electrodialysis, and hydrovoltaic systems. The most recent advancements in cogeneration hybrid systems for water and electricity were taken into account, particularly highlighting their electric generating approaches, system configurations, and efficiency. It was also found that using a hydrovoltaic nanogenerators with solar-driven interfacial evaporation systems will be beneficial. Alhajeri et al. (2024) presented a methodical approach to optimize the water–energy–food nexus by using a combination of water desalination, energy generation, and agricultural production systems. In this regard, the problem was approached as a multitemporal complex optimization problem. It deployed data-driven models obtained from field research. The report also used a technique for including agricultural activities, water systems, energy-resource options, carbon emissions, and economic considerations. The developed problem was then solved to provide a trade-off between economic and environmental objectives. An integrated seawater hub (ISH) was designed by Sajna et al. (2024), using seawater desalination, to not only supply water but also produce electrical power and extract minerals. This technique aimed to mitigate resource limitations, particularly in arid locations. The ISH employed the saltwater reverse osmosis technique, together with an inventive brine management approach, for extracting minerals and generating electrical power. The proposed method not only raises the amount of freshwater available but also supports sustainability by alleviating the downsides of disposing of brine and enhancing economic feasibility via the extraction of valuable resources. Furthermore, utilizing RESs to feed the processes was also taken into consideration. The method can be employed to alleviate carbon footprints and promote ecological merits. Babel et al. (2023) investigated the Karnafuli River Basin in Bangladesh and deployed an indicator-based method to effectively allocate water resources across several sectors, comprising residential, agricultural, energy, industrial, and environmental. The LINDO optimization tool was utilized for maximizing economic returns. The water–energy–food nexus model analyzes the water demands for energy generation and agriculture, addressing the inefficiencies such as large unirrigated regions resulting from inappropriate management. The papers indicated that giving priority to agriculture has the potential to provide a maximum economic benefit of $30.3 million. Nevertheless, this method may adversely impact the environment. Thus, it is essential to employ integrated resource management methods for achieving sustainable socioeconomic growth. For the sake of optimizing resource allocation and providing a healthy and sustainable diet within environmental planetary boundary restrictions and economic budgets, (Mansour et al., 2023) proposed a holistic water–energy–food nexus management framework. The mentioned framework seeks to maximize food production while minimizing energy and water by deploying an effective multiobjective optimization model. Economic, environmental, and dietary limitations have all been considered. The obtained results indicated how priorities impact resource allocation and meeting the sustainable development targets. In the food, energy, and water nexus, (Agrawal et al., 2024) suggests a multiobjective optimization method for allocating land to various crops. The objectives include maximization of food and bioenergy generation and minimization of water usage. During the winter and monsoon seasons, three areas of India have been studied. The results indicated that efficacious land use will lead to remarkable gains in the generation of bioenergy and mitigating the use of water.

China's South-to-North Water Transfer Project has been studied in the research work by Liu et al. (2023) aimed at minimizing energy usage. The project is the biggest water transfer project in the world and plays a crucial role in improving water security in dry areas. The paper presented an optimal operational framework for energy consumption minimization by effectively scheduling the water-pumping process along the transfer route. The derived solutions depicted that the developed operating strategies have the potential to alleviate water pumping and energy use. Moreover, the use of water supplied by nearby midstream lakes and the elevation of water levels in southern lakes mat efficiently mitigate energy consumption. The paper also highlights the best locations for energy conservation along the eastern path of the project. A broad study was carried out on the optimal integration of rural water–energy supply networks by Fuentes-Cortés et al. (2023), emphasizing the water–energy–environment nexus and energy security. In this respect, a normalization and subsetting technique was presented for multiobjective optimization to handle several design objectives concurrently and visualize trade-offs in the system with conflicting objectives. Besides, a rural indigenous community in central Mexico was investigated, while optimizing a system balancing local water consumption, energy efficiency, land use, and local CO_2 balance. The paper deployed a multiobjective optimization method, using the Shannon–Wiener index for evaluating energy security, increasing with the diversification of primary energy sources. Nonetheless, diversification also results in higher system costs and effects on land use and CO_2 emissions.

The optimization of distribution systems for water and energy in isolated areas has been done by Rojas-Punzo et al. (2023), deploying a multiobjective optimization technique involving open cycle ocean thermal energy conversion (OC-OTEC) technologies. The paper addressed resource scarcity and decreasing environmental effects in regions such as islands, where conventional energy and water sources are restricted and costly. By integrating OC-OTEC into Tahiti's existing systems, the study aimed to optimize the balance between economic viability, environmental sustainability, and social benefits. The method included a MINLP framework for evaluating various performance indexes such as CO_2 emissions, water usage from aquifers, and job creation potential. The obtained results showed that using OC-OTEC substantially enhances energy and water self-sufficiency, mitigates reliance on imported fuels, and supports local economies through job creation, which makes it a practical solution. Abidi et al. (2023) designed a water–energy nexus utilizing a multipump PV water station. The study investigates the hydraulic efficiency and water pumping capacity of two different water station configurations known as "monopump water station" and "multipump water station." The multipump water station is associated with higher hydraulic efficiency. In this regard, the unit commitment problem of the grid-tied PV unit has also been addressed. The problem was formulated as a MILP optimization problem aimed at maximizing water generation during a whole year. Chen et al. (2022) has been devoted to creating and implementing a graph theory-based bilevel water network planning model, addressing the water–energy–carbon nexus in the context of integrated regional water network planning. By incorporating the graph theory and bilevel programming, the model sought to optimize the planning of water systems by reconciling the conflicting objectives, that is, energy-saving maximization and cost minimization. The system deployed a hierarchical decision-making method at the regional level, taking into consideration the uncertainties associated with water demands. The efficacy of the presented framework was assessed and verified through different case studies, demonstrating the effect of different carbon emission control scenarios on the system's decision-making, spatial layout, and efficiencies. Utilizing multiobjective programming and machine learning algorithms, particularly soft clustering and fuzzy C-means, (Valencia-Marquez et al., 2023) depicts a through plan to manage the water–energy–waste nexus. It looks at how to build and run biogas-fired CHP systems and other distributed energy technologies in an optimal manner, considering environmental, economic, and practical objectives. The proposed framework found the Pareto frontier for the multiobjective optimization of the problem. This gives a deeper knowledge of how the nexus operates with distributed generation. Deploying clusters to see and understand how different objectives are interrelated and influenced by each other enhances the decision-making process in planning the energy system. Using an hourly OSeMOSYS model, (Groppi et al., 2023) investigated how to optimally plan the energy and water networks for a small island. In this respect, the long-term planning of supplying water and energy to the Favignana island in Italy was carried out and the way the reverse osmosis desalination and water storage may operate synchronously was assessed. Moreover, the impacts of carbon tax, emission limits, and no policy on the energy strategy of Favignana were evaluated. The results derived depicted that the minimum amount of emission is achieved by employing the carbon tax covering also shipping. Additionally, the paper looked also into the decision-making procedure for installing a desalination system. The research work by Cansino-Loeza et al. (2022) has been dedicated to developing an efficient model for planning and managing resources in a water–energy–food nexus. A mathematical model with different allocation mechanisms including social welfare, Rawlsian, Nash, and Rawlsian–Nash was also used to optimally operate resources and enhance global security concerns for water, energy, and food availability and adequacy. In this regard, issues like climate change, urbanization, and increasing demand were all taken into account in a case study from a Mexican state, where it was shown how the social aid plan has resulted in considerable changes in the water, energy, and food security. A deep reinforcement learning model was introduced in Riebel et al. (2024) for the sake of improving the water heating system control harnessing solar energy and heat recovery. The mentioned model trains intelligent agents to control a complicated water heating system in an institutional building. The system is equipped with solar power generation assets and a system for using waste heat from a water chiller to supply the required energy. In addition, the agents are trained not only to optimize energy use during hot water delivery but also to enable the chiller activation in order to avoid early degradation of the system and supply uninterrupted hot water once the heating device fails. Afsari Mamaghani et al. (2023) proposed a detailed model for water–energy–emission nexus planning, considering the interconnections between water, energy, and emissions. The model seeks to optimize the costs while taking into consideration technological and economic constraints and socioenvironmental requirements. A novel multidimensional measure, known as the water–electricity nexus measure, is deployed for the interconnectedness of the water and power industries assessment. The simulation results indicate that scenarios of high economic expansion would not be viable in the long run. To mitigate the nexus index, key factors involve enhancing renewable power production capacity, promoting technical efficiency, and introducing carbon and water tariffs.

Briefly stated, it is evident that the interaction between water and energy systems is vital for system efficiency maximization and getting environmental sustainability ensured in different sectors either residential or industrial and commercial. The reviewed documents proposed effective frameworks and methods to handle the complicated dynamics of the water–energy nexus. Each approach not only promotes the technical and economic performance of the systems but also addresses the importance of unified solutions in meeting sustainability objectives. Overall, it is essential to continuously enhance these frameworks and their applications to efficiently investigate the difficulties of resource management in a world that is becoming more interconnected and associated with resource limitations.

Contributions

The main contributions of the paper are as follows:

1. Proposing an integrated scheduling framework for the water–energy nexus considering renewable energy, desalination, and storage systems.
2. Presenting an efficient model with low computational burden for the short-term scheduling of water-energy systems.

Organization

The chapter is organized as follows, where Section 4.2 includes the mathematical modeling of the problem. The obtained results from simulating the problem in GAMS have been presented and discussed in Section 4.3. Lastly, Section 4.4 is dedicated to drawing the relevant concluding remarks.

Mathematical modeling

The proposed mathematical programming problem is formulated within a single-objective optimization framework where the objective is to minimize the total operating costs of the EH as shown in (4.1) (Esmaeel Nezhad et al., 2022; Javadi et al., 2019a, 2019b). The mentioned EH is a grid-tied one with the capability of bidirectional power flow resulting in reducing the costs which in turn increases the profit by transacting energy with the utility grid as addressed in (4.1). It is worth noting that the subscripts s and t denote the scenarios and time intervals, respectively. The objective function of the problem stated in (4.1) includes three items where the first item relates to the cost of transacting energy with the utility grid. The other two items indicate the cost of electrical energy generation by the EH resources and the cost of heat generation by the boiler, respectively. It is also noteworthy that the presented model is flexible enough to calculate the seasonal costs using the data of different seasons.

$$\text{Min}\left[\sum_{s=1}^{N_s}\omega_s\left[\sum_{t=1}^{N_T}\left[\left(PG_{s,t}^{G2H}\lambda_t^{\text{Buy}}-PG_{s,t}^{H2G}\lambda_t^{\text{Sell}}\right)+F_{s,t}^{\text{CHP}}+F_{s,t}^{\text{Boiler}}\right]\right]\right] \tag{4.1}$$

The last two cost terms of the objective function have been stated in (4.2) and (4.3), respectively. It is worth mentioning that f, $PG_{s,t}^{G2H}$, and $PG_{s,t}^{H2G}$ indicate the operating cost, the imported energy from the utility grid, and the energy exported to the grid, respectively. The problem is approached using a scenario-based stochastic programming problem, where ω denotes the probability of each scenario occurrence. Furthermore, λ_t^{sell}, λ_t^{buy}, and λ_t^{NG} denote the hourly energy selling price, energy buying price, and natural gas price, respectively.

$$F_{s,t}^{\text{CHP}}=\left(\frac{PG_{s,t}^{\text{CHP}}}{\eta_E^{\text{CHP}}}+\frac{PH_{s,t}^{\text{CHP}}}{\eta_H^{\text{CHP}}}\right)\lambda_t^{NG} \tag{4.2}$$

$$F_{s,t}^{\text{Boiler}}=\left(\frac{PH_{s,t}^{\text{Boiler}}}{\eta_H^{\text{Boiler}}}\right)\lambda_t^{NG} \tag{4.3}$$

The cost of generating electricity and heat by the CHP unit is denoted in (4.2), while (4.3) includes the cost of generating heat by the boiler. The electrical and heat generation of the CHP unit are denoted by $PG_{s,t}^{\text{CHP}}$ and $PH_{s,t}^{\text{CHP}}$, respectively while $PH_{s,t}^{\text{Boiler}}$ indicates the heat generation of the boiler. The proposed energy hub is equipped with an electrical energy storage (EES) unit operating at a negligible cost. The constraints of the system have been presented in (4.4) to (4.52).

$$PS_{s,t}^{\text{CHP,min}}I_{s,t}^{\text{CHP}}\le PG_{s,t}^{\text{CHP}}+PH_{s,t}^{\text{CHP}}\le PS_{s,t}^{\text{CHP,max}}I_{s,t}^{\text{CHP}} \tag{4.4}$$

$$PG_{s,t}^{\text{CHP,min}}I_{s,t}^{\text{CHP}}\le PG_{s,t}^{\text{CHP}}\le PG_{s,t}^{\text{CHP,max}}I_{s,t}^{\text{CHP}} \tag{4.5}$$

$$PH_{s,t}^{\text{CHP,min}}I_{s,t}^{\text{CHP}}\le PH_{s,t}^{\text{CHP}}\le PH_{s,t}^{\text{CHP,max}}I_{s,t}^{\text{CHP}} \tag{4.6}$$

$$PH_{s,t}^{\text{Boiler,min}}I_{s,t}^{\text{Boiler}}\le PH_{s,t}^{\text{Boiler}}\le PH_{s,t}^{\text{Boiler,max}}I_{s,t}^{\text{Boiler}} \tag{4.7}$$

$$PC^{\text{Chiller,min}}I_{s,t}^{\text{Chiller}}\le PC_{s,t}^{\text{Chiller}}\le PC^{\text{Chiller,max}}I_{s,t}^{\text{Chiller}} \tag{4.8}$$

$$PC_{s,t}^{\text{Chiller}}=PH_{s,t}\text{COP}^{\text{Chiller}} \tag{4.9}$$

$$PC^{\text{EHP,min}}I_{s,t}^{\text{EHP,Cooling}}\le PC_{s,t}^{\text{EHP,Cooling}}\le PC^{\text{EHP,max}}I_{s,t}^{\text{EHP,Cooling}} \tag{4.10}$$

$$PH^{\mathrm{EHP,min}} I_{s,t}^{\mathrm{EHP,Heating}} \leq PH_{s,t}^{\mathrm{EHP,Heating}} \leq PH^{\mathrm{EHP,max}} I_{s,t}^{\mathrm{EHP,Heating}} \tag{4.11}$$

$$0 \leq I_{s,t}^{\mathrm{EHP,Heating}} + I_{s,t}^{\mathrm{EHP,Cooling}} \leq 1 \tag{4.12}$$

It is noteworthy I is a binary variable used to specify the operating status of the equipment. The CHP system is modeled in (4.4–4.6) (Geidl & Andersson, 2007; Kienzle et al., 2011), where a convex feasible operating region (FOR) is employed to characterize the CHP operation. $I_{s,t}^{\mathrm{CHP}}$ specifies the operation of the CHP unit within the FOR. The operation of the boiler and the chiller to produce heat is modeled in (4.7) and (4.8), respectively. It is noted that the chiller operation is based on absorbing heat and producing cooling power. In this respect, $\mathrm{COP}^{\mathrm{Chiller}}$ shows the energy conversion factor of the chiller in (4.9). The electric heat pump (EHP) also operates in either cooling to produce cooling power denoted by $PC_{s,t}^{\mathrm{EHP}}$, or heating mode to produce heat denoted by $PH_{s,t}^{\mathrm{EHP}}$. The operation of the EHP is modeled in (4.10–4.12).

$$PG_{s,t}^{\mathrm{EHP,Cooling}} = PG_{s,t}^{\mathrm{EHP}} \mathrm{COP}^{\mathrm{EHP,Cooling}} \tag{4.13}$$

$$PH_{s,t}^{\mathrm{EHP,Heating}} = PG_{s,t}^{\mathrm{EHP}} \mathrm{COP}^{\mathrm{EHP,Heating}} \tag{4.14}$$

$$0 \leq PH_{s,t}^{\mathrm{Heater}} \leq PH_{s,t}^{\mathrm{Heater,max}} I_{s,t}^{\mathrm{Heater}} \tag{4.15}$$

$$PH_{s,t}^{\mathrm{Heater}} = PG_{s,t}^{\mathrm{Heater}} \mathrm{COP}^{\mathrm{Heater}} \tag{4.16}$$

$$\mathrm{Eng}_{s,t}^{\mathrm{EES}} = \mathrm{Eng}_{s,t-1}^{\mathrm{EES}} + PG_{s,t}^{\mathrm{EES,Ch}} \eta^{\mathrm{EES,Ch}} \Delta t - PG_{s,t}^{\mathrm{EES,Dis}} \Delta t / \eta^{\mathrm{EES,Dis}} \ \forall\, t > 1 \tag{4.17}$$

$$\mathrm{Eng}^{\mathrm{EES,min}} \leq \mathrm{Eng}_{s,t}^{\mathrm{EES}} \leq \mathrm{Eng}^{\mathrm{EES,max}} \tag{4.18}$$

$$\mathrm{Eng}_{s,t=0}^{\mathrm{EES}} = \mathrm{Eng}_{s,t=24}^{\mathrm{EES}} \tag{4.19}$$

$$0 \leq PG_{s,t}^{\mathrm{EES,Ch}} \leq PG_{s,t}^{\mathrm{EES,Ch,max}} I_{s,t}^{\mathrm{EES,Ch}} \tag{4.20}$$

$$0 \leq PG_{s,t}^{\mathrm{EES,Dis}} \leq PG_{s,t}^{\mathrm{EES,Dis,max}} I_{s,t}^{\mathrm{EES,Dis}} \tag{4.21}$$

$$0 \leq I_{s,t}^{\mathrm{EES,Ch}} + I_{s,t}^{\mathrm{EES,Dis}} \leq 1 \tag{4.22}$$

$$PG_{s,t}^{\mathrm{EES}} = PG_{s,t}^{\mathrm{EES,Dis}} - PG_{s,t}^{\mathrm{EES,Ch}} \tag{4.23}$$

$$0 \leq PG_{s,t}^{G2H} \leq PL^{\max} I_{s,t}^{G2H} \tag{4.24}$$

$$0 \le PG_{s,t}^{H2G} \le PL^{\max} I_{s,t}^{H2G} \tag{4.25}$$

$$0 \le I_{s,t}^{G2H} + I_{s,t}^{H2G} \le 1 \tag{4.26}$$

Eq. (4.13) models the cooling energy conversion of the EHP, while its heat conversion model is presented in (4.14). The conversion factors in the cooling and heating modes are denoted by $\mathrm{COP}^{\mathrm{EHP,Cooling}}$ and $\mathrm{COP}^{\mathrm{EHP,Heating}}$, respectively. Constraint (4.15) shows that the heat generated by the heater must be within the feasible operating interval and the electricity-to-heat conversion model is shown in (4.16). The model represented in (4.17) to (4.23) also indicate the operational model of the EES (Javadi et al., 2017; Javadi, Lofti et al., 2020). The works published by Delfino et al. (2017, 2019) include detailed descriptions of the EES model. As constraints (4.24) and (4.25) depict, the power imported from the grid and the power exported to the grid must be within less than the capacity of the feeder. $\mathrm{PL}^{\mathrm{Max}}$ is the capacity of the feeder connecting the EH to the utility grid.

$$W_{s,t}^{\mathrm{Des}} = K^{\mathrm{Des}} PG_{s,t}^{\mathrm{Des}} \tag{4.27}$$

$$W_{s,t}^{\mathrm{Tank}} = W_{s,t-1}^{\mathrm{Tank}} + W_{s,t}^{\mathrm{Des}} - W_{s,t}^{L} \; \forall t > 1 \tag{4.28}$$

$$W^{\min} \le W_{s,t}^{\mathrm{Des}} \le W^{\mathrm{Max}} \tag{4.29}$$

$$W_{s,t=0}^{\mathrm{Tank}} = W_{s,t=24}^{\mathrm{Tank}} \tag{4.30}$$

$$W_{\min}^{\mathrm{Tank}} \le W_{s,t}^{\mathrm{Tank}} \le W_{\max}^{\mathrm{Tank}} \tag{4.31}$$

$$W_{s,t}^{\mathrm{Des}} - W_{s,t-1}^{\mathrm{Des}} \le RD^{\mathrm{Des}} \; \forall t > 1 \tag{4.32}$$

$$W_{s,t-1}^{\mathrm{Des}} - W_{s,t}^{\mathrm{Des}} \le RU^{\mathrm{Des}} \; \forall t > 1 \tag{4.33}$$

Eq. (4.27) states the demand power of desalination system (Jalili et al., 2021). The coefficient factor is selected as 2.66 in this study (Jalili et al., 2021). Eq. (4.28) shows that the water available in the tank of the desalination system in scenario s and time t is determined with respect to the water available in the system in scenario s and time $t-1$ plus the water desalinated minus the water discharged from the tank in scenario s and time t. Constraint (4.29) indicates the maximum and minimum water demand of the desalination system. It is also emphasized in (4.30) that the water available in the desalination tank at the end of the scheduling period should be equal to the one at the beginning of the scheduling period. The limitation of the water tank capacity has also been expressed in (4.31), while (4.32) and (4.33) state the rampdown and rampup limits of the water demand of the desalination system. Energy flow equations have also been indicated in (4.34–4.52). Eq. (4.34) shows that the power generated by the CHP unit will be used by the electrolyzer, delivered to the grid, absorbed by the EES system, consumed by the EH and EHP, and also used for desalination. Eq. (4.35) indicates that the solar power generation by the PV unit is delivered to the electrolyzer, grid, EES system, EH, and EHP and also used for desalination. As can be observed in Eq. (4.36), the power demanded by the heater is supplied by absorbing solar power generation, the power supplied by the utility grid, the power produced by the CHP unit, and also the power injected by the EES system. Eq. (4.37) indicates that the power discharged to the system by the EES is used by the electrical load demand, exported to the grid, consumed by the EH and EHP, and also will be utilized for desalination. Eq. (4.38) includes the energy flow of the battery in the charging mode where the charge power is supplied by using the CHP, solar PV unit, and also the grid. Eq. (4.39) shows the energy flow of the EHP where the required power is procured by the CHP, PV system, and also the utility grid. The amount of power needed for desalination is provided by the CHP unit, PV system, utility grid, and also battery. It is also noteworthy that the EES system, CHP unit, and PV system can supply the power exported to the utility grid, as shown in (4.41). The power imported from the utility grid will be also used by the electrical load demand, the EES system to charge, the EHP, and for desalination, as indicated in (4.42). Eq. (4.43) indicates that the heat generated by the boiler can be used to supply the heating load and by the chiller. Likewise, the heat produced by the CHP is also used to feed heating loads and chiller. The heat proposed by the heater and the EHP is also used to meet the heating load demand, as shown in (4.45) and (4.46), respectively. Eq. (4.47) states that the power demanded by the chiller is supplied by the CHP unit and the boiler as the chiller can use the heat generated by these two assets only. Accordingly, the cooling power generated by the chiller will be used to meet the cooling power, as stated in (4.48). Similarly, the cooling power produced by the EHP can also be used for supplying the cooling load demand, as expressed in (4.49). Eq. (4.50) indicates that the electrical load can be supplied using the CHP unit, PV system, utility grid, and EES system. The heating load demand can also be met by the CHP, boiler, heater, and EHP, as indicated in (4.51). Lastly, the cooling load can also be supplied using the EHP and chiller as emphasized in (4.52) (Fig. 4.4, Fig. 4.5).

$$PG_{s,t}^{\mathrm{CHP}} = P_{s,t}^{\mathrm{CHP}\to\mathrm{EL}} + P_{s,t}^{\mathrm{CHP}\to H2G} + P_{s,t}^{\mathrm{CHP}\to\mathrm{EES}} + P_{s,t}^{\mathrm{CHP}\to\mathrm{EH}} + P_{s,t}^{\mathrm{CHP}\to\mathrm{EHP}} + P_{s,t}^{\mathrm{CHP}\to\mathrm{Des}} \tag{4.34}$$

$$PG_{s,t}^{\mathrm{PV}} = P_{s,t}^{\mathrm{PV}\to\mathrm{EL}} + P_{s,t}^{\mathrm{PV}\to H2G} + P_{s,t}^{\mathrm{PV}\to\mathrm{EES}} + P_{s,t}^{\mathrm{PV}\to\mathrm{EH}} + P_{s,t}^{\mathrm{PV}\to\mathrm{EHP}} + P_{s,t}^{\mathrm{PV}\to\mathrm{Des}} \tag{4.35}$$

$$PG_{s,t}^{\mathrm{Heater}} = P_{s,t}^{\mathrm{PV}\to\mathrm{Heater}} + P_{s,t}^{G2H\to\mathrm{Heater}} + P_{s,t}^{\mathrm{CHP}\to\mathrm{Heater}} + P_{s,t}^{\mathrm{EES}\to\mathrm{Heater}} \tag{4.36}$$

$$PG_{s,t}^{\mathrm{EES,Dis}} = P_{s,t}^{\mathrm{EES}\to\mathrm{EL}} + P_{s,t}^{\mathrm{EES}\to H2G} + P_{s,t}^{\mathrm{EES}\to\mathrm{EH}} + P_{s,t}^{\mathrm{EES}\to\mathrm{EHP}} + P_{s,t}^{\mathrm{EES}\to\mathrm{Des}} \tag{4.37}$$

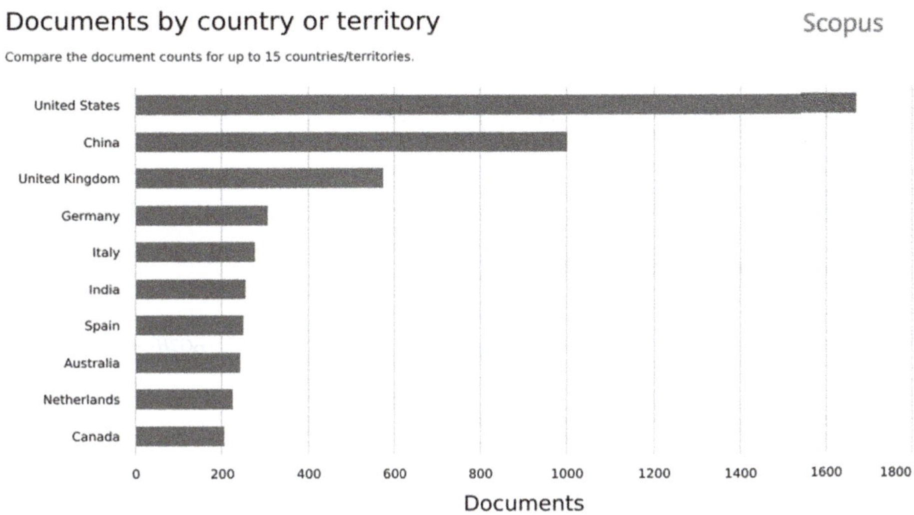

Figure 4.4 Number of documents by country.

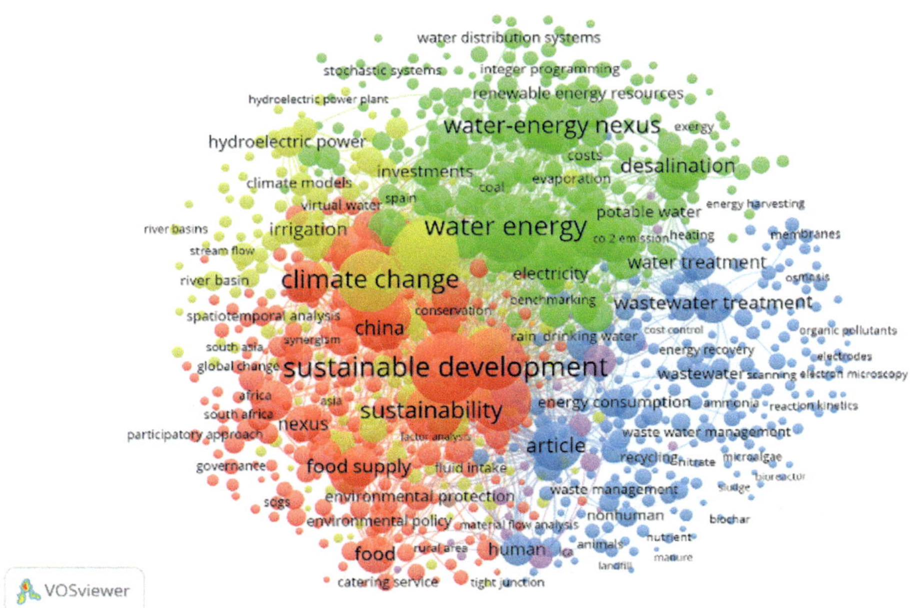

Figure 4.5 Most prevalent keywords.

$$PG_{s,t}^{\text{EES,Ch}} = P_{s,t}^{\text{CHP}\rightarrow\text{EES}} + P_{s,t}^{\text{PV}\rightarrow\text{EES}} + P_{s,t}^{G2H\rightarrow\text{EES}} \tag{4.38}$$

$$PG_{s,t}^{\text{EHP}} = P_{s,t}^{\text{CHP}\rightarrow\text{EHP}} + P_{s,t}^{PV\rightarrow\text{EHP}} + P_{s,t}^{G2H\rightarrow\text{EHP}} \tag{4.39}$$

$$PG_{s,t}^{\text{Des}} = P_{s,t}^{\text{CHP}\rightarrow\text{Des}} + P_{s,t}^{\text{PV}\rightarrow\text{Des}} + P_{s,t}^{G2H\rightarrow\text{Des}} + P_{s,t}^{\text{EES}\rightarrow\text{Des}} \tag{4.40}$$

$$PG_{s,t}^{H2G} = P_{s,t}^{\text{EES}\rightarrow H2G} + P_{s,t}^{\text{CHP}\rightarrow H2G} + P_{s,t}^{\text{PV}\rightarrow H2G} \tag{4.41}$$

$$PG_{s,t}^{G2H} = P_{s,t}^{G2H\rightarrow\text{EL}} + P_{s,t}^{G2H\rightarrow\text{EES}} + P_{s,t}^{G2H\rightarrow\text{EH}} + P_{s,t}^{G2H\rightarrow\text{EHP}} + P_{s,t}^{G2H\rightarrow\text{Des}} \tag{4.42}$$

$$PH_{s,t}^{\text{Boiler}} = H_{s,t}^{\text{Boiler}\rightarrow\text{HL}} + H_{s,t}^{\text{Boiler}\rightarrow\text{Chiller}} \tag{4.43}$$

$$PH_{s,t}^{\text{CHP}} = H_{s,t}^{\text{CHP}\rightarrow\text{HL}} + H_{s,t}^{\text{CHP}\rightarrow\text{Chiller}} \tag{4.44}$$

$$PH_{s,t}^{\text{Heater}} = H_{s,t}^{\text{Heater}\rightarrow\text{HL}} \tag{4.45}$$

$$PH_{s,t}^{\text{EHP}} = H_{s,t}^{\text{EHP,Heating}\rightarrow\text{HL}} \tag{4.46}$$

$$PC_{s,t}^{\text{Chiller}} = \left(H_{s,t}^{\text{CHP}\rightarrow\text{Chiller}} + H_{s,t}^{\text{Boiler}\rightarrow\text{Chiller}}\right)\text{COP}^{\text{Chiller}} \tag{4.47}$$

$$PC_{s,t}^{\text{Chiller}} = C_{s,t}^{\text{Chiller}\rightarrow\text{CL}} \tag{4.48}$$

$$PC_{s,t}^{\text{EHP,Cooling}} = C_{s,t}^{\text{EHP,Cooling}\rightarrow\text{CL}} \tag{4.49}$$

$$PL_{s,t}^{\text{EL}} = P_{s,t}^{\text{CHP}\rightarrow\text{EL}} + P_{s,t}^{\text{PV}\rightarrow\text{EL}} + P_{s,t}^{G2H\rightarrow\text{EL}} + P_{s,t}^{\text{EES}\rightarrow\text{EL}} \tag{4.50}$$

$$PH_{s,t}^{\text{HL}} = H_{s,t}^{\text{CHP}\rightarrow\text{HL}} + H_{s,t}^{\text{Boiler}\rightarrow\text{HL}} + H_{s,t}^{\text{Heater}\rightarrow\text{HL}} + H_{s,t}^{\text{EHP}\rightarrow\text{HL}} \tag{4.51}$$

$$PC_{s,t}^{\text{CL}} = C_{s,t}^{\text{EHP,Cooling}\rightarrow\text{CL}} + C_{s,t}^{\text{Chiller}\rightarrow\text{CL}} \tag{4.52}$$

Simulation results

This section includes the results obtained from simulating the EH with the desalination system. Ten different scenarios have been considered for simulating the problem. In this regard, the NG price is shown in Fig. 4.6. The data are hourly based and dynamic for 24 hours of the day.

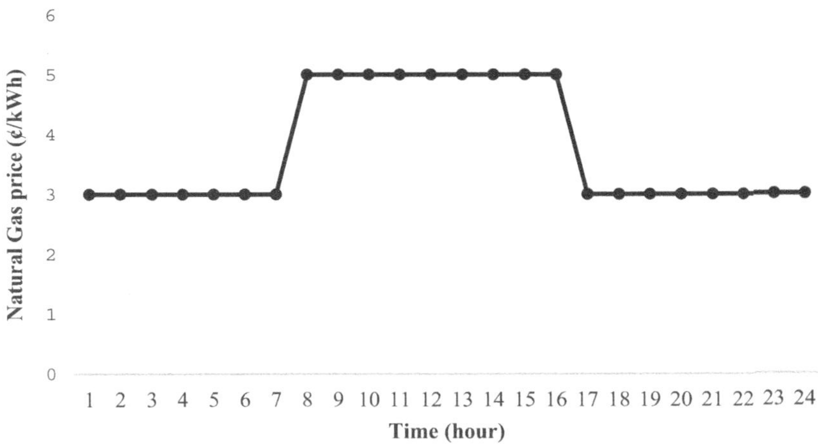

Figure 4.6 The hourly natural gas price.

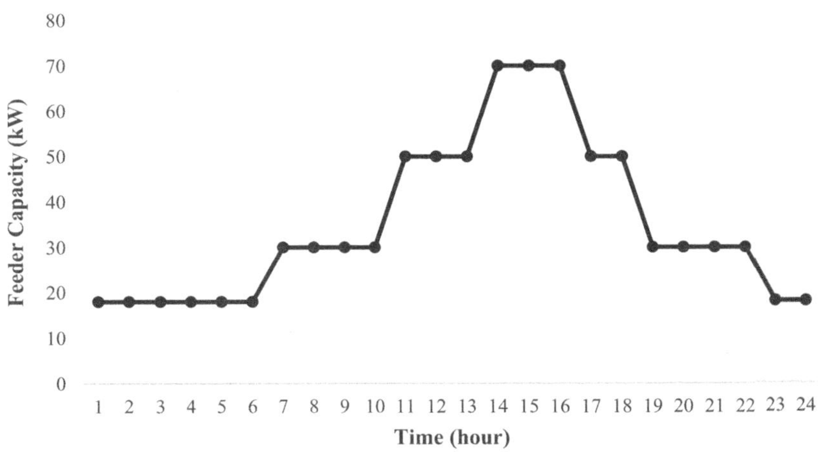

Figure 4.7 Capacity of the feeder connecting the EH to the utility grid.

The capacity of the feeder has also been depicted in Fig. 4.7. It is assumed that the capacity of the feeder is the same in the 10 scenarios simulated as the cable connecting the hub to the utility grid remains the same in all scenarios and it is not affected by the weather changes.

The power generated by the PV panel is affected by the weather changes addressed through the 10 scenarios studied in this chapter. In this respect, solar power generation in the 10 scenarios simulated in this chapter can be found in Fig. 4.8.

The electrical load demand, the water load demand, the heat load demand, and the cooling load demand are indicated in Figs. 4.9–4.12.

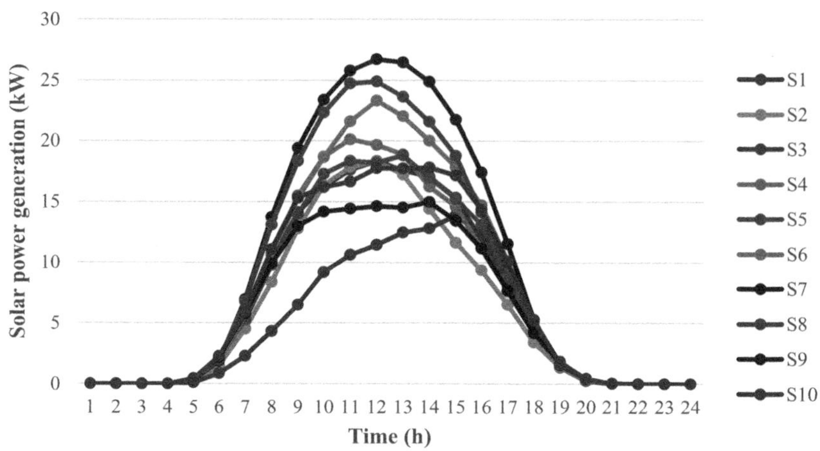

Figure 4.8 Solar power generation in different scenarios.

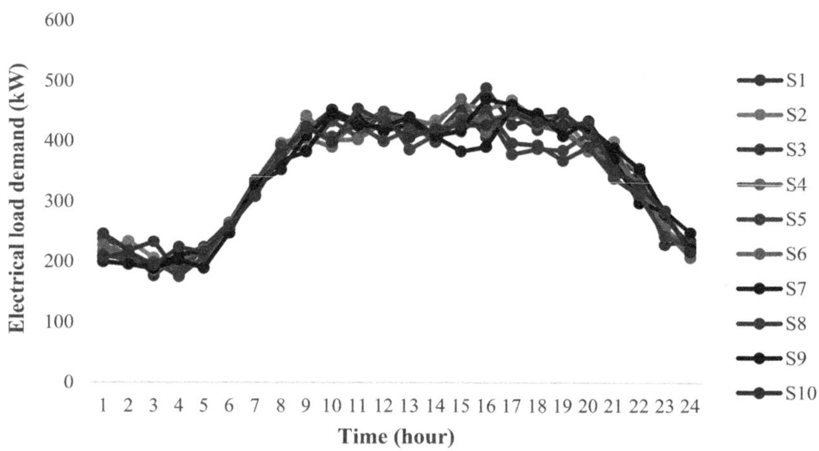

Figure 4.9 Electrical load demand of the EH in different scenarios.

The proposed MIP-based day-ahead scheduling of EH with a desalination system has been simulated in GAMS and solved using the CPLEX solver on an AMD Ryzen7 laptop with 16 GB RAM. In this respect, the optimal daily schedule of the CHP unit to produce electrical power in different scenarios can be observed in Fig. 4.13. This figure shows that the power generation of the CHP unit remains constant in all scenarios.

The energy stored in the EES system has also been shown in Fig. 4.14, where the optimal dispatch of this system has been given for the 10 scenarios.

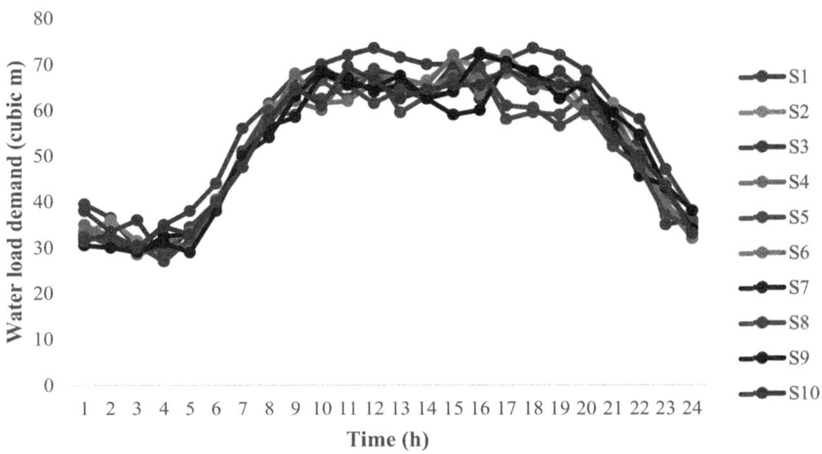

Figure 4.10 Water demand of the hub in different scenarios.

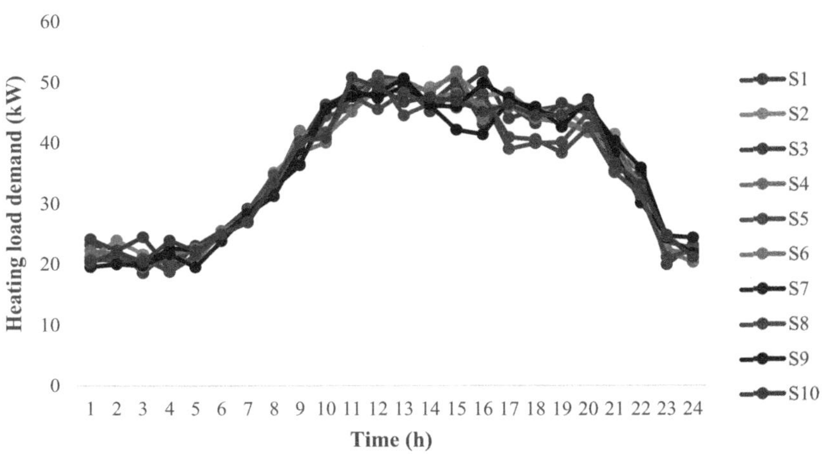

Figure 4.11 The heating load demand of the hub in different scenarios.

This figure shows that the EES system has operated within the permitted operating range, and it reaches the final value specified in constraint (19). The optimal dispatch of the boiler to produce heat in different scenarios and different hours of the day is demonstrated in Fig. 4.15.

The cooling power of the chiller has also been indicated in Fig. 4.16, and the cooling power of the EHP is shown in Fig. 4.17. Besides, the hourly power transacted with the utility grid has been depicted in Fig. 4.18.

The desalinated water available in the tank has also been optimally obtained and shown in Fig. 4.19. Also, the desalination power is shown in Fig. 4.20.

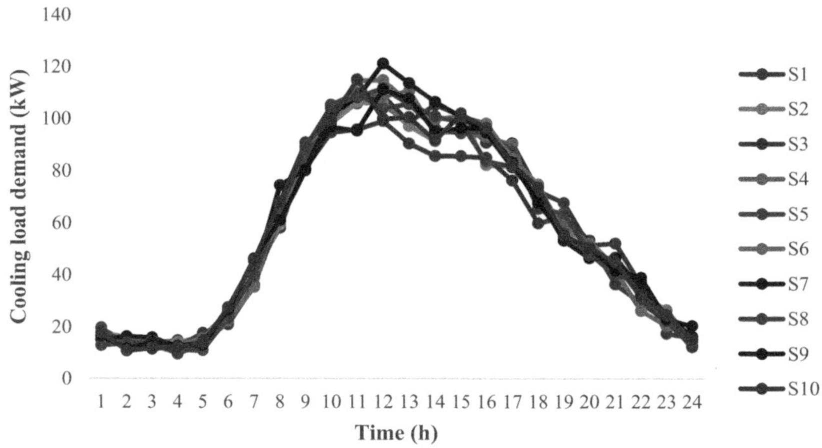

Figure 4.12 Cooling load demand of the hub in different scenarios.

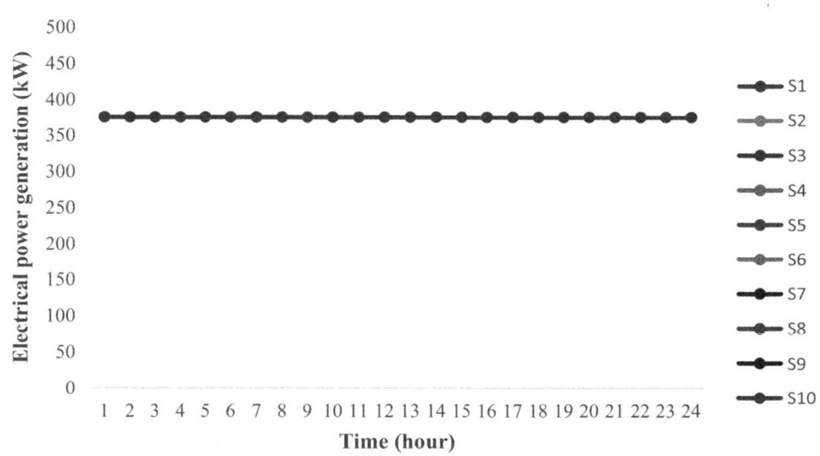

Figure 4.13 Electrical power generation of the CHP unit in different scenarios.

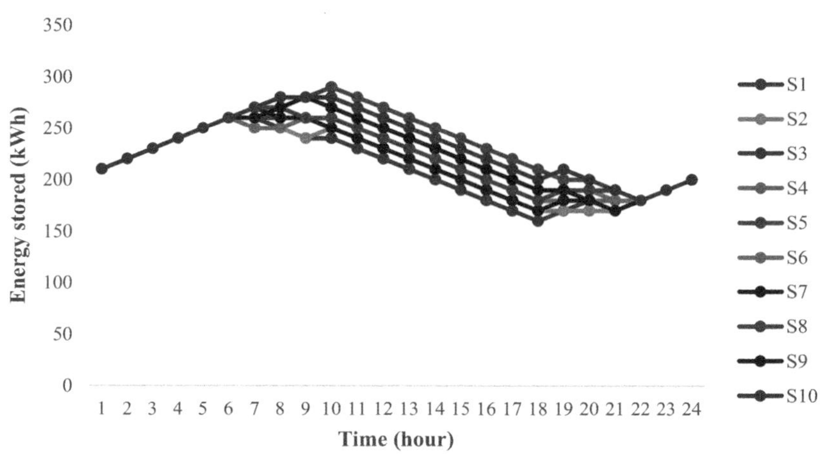

Figure 4.14 Optimal dispatch of the EES unit in different scenarios.

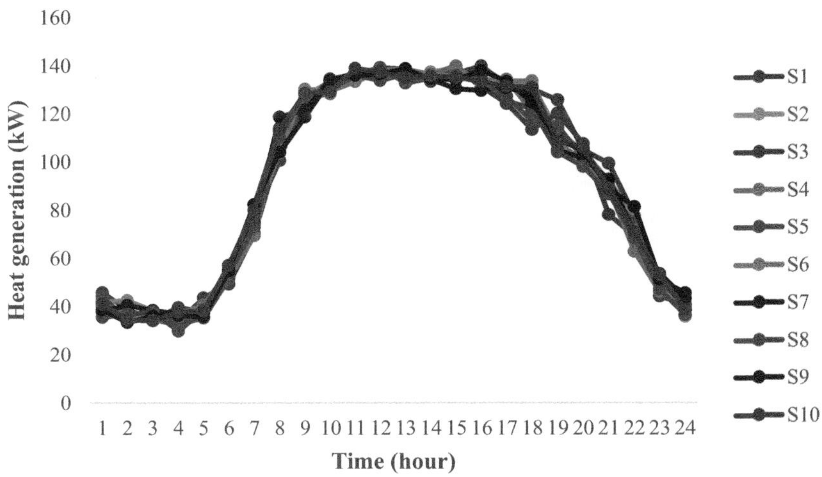

Figure 4.15 Heat generation of the boiler in different scenarios.

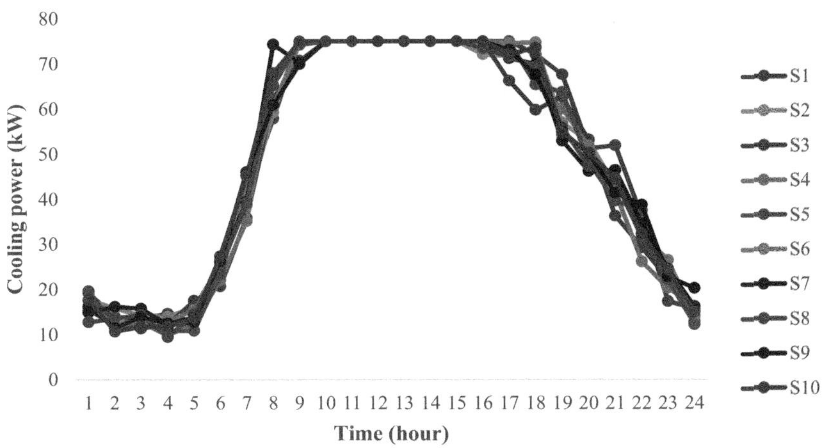

Figure 4.16 Cooling power of the chiller in different scenarios.

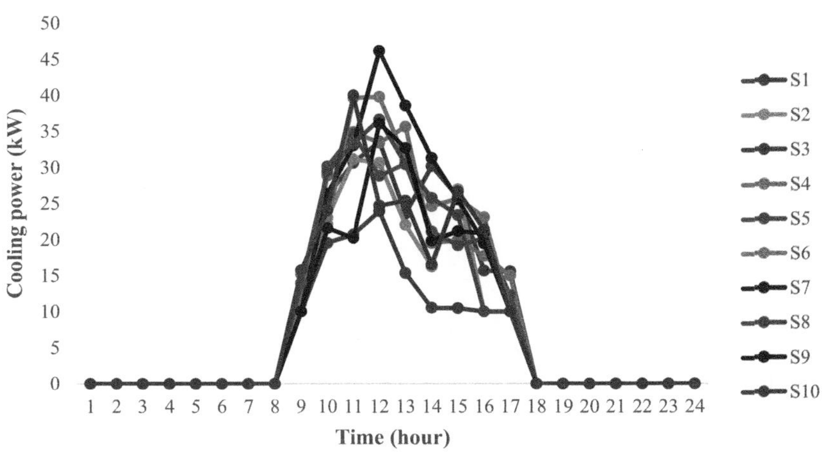

Figure 4.17 Cooling power of the EHP in different scenarios.

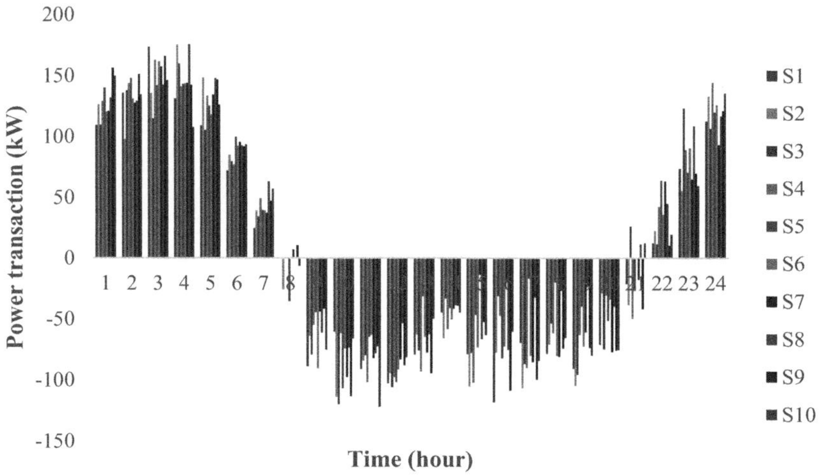

Figure 4.18 Power transaction between the EH and utility grid in different scenarios.

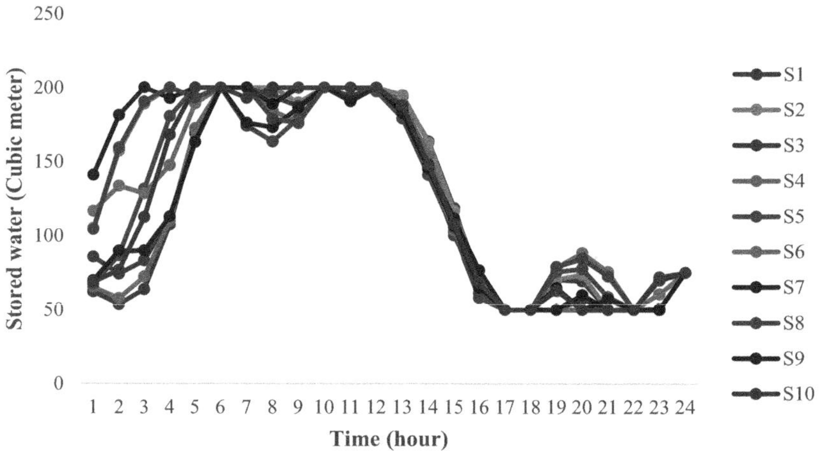

Figure 4.19 Desalinated water stored in the tank in different scenarios.

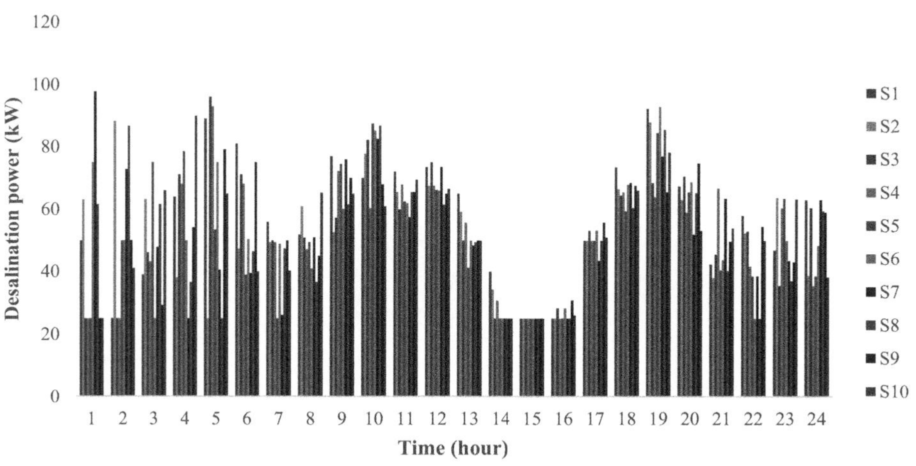

Figure 4.20 Desalination power in different scenarios.

Conclusion

This chapter proposed a comprehensive framework for scheduling energy and water systems within an energy hub. The framework effectively combined different technologies for converting and storing energy to meet different load types including electricity, heat, cooling, and water. The incorporation of a CHP unit, EHP, boiler, and heater in the hub guarantees a versatile and effective solution to electrical and heating energy demand. Installing an absorption chiller utilizing waste heat from the CHP unit demonstrates the strategic exploitation of byproduct energy to improve the overall efficiency of the system. The integration of battery-based electrical storage systems with solar PV panels will aid the system with local power generation at a lower cost and emission. Operating a saltwater desalination system will provide the EH with a reliable and sustainable source of freshwater, essential for dry areas. This indicates the interconnection between managing water and energy networks. The use of both renewable and thermal energy sources in this system maximizes resource efficiency and minimizes operating costs, showing remarkable progress in the design of sustainable energy systems. Deploying a MILP model coded in GAMS and solved by the CPLEX solver, the mathematical modeling of the day-ahead scheduling problem proposes a realistic and efficient methodology for operational purposes of such water-energy systems. This chapter not only proposed an integrated energy-water management system, but also provided a standard and efficient model for future studies in this area.

Acknowledgment

This paper is partly supported by (1) Research Council of Finland via: (a) EnergyNet Fellowship n.321265/n.328869/n.352654, and (b) X-SDEN project n.349965; and (2) Marie Skłodowska-Curie Actions (MSCA) Staff Exchanges via COALESCE project. Mohammad Sadegh Javadi acknowledges FCT for his contract funding provided through 2021.01052. CEECIND, 10.54499/2021.01052.CEECIND/CP1682/CT0001.

References

Abidi, M., Rhouma, A. B., & Belhadj, J. (2023). Optimal coordinated planning of water-energy system-based MILP algorithm of a multipump PV water station by deeming power commitment. *Electric Power Systems Research*, *220*, 109343. Available from https://doi.org/10.1016/j.epsr.2023.109343.

Afsari Mamaghani, F., Omidvar, B., Avami, A., & Nabi-Bidhendi, G. (2023). An optimal integrated power and water supply planning model considering Water-Energy-Emission nexus. *Energy Conversion and Management*, *277*, 116595. Available from https://doi.org/10.1016/j.enconman.2022.116595.

Agrawal, A., Bakshi, B. R., Kodamana, H., & Ramteke, M. (2024). Multi-objective optimization of food-energy-water nexus via crops land allocation. *Computers & Chemical Engineering*, *183*, 108610. Available from https://doi.org/10.1016/j.compchemeng.2024.108610.

Akbar, H., Nilsalab, P., Silalertruksa, T., & Gheewala, S. H. (2023). An inclusive approach for integrated systems: Incorporation of climate in the water-food-energy-land nexus index. *Sustainable Production and Consumption, 39*. Available from https://doi.org/10.1016/j.spc.2023.05.005.

Alhajeri, N. S., Al-Fadhli, F. M., Deshpande, A. A., & El-Halwagi, M. M. (2024). Optimization of water-energy-food nexus via an integrated system of solar-assisted desalination and farming. *Journal of Cleaner Production, 434*, 140362. Available from https://doi.org/10.1016/j.jclepro.2023.140362.

Babel, M. S., Rahman, M., Budhathoki, A., & Chapagain, K. (2023). Optimization of economic return from water using water-energy-food nexus approach: A case of Karnafuli Basin, Bangladesh. *Energy Nexus, 10*, 100186. Available from https://doi.org/10.1016/j.nexus.2023.100186.

Cansino-Loeza, B., Munguía-López, A. D. C., & Ponce-Ortega, J. M. (2022). A water-energy-food security nexus framework based on optimal resource allocation. *Environmental Science & Policy, 133*. Available from https://doi.org/10.1016/j.envsci.2022.03.006.

Chen, C., Zhang, X., Zhang, H., Cai, Y., & Wang, S. (2022). Managing water-energy-carbon nexus in integrated regional water network planning through graph theory-based bi-level programming. *Applied Energy, 328*, 120178. Available from https://doi.org/10.1016/j.apenergy.2022.120178.

Delfino, F., Ferro, G., Minciardi, R., Robba, M., Rossi, M., & Rossi, M. (2017). Identification and management of an electrical storage system for application in photovoltaic installations. *13th IEEE International Conference on Control & Automation (ICCA)*. Available from https://doi.org/10.1109/icca.2017.8003178.

Delfino, F., Ferro, G., Minciardi, R., Robba, M., Rossi, M., & Rossi, M. (2019). Identification and optimal control of an electrical storage system for microgrids with renewables. *Sustainable Energy, Grids and Networks, 17*, 100183. Available from https://doi.org/10.1016/j.segan.2018.100183.

Esmaeel Nezhad, A., Nardelli, P. H. J., Sahoo, S., & Ghanavati, F. (2022). Scheduling of energy hub resources using robust chance-constrained optimization. *IEEE Access, 10*. Available from https://doi.org/10.1109/access.2022.3228388.

Fuentes-Cortés, L. F., Rodríguez-Gutiérrez, J. E., López-Ramírez, M. D., & Martínez-Gutiérrez, N. (2023). Involving energy security and a Water—Energy-Environment nexus framework in the optimal integration of rural water—energy supply systems. *Energy Conversion and Management, 293*, 117452. Available from https://doi.org/10.1016/j.enconman.2023.117452.

Geidl, M., & Andersson, G. (2007). Optimal power flow of multiple energy carriers. *IEEE Transactions on Power Systems, 22*(1). Available from https://doi.org/10.1109/tpwrs.2006.888988.

Groppi, D., Kumar Pinayur Kannan, S., Gardumi, F., & Astiaso Garcia, D. (2023). Optimal planning of energy and water systems of a small island with a hourly OSeMOSYS model. *Energy Conversion and Management, 276*, 116541. Available from https://doi.org/10.1016/j.enconman.2022.116541.

Gui, J., Li, C., Cao, Y., Liu, Z., Shen, Y., Huang, W., & Tian, X. (2023). Hybrid solar evaporation system for water and electricity cogeneration: Comprehensive utilization of solar and water energy. *Nano Energy, 107*, 108155. Available from https://doi.org/10.1016/j.nanoen.2022.108155.

Jalili, M., Sedighizadeh, M., & Sheikhi Fini, A. (2021). Optimal operation of the coastal energy hub considering seawater desalination and compressed air energy storage system.

Thermal Science and Engineering Progress, *25*, 101020. Available from https://doi.org/10.1016/j.tsep.2021.101020.

Javadi, M. S., Anvari-Moghaddam, A., & Guerrero, J. M. (2017). Optimal scheduling of a multicarrier energy hub supplemented by battery energy storage systems. *IEEE international conference on environment and electrical engineering and 2017 IEEE industrial and commercial power systems Europe (EEEIC/I&CPS Europe). Milan, Italy: IEEE.* Available from https://doi.org/10.1109/eeeic.2017.7977520.

Javadi, M. S., Lotfi, M., Nezhad, A. E., Anvari-Moghaddam, A., Guerrero, J. M., & Catalao, J. P. S. (2020). Optimal operation of energy hubs considering uncertainties and different time resolutions. *IEEE Transactions on Industry Applications*, *56*(5). Available from https://doi.org/10.1109/tia.2020.3000707.

Javadi, M. S., Anvari-Moghaddam, A., Guerrero, J. M., Esmaeel Nezhad, A., Lotfi, M., & Catalao, J. P. S. (2019). Optimal operation of an energy hub in the presence of uncertainties. *IEEE International Conference on Environment and Electrical Engineering and 2019 IEEE Industrial and Commercial Power Systems Europe (EEEIC/I& CPS Europe)*. Available from https://doi.org/10.1109/eeeic.2019.8783452.

Javadi, M. S., Firuzi, K., Rezanejad, M., Lotfi, M., Gough, M., & Catalao, J. P. S. (2019). Optimal sizing and siting of electrical energy storage devices for smart grids considering time-of-use programs. *IECON 2019—45th Annual Conference of the IEEE Industrial Electronics Society*. Available from https://doi.org/10.1109/iecon.2019.8927263.

Kienzle, F., Ahcin, P., & Andersson, G. (2011). Valuing investments in multi-energy conversion, storage, and demand-side management systems under uncertainty. *IEEE Transactions on Sustainable Energy*, *2*(2). Available from https://doi.org/10.1109/tste.2011.2106228.

Liu, Y., Zheng, H., Wan, W., & Zhao, J. (2023). Optimal operation toward energy efficiency of the long-distance water transfer project. *Journal of Hydrology*, *618*, 129152. Available from https://doi.org/10.1016/j.jhydrol.2023.129152.

Mansour, F., Al-Hindi, M., Abou Najm, M., & Yassine, A. (2023). Multi-objective optimization for comprehensive water, energy, food nexus modeling. *Sustainable Production and Consumption*, 38. Available from https://doi.org/10.1016/j.spc.2023.04.013.

Mirzaei, M. A., Mehrjerdi, H., & Mansour Saatloo, A. (2023). Look-ahead scheduling of energy-water nexus integrated with Power2X conversion technologies under multiple uncertainties. *Sustainable Cities and Society*, *99*, 104902. Available from https://doi.org/10.1016/j.scs.2023.104902.

Nakhaei, M., Akrami, M., Gheibi, M., Daniel Urbina Coronado, P., Hajiaghaei-Keshteli, M., & Mahlknecht, J. (2022). A novel framework for technical performance evaluation of water distribution networks based on the water-energy nexus concept. *Energy Conversion and Management*, *273*, 116422. Available from https://doi.org/10.1016/j.enconman.2022.116422.

Perianes-Rodriguez, A., Waltman, L., & Eck, N. J. van (2016). Constructing bibliometric networks: A comparison between full and fractional counting. *Journal of Informetrics*, *10*(4). Available from https://doi.org/10.1016/j.joi.2016.10.006.

Rezaei, N., Pezhmani, Y., & Pir Mohammadiani, R. (2023). Optimal stochastic self-scheduling of a water-energy virtual power plant considering data clustering and multiple storage systems. *Journal of Energy Storage*, *65*, 107366. Available from https://doi.org/10.1016/j.est.2023.107366.

Riebel, A., Cardemil, J. M., & López, E. (2024). Multi-objective deep reinforcement learning for a water heating system with solar energy and heat recovery. *Energy*, *291*, 130296. Available from https://doi.org/10.1016/j.energy.2024.130296.

Rojas-Punzo, A. A., Hernández-Romero, I. M., Tovar-Facio, J., & Nápoles-Rivera, F. (2023). Optimization of distribution networks for water and energy in isolated regions: A multi-objective approach incorporating ocean thermal energy conversion technologies. *Sustainable Production and Consumption*, *40*. Available from https://doi.org/10.1016/j.spc.2023.07.016.

Sajna, M. S., Elmakki, T., Schipper, K., Ihm, S., Yoo, Y., Park, B., Park, H., Shon, H. K., & Han, D. S. (2024). Integrated seawater hub: A nexus of sustainable water, energy, and resource generation. *Desalination*, *571*, 117065. Available from https://doi.org/10.1016/j.desal.2023.117065.

Saskia, A., Putri, F., Moazeni, J., & Khazaei. (2023). Predictive control of interlinked water-energy microgrids. *Applied Energy*, *347*, 121455. Available from https://doi.org/10.1016/j.apenergy.2023.121455.

Shin, H. H., Kim, K., Lee, M., Han, C., & Kim, Y. (2024). Maximized thermal energy utilization of surface water-source heat pumps using heat source compensation strategies under low water temperature conditions. *Energy*, *288*, 129700. Available from https://doi.org/10.1016/j.energy.2023.129700.

Sun, J., Zhang, J., Wang, X., Jaramillo, F., Tang, Z., & Chen, L. (2023). Water-energy-food integrated management model under uncertainty-a case study of yulin city. *Journal of Hydrology*, *625*, 130059. Available from https://doi.org/10.1016/j.jhydrol.2023.130059.

Valencia-Marquez, D., Arturo Ortiz-Munguia, J., Maldonado-López, E., Alberto Quintana-Hernández, P., Francisco Louvier-Hernández, J., & Fabián Fuentes-Cortés, L. (2023). Multi-objective and machine learning strategies for addressing the Water–Energy–Waste nexus in the design of energy systems. *Sustainable Energy Technologies and Assessments*, *60*, 103445. Available from https://doi.org/10.1016/j.seta.2023.103445.

Xie, S., Zhang, J., Shi, D., & Zhang, Q. (2023). Multi-energy optimization under the water-energy-carbon nexus in a typical iron and steel plant. *Applied Thermal Engineering*, *224*, 120086. Available from https://doi.org/10.1016/j.applthermaleng.2023.120086.

Yu, D., Li, Z., Fan, S., & Sun, T. (2023). Optimal management of an energy-water-carbon nexus employing carbon capturing and storing technology via downside risk constraint approach. *Journal of Cleaner Production*, *418*, 138168. Available from https://doi.org/10.1016/j.jclepro.2023.138168.

Biomass for heating

5

Daniel Sánchez-Lozano, Antonio Escámez, Roque Aguado, David Vera and Francisco Jurado Melguizo
Department of Electrical Engineering, University of Jaén, Jaén, Spain

Introduction

Biomass is regarded as an essential component in the global transition toward sustainable energy solutions, particularly in the production of heat. Biomass has been utilized for heating purposes for centuries, yet its role in modern energy systems has expanded remarkably due to advancements in technology and a growing emphasis on renewable energy sources. As the largest renewable energy source globally, biomass has the potential to mitigate climate change and reduce dependency on fossil fuels. The versatility of biomass as an energy source is evident in its diverse feedstock, technology pathways, and end products (World Bioenergy Association, 2022c).

According to the World Bioenergy Association (2022c), 1.26 EJ of renewable heat was produced in both heat-only plants as well as combined heat and power (CHP) plants in 2020. Biomass accounted for 96% of all renewable heat produced, with minor contributions from geothermal and solar thermal technologies (World Bioenergy Association, 2022c). In other words, biomass contributed with 1.2 exajoules (EJ) to global heat production. Of this, 52% originated from solid biomass, while 25% came from municipal solid waste (World Bioenergy Association, 2022c). Europe has been at the forefront of biomass use for heating applications, producing 88% of the world's bioheat, predominantly through district heating networks.

The primary methods of converting biomass into heat include heat-only plants and CHP plants. Heat-only plants are dedicated solely to heat production for residential, commercial, or industrial use, while CHP plants generate both heat and electricity, optimizing energy efficiency. In 2020 CHP plants utilized 3.4 EJ of biomass for heat production, with solid biofuels making up the majority of the input feedstock (World Bioenergy Association, 2022c).

The global energy landscape has seen a significant shift toward renewable energy sources over the past two decades. Renewable technologies, including biomass, have doubled their share in global heat production. Biomass alone accounted for 96% of all renewable heat generated, underscoring its essential role in the energy sector (World Bioenergy Association, 2022c). Wood pellets, wood chips, and other forms of processed biomass provide renewable burning fuel options, facil-

Towards Future Smart Power Systems with High Penetration of Renewables. DOI: https://doi.org/10.1016/B978-0-443-29871-4.00005-1

itating widespread adoption across different regions. In 2021, the global production of wood pellets reached 44.3 million tons, with Europe accounting for more than half of this production.

This chapter delves into the various aspects of biomass for heating, exploring its potential, available technologies, and its role in the global energy system.

Types of biomass

Definition and classification of biomass

Biomass encompasses a wide range of organic materials derived from plants and animals. This organic matter can be utilized as a sustainable energy source due to its renewable nature. Understanding the classification of biomass is essential for effectively harnessing its energy potential.

1. **Primary biomass:** This category includes biomass derived directly from living or recently living organisms. Examples of primary biomass include wood from forests, agricultural crops like corn and sugarcane, and animal waste such as manure. Primary biomass is typically used for various energy applications, including heating, electricity generation, and biofuel production (Basu, 2018; Galadima & Muraza, 2020; Vukovic & Makogon, 2022; Zhao et al., 2012).
2. **Secondary biomass:** Secondary biomass consists of organic materials that are derived from industrial and municipal waste streams (Funk et al., 2020). These materials may include food scraps, paper and cardboard waste, yard trimmings, and sewage sludge. Secondary biomass can be processed and converted into energy through methods like anaerobic digestion, composting, and thermochemical conversion (Uddin & Wright, 2023; Vukovic & Makogon, 2022).

   Biomass can also be categorized as either renewable or nonrenewable, depending on the period of its formation process (Fig. 5.1).
3. **Renewable biomass:** Biomass is considered renewable when it can be replenished within a relatively short period, typically within a human lifespan. Renewable biomass sources include fast-growing energy crops like switchgrass and willow trees, as well as organic waste materials that can be continuously generated, such as agricultural residues and forestry byproducts. Utilizing renewable biomass helps reduce dependence on fossil fuels and mitigates greenhouse gas (GHG) emissions (Balaman, 2019a).
4. **Nonrenewable biomass:** Some biomass sources, such as fossilized organic materials like peat, are considered nonrenewable because their formation process occurs over geological timescales (Schilstra, 2001). While these materials can still be used as energy sources, their extraction and combustion contribute to carbon dioxide emissions and are not sustainable in the long term.

Different forms of biomass suitable for heating

Biomass is a versatile energy source that can be used for heating purposes in various forms (Fig. 5.2). A diverse array of primary sources contributes to biomass fuels, broadly categorized as either "wet" or "dry" sources. Within these classifications, the sources can be further grouped into five distinct categories (Smith, 2009).

Figure 5.1 General classification of biomass.

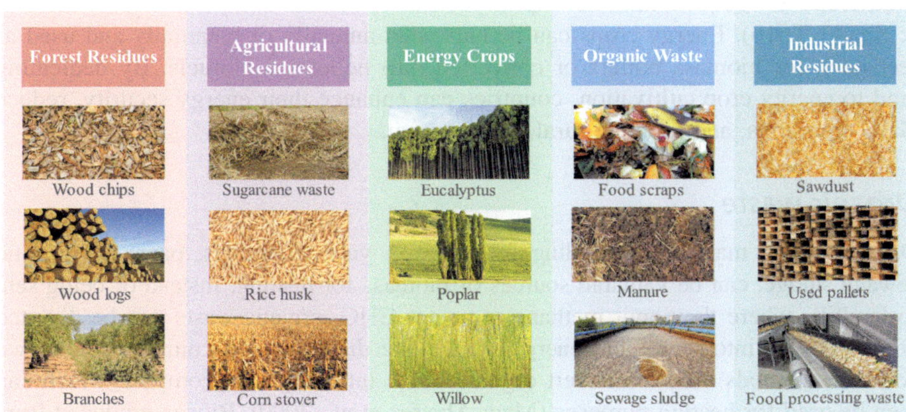

Figure 5.2 Categories of biomass for heating.

Forest residues

Forest residues refer to the organic materials left behind in forests after logging operations, including branches, bark, tops of trees, and other woody debris. These residues are a valuable source of biomass for heating due to their abundance and accessibility. After harvesting timber, the leftover forest residues can be collected and processed for use in biomass boilers or stoves. They are often chipped or ground into smaller pieces to improve combustion efficiency. Forest residues offer a sustainable alternative to fossil fuels for heating applications, helping to reduce GHG emissions and reliance on nonrenewable energy sources (Titus et al., 2021).

Agricultural residues

Agricultural residues are the organic materials left behind in fields after harvesting crops, including straw, husks, stalks, and other agricultural waste (Alakangas, 2016). These residues are a significant source of biomass for heating, especially in agricultural regions. Instead of being left to decompose in the fields, agricultural residues can be collected and utilized as a renewable energy source (Parfitt et al., 2021). They can be burned directly in biomass boilers or processed into pellets or briquettes for more efficient combustion (Bäfver et al., 2011). By using agricultural residues for heating, farmers can reduce waste disposal costs and generate additional income while contributing to renewable energy production.

Energy crops

Energy crops are specifically grown for their biomass energy potential, providing a renewable and sustainable source of fuel for heating applications (Whittaker & Shield, 2016). Examples of energy crops include willow (Šyc et al., 2012), miscanthus (Głowacka, 2011), switchgrass (Özenen-Kavlak & Çabuk, 2022), and poplar trees. These fast-growing plants are cultivated on marginal lands unsuitable for food crops and require minimal inputs such as fertilizers and pesticides (Whittaker & Shield, 2016). Energy crops can be harvested annually or biennially and used as feedstock for biomass boilers or converted into pellets or biofuels. By dedicating land to energy crop cultivation, countries can enhance their energy security, reduce GHG emissions, and promote rural development.

Organic waste

Organic waste materials, including food scraps, yard trimmings, paper waste, and sewage sludge, can be valuable sources of biomass for heating. Instead of being sent to landfills where they emit methane, a potent GHG, organic waste can be diverted and processed into renewable energy. Anaerobic digestion and composting are two common methods used to convert organic waste into biogas or compost, which can then be used for heating purposes (Mata-Alvarez et al., 2000). Biogas produced from anaerobic digestion can be burned in biomass boilers or used to generate electricity and heat through CHP systems (Hakawati et al., 2017). By harnessing the energy potential of organic waste, communities can reduce their environmental footprint, improve waste management practices, and promote a circular economy.

Industrial residues

Industrial residues comprise various organic materials generated as byproducts or waste during industrial processes (Naik et al., 2023). These residues may include wood chips, sawdust, nut shells, and other materials produced by industries such as forestry, woodworking, and food processing (Naik et al., 2023). In biomass heating plants, both treated and untreated wood residues can be used. Untreated residues such as off-cuts and sawdust from sawmills and timber yards are common sources. Treated residues, found in construction or furniture manufacturing, require careful consideration due to their potential adverse impact on plant performance (Bibi et al., 2023). Proper commissioning is necessary, especially for materials like extremely dry wood chip, to avoid detrimental effects on plant performance.

Advantages of biomass heating systems

Biomass is a cost-effective and practical solution for providing space heating, hot water, and process heating/steam from a low-carbon source (Smith, 2009). Biomass heating systems offer several compelling advantages over traditional heating methods powered by fossil fuels:

1. **Renewable energy source**. Biomass is derived from organic materials such as wood, agricultural residues, and organic waste, which can be replenished through natural processes. Unlike finite fossil fuels, which are being depleted at an alarming rate, biomass is considered a renewable energy source (Balaman, 2019b). This means that biomass can be sustainably harvested and utilized without depleting the Earth's finite resources, contributing to long-term environmental sustainability and reducing reliance on nonrenewable energy sources (Mbiankeu Nguea & Fotio, 2024).
2. **Reduced carbon emissions**. When biomass is burned for heating, it releases carbon dioxide (CO_2) into the atmosphere, just like burning fossil fuels. However, unlike fossil fuels, which release carbon dioxide that has been sequestered underground for millions of years, biomass is part of the natural carbon cycle. The plants used to produce biomass absorb CO_2 during photosynthesis, effectively offsetting the carbon emissions released during combustion (Fernández-Lobato et al., 2022). As a result, biomass heating systems are considered carbon-neutral or even carbon-negative, making them environmentally friendly options for reducing GHG emissions and mitigating climate change (Fernández-Lobato et al., 2022).
3. **Local resource utilization**. Biomass heating systems can utilize locally available biomass feedstocks, such as wood chips, agricultural residues, and organic waste. By harnessing local resources for heating, communities can reduce dependence on imported fossil fuels and support local economies (Balaman, 2019b; Vijay et al., 2022). This reduces reliance on distant energy sources and associated transportation-related emissions. Additionally, utilizing local biomass resources creates opportunities for local employment and economic development in rural areas (Vijay et al., 2022). Although biomass fuels may experience fluctuations in price over time and be influenced to some extent by international fossil fuel and commodity prices, these fluctuations are expected to be less severe than those affecting fossil fuels. Moreover, if biomass is obtained locally or from established suppliers, these fluctuations are likely to be more manageable and predictable.
4. **Energy independence**. Using biomass for heating can enhance energy security and independence by diversifying the energy supply. Unlike fossil fuels, which are subject to price volatility and geopolitical risks, biomass resources are often locally sourced and more stable in price. This reduces vulnerability to fluctuations in global energy markets and geopolitical tensions, enhancing energy resilience and stability (IEA Bioenergy, 2022).
5. **Waste management**. Using certain biomass resources as fuels for heating systems can divert them from becoming wastes and being disposed of in landfills. By converting

agricultural residues, forestry byproducts, and organic waste into energy, biomass heating systems can reduce the need for landfill disposal and mitigate environmental pollution. This helps address waste management challenges while producing valuable renewable energy, contributing to a circular economy and resource efficiency (Balaman, 2019b).

6. **Financial incentives**. In many regions, there are financial incentives and subsidies available to support the adoption of biomass heating systems. These incentives may include tax credits, grants, rebates, and low-interest loans, making biomass heating an economically viable option for homeowners, businesses, and institutions. Additionally, biomass heating systems can provide long-term cost savings compared to fossil fuel-based heating systems, especially as fuel prices fluctuate. By investing in biomass heating, individuals and organizations can realize both environmental and financial benefits.

Fuel properties of biomass for heating applications

Heating value

The heating value is an essential property of fuels that indicates their heating potential and is a measure of its energy content (Smith, 2009). The higher heating value (HHV), also known as gross calorific value, is defined as the heat released to a specified amount of water by a unit mass or volume of fuel (initially at 25°C) postcombustion with oxygen in a bomb calorimeter and subsequent return of the combustion products to standard ambient temperature conditions of 25°C (Basu, 2018). The heat released in this process represents the utmost energy potential achievable from fuel combustion. As liquid water forms during the test, the bomb calorimeter provides a measurement of the HHV. Nonetheless, the truly usable energy is what remains after the water produced during combustion evaporates. This residual energy constitutes the lower heating value (LHV), also termed net calorific value, and can be derived from the HHV. The relationship between HHV and LHV is given by:

$$\mathrm{LHV} = \mathrm{HHV} - \Delta h_g(\mathrm{H} - W)$$

where H and W denote, respectively, the mass fraction of hydrogen and moisture content of the fuel on a wet basis (expressed as ${}^{\mathrm{kg\ H_2O}}/_{\mathrm{kg\ mixture}}$ in SI units). In addition, Δh_g represents the specific enthalpy of water vaporization in the same units and reference temperature as the HHV (typically, 25°C). In numerous European nations, the efficiency of a thermal system is commonly defined with respect to the LHV. Consequently, efficiency values expressed in terms of the LHV may appear higher compared to those referenced to the HHV, which is customary in other regions such as the United States and Canada (Basu, 2018). Therefore, it is imperative to specify the basis of efficiency calculation to ensure accurate comparisons.

Moisture content

Moisture content denotes the proportion of moisture within a fuel sample upon being heated slightly beyond the boiling point of water (105°C) (Miller, 2013). It is determined by computing the weight loss, expressed as a percentage of the initial material, once the weight reaches stability with no further alterations. The moisture content, typically reported on a wet basis (W), is calculated as follows:

$$W = \frac{m_{\text{wet}} - m_{\text{dry}}}{m_{\text{wet}}}$$

The majority of the biomass industry commonly reports the moisture content on wet basis, as it provides a more straightforward insight into the moisture content present in biomass fuels (Smith, 2009). However, moisture content is occasionally presented on a dry weight basis (X), wherein:

$$X = \frac{m_{\text{wet}} - m_{\text{dry}}}{m_{\text{dry}}}$$

The dry weight basis may yield a moisture percentage exceeding 100% for highly saturated samples, potentially causing confusion. Therefore, it is essential to always specify the moisture basis. Moisture content values provided in one format can be converted to the other through the following relationships:

$$W = \frac{X}{1 + X}$$

$$X = \frac{W}{1 - W}$$

A higher moisture content leads to a reduced heating value, since each unit mass of fuel contains a reduced amount of oven-dry biomass, which constitutes the fraction of the fuel that actively undergoes combustion to generate heat. This effect is particularly pronounced in most biomass heating systems where the water vapor in the combustion byproducts remains uncondensed. This is due to the fact that the moisture in the fuel must also be vaporized before combustion can take place, requiring an energy input that cannot be recovered afterward.

Bulk density

The true density is defined as the ratio between the mass and the initial volume of a fuel, expressed as follows:

$$\rho_f = \frac{m_f}{V_f}$$

However, for powdered or granular solid fuels with a heterogeneous composition containing empty spaces or interstices filled with air or other substances, the concept of bulk density becomes pertinent. The bulk density of a solid fuel is defined as the mass of the fuel particles divided by the total volume they occupy, expressed by:

$$\rho_f^* = \frac{m_{\text{bulk}}}{V_{\text{bulk}}} = \frac{m_f + m_{\text{air}}}{V_f + V_{\text{air}}}$$

The total volume includes the volume occupied by particles, the void volume between particles, and the internal pore volume. As a result, the bulk density of a specific volume of particulate solid fuel is lower than its true density if it were compacted. Accordingly, it always holds true that $\rho^* \leq \rho$.

Bulk density, in contrast to true density, is not inherent to a material (Smith, 2009). For instance, an identical piece of wood may exhibit varying bulk densities if processed into wood chips, pellets, or logs. Moisture content also affects bulk density, since each particle may have a higher mass without necessarily occupying more space.

Energy density

Energy density is derived from the bulk density of a fuel and quantifies the energy held within a unit of fuel. Typically denoted in MJ/m^3, it can be calculated by multiplying the calorific value (MJ/kg), either lower or heating value, by the bulk density (kg/m^3).

$$\rho_{\text{LHV}} = \rho_f^* \text{LHV}_f$$

Energy density allows determining volumetric fuel consumption rates and the necessary fuel storage capacity, which allows estimating the frequency of deliveries and assessing the total annual fuel requirement.

Particle size

Biomass heating systems require mechanisms for physically transferring fuel from storage to the combustion area within the plant. Oversized fuel particles can cause blockages in specific fuel feed systems, such as augers, making particle size a critical aspect to consider regarding biomass fuel suitability. Biomass fuels exhibit a wide array of shapes and sizes, with particle size ranking alongside moisture content as a primary consideration when aligning system design with available fuel (Smith, 2009). Some fuel feed systems such as walking floors can accommodate a broader spectrum of particle sizes, whereas systems designed for pellet fuels have narrower tolerances and can only handle a more limited range of particle sizes.

Mechanical durability

If the biomass heating system relies exclusively on pellet fuels, then the mechanical durability of the pellets (i.e., their ability to remain intact during handling) becomes a critical consideration to be specified (Smith, 2009). A common issue with pellet fuels is their tendency to disintegrate during the handling process. High-quality pellets should exhibit a mechanical durability of at least 97.5%, meaning that less than 2.5% of the pellets should break down after delivery (Smith, 2009). Additionally, many pellet fuels require additives to act as binding agents, which should be provided by the manufacturer.

Ash content

Ash content refers to the residue left behind following the incineration of biomass at a temperature of 550°C ± 10°C for a minimum of 60 minutes, during which the moisture content evaporates and the organic matter is combusted. This residual material lacks any energy value and comprises primarily inorganic substances. Its major components include silica, aluminum, iron, and calcium, with minor quantities of magnesium, titanium, sodium, and potassium possibly present (Basu, 2018).

Even though the amount of ash generated depends on the type and efficiency of the biomass plant in operation, it also constitutes an intrinsic characteristic of the fuel itself, specified as a fuel property. For instance, wood chip or pellet fuels typically exhibit an ash content of approximately 1% by weight (equivalent to 1%−3% by volume) of the fuel consumed, whereas energy crops like miscanthus and straw tend to have higher ash content (Smith, 2009).

Biomass heating technologies

Small-scale heating systems

Firewood stoves

Firewood stoves for biomass heating represent a traditional and effective option for heating homes using firewood as fuel (Brandelet et al., 2018). These stoves are designed to burn firewood efficiently, generating heat that can warm a room or even an entire house. What sets them apart is their ability to use firewood as a form of renewable biomass, making them a more sustainable option compared to heating systems that rely on fossil fuels. In Europe, the residential sector (excluding pellets) accounts for the largest share of wood energy consumption, comprising 27% of the total usage (Brandelet et al., 2018; European Biomass Association, 2016).

A key characteristic of firewood stoves for biomass heating is their simplicity and reliability. These stoves are often constructed with durable and heat-resistant materials such as cast iron or steel, allowing them to withstand the rigorous process of wood combustion. Additionally, many stoves are designed with ventilation systems and airflow controls that optimize combustion and maximize the efficiency of the generated heat. Another important aspect to consider is safety (Agurto Adrianzén, 2013; Brandelet et al., 2021, 2017). Firewood stoves must be installed and used correctly to avoid risks such as fires, carbon monoxide (CO) poisoning, or particles (Brandelet et al., 2017, 2021). It is essential to follow the manufacturer's instructions for stove installation and maintenance, as well as to maintain an adequate supply of fresh air in the room where the stove is located. The thermal efficiency of modern firewood stoves is reported to be as high as 80% (World Bioenergy Association, 2012b).

Despite their simplicity, firewood stoves for biomass heating remain a popular choice in many rural and suburban areas (Agarwal, 1983; Demurger & Fournier, 2011). In addition to providing heat, the firewood used in these stoves is considered a renewable energy source, as long as it is managed sustainably. This entails harvesting firewood from responsibly managed forests or using wood waste from the timber industry.

Firewood boilers

Firewood boilers represent a traditional yet highly effective form of biomass heating technology, harnessing the energy potential of firewood to provide warmth for residential, commercial, and industrial spaces. These boilers operate by burning seasoned firewood within a controlled combustion chamber, where the heat generated is transferred to a heat exchanger. This heat exchanger then transfers the thermal energy to water or another heat transfer fluid, which is circulated through the building's heating system via pipes or radiators.

Firewood boilers, utilizing seasoned firewood for heating, are efficient and traditional biomass heating systems. They burn firewood in controlled chambers, transferring heat to water or fluid circulated through buildings. Wood pellet and chip boilers boast 90% efficiency, but emit CO_2, CO, and particulates, mitigated by emission control tech. Accessible and affordable, firewood is sourced sustainably and locally, reducing transport emissions. Available in various sizes, some models automate fuel feeding. Properly stored and dried firewood ensures optimal boiler performance and efficiency, minimizing emissions and maximizing heat output (Chen et al., 2023; Niklitschek et al., 2020).

Wood chip-fueled boilers

Wood chips are commonly used as biomass fuel for larger installations with capacities exceeding 100 kW (World Bioenergy Association, 2012b). However, they are also occasionally used for small-scale heating in agricultural or forest settings (World Bioenergy Association, 2012b).

Wood chip-fueled boilers are widely utilized for energy production, predominantly through combustion processes (Mižakova et al., 2021). Wood chips, derived notably from waste wood during extraction and processing activities, emerge as a compelling fuel source due to their economic viability. However, optimizing the combustion process is pivotal for reducing emissions, particularly CO (Quintero-Marquez et al., 2014). Enhanced control mechanisms targeting parameters such as excess air ratio, primary and secondary air distribution systems, and fuel feed rates are essential for maintaining stable combustion, thereby mitigating emissions and maximizing efficiency (Quintero-Marquez et al., 2014).

Despite its environmental benefits, wood combustion may lead to elevated levels of particle and gaseous emissions in comparison to fossil fuels like oil or natural gas. Ensuring proper fuel quality is imperative for both uninterrupted operation and minimizing emissions in wood chip boilers, especially in smaller units with heat outputs below 100 kW (Schön et al., 2019). Challenges such as clogging of screw conveyors, bridging, or slagging can be averted by utilizing homogeneous and high-quality fuels. Additionally, adherence to relevant fuel parameters including chemical composition, moisture content, and fines content within the wood chips is essential (Schön et al., 2019).

Wood pellet stoves

Wood pellet stoves offer a contemporary alternative to traditional wood stoves (Bäfver et al., 2011). Wood pellets are a densified fuel with uniform physical properties suitable for various domestic and industrial furnace applications. This type of fuel has seen significant growth in recent years, becoming a noteworthy option globally (Toscano et al., 2014).

Within the framework of energy sustainability, renewable energy systems are gaining ground. Wood pellets have emerged as a valuable renewable energy source, with a noticeable increase in consumption across Europe in the past decade (Bäfver et al., 2011; Lee et al., 2011; Toscano et al., 2014; Venturini et al., 2018). Pellet stoves offer advantages such as carbon neutrality and reliance on renewable resources. Recent research has underscored the significance of pellet quality, as lower-quality pellets have been linked to higher emissions (Venturini et al., 2018). Regarding particle concentrations in flue gas, levels ranging from 13 to 81 mg/MJ have been reported, notably lower than those from wood stoves (86–780 mg/MJ) (Bäfver et al., 2011).

Designing a wood pellet stove requires a deep understanding of solid fuel combustion, gas flow mixing, and reactions (Venturini et al., 2018). It is imperative to emphasize the importance of avoiding errors in furnace design and exhaust system installation, as these can lead to excessive CO emissions or diminished energy efficiency (Lee et al., 2011).

Wood pellet boilers

In Europe, the demand for wood pellet heating is on the rise due to increased government backing for renewable energies, coupled with elevated prices of fossil fuels and environmental apprehensions (Carvalho et al., 2013; Thomson & Liddell, 2015). Despite this surge, the suitability of wood pellet boilers for household use has been relatively overlooked compared to alternative renewable heating technologies or commercial biomass heating practices.

The benefits of wood pellet heating include mitigating CO_2 emissions, potential financial savings, and enhanced comfort (Thomson & Liddell, 2015). Nonetheless, there are concerns regarding risks across the entire supply chain, as well as the reliance on well-designed and properly maintained systems for optimal performance and safety (Thomson & Liddell, 2015). Furthermore, the substantial initial costs associated with wood pellet heating may discourage adoption, particularly among low-income households (Thomson & Liddell, 2015). Although some households may view wood pellet heating as a feasible alternative to fossil fuel systems, challenges faced by end-users, uncertainties regarding pellet supply and pricing, and maintenance needs presently constrain its suitability for residential settings (Carvalho et al., 2013; Thomson & Liddell, 2015). Nevertheless, there is considerable promise for the use of wood pellet fuel in urban heating networks (Carvalho et al., 2013; Thomson & Liddell, 2015), as well as within industrial contexts (Carvalho et al., 2013).

The basic schematic of a wood pellet boiler is depicted in Fig. 5.3. The fuel hopper with pellets can be loaded automatically or manually. Pellets fall to the bottom of the hopper, from where they are conveyed by a screw conveyor to the burner. The rotation of the conveyor is controlled by an electronic unit, as well as the airflow of the fan (Horák et al., 2022). This schematic provides an overview of the internal design of the boiler, although each manufacturer may offer specific solutions for this technology (Horák et al., 2022).

Medium- and large-scale heating systems

A boiler and a furnace serve different heating functions and operate distinctly. A boiler is a closed vessel that heats water or another fluid to generate steam or hot water, which is then used for heating, power generation, or industrial processes. It distributes heat via radiators, baseboard heaters, or radiant floor systems and can use various fuels such as natural gas, oil, electricity, coal, biomass, or alternative fuels. Boilers are more efficient for generating large amounts of heat and can be environmentally friendly when using renewable energy sources. They are used in residential heating, commercial buildings, and industrial processes, with a longer

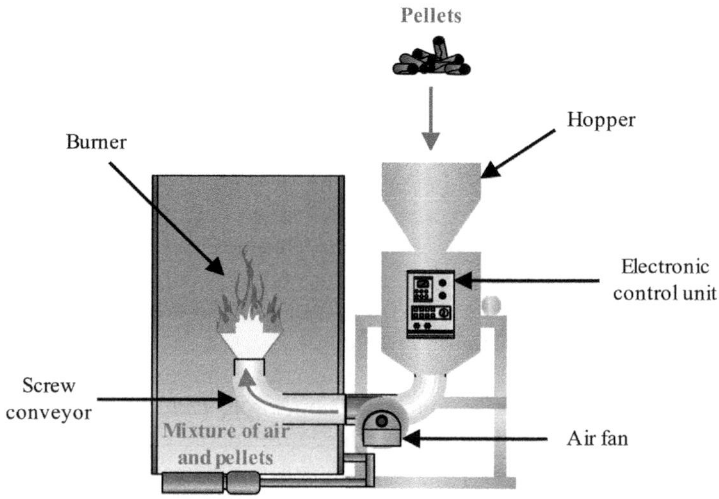

Figure 5.3 Diagram of a small-scale wood pellet boiler.

lifespan if properly maintained, although they have higher initial installation costs. On the other hand, a furnace heats air directly and uses a blower to distribute the heated air through ducts for space heating. It typically uses natural gas, oil, or electricity and is less versatile in fuel types compared to boilers. Furnaces are commonly found in residential and commercial heating systems, with applications in industrial heat treatment processes. They vary in efficiency, can have higher emissions depending on the fuel source, and generally have lower installation costs but a shorter lifespan. Regular maintenance is required for both systems to ensure efficient operation and longevity.

Industrial boilers

Biomass boilers (see Fig. 5.4) have become a sustainable solution for heating purposes, particularly in residential, commercial, and industrial settings (Saidur et al., 2011). These boilers use organic materials such as wood pellets, chips, or logs as fuel to produce heat, in contrast to traditional fossil fuel boilers. In the European Union (EU28), more than half of the biomass is eventually dedicated to residential heating. Among the prevalent heating technologies adopted in these regions are logwood stoves, logwood boilers, pellet boilers, and wood chip boilers (Jaworek et al., 2021; MBtek Inc., 2023). Logwood stoves are common in residential settings, providing direct heating through burning logs, while logwood boilers are larger systems suitable for heating water for both domestic and heating purposes. Pellet boilers, on the other hand, are automated systems burning compressed wood pellets, offering convenience and efficiency for residential and commercial applications. For larger-scale heating needs, wood chip boilers prove to be ideal, commonly used

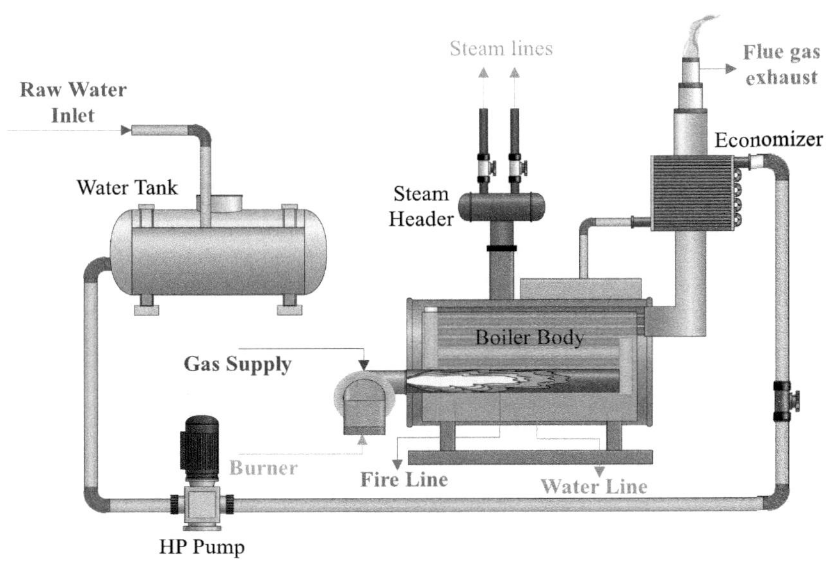

Figure 5.4 Schematic diagram of a fire tube boiler.

in district heating systems (DHSs) or industrial facilities (BECoop, 2022; Sipilä & Wiltshire, 2016). The advantages of biomass boilers are noteworthy, including their utilization of renewable energy sources, carbon neutrality due to carbon offsetting, cost-effectiveness compared to fossil fuels, and promotion of local energy production. However, several considerations must be taken into account when deploying biomass boilers, such as fuel availability, efficiency, maintenance requirements, emissions control, and regulatory compliance. Despite these challenges, biomass boilers are poised to play a significant role in meeting heating needs globally, driven by advancements in technology and growing awareness of the importance of renewable energy sources.

Industrial furnaces

Various technologies exist for utilizing biomass in furnaces, providing sustainable heating solutions across residential, commercial, and industrial sectors. One common approach is the integration of biomass combustion systems into traditional furnace designs, where biomass materials such as wood pellets, chips, or logs are burned to generate heat (see Fig. 5.5). These biomass-fired furnaces can operate similarly to

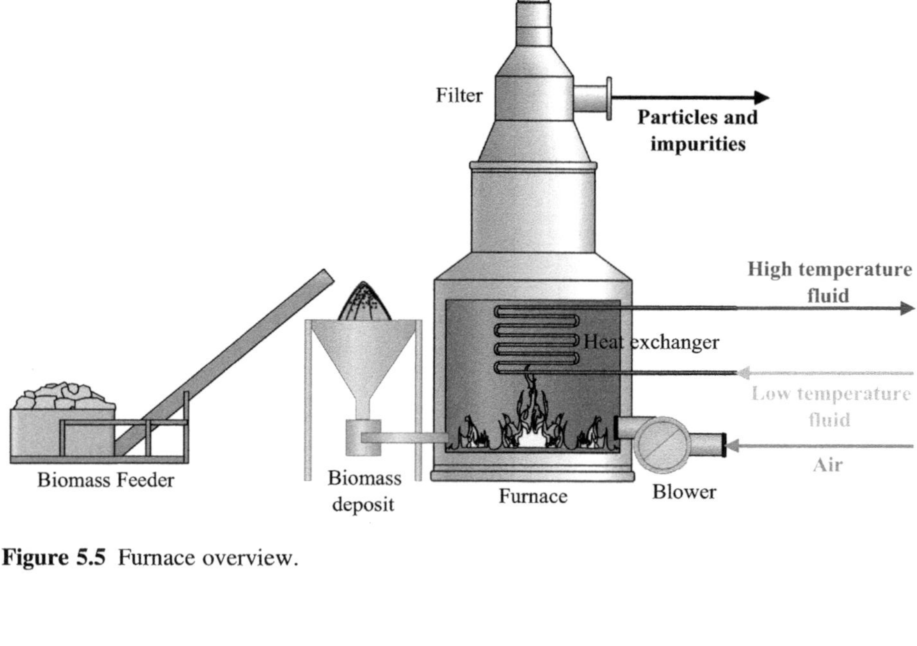

Figure 5.5 Furnace overview.

conventional furnaces, utilizing combustion chambers, heat exchangers, and distribution systems to deliver warmth throughout the building. Advanced biomass combustion technologies, such as fluidized bed combustion or gasification, offer enhanced efficiency and reduced emissions compared to conventional biomass combustion (Basu, 2018). Additionally, biomass can be integrated into CHP systems, where heat generated from biomass combustion is used for space heating or industrial processes, while electricity is generated through turbine-driven generators (Milani & Khalilpour, 2019; U.S. Environmental Protection Agency, 2024). Overall, the integration of biomass technologies into furnaces presents a versatile and sustainable approach to meeting heating needs while reducing reliance on fossil fuels and promoting environmental stewardship (Basu, 2018).

Combined heat and power systems

CHP refers to the concurrent generation and utilization of both heat and electricity. CHP production is based on a wide range of established, dependable, and economically efficient technologies that play a significant role in fulfilling worldwide requirements for heat and electricity (U.S. Environmental Protection Agency, 2024).

In contrast to standalone power plants, CHP facilities provide both electricity and usable heat (see Fig. 5.6), resulting in significantly higher efficiency in terms of usable energy output (Milani & Khalilpour, 2019; U.S. Environmental Protection Agency, 2024). When planning for a CHP facility, it is essential to thoroughly assess the heating and cooling demands of the surrounding area, whether for industrial or residential purposes. The heat energy supplied by a CHP plant comes from excess heat released after condensing and cooling the steam generated through fuel combustion (Milani & Khalilpour, 2019; U.S. Environmental Protection Agency, 2024).

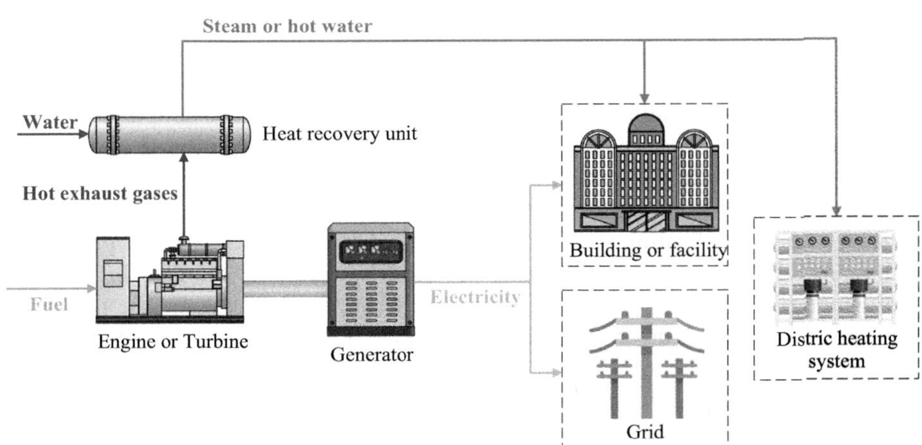

Figure 5.6 CHP system overview. *CHP*, Combined heat and power.

The generation of heat, distributed to industries or for district heating, allows for a slight reduction in potential electricity output compared to running the plant solely in condensing mode, that is, focused solely on electricity production. By using the CHP mode with a modern combustion steam cycle and a relatively large CHP unit ($>50\ MW_e$), approximately 35% of electricity and 55% of useful heat energy could be produced (World Bioenergy Association, 2012a). Accordingly, the energy loss within this CHP unit amounts to only 10% (World Bioenergy Association, 2012a)

District heating systems

A DHS is a centralized heating infrastructure that supplies heat to multiple buildings and users within a defined geographical area, such as a neighborhood, an urban district, or an entire city (BECoop, 2022; Cadham et al., 2016; Imran et al., 2017; International Energy Agency, 2023). In this system, a central power plant or heat source, such as a boiler or CHP plant, generates heat, which is then distributed through a pipe network system that transports a thermal fluid to customers to cover their heat demands (BECoop, 2022; Ericsson & Werner, 2016; Sipilä & Wiltshire, 2016). Due to the investments required in distribution infrastructure and resulting distribution losses, district heating is more cost-effective in densely populated urban areas, while it is less economically viable in sparsely populated regions with single-family homes or small villages (Ericsson & Werner, 2016).

Instead of each building having its own individual heating system, DHS enables greater energy efficiency by leveraging economies of scale (Dou et al., 2021; Imran et al., 2017), as well as the ability to utilize more diverse and efficient energy sources, such as biomass, geothermal, cogeneration, or waste heat from industrial processes (BECoop, 2022; Sipilä & Wiltshire, 2016). This approach can contribute to reducing GHG emissions and improving the energy sustainability of urban communities toward a low-carbon transition (BECoop, 2022; Dou et al., 2021).

A DHS comprises various essential components; see Fig. 5.7. Firstly, a generation plant centrally produces heat to meet the demands of multiple consumers (BECoop, 2022). Secondly, a distribution piping network facilitates the transport of fluids, typically water at 70°C (BECoop, 2022; Vallios et al., 2009), with a maximum drop of 45°C (Vallios et al., 2009), via insulated pipes to minimize thermal losses, which range from 5% to 10% (BECoop, 2022). These pipes are often laid in subway trenches following urban street layouts. Thirdly, substations facilitate heat transfer between the distribution network and consumers (BECoop, 2022).

District biomass heating systems commonly employ a variety of heat production techniques. These include CHP plants powered by biomass, biomass boilers, and heat generation facilities that convert biomass waste into energy and the utilization of surplus heat from industrial processes like bone boilers in oil mills, which supply hot water for oil extraction.

According to the data provided by International Energy Agency (2023), in 2022, approximately 17 EJ of heat was produced for district heating networks globally, a

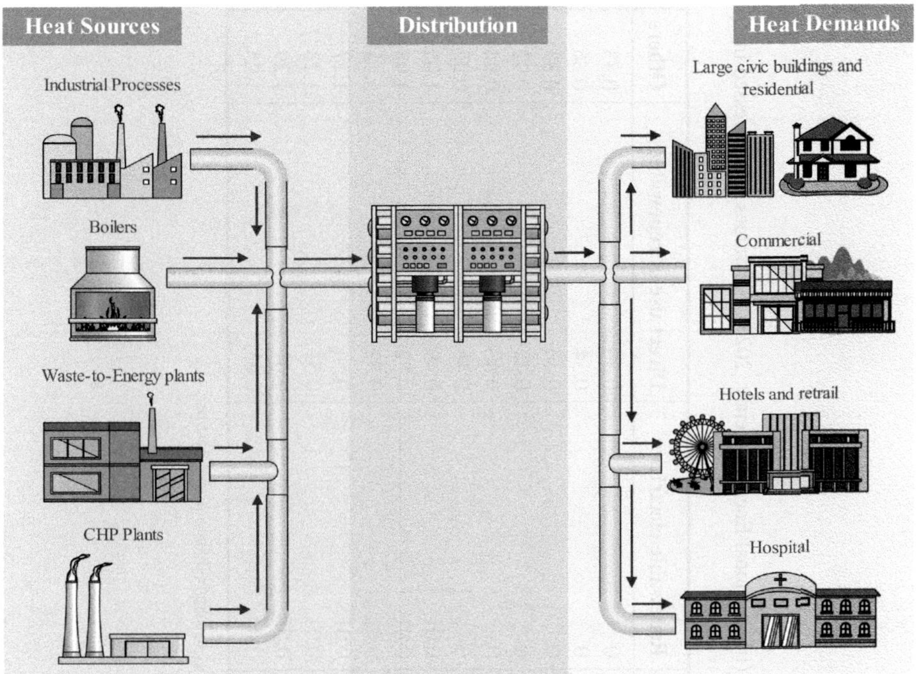

Figure 5.7 Schematic diagram of a biomass district heating system.

figure relatively consistent with 2021 but roughly 10% higher than 2020 and 17% higher than a decade earlier. China, Russia, and Europe collectively account for over 90% of global district heat production, with China leading in the fastest growth of district network connections. As shown in Table 5.1, about 90% of global district heat in 2022 came from fossil fuels, primarily coal (over 48%, especially in China), natural gas (about 38%, particularly in Russia), and oil (3%), marking a decrease from approximately 95% in 2000. The remaining share is sourced from municipal waste, waste heat recovery, nuclear energy, and renewables. Renewables comprised only about 5% of district heat supplies globally, although certain countries have substantially higher proportions. While bioenergy and renewable municipal waste dominate renewable district network supplies, there is growing interest in large-scale solar thermal systems, geothermal energy, and heat pumps. Europe leads in integrating renewables into district heating, with approximately 25% of its district heat supplies coming from renewable sources. Some countries, including Sweden, Denmark, Austria, Estonia, Lithuania, Latvia, and Iceland, exceed 50% renewable district heat. Globally, slightly over 40% of the heat generated in district heating plants is consumed by the buildings sector, with another 40% going to industry. The remainder is self-consumed on production sites, used in agriculture, or lost during distribution. District heating supplies account for only 9% of total

Table 5.1 Distribution of fuel types in district heating networks (2010–22) (International Energy Agency, 2023). All values are expressed in exajoules (EJ).

Year	Coal	Natural gas	Oil	Nonrenewable electricity	Renewable electricity	Direct use of renewables	Others
2010	5.37	6.78	0.85	0.01	0	0.36	0.83
2011	5.62	6.87	0.81	0.01	0	0.34	0.86
2012	5.86	6.92	0.78	0.02	0	0.40	0.89
2013	5.81	6.39	0.65	0.01	0	0.42	0.92
2014	5.55	6.24	0.64	0.02	0	0.43	0.94
2015	5.77	5.89	0.59	0.02	0	0.46	0.95
2016	6.00	6.23	0.59	0.03	0.01	0.49	1.05
2017	6.22	6.17	0.53	0.03	0.01	0.52	1.08
2018	6.82	6.28	0.55	0.03	0.01	0.52	1.13
2019	6.85	6.28	0.51	0.03	0.01	0.54	1.17
2020	7.04	6.23	0.54	0.03	0.01	0.51	1.23
2021	8.04	6.77	0.56	0.03	0.01	0.61	1.29
2022	8.39	6.58	0.54	0.00	0.00	0.63	1.25

Source: Data from International Energy Agency. (2023). District heating. https://www.iea.org/energy-system/buildings/district-heating.

final heat consumption globally. Despite stable weather conditions in key markets during the 2022 heating seasons, this stability may obscure an increase in district network connections (International Energy Agency, 2023).

Environmental and economic aspects

Environmental advantages of using biomass for heating

One of the primary advantages of biomass is the reduction of GHG emissions. Unlike fossil fuels, biomass is considered carbon-neutral over its lifecycle since the carbon dioxide (CO_2) released during combustion is offset by the CO_2 absorbed by plants during their growth. According to Cherubini and Strømman (2011), utilizing biomass can result in a considerable decrease in GHG emissions, especially when agricultural and forestry residues are used, which would otherwise release CO_2 through decomposition.

Biomass for heating can include a variety of organic materials such as agricultural residues, forestry residues, and industrial wood waste. The use of these residues contributes to sustainable waste management by preventing these materials from ending up in landfills where they could generate methane, a much more potent GHG than CO_2 (Siwal et al., 2021).

The collection of biomass from perennial crops, such as miscanthus and willow, can have positive effects on biodiversity and soil health. These crops not only provide habitats for wildlife but also improve soil structure and fertility by increasing organic matter and microbial activity. A study by Tilman et al. (2006) found that well-managed biomass cropping systems can enhance biodiversity and the resilience of agricultural ecosystems.

Economic benefits of biomass heating

One of the most significant economic benefits of biomass heating is job creation. The biomass industry encompasses a wide range of activities, from agriculture and forestry to manufacturing and maintenance of biomass heating systems. According to a report by the International Renewable Energy Agency (2020a, 2020b), the bioenergy sector employed around 3.58 million people worldwide in 2019. This sector provides jobs in rural areas, where employment opportunities are often limited, thus supporting local economies and reducing urban migration.

Another key advantage of biomass heating is enhanced energy independence. Many countries rely heavily on imported fossil fuels, which exposes them to price volatility and geopolitical risks (Hille, 2023). By investing in domestic biomass resources, countries can reduce their dependence on foreign energy supplies and increase their energy security. A study found that expanding biomass energy could significantly decrease Europe's reliance on imported natural gas and oil, thus enhancing the continent's energy autonomy.

In terms of cost, biomass is often more economical than fossil fuels (Antar et al., 2021). The price of biomass fuels, such as wood pellets and agricultural residues, tends to be more stable compared to fossil fuels, which are subject to global market fluctuations. Additionally, biomass can be sourced locally, reducing transportation costs and supporting local economies.

Furthermore, the development of the biomass heating sector can stimulate technological innovation and infrastructure development. Investments in biomass processing facilities, distribution networks, and advanced combustion technologies can lead to economic growth and increased competitiveness of the renewable energy sector.

Conclusions and recommendations

Biomass heating systems offer a sustainable alternative to conventional fossil fuel-based heating methods. The diverse types of biomass, including forest residues, agricultural residues, energy crops, organic waste, and industrial residues, provide a renewable source of energy that can be harnessed locally. This local sourcing reduces dependence on imported fossil fuels, promotes energy security, and supports rural economies by creating jobs and stimulating economic development.

The environmental benefits of biomass heating are significant. Biomass is considered carbon-neutral because the carbon dioxide released during combustion is offset by the carbon dioxide absorbed during the growth of the biomass. This natural carbon cycle helps mitigate climate change by reducing GHG emissions. Moreover, utilizing biomass for heating can improve waste management practices by converting agricultural, forestry, and organic waste into valuable energy, thus reducing landfill use and preventing environmental pollution.

The combustion efficiency and fuel properties of biomass, such as heating value, moisture content, bulk density, energy density, particle size, mechanical durability, and ash content influence the performance and environmental impact of biomass heating systems. Advances in technology and improved combustion techniques can further enhance the efficiency and sustainability of these systems.

In terms of economic impact, biomass heating systems are cost-effective due to the stability of biomass fuel prices compared to the volatility of fossil fuel markets. Financial incentives and subsidies available in many regions make biomass heating systems an economically viable option for homeowners, businesses, and institutions. These systems also provide long-term savings and contribute to the overall viability of renewable energy investments.

To maximize the benefits of biomass heating, it is essential to continue investing in advanced conversion technologies and ensuring the sustainable sourcing of biomass. Policies that support the development of biomass infrastructure and promote education and outreach about the advantages and best practices of biomass heating will be crucial in achieving widespread adoption and realizing the full potential of this renewable energy source.

References

Agarwal, B. (1983). Diffusion of rural innovations: Some analytical issues and the case of wood-burning stoves. *World Development, 11*(4). Available from https://doi.org/10.1016/0305-750x(83)90047-5.

Agurto Adrianzén, M. (2013). Improved cooking stoves and firewood consumption: Quasi-experimental evidence from the Northern Peruvian Andes. *Ecological Economics, 89*. Available from https://doi.org/10.1016/j.ecolecon.2013.02.010.

Alakangas, E. (2016). Biomass and agricultural residues for energy generation. Elsevier BV. Available from https://doi.org/10.1016/b978-1-78242-378-2.00003-1.

Antar, M., Lyu, D., Nazari, M., Shah, A., Zhou, X., & Smith, D. L. (2021). Biomass for a sustainable bioeconomy: An overview of world biomass production and utilization. *Renewable and Sustainable Energy Reviews, 139*, 110691. Available from https://doi.org/10.1016/j.rser.2020.110691.

Bäfver, L. S., Leckner, B., Tullin, C., & Berntsen, M. (2011). Particle emissions from pellets stoves and modern and old-type wood stoves. *Biomass and Bioenergy, 35*(8). Available from https://doi.org/10.1016/j.biombioe.2011.05.027.

Balaman, Ş. Y. (2019a). Basics of decision-making in design and management of biomass-based production chains. Elsevier BV. Available from https://doi.org/10.1016/b978-0-12-814278-3.00006-6.

Balaman, Ş. Y. (2019b). Sustainability issues in biomass-based production chains. Elsevier BV. Available from https://doi.org/10.1016/b978-0-12-814278-3.00004-2.

Basu, P. (2018). Biomass gasification, pyrolysis and torrefaction (3rd ed.), Academic Press. Available from https://doi.org/10.1016/C2016-0-04056-12018.

BECoop. (2022). *Technical catalogue on biomass district heating*.

Bibi, F., Ilyas, N., Saeed, M., Shabir, S., Shati, A. A., Alfaifi, M. Y., Amesho, K. T. T., Chowdhury, S., & Sayyed, R. Z. (2023). Innovative production of value-added products using agro-industrial wastes via solid-state fermentation. *Environmental Science and Pollution Research, 30*(60). Available from https://doi.org/10.1007/s11356-023-28765-6, https://www.springer.com/journal/11356.

Brandelet, B., Rose, C., Rogaume, C., & Rogaume, Y. (2017). Investigation of the organic carbon ratio analysis on particles from biomass combustion and its evolution in three generations of firewood stoves. *Biomass and Bioenergy, 99*. Available from https://doi.org/10.1016/j.biombioe.2017.03.002.

Brandelet, B., Rose, C., Landreau, J., Druette, L., & Rogaume, Y. (2021). Toward a cleaner domestic wood heating by the optimization of firewood stoves? *Journal of Cleaner Production, 325*, 129338. Available from https://doi.org/10.1016/j.jclepro.2021.129338.

Brandelet, B., Rose, C., Rogaume, C., & Rogaume, Y. (2018). Impact of ignition technique on total emissions of a firewood stove. *Biomass and Bioenergy, 108*. Available from https://doi.org/10.1016/j.biombioe.2017.10.047.

Cadham, W., Van Dyk, J. S., Linoj Kumar, J. S., Saddler, J. N., Thiffault, E., Berndes, G., Junginger, M., Saddler, J. N., Smith, C. T. (2016). Chapter 7—Challenges and opportunities for the conversion technologies mobilisation of forest bioenergy. Mobilisation of forest bioenergy in the boreal and temperate biomes. Academic Press. Available from https://doi.org/10.1016/B978-0-12-804514-5.00007-X.

Carvalho, L., Wopienka, E., Pointner, C., Lundgren, J., Verma, V. K., Haslinger, W., & Schmidl, C. (2013). Performance of a pellet boiler fired with agricultural fuels. *Applied Energy, 104*. Available from https://doi.org/10.1016/j.apenergy.2012.10.058, http://www.elsevier.com/inca/publications/store/4/0/5/8/9/1/index.htt.

Chen, F., Filipski, M., Luo, Y., & Qiu, H. (2023). Why is firewood use so persistent? The role of income uncertainty. *Energy for Sustainable Development*, *76*, 101285. Available from https://doi.org/10.1016/j.esd.2023.101285.

Cherubini, F., & Strømman, A. H. (2011). Life cycle assessment of bioenergy systems: State of the art and future challenges. *Bioresource Technology*, *102*(2). Available from https://doi.org/10.1016/j.biortech.2010.08.010.

Demurger, S., & Fournier, M. (2011). Poverty and firewood consumption: A case study of rural households in China. *Economic Review*, *22*(4). Available from https://doi.org/10.1016/j.chieco.2010.09.009.

Dou, Y., Sun, L., Fujii, M., Kikuchi, Y., Kanematsu, Y., Ren, J., Ren, J. (2021). Chapter 10—Towards a renewable-energy-driven district heating system: Renewable-energy-driven future. Academic Press. Available from https://doi.org/10.1016/B978-0-12-820539-6.00010-8.

Ericsson, K., & Werner, S. (2016). The introduction and expansion of biomass use in Swedish district heating systems. *Biomass and Bioenergy*, *94*. Available from https://doi.org/10.1016/j.biombioe.2016.08.011.

European Biomass Association (2016). *Aebiom statistical report-key findings*. http://www.aebiom.org/statistical-report-2016/.

Fernández-Lobato, L., Aguado, R., Jurado, F., & Vera, D. (2022). Biomass gasification as a key technology to reduce the environmental impact of virgin olive oil production: A life cycle assessment approach. *Biomass and Bioenergy*, *165*, 106585. Available from https://doi.org/10.1016/j.biombioe.2022.106585.

Funk, K., Milford, J., Simpkins, T. (2020). Waste not, want not: Analyzing the economic and environmental viability of waste-to-energy technology for site-specific optimization of renewable energy options. Bioenergy: Biomass to biofuels and waste to energy. Elsevier. Available from https://www.elsevier.com/books/bioenergy/dahiya/978-0-12-815497-7, https://doi.org/10.1016/B978-0-12-815497-7.00019-1.

Galadima, A., & Muraza, O. (2020). Waste materials for production of biodiesel catalysts: Technological status and prospects. *Journal of Cleaner Production*, *263*, 121358. Available from https://doi.org/10.1016/j.jclepro.2020.121358.

Głowacka, K. (2011). A review of the genetic study of the energy crop Miscanthus. *Biomass and Bioenergy*, *35*(7). Available from https://doi.org/10.1016/j.biombioe.2011.01.041.

Hakawati, R., Smyth, B. M., McCullough, G., De Rosa, F., & Rooney, D. (2017). What is the most energy efficient route for biogas utilization: Heat, electricity or transport? *Applied Energy*, *206*. Available from https://doi.org/10.1016/j.apenergy.2017.08.068, http://www.elsevier.com/inca/publications/store/4/0/5/8/9/1/index.htt.

Hille, E. (2023). Europe's energy crisis: Are geopolitical risks in source countries of fossil fuels accelerating the transition to renewable energy? *Energy Economics*, *127*, 2023.

Horák, J., Hopan, F., Kremer, J., Kuboňová, L., Polcar, L., Molchanov, O., Ryšavý, J., Krpec, K., Kubesa, P., Dej, M., & Garba, M. (2022). Real measurement of carbon monoxide, total suspended particulate, and thermal efficiency in modern biomass household boilers. *Biomass Conversion and Biorefinery*, *12*(10). Available from https://doi.org/10.1007/s13399-022-02657-0, http://www.springer.com/engineering/energy + technology/journal/13399.

IEA Bioenergy. (2022). *The dawn of greater energy independence*?.

Imran, M., Usman, M., Im, Y. H., & Park, B. S. (2017). The feasibility analysis for the concept of low temperature district heating network with cascade utilization of heat between networks. *Energy Procedia*, *116*(4), 12. Available from https://doi.org/10.1016/j.egypro.2017.05.050, http://www.sciencedirect.com/science/journal/18766102.

International Energy Agency. (2023). District heating. https://www.iea.org/energy-system/buildings/district-heating.

International Renewable Energy Agency. (2020a). Renewable energy and jobs. https://www.irena.org/-/media/Files/IRENA/Agency/Publication/2020/Sep/IRENA_RE_Jobs_2020.pdf?rev = db153791a7744a33913b553e02a1e5b0.

International Renewable Energy Agency. (2020b). Renewable energy and jobs.

Jaworek, A., Sobczyk, A. T., Marchewicz, A., Krupa, A., & Czech, T. (2021). Particulate matter emission control from small residential boilers after biomass combustion. A review. *Renewable and Sustainable Energy Reviews, 137*, 110446. Available from https://doi.org/10.1016/j.rser.2020.110446.

Lee, Y. W., Ryu, C., Lee, W. J., & Park, Y. K. (2011). Assessment of wood pellet combustion in a domestic stove. *Journal of Material Cycles and Waste Management, 13*(3). Available from https://doi.org/10.1007/s10163-011-0014-0. http://link.springer.de/link/service/journals/10163/index.htm.

Mata-Alvarez, J., Macé, S., & Llabrés, P. (2000). Anaerobic digestion of organic solid wastes. An overview of research achievements and perspectives. *Bioresource Technology, 74*(1). Available from https://doi.org/10.1016/S0960-8524(00)00023-7, http://www.elsevier.com/locate/biortech.

Mbiankeu Nguea, S., & Fotio, H. K. (2024). Synthesizing the role of biomass energy consumption and human development. *Energy, 293*, 130500. Available from https://doi.org/10.1016/j.energy.2024.130500.

MBtek Inc. (2023). MBtek Inc. https://www.mbtek.com/.

Milani, D., Khalilpour, K. R. (2019). Chapter 14—Renewable energy integration in combined cooling, heating, Polygeneration with polystorage for chemical and energy hubs. Academic Press. Available from https://doi.org/10.1016/B978-0-12-813306-4.00014-8.

Miller, K. (2013). Coal analysis. The coal handbook: Towards cleaner production. vol **1**, Elsevier Inc., Australia. Available from http://www.sciencedirect.com/science/book/9780857094223, https://doi.org/10.1533/9780857097309.1.151.

Mižakova, J., Piteľ, J., Hošovský, A., Pavlenko, I., Ochowiak, M., & Khovanskyi, S. (2021). Biomass combustion control in small and medium-scale boilers based on low. *Processes, 9*(11). Available from https://doi.org/10.3390/pr9112030.

Naik, B., Kumar, V., Rizwanuddin, S., Chauhan, M., Gupta, A. K., Rustagi, S., Kumar, V., & Gupta, S. (2023). Agro-industrial waste: A cost-effective and eco-friendly substrate to produce amylase. *Food Production, Processing and Nutrition, 5*(1). Available from https://doi.org/10.1186/s43014-023-00143-2, https://www.springer.com/journal/43014.

Niklitschek, M. E., Labbé, R., & Guerrero, J. (2020). Heating and hot water with wood chips. Is it convenient to replace firewood boilers in southern Chile? *Energy for Sustainable Development, 55*. Available from https://doi.org/10.1016/j.esd.2019.12.001, http://www.elsevier.com.

Özenen-Kavlak, M., & Çabuk, S. N. (2022). Geospatial suitability and cost analysis modeling for energy crop cultivation: Switchgrass case for Eskisehir. *Renewable Energy, 194*. Available from https://doi.org/10.1016/j.renene.2022.05.144.

Parfitt, J., Croker, T., & Brockhaus, A. (2021). Global food loss and waste in primary production: A reassessment of its scale and significance. *Sustainability, 13*(21), 12087. Available from https://doi.org/10.3390/su132112087.

Quintero-Marquez, A., Bernard, C., Zoulalian, A., & Rogaume, Y. (2014). Improving the operation of an automatic wood chip boiler by optimizing CO emissions. *Energy & Fuels, 28*(3). Available from https://doi.org/10.1021/ef402021c.

Saidur, R., Abdelaziz, E. A., Demirbas, A., Hossain, M. S., & Mekhilef, S. (2011). A review on biomass as a fuel for boilers. *Renewable and Sustainable Energy Reviews, 15*(5). Available from https://doi.org/10.1016/j.rser.2011.02.015.

Schilstra, A. J. (2001). How sustainable is the use of peat for commercial energy production? *Ecological Economics*, *39*(2). Available from https://doi.org/10.1016/S0921-8009(01)00216-6.

Schön, C., Kuptz, D., Mack, R., Zelinski, V., Loewen, A., & Hartmann, H. (2019). Influence of wood chip quality on emission behaviour in small-scale wood chip boilers. *Biomass Conversion and Biorefinery*, *9*(1). Available from https://doi.org/10.1007/s13399-017-0249-7.

Sipilä, K., & Wiltshire, R. (2016). 3—Cogeneration, biomass, waste to energy and industrial waste heat for district heating Advanced District Heating and Cooling (DHC) systems. Woodhead Publishing, Oxford. Available from https://doi.org/10.1016/B978-1-78242-374-4.00003-3.

Siwal, S. S., Zhang, Q., Devi, N., Saini, A. K., Saini, V., Pareek, B., Gaidukovs, S., & Thakur, V. K. (2021). Recovery processes of sustainable energy using different biomass and wastes. *Renewable and Sustainable Energy Reviews*, *150*. Available from https://doi.org/10.1016/j.rser.2021.111483. https://www.journals.elsevier.com/renewable-and-sustainable-energy-reviews.

Smith, (2009). Biomass heating: A practical guide for potential users.

Šyc, M., Pohořelý, M., Kamenikova, P., Habart, J., Svoboda, K., & Punčochař, M. (2012). Willow trees from heavy metals phytoextraction as energy crops. *Biomass and Bioenergy*, *37*. Available from https://doi.org/10.1016/j.biombioe.2011.12.025.

Thomson, H., & Liddell, C. (2015). The suitability of wood pellet heating for domestic households: A review of literature. *Renewable and Sustainable Energy Reviews*, *42*. Available from https://doi.org/10.1016/j.rser.2014.11.009.

Tilman, D., Hill, J., & Lehman, C. (2006). Carbon-negative biofuels from low-input high-diversity grassland biomass. *Science*, *314*(5805). Available from https://doi.org/10.1126/science.1133306.

Titus, B. D., Brown, K., Helmisaari, H. S., Vanguelova, E., Stupak, I., Evans, A., Clarke, N., Guidi, C., Bruckman, V. J., Varnagiryte-Kabasinskiene, I., Armolaitis, K., de Vries, W., Hirai, K., Kaarakka, L., Hogg, K., & Reece, P. (2021). Sustainable forest biomass: A review of current residue harvesting guidelines. *Energy, Sustainability and Society*, *11*(1). Available from https://doi.org/10.1186/s13705-021-00281-w. https://energsustainsoc.biomedcentral.com/.

Toscano, G., Duca, D., Amato, A., & Pizzi, A. (2014). Emission from realistic utilization of wood pellet stove. *Energy*, *68*. Available from https://doi.org/10.1016/j.energy.2014.01.108.

Uddin, M. M., & Wright, M. M. (2023). Anaerobic digestion fundamentals, challenges, and technological advances. *Physical Sciences Reviews*, *8*(9). Available from https://doi.org/10.1515/psr-2021-0068. https://www.degruyter.com/journal/key/psr/html.

U.S. Environmental Protection Agency. (2024). What is CHP? https://www.epa.gov/chp/what-chp.

Vallios, I., Tsoutsos, T., & Papadakis, G. (2009). Design of biomass district heating systems. *Biomass and Bioenergy*, *33*(4). Available from https://doi.org/10.1016/j.biombioe.2008.10.009.

Venturini, E., Vassura, I., Agostini, F., Pizzi, A., Toscano, G., & Passarini, F. (2018). Effect of fuel quality classes on the emissions of a residential wood pellet stove. *Fuel*, *211*. Available from https://doi.org/10.1016/j.fuel.2017.09.017.

Vijay, V., Kapoor, R., Singh, P., Hiloidhari, M., & Ghosh, P. (2022). Sustainable utilization of biomass resources for decentralized energy generation and climate change mitigation: A regional case study in India. *Environmental Research*, *212*, 113257. Available from https://doi.org/10.1016/j.envres.2022.113257.

Vukovic, N., & Makogon, E. (2022). Waste-to-energy generation: Complex efficiency analysis of modern technologies. *Sustainability*, *14*(21), 13814. Available from https://doi.org/10.3390/su142113814.

Whittaker, C., & Shield, I. (2016). Short rotation woody energy crop supply chains. Biomass supply chains for bioenergy and biorefining. Elsevier Inc. Available from http://www.sciencedirect.com/science/book/9781782423669, https://doi.org/10.1016/B978-1-78242-366-9.00010-1.

World Bioenergy Association. (2012a). Biomass combined heat and power (CHP). https://www.worldbioenergy.org/uploads/Fact_sheet_CHP.pdf.

World Bioenergy Association. (2012b). Small-scale biomass heating. https://www.worldbioenergy.org/uploads/Factsheet%20-%20Small%20scale%20biomass%20heating.pdf.

World Bioenergy Association. (2022c). https://www.worldbioenergy.org/uploads/221223%20WBA%20GBS%202022.pdf.

Zhao, X. Q., Zi, L. H., Bai, F. W., Lin, H. L., Hao, X. M., Yue, G. J., & Ho, N. W. Y. (2012). *Bioethanol from lignocellulosic biomass*. Berlin Heidelberg: Springer. Available from https://doi.org/10.1007/10_2011_1292012.

Part 2

New markets, businesses, and structures

Self-consumption and energy communities

6

Jose Villar[1] *and João Mello*[1,2]
[1]Institute for Systems and Computer Engineering, Technology and Science, Porto, Portugal, [2]Faculty of Engineering, University of Porto, Porto, Portugal

Introduction

Individual and collective self-consumption (ISC and CSC) structures, alongside energy communities (EC), are expected to be an important driver for a sustainable energy transition and decarbonization of the energy system. Operating on a decentralized model, these frameworks empower end-users by equipping them with the tools to actively participate in the energy system. By becoming self-producers of renewable energy and optimizing the potential flexibility of their energy resources, they will be able to reduce their environmental footprint, but also offer valuable flexibility services for a more cost-effective and resilient operation of the power system.

While ISC refers to the energy activity of individuals generating their own electricity from renewable energy sources and consuming it to meet their energy needs, CSC refers to the energy activity where multiple energy consumers collectively produce local renewable generation for SC and for sharing with other members of the CSC in need of supply. On the other hand, EC can be of two main types, renewable energy communities (REC) and citizen energy communities (CEC). Their primary purpose is to provide environmental, economic, or social community benefits to its members or shareholders or to the local areas where they operate, rather than to generate financial profits, and both types of EC can engage in CSC. However, while REC focus on local community premises and are limited to renewable generation, CEC may include generation from renewable and nonrenewable sources, and may also develop other activities such as distribution, supply, or aggregation.

To promote SC schemes, the European Union (EU) provides guidelines to be transposed by its members states (Ms) to their national regulations, defining the rules for its implementation, participation of stakeholders, and integration with the wholesale market. However, SC and EC regulation is relatively recent in the EU and is still evolving.

This chapter lists the main EU regulatory documents related to SC and EC, describes the new power system actors and roles that stem from these relatively new energy activities, and describes in more detail ISC, CSC, energy sharing and EC concepts, the energy services EC can provide to their members and third parties, and the tools needed to manage EC. Some conclusions are provided at the end of the chapter.

Towards Future Smart Power Systems with High Penetration of Renewables. DOI: https://doi.org/10.1016/B978-0-443-29871-4.00006-3

European Union regulatory context

SC and EC are one of the focuses of the EU energy strategy, where Ms are transposing the EU directives to their national regulations. The EU Parliament and Counsel published four main documents related to SC:

- Directive 2018/2001 (D2018) "Directive EU, 2018/2001 of the European Parliament and of the Council of 11 December 2018 on the promotion of the use of energy from renewable sources (Text with EEA relevance, 2018)", with the main guidelines and definitions regarding renewable self-consumption and REC, including rights and obligations and integration with energy markets.
- Directive 2019/944 (Directive EU, 2019), which is focused on electricity markets, with the definition of CEC, and the clear purpose of enforcing the integration of self-generation with energy and flexibility markets.
- Regulation 2019/943 (Regulation EU, 2019), which regulates specific topics regarding CSC and REC, such as the remuneration of renewable sources, energy storage and demand response, and the grid charges to be applied to self-consumption.
- Amendment proposals of the European Parliament and of the Council (European Union, 2023) with amendments to the three abovementioned documents and broadens the rights of self-consumers.
- Directive 2023/2413 (D2023) (European Union, 2023) amending D2018 and including recommendations and mandatory points focused, for example, in increasing the speed of the licensing processes of ISC and CSC.

The following table provide definitions of the main concepts used along this text that can be deduced from the above EU regulatory texts (Table 6.1).

New power system actors under the self-consumption paradigm

A role is usually defined as an intended behavior of a business party, which is unique, cannot be shared, and intends to satisfy a service or a transaction (Rodrigues et al., 2023). Roles are played by actors, and each actor can play one or several roles at the same time. While a role corresponds to a set of skills, competencies, and responsibilities, being more associated with business processes and tasks, an actor is usually associated with a person or an organization. However, and even if for business models and regulatory issues the distinction between roles and actor may be very relevant, in this text, we will not make this strict differentiation, since for simplicity, we can assume actors that play a unique role. With these considerations, in the traditional power system model, the main actors related to basic activity of residential energy consumption, supported by the activities of energy production, transportation, distribution and commercialization, are the following ones:

- *Consumers*: Individuals that contract their electricity supply with a supplier at a regulated tariff (last resort tariff) or on the free retail market. Traditional consumers have a passive

Table 6.1 Definitions of the main concepts associated with collective self-consumption.

ISC	Individual self-consumption: the individual energy activity of producing local renewable generation for self-consumption.
CSC	Collective self-consumption: the collective energy activity of producing local renewable generation for self-consumption and for sharing with those members of the CSC in need of supply.
SC	Self-consumption, either individual or collective.
EC	Energy community: a legal person that empower citizens, small businesses, and local authorities to produce, manage, and consume their own energy. Can be REC or CEC.
REC	Renewable energy community: a legal entity of members or shareholders that are natural persons, small, or medium companies, or local authorities, including municipalities, located in the proximity of the renewable projects it owns and develops, whose main purpose is to provide environmental, economic, or social community benefits for its shareholders or members or the local area where it operates, rather than financial profits. They can produce, consume, store, sell, and share renewable energy.
CEC	Citizens energy community: a legal entity of members or shareholders that are natural persons, local authorities, including municipalities, or small enterprises, with the primary purpose of providing environmental, economic, or social community benefits to its members or shareholders, or to the local areas where it operates, rather than to generate financial profits. They can produce, store, distribute, share, supply, provide energy efficiency services or charging service for electric vehicles, and aggregate and consume energy to its members or shareholders.
AC	Allocation coefficients: AC are factors that, when multiplied by the net energy locally generated by the members of a CSC, provide the energy allocated to the CSC members with net consumption.

role and do not react significantly to external signals. They pay the cost of the energy supplied and the corresponding grid access charges to the suppliers they have contracted.

- *Suppliers:* Companies that sell electricity to consumers. They purchase the energy on the wholesale market to meet the expected consumption of their customers' portfolio and are responsible for the imbalances they cause when the energy purchased differs from that finally supplied. Suppliers usually bill their customers for the full costs of supply and pass on the grid access charges to the grid operators, whether distribution system operators (DSO) or transmission system operators (TSO).
- *Producers:* Companies that sell energy on the wholesale market to suppliers and large industrial consumers.
- *DSO:* Company responsible for the physical delivery of energy to the final customer by operating and maintaining the distribution grid and metering the energy delivered.
- *TSO:* Company responsible for the physical delivery of energy from the producers connected to the transmission network to the distribution networks (operated by the DSO) by operating and maintaining the transmission network and metering the energy delivered to that grid.

The decentralization of the energy system with the activities of ISC and CSC and the participation of distributed energy resources (DER) in the provision of flexibility services are changing the traditional roles of the energy role model. The existing players are changing their roles and responsibilities, and new players appear.

- Consumers become *prosumers* with an active energy role, generating their own energy and managing load and storage systems. They may also share energy locally with other prosumers by engaging in CSC and provide grid services by contracting an aggregator. They still must contract a supplier to supply the energy not produced locally, and their active role, including self-consumption and energy sharing activities, may entail changes in their consumption profiles and increase the supplier balancing risks.
- *Aggregators:* Companies that intermediate prosumers and wholesale agents. They buy the energy from the consumers in their portfolio and sell it to the wholesale energy markets. In addition to producing, prosumers may also adapt their consumption profile by reacting to the economic signals, bringing additional flexibility to the system that is also managed by their aggregator. This flexibility, of great importance in a context where traditional flexibility providers are being progressively replaced by generation based on nondispatchable renewable sources, may be traded in flexibility markets, whether intraday (where market agents can manage their portfolios to minimize imbalances), TSO's flexibility markets (for overall system balancing or to solve constraints in the transmission network), or DSO's local flexibility markets (to solve constraints in the distribution networks), contributing to a more cost-efficient operation of the power system.
- The integration of DER in energy and flexibility markets adds complexity to the system operation, requiring new functions to TSOs and DSOs, whose roles and responsibilities must be redefined, also incorporating greater *TSO–DSO coordination* (Silva et al., 2021).

They represent some of these new actors and responsibilities, with the focus on SC activities, and these new activities and actors are highlighted in yellow arrows.

SC also adds responsibilities to DSOs, such as measuring the net consumption or surplus injection of prosumers or applying the energy sharing mechanisms selected by CSCs. DSOs also have to inform prosumers of the grid access charges to pay for the self-consumed energy that used the public grid, suppliers of how much they have supplied to their consumers, and aggregators of how much surplus of their prosumers they have to sell in other markets.

Self-consumption and energy communities

Individual self-consumption

More formally, *individual self-consumption* (Fig. 6.1) is a regulated energy activity where a consumer becomes a prosumer (or self-consumer) by installing energy assets and generating part of its own supply.

As shown in Fig. 6.1, energy assets can be behind-the-meter (BTM), that is, connected directly to the internal prosumer grid behind the DSO meter, whose generation or consumption is internally balanced before being metered by the DSO, or in front-of-the-meter (FTM), connected elsewhere to the distribution grid and with a dedicated

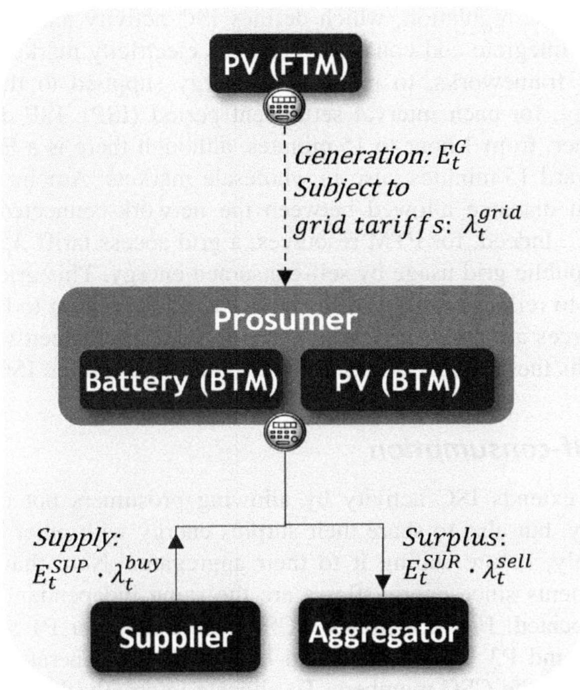

Figure 6.1 Individual self-consumption, where prosumers may generate their own energy, and have a contract with a supplier for the energy they still need from the main grid, and with an aggregator to sell their energy surplus.

DSO meter to account for its generated energy E_t^G. The prosumer has a contract with a supplier who supplies, at the agreed price λ_t^{buy}, the energy needs E_t^{SUP} not supplied locally. The prosumer can also contract an aggregator to sell his surplus E_t^{SUR}, at the agreed price λ_t^{sell}, if his generation is greater than his consumption.

National transpositions of the EU directives related to SC activities define ISC, CSC, and EC concepts and rules to comply with the following:

- ISC, CSC, REC, CEC, and entities definitions, rights, and obligations.
- Generation and consumption assets proximity rules.
- CSC sharing mechanisms, allocation coefficients, and hierarchical structures.
- Energy surplus treatment and market access.
- Grid access tariffs for self-consumption.
- Operation of private distribution grids.
- Storage definitions, rules, and charges.
- Implicit or explicit regulation of local markets.
- Licensing procedures for resources and CSC structures.
- Grid connecting procedures, metering data sharing, smart meters distribution, and so forth.
- Digital platforms support from administration, DSOs, and so forth.
- Consideration of other energy resources (such as cogeneration units).

In particular, ISC regulation, which defines ISC activity and rules, is in fact a powerful tool to integrate end-consumers into the electricity market by allowing, in most regulatory frameworks, to reduce the energy supplied to them in the self-consumed energy, for each interval settlement period (ISP). ISP differs from one country to another, from 1 hour to 15 minutes, although there is a EU general trend to converge toward 15 minutes, also in wholesale markets. Among the above rules is the maximum distance allowed between the network-connected resources that make up the ISC. Indeed, for FTM resources, a grid access tariff λ_t^{grid} is charged to account for the public grid usage by self-consumed energy. This grid access tariff is usually reduced to reflect a partial local use of the grid, limiting to the voltage level where the resources are connected, which is somehow guaranteed with the proximity rule that limits the physical or electrical distance between the ISC resources.

Collective self-consumption

CSC (Fig. 6.2) extends ISC activity by allowing prosumers not only to produce their own energy, but also to share their surplus energy with other local prosumers in need of supply, before selling it to their aggregator. Note that these are only financial agreements since energy flows are the same independently of where the energies are allocated. Fig. 6.2 shows a CSC where member P1 shares its energy surplus with P2 and P3. Power generated by the CSC's generation facilities may also be shared with the CSC members. To allocate internally the energy shared, the CSC must define the rules to be applied, in accordance with current regulations, and usually compensate producers for the energy they supply locally.

Note that each CSC member i can have its own opportunity costs, that is, the price $\lambda_{i,t}^{\text{buy}}$ of buying from the grid and the price $\lambda_{i,t}^{\text{sell}}$ of selling to the grid for ISP t.

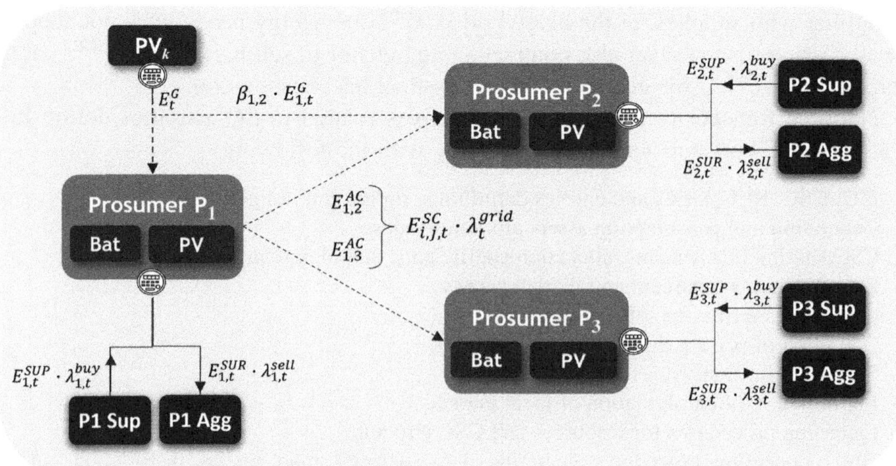

Figure 6.2 Collective self-consumption, where prosumers join to share the energy among them before injecting it to the main grid.

The final surpluses can be sold individually by the members, or collectively by the CSC, depending on the regulatory framework. For instance, Portugal defines that the surpluses must be aggregated and sold by the CSC manager, which may later distribute the revenues among the members according to the CSC internal rules.

As for the ISC, CSC regulation also defines the rules this activity must comply with. Again, a maximum physical or electrical distance between CSC members is commonly defined, as well as the energy sharing mechanisms allowed to allocate energy among the CSC members, and to compute the energy self-consumed and the grid access tariffs to be applied when the public grid is used.

Energy allocation mechanisms and allocation coefficients

Energy allocation mechanisms are the rules that define how the energy can be shared among the members of a CSC. Many of these mechanisms can be explained with the so-called *allocation coefficients (AC)*, which are factors that, when multiplied by the locally injected energies, give back the energies allocated to the CSC members. For example, in Spain, France, Portugal, and some other EU countries, allocation mechanisms can be explained with AC. However, as mentioned below, other regulations exist, such as in Italy, that apply different allocation rules.

Let us assume that we depart from the ISC represented in Fig. 6.3. As can be seen, consumers P1 and P2 are net producers (in green) and sell their surplus E_n^{SUR} to their contracted aggregators with the corresponding profit, while consumers P3 and P4 are net producers (in brown) and buy their consumption E_n^{SUP} from their contracted suppliers with the corresponding supply cost. If all profits and costs are added, there is a total net energy cost $C_{\mathrm{tot}}^{\mathrm{ind}}$ of 1.55€.

Since in general the surplus sold to the grid is paid at prices below supply (also with grid access tariffs for sellers usually lower than for buyers), P3 and P4 would be willing to buy local supply from P1 and P2 at prices lower than 0.55 and 0.4 /kWh, respectively. Similarly, P1 and P2 can sell their surplus to their aggregator at 0.08 and 0.06 /kWh, respectively, so they would be willing to sell them first to P1 and P2 considering the prices they would be willing to pay.

The CSC energy allocation mechanisms define the possible rules that can be applied to allocate P1 and P2 surpluses to P3 and P4 before selling it back to the grid. Fig. 6.4 shows these possible rules.

The first alternative shown in Fig. 6.4A is borrowed from the Portuguese regulation and defines that the energy injected by the CSC members (or resources) producing can be allocated to the consuming members proportionally to their consumption. We call this proportional AC, and it can be easily seen that this approach minimizes the total energy surplus of the CSC. In the example considered, applying proportional ACs leads to a total CSC cost $C_{\mathrm{tot}}^{\mathrm{prop}}$ of 0.44, compared to the 1.55€ of the ISC, as shown in Fig. 6.3.

A more flexible mechanism is based on the so-called dynamic AC (allowed, for example, by the Portuguese and French regulations). Dynamic AC are AC computed by the CSC manager according to rules that the CSC members have previously agreed upon and after the DSO provide the measured energies. This makes

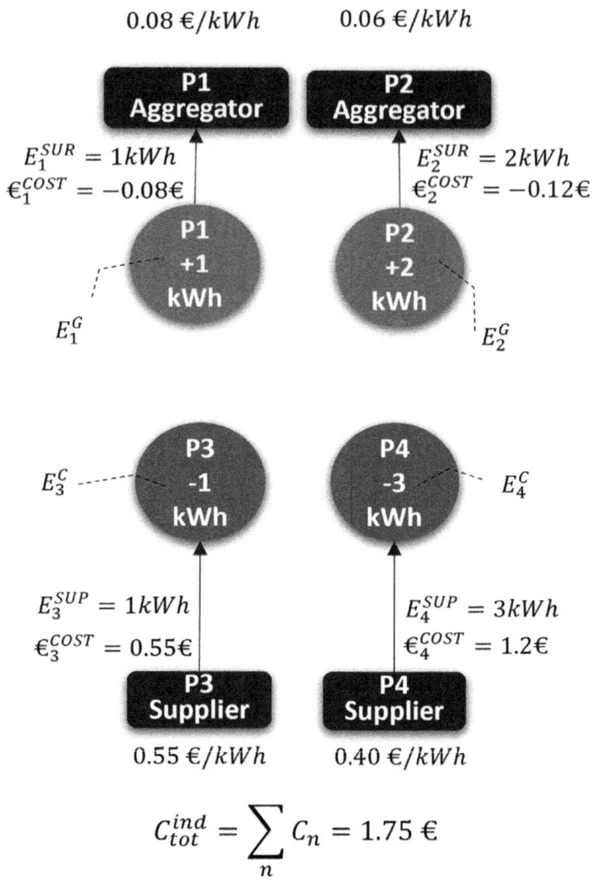

Figure 6.3 ISC with four members, some selling their energy surplus to their aggregators, and some still needing supply from their suppliers. *ISC*, Individual self-consumption.

them very flexible, and allows, for example, to minimize the total CSC cost according to the opportunity costs of its members, as Fig. 6.4B shows. In this case, the final CSC cost $C_{\text{tot}}^{\text{dyn}}$ is 0.40, lower than the cost obtained with proportional AC.

Whatever the AC type, consuming members (such as P1 and P2) should compensate those members supplying them locally (such as P3 and P4), for example, by computing a local price ^{P2P} that benefits both buyers and sellers. For example, in Fig. 6.6, this internal price could be set to $^{P2P} = 0.15°/\text{kWh}$ to compute these internal compensations.

Reference Mello, de Lorenzo et al. (2023) and Mello and Villar (2023) described the *postdelivery pool market*, which is a simple local market design to determine both the energies to be allocated and the internal compensation price based on the opportunity costs of the members. In this design, the aggregated selling and buying curves can be built after delivery by using, for each CSC member ("Directive EU, 2018/2001 of the European Parliament and of the Council of 11 December 2018 on

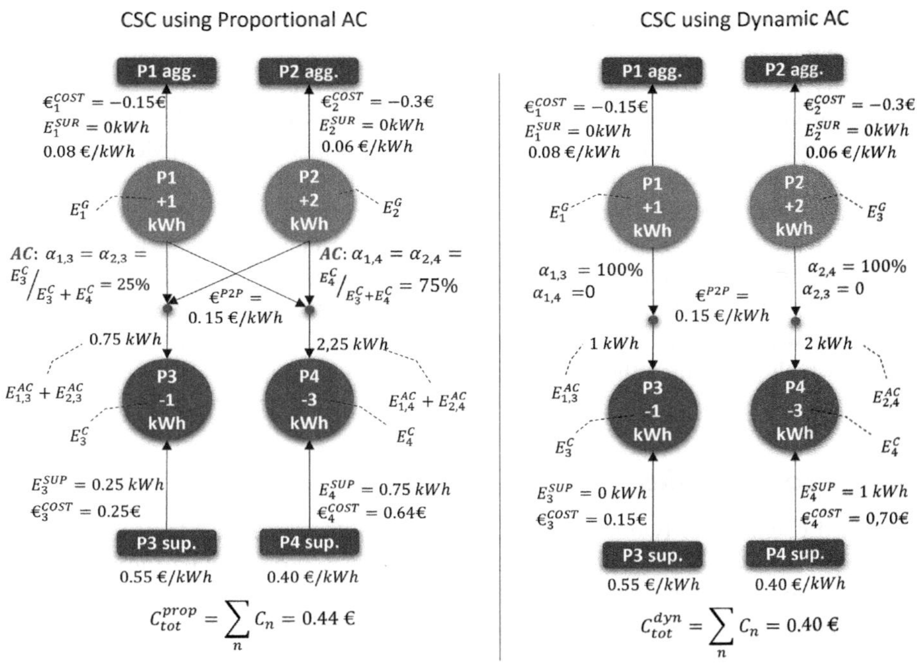

Figure 6.4 CSC with (A) proportional AC and (B) dynamic AC, two different energy sharing mechanisms. *AC*, Allocation coefficients; *CSC*, collective self-consumption.

the promotion of the use of energy from renewable sources Text with EEA relevance," 2018) the energy produced and the corresponding retailer buying price to make its selling bid and (Directive EU, 2019) the energy consumed and the corresponding retailer selling price to make its buying bids. For each ISP, a CSC member will only have either a buying or a selling bid. The market equilibrium is computed by intersecting the aggregated supply and demand curves, providing energies traded and the market price.

Note that, in general, as described in Mello et al. (2024), AC can be classified as *predelivery AC* and *postdelivery AC*. Predelivery AC must be defined a priori, that is, without knowing the energies finally supplied or injected measured in the smart meters of the CSC members. They are usually communicated at the CSC creation time and can only be changed when the CSC structure changes, for example, to the entry or exit of members. Examples are the so-called fixed AC of the Portuguese regulation, or the AC proportional to the contracted power of the Spanish regulation. Although fixed AC could be estimated to minimize the expected CSC energy cost, since they are defined before metering, they will not allow to reach optimal allocations. Postdelivery AC can be computed after the DSO energies measurements are made available to the CSC. Both Portugal and France allow for dynamic allocation, while Spain still does not. Note also that dynamic AC are essential to reflect the actual internal transactions of a local electricity market, for example, to postdelivery pool market previously described, being a prerequisite for their implementation.

Finally, let us make two remarks. The first one is that for the AC-based energy sharing mechanism, some constraints apply such as they must add 1 and not be negative (see paper IEEE for a regulatory discussion on this topics or [Mello & Villar, 2023] for examples where these constraints limit energy and flexibility integration). The second one is that other energy sharing mechanisms not based on AC can be defined, for example, in Italy, the use of the virtual energy-sharing model, where the energy shared in a CSC is computed by the energy regulator, a state-owned company, as the minimum between total net surplus and net supply of all CSC members, who pays a financial incentive to the CSC proportional to the energy shared, which predefined the rules to share it among its members.

Energy communities

As said, EC can be of two main types, REC and CEC. According to EU regulation, both types of EC should have, as a primary purpose, provide environmental, economic, or social community benefits to its members or shareholders or to the local areas where they operate, rather than generating financial profits.

The link between EC and self-consumption comes from the fact that both types of ECs can engage in CSC activities. In addition, although many regulations allow CSC structures for a set of neighbors in a condominium to be licensed, EC implementing CSC are intended for a more complex organization, with different types of members, including, for example, commercial or industrial business, or administrations facilities, and must have legal personality.

When comparing REC with CEC, while RECs focus on local community premises and are limited to renewable generation, CECs may include generation from renewable and nonrenewable sources and may also develop other activities such as distribution, supply, or aggregation. Indeed, in the EU regulation, it appears that CEC are structures that manage resources located anywhere in the system, as a tool to develop large-scale aggregators or virtual power plants with largely distributed resources, which are able to provide services to the grid by managing distributed assets in coordination. However, national transpositions do not always respect EU principles. For example, limiting the distance of CEC members also limits aggregators' capacity to efficiently operate these assets, as it is the case of Portugal, where the same proximity criteria apply to RECs and CECs.

In ECs with CSC, business models vary regarding the objectives of the stakeholders, the way energy is shared and financially compensated among the members, their activity sectors, and their energy behavior. Members with an investor profile may be willing to invest in generation assets to monetize them by saving energy costs and by profiting from the extra revenues obtained by sharing their surplus locally with members with more passive profiles, who may want to reduce their energy bill without investing in generation facilities.

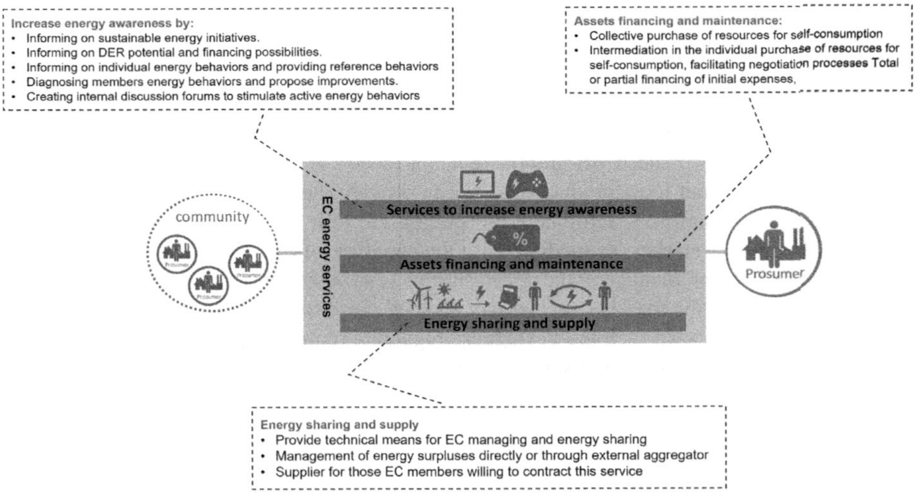

Figure 6.5 Energy services provided by EC to their members. Energy services provided by EC to their members, to increase awareness, help to finance and maintain assets, and manage the collective self-consumption activity. *EC*, Energy communities.
Source: Adapted from USEF, White Paper. (2018). Flexibility value chain.

Energy services offered by energy communities

EC have the capability of providing several energy services to their members that improve the value they get from belonging to it and incentivizing their engagement in the energy system. Fig. 6.5 adapted from (USEF, White Paper, 2019) summarizes these services.

- *Services to increase energy awareness:* EC, defined as structures whose primary purpose is to provide environmental, economic, or social benefits to its members or shareholders, or to the local areas where they operate, rather than to generate financial profits, can provide energy services to enhance these objectives (Rocha et al., 2019). Some are as follows:
- Inform its member about sustainable energy initiatives or alternative to their current practices.
- Inform about the potential of DER according to the member financing possibilities.
- Assess individual energy behaviors and provide references of efficient behaviors aligned with the system needs from reference patterns or from other members.
- Provide diagnosis on members energy behaviors and facilities to enhance improvements.
- Create internal discussion forums to stimulate active energy behaviors and improve energy literacy and sustainable attitudes.
- *Assets financing and maintenance:* EC, with a better capability to negotiate financing than individual members, can contribute to offering the following services:
- Collective purchase of resources for self-consumption (solar panels, insulation materials, batteries, etc.) for shared resources. Collective investment for resources sharing makes it possible to invest in more advanced resources with greater capacity than individual investments.
- Intermediation in the individual purchase of resources for self-consumption, facilitating negotiation processes with suppliers to obtain better financing conditions, installation, and maintenance of assets.

- Total or partial financing of initial expenses, complemented by reimbursement schemes that can be integrated into the REC's own business model.
- *Energy sharing and supply:*
- One of the main services EC may offer to their members is the sharing of locally generated energy, so that the individual energy surpluses of some members can be shared and traded with others in need of supply, reducing costs, and improving the local balance between demand and local production. EC can provide the technical means (management and transactions platforms, settlement procedures, etc.) for such sharing and for the implementation of local markets or mediate with the suppliers of these systems.
- EC, responsible for managing the local surplus, can also act as an aggregator and sell these surpluses on the markets, or contract and sell these surpluses to aggregators who sell them on the markets. In fact, the Portuguese regulation enforces the EC manager to aggregate and manage the individual surpluses after the local energy sharing.
- ECs could also adopt the role of marketer for its members, replacing traditional marketers, but with similar obligations.

Flexibility services offered by energy communities

Flexibility refers to the modification of the generation or consumption profile in reaction to an external signal to contribute to the power system (Villar, Bessa, & Matos, 2018). When the external signals are prices or tariffs, flexibility is known as implicit flexibility. When there is an explicit activation signal, then the flexibility provided is known as explicit flexibility (USEF, White, Paper, 2019). The potential flexibility an EC could provide has multiple uses, as shown in Fig. 6.6 (Mello & Villar, 2023),

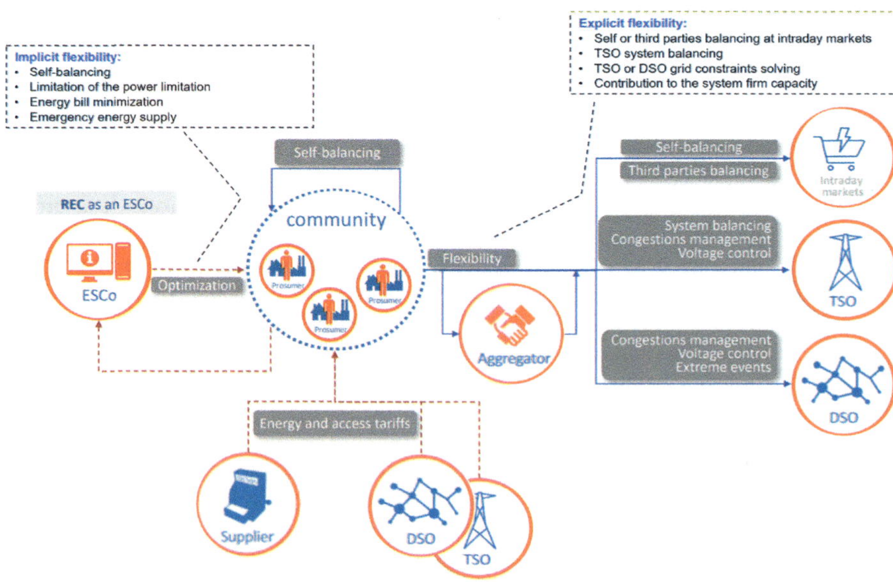

Figure 6.6 EC flexibility services. Flexibility services an energy community can provide to its members for local balancing, or to third parties such as balance responsible parties, or grid operators in need of flexibility. *EC*, Energy communities.

where both implicit and explicit flexibility are represented. These uses go from the internal balance and cost minimization of the EC, to third parties such as BRPs to balance their portfolio in energy adjustment markets (intraday markets), or as TSO or DSO services markets to provide system services (see (Silva et al., 2021) for a review of TSO and DSO flexibility needs and flexibility coordination schemes for distributed flexibility).

As a concrete example on how to deal with EC flexibility for third parties, Rodrigues et al. (2024) proposed a mechanism to compute flexibility bidding curves to bid in DSO flexibility markets. Since DSO publish their needs but not the flexibility price, bidding curves allow to determine, for each possible price, the optimal flexibility to be provided to deal with price uncertainty. To do so, this proposal computes, first, the EC baseline as its energy profile when optimizing the schedules of the flexible assets to minimize the EC energy cost without providing flexibility. Then, by discretizing the expected flexibility price range, the optimal flexibility that would minimize the REC cost for each flexibility price provides a point of the bidding curve. This process is illustrated in Fig. 6.7.

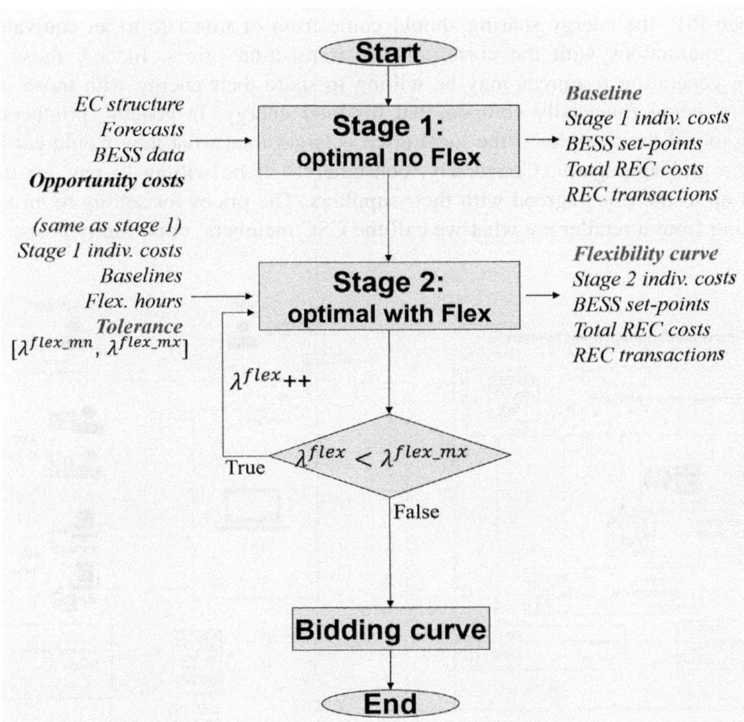

Figure 6.7 Computing the flexibility bidding curve of a REC by computing the flexibility that maximizes the community profit (or minimizes its energy cost) for different values of the flexibility price. *REC*, Renewable energy communities.

Tools for energy community management

The practical implementation of CSC structures and EC requires digital systems to connect new entities, business models, and technologies in a secure and efficient manner. This section outlines an innovative digital platform (Tavares et al., 2023) developed by INESC TEC (Fig. 6.8) to manage REC for self-consumption and energy trading within a LEM, adapted to the Portuguese regulatory framework (Erse, Regulamento, 2021; Portuguese Gov, 2022), but easily replicable in similar frameworks. The platform was designed to integrate functionalities as microservices and includes, among others, an energy transactions service, a settlement service that computes energy allocations and financial compensations, and a distributed resources sizing service and provides efficient and secure management of the community and the internal energy transactions. Transactions can be generated automatically either from a proportional sharing as defined in the Portuguese regulation, by implementing a postdelivery pool market as described in the section devoted to the energy allocation mechanisms, or other more complex designs.

Platforms for EC management should include several relevant modules, such as:

- The *transactions module* to support the energy transactions among community members. For each ISP, the energy sharing should come from or translate to an equivalent set of energy transactions with the corresponding transaction prices. Indeed, those members owning generation resources may be willing to share their energy with those in need of supply if being financially compensated for their energy. In general, producers will be willing to sell their surplus if the local price is larger than what they would get by selling it to the grid aggregator. Conversely, consumers will be willing to pay for the energy shared up to the price agreed with their suppliers. The prices for selling to an aggregator or buying from a retailer are what we call the CSC members' opportunity costs.

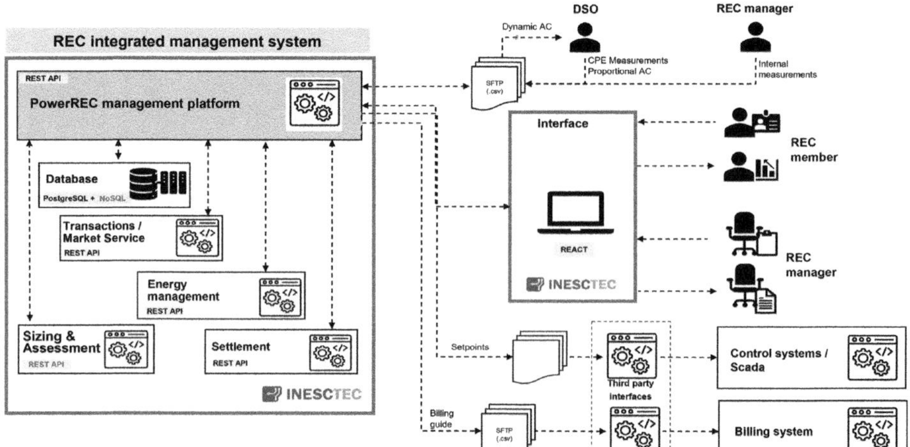

Figure 6.8 INESC TEC platform for the management of energy communities, with the main functional modules, agents, and interactions.

- The *settlement module*: Using as inputs the energy transactions and prices provides the internal financial compensations among the CSC members and, in the case of the Portuguese regulation, also computes the grid access tariffs to be paid to the DSO for the energy self-consumed that uses the public distribution grid. Fixed, proportional, and dynamic AC (Mello et al., 2022) are supported.
- To further enhance the management of energy resources, the platform includes an *energy management module*, responsible for dispatching flexible resources within the community, such as energy storage systems.
- The *sizing and assessment module* is responsible for sizing new energy resources to optimally meet the community's energy needs under different business models.

Inputs to the platform are sourced from the interface service, which provides a user-friendly interface for EC managers and members to interact with the system. Inputs are also sourced from the DSO that provides the energy allocation according to the selected energy management mechanism and the verified measurements from the smart meters, needed when dynamic allocation (which, as explained in the section devoted to the energy allocation mechanisms, is computed after delivery) is in place, from Iberian Market Operator for Electricity (OMIE) where day-ahead electricity market prices are obtained to compute those opportunity costs of the CSC members that are a function of them and from internal meters when connected to the CSC resources.

The main outputs of the platform include the set points of the flexible resources when the energy management system is used for the EC management, the billing guide, and individual and collective performance indicators. Flexible resource set points provide third-party systems (i.e., SCADA) with energy resources optimal dispatches, while the billing guide ensures that members are accurately and fairly billed for their energy consumption and production.

The EC management platform is supported by a data model that represents the different components and interactions involved, with a very simplified version represented in Fig. 6.9, which includes the members that can partially own resources connected to either DSO or private meters, with each DSO meter being the responsibility of a member with a supply contract (and possibly a surplus selling contract). Transactions are between resources to share local energy and properly account for the self-consumption grid access tariffs. Resources can be consumption facilities (that can both consume and generate), generation units for self-consumption (UPAC under the Portuguese terminology) such as PV panels, and storage systems for energy management energy surplus minimization.

Fig. 6.10 exemplifies a possible EC configuration that helps to understand members responsibilities. Member A is responsible for DSO meter A, which has three private meters measuring three BTM resources, with member A1 owning the resource associated to the first private meter and member A2 owning the resources associated to the second and third ones. Member B is responsible for DSO meter B and owns a resource BTM. Transactions occur among resources, but the settlement is be done at the DSO meter level, at which the DSO performs the energy allocation. An important point is that the EC platform must be able to represent such structures, and the settlement module designed to compute the financial compensations for the energy shared considering all these interactions and according to the in place regulation.

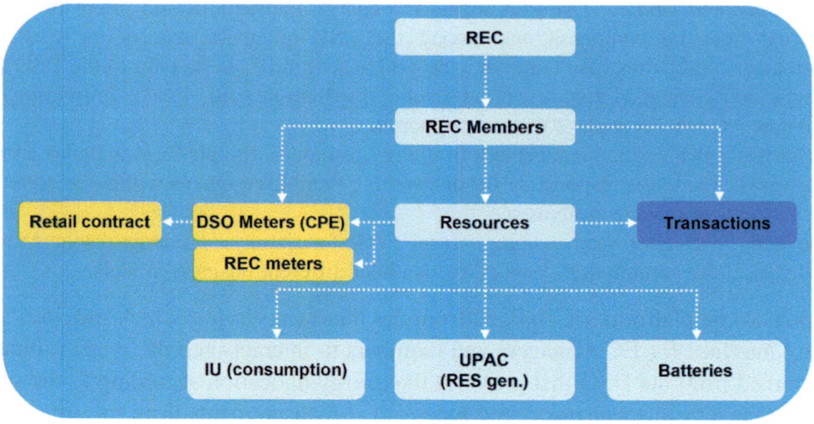

Figure 6.9 EC management platform data model with members, resources, meters and supply contracts, and transactions. *EC*, Energy communities.

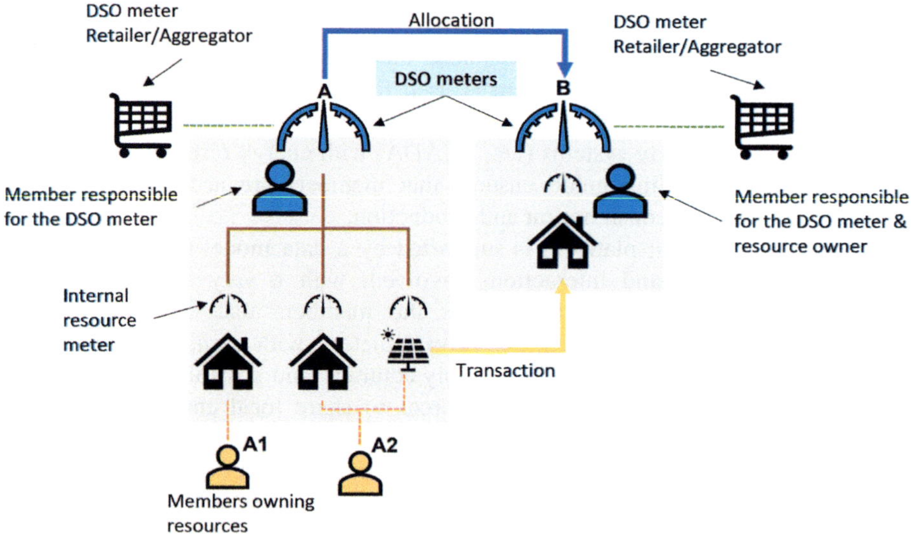

Figure 6.10 EC configuration example. Example of a possible configuration of an energy community, with meters, resources, supply contracts, and transactions among members resources. *EC*, Energy communities.

According to the Portuguese regulation, but similar to other regulatory approaches, the EC platform offers two operation modes based on the energy sharing mechanism: fixed/proportional (FP) mode and dynamic (DY) mode. Fig. 6.11A, B represents these modes of operation, respectively.

In the FP mode, the EC specifies the energy sharing mechanism at licensing time. The DSO applies it daily, for each ISP, to compute the energy allocated

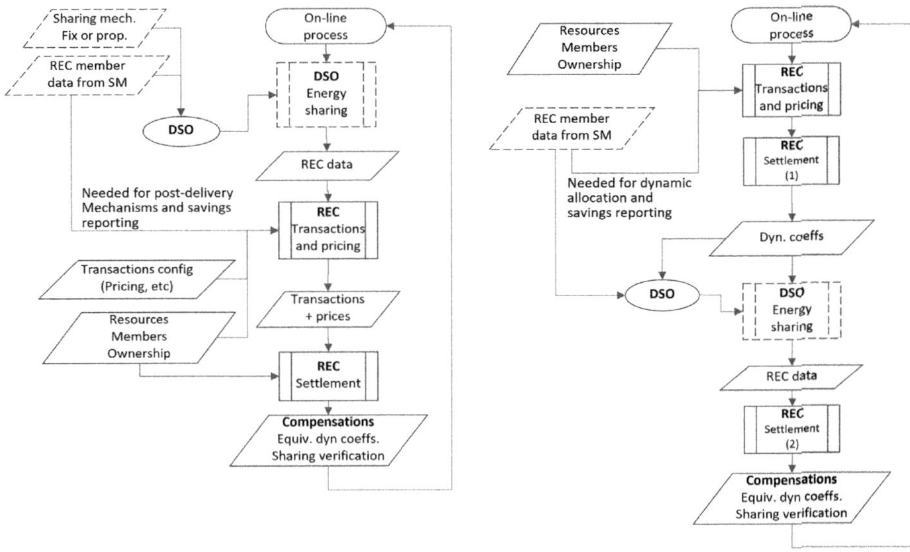

Figure 6.11 EC operation with (A) fixed or proportional AC (FP mode) and (B) dynamic AC (DY mode). Operation of an energy community according to the proportional sharing mechanisms, or to the use of dynamic allocation coefficients. *AC*, Allocation coefficients; *DY*, dynamic; *EC*, energy communities; *FP*, fixed/proportional.

among the consuming members (using fixed or proportional allocation), self-consumed, supplied, and surplus. Then, the REC transactions service computes the transactions that result in the same allocations, and the settlement service computes the financial compensations from the transactions, the pricing mechanism predefined (Mello, de Lorenzo, Campos, et al., 2023), and the energy ownerships.

The DY mode allows the EC to decide the energy allocation, using, for example, a local market approach. The settlement service computes the dynamic AC that represent the desired energy allocation and sends them to the DSO, which verifies and applies them to return the energy allocated, self-consumed, supplied, and surplus. Finally, the settlement service computes the financial compensations based on the transactions and prices and energy ownerships (Mello et al., 2022). The DY mode allows for more flexible and granular approaches, enabling the REC to make decisions based on real-time data, whereas the FP mode is more suitable for simple and predictable energy sharing scenarios.

Conclusions

This chapter has provided an overview on individual and CSC energy activities and basic regulation, explaining some of the most common energy sharing mechanisms. EC are also described, as well as the energy and flexibility services EC can provide to their members and third parties. The chapter ends with the description of the tools needed to manage EC and their main modules and process flows depending on the energy sharing mechanisms selected.

While the EU regulation seems to be strongly committed to the development of self-consumption and EC, national regulatory and administrative improvements are needed to achieve their full potential. For example, while EU directives state that grid operators and regulators should facilitate the procedures to connect self-consumption assets to the grid, DSOs, responsible for many of the technical processes involved, are often reluctant to changes and slow in implementing the new procedures needed, creating barriers to its full deployment. Smart meters are also necessary to implement the energy sharing procedures, and in some countries, there are still in the preliminary deployment phase. National energy agencies and administrations should also simplify, digitalize, and accelerate the procedures for EC licensing to limit bureaucratic barriers.

Finally, low-cost tools need to be developed to help manage EC and implement energy sharing procedures with automated systems to really engage consumers into collective structures financially sustainable. CSC management may impose software and operational cots that can only be paid with gains of scale through successfully spreading EC with CSC.

References

Directive (EU) (2018). 2018/2001 of the European Parliament and of the Council of 11 December 2018 on the promotion of the use of energy from renewable sources (Text with EEA relevance). 32018L2001.

Directive (EU) (2012). 2019/944 of the Parliament and the Council of 5 June 2019 on common rules for the internal market of electricity and amending directive

ERSE, Regulamento n.º 373/2021.

European Union. (2023). Regulation of the european parliament and the council. Accessed: Aug. 15, 2024. [Online]. Available: https://eur-lex.europa.eu/legal-content/EN/TXT/?uri = celex:52023PC0148.

Mello, J., de Lorenzo, C., Campos, F. A., & Villar, J. (2023). Pricing and simulating energy transactions in energy communities. *Energies*, *16*(4). Available from https://doi.org/10.3390/en16041949, http://www.mdpi.com/journal/energies/.

Mello, J., & Villar, J. (2023). Integrating flexibility and energy local markets with wholesale balancing responsibilities in the context of renewable energy communities. *Energy.*, *282*, 128853. Available from https://doi.org/10.1016/j.energy.2023.128853.

Mello, J., Villar, J., & Saraiva, J. T. (2022). Conciliating the settlement of local energy markets with self-consumption regulations. *SSRN, Portugal SSRN*. Available from https://doi.org/10.2139/ssrn.4097357, https://www.ssrn.com/index.cfm/en/.

Mello, J., Villar, J., Bessa, R. J., Antunes, A., & Sequeira, M. (2024). *Decarbonized and inclusive energy: A two-fold strategy for renewable energy communities.*

Portuguese Gov. (2022). *15/2022: Organisation and functioning of the national electricity system.*

Regulation (EU) (2019). *2019/943 of the European parliament and of the council of 5 June 2019 on the internal market for electricity.*

Rocha, R., Villar, J., & Bessa, R. J. (2019). *Business models for peer-to-peer energy markets.* International Conference on the European Energy Market, EEM. IEEE. Available from https://doi.org/10.1109/EEM.2019.8916543, http://ieeexplore.ieee.org/xpl/mostRecentIssue.jsp?punumber = 65951682019.

Rodrigues, L., Faria, D., Coelho, F., Mello, J., Saraiva, J. T., Villar, J., & Bessa, R. J. (2023). *Analysis of flexibility-centric energy and cross-sector business models. International Conference on the European Energy Market, EEM.* IEEE. Available from https://doi.org/10.1109/EEM58374.2023.10161816, http://ieeexplore.ieee.org/xpl/mostRecentIssue.jsp?punumber = 65951682023.

Rodrigues, L., Ganesan, J., José, S. (2024). Building flexibility bidding curves for energy communities.

Silva, R., Alves, E., Ferreira, R., Villar, J., & Gouveia, C. (2021). Characterization of TSO and DSO grid system services and TSO-DSO basic coordination mechanisms in the current decarbonization context. *Energies, 14*(15). Available from https://doi.org/10.3390/en14154451, https://www.mdpi.com/1996-1073/14/15/4451/pdf.

Tavares, T., Mello, J., Silva, R., Moreno, A., Garcia, A., Pacheco, J., Pereira, C., Amorim, M., Gouveia, C., & Villar, J. (2023). Operation and simulation of a renewable energy community based on a local postdelivery pool market. *International Conference on the European Energy Market, (EEM). IEEE.*. Available from https://doi.org/10.1109/EEM58374.2023.10161761, http://ieeexplore.ieee.org/xpl/mostRecentIssue.jsp?punumber = 65951682023.

USEF, White Paper. (2018). Flexibility value chain.

USEF, WhitePaper. (2019). Energy and flexibility services for citizens energy communities.

Villar, J., Bessa, R., & Matos, M. (2018). Flexibility products and markets: Literature review. *Electric Power Systems Research., 154.* Available from https://doi.org/10.1016/j.epsr.2017.09.005.

Microgrids and clusters of microgrids

7

Guillermo Gutiérrez-Alcaraz[1], *Danny Espín-Sarzosa*[2], *Víctor Hinojosa*[2], *Felipe Valencia-Arroyave*[3], *Patricio Mendoza-Araya*[4] *and Andrés Montenegro*[5]
[1]Department of Electrical Engineering, Tecnológico Nacional de México/I.T. Morelia, Morelia, Michoacán, Mexico, [2]Department of Electrical Engineering, Universidad Técnica Federico Santa María, Valparaíso, Chile, [3]Electricity and Electronics Insitute, Faculty of Engineering Sciences, Universidad Austral de Chile, Valdivia, Los Ríos, Chile, [4]Department of Electrical Engineering, Universidad de Chile, Santiago, Chile, [5]Department of Electrical and Electronic Engineering, Universidad Nacional de Colombia, Bogotá, Colombia

Microgrid

According to the European Union research projects, a microgrid (MG) comprises low-voltage distribution systems with distributed energy resources (DER) (microturbines, fuel cells, wind turbines, photovoltaics, etc.) together with storage devices (flywheels, energy capacitors, and batteries) and flexible loads (Microgrids—Large Scale Integration of Micro-Generation to Low Voltage Grids, 2002). Such systems can be operated in a nonautonomous way, if interconnected to the grid, or in an autonomous way, if disconnected from the main grid. The operation of microsources in the network can provide distinct benefits to the overall system performance if managed and coordinated efficiently (Summary report: 2012 doe microgrid workshop, 12 doe microgrid workshop, 2012).

The DER generate and distribute electricity within clearly defined electrical boundaries that act as a single controllable entity concerning the grid (Hatziargyriou, 2014). MGs have emerged as a promising potential solution to address energy challenges, thanks to their ability to provide reliable, clean, and cost-effective energy.

Microgrid clusters

The interconnection among MGs promotes the flexible and optimized operation of the MG cluster by increasing the proportion of distributed generation (DG) penetration in autonomous MGs, sharing power with other interconnected MGs in the island mode via the advanced control system strategies and increasing

Towards Future Smart Power Systems with High Penetration of Renewables. DOI: https://doi.org/10.1016/B978-0-443-29871-4.00007-5

communication connections. The interconnection of multiple autonomous MGs with individual goals forms the MG cluster.

Microgrid power coordination

In the MG cluster context, one or more MGs exhibit excess generation utilizing green energy from DER for a given period. In contrast, one or more MGs exhibit a power generation shortage. In such cases, extra power can be shared among MGs to overcome the MG cluster's power shortage, reducing the grid dependency of MGs. Hence, cooperative operation among MGs can increase customer satisfaction through economic benefits by minimizing operation costs, maximizing DER utilization, minimizing transmission losses, and improving reliability, resiliency, system flexibility, and stability.

The MG's cooperative operation brings with it several challenges. For instance, tie-line power fluctuation and frequency deviation for interconnected MGs are more complex than traditional large-scale power systems due to renewable generation's intermittent and stochastic output power and lower equivalent inertia (Khooban et al., 2018). To deal with this issue, an energy storage system (ESS) offers an alternative and fast method of smoothing power fluctuations in the MG cluster (Yang & Hu, 2021). In addition, the ESS can effectively provide frequency support in MG clusters. Communication between MGs is performed to share the tie-line power status and the scheduled tie-line reference (Islam, 2021). MGs communicate with one another to calculate the tie-line power reference of an MG cluster and for real-time adjustment of the reference power of tie-line converters based on the information of instantaneous power drawn by other MGs in the cluster.

Market models and actors

In electrical systems, two major markets can be distinguished in the MG: the wholesale market and the retail market. These markets can interact via a pool or/ and bilateral transactions. Traditionally, mandatory open transmission access to generators and energy importers has created more competitive wholesale power markets (Hatziargyriou, 2014).

Due to their small sizes, MGs face challenges in participating directly in wholesale or retail energy markets. However, they can potentially integrate into the market through partnerships with retail suppliers or energy service companies (ESCOs). Another model involves direct control of DERs by distribution system operators (DSOs). The successful adoption of MGs is expected to rely on fair competition in the retail sector. This fact assumes a conducive regulatory environment and a robust communication infrastructure that supports market operations, including electronic/smart meters and standardized interactions between retail companies and DSOs.

The relevant key actors in the energy market are as follows (Hatziargyriou, 2014):

1. **Consumer:** Consumers within an MG can encompass households or medium to small enterprises. Generally, these consumers establish contractual agreements with retail companies to fulfill their energy supply needs. Additionally, consumers must pay fees to the distribution network owner for utilizing its services. Demand aggregators consolidate the electricity demand from various sources, including residential homes, commercial buildings, and industrial facilities. They play a crucial role in optimizing energy consumption in the MG. Consequently, demand aggregators could negotiate favorable energy agreements, optimize usage patterns, and contribute significantly to overarching demand management strategies.
2. **DG owner/operator:** DG owners, typically individuals or entities, are responsible for managing the operation of their units. They supply generated power to the network, having priority dispatch and fixed feed-in tariffs, especially for DER-based DG. DG owners may also form contracts with retail companies and may be subject to distribution network charges.
3. **Prosumer:** They are consumers with small DG units on their networks. The local DG production can either entirely or partially fulfill the owner's consumption, with any surplus exported to the main power grid. Moreover, the entirety of local production may be directly sold to the main power grid, benefiting from favorable feed-in tariffs. Consumers, DG, and prosumers are referred to as customers in MG.
4. **Market regulator:** A regulatory authority for energy is an independent company ensuring fair and transparent market operations, open network access, and efficient allocation of network costs. It also approves network usage charges and, in some cases, end-user prices based on local conditions.
5. **Retail supplier, ESCO:** A supplier in the energy sector engages directly with customers, providing electricity and other energy services. In MGs, suppliers optimize DER participation, enhancing the MG's value. They simplify retail market complexities for customers, offering understandable and attractive energy products. Suppliers may expand services based on regulations, providing reliability, green energy, and convenience. In the future, MGs may become part of an ESCO portfolio, offering comprehensive energy solutions to reduce overall energy costs.
6. **Distribution system operator:** The DSO manages and operates the distribution network, including the MG. Their tasks involve ensuring the safe and reliable delivery of electricity to end-users within the grid. DSOs coordinate with various entities, monitor grid conditions, and manage the flow of electricity in the distribution network. In the era of smart grids, DSOs become neutral market facilitators, providing a platform for suppliers to introduce innovative products. They play a vital role in metering operations for efficient energy service delivery. Embracing the flexibility of MGs, DSOs engage in contracts with suppliers and individual generators for local network balancing.
7. **Microgrid operator:** The MG operator is tasked with operating, maintaining, and developing the local low-voltage distribution network that constitutes the MG. Their responsibilities include optimizing the operation of local generation sources, ESSs, and load balancing within the MG to enhance the MG's efficiency, resilience, and reliability. In some situations, this role can be fulfilled by the local DSO.

Various agents play integral roles in the MG, collectively ensuring efficient energy generation, storage, and distribution. MGs are especially beneficial when a self-sufficient energy infrastructure is desired, offering advantages such as enhanced energy security, minimized transmission losses, and seamless integration of renewable energy sources.

Energy management strategies

Modes of microgrid operation

An MG can operate either in grid-connected (on-grid) mode through a point of common coupling (PCC), as illustrated in Fig. 7.1, or isolated mode (off-grid) and manage the transitions between these two modes of operation (Li et al., 2014). In the grid-connected mode, the main grid can supply the power deficit to maintain the balance between generation and demand, and the excess power generated in the MG can be exchanged with the main grid and provide ancillary services. The islanding, that is, the disconnection of the MG from the main grid, can be intentional (scheduled) or unintentional. Intentional islanding may occur in situations such as scheduled maintenance or when degradation of the main grid power quality may cause the operation of the MG to be compromised. Unintentional islanding can occur due to faults and other unexpected events unknown to the MG; proper detection of such tripping is critical for the safety of personnel, the appropriate operation of the MG, and the implementation of necessary changes in the control strategy. MG transitions from interconnected to islanded mode of operation and vice versa will likely cause large mismatches between generation and loads, causing severe frequency and voltage variations (Hatziargyriou, 2014). The specialized literature offers many islanding detection algorithms, which can be classified into active, passive, and remote methods (Li et al., 2014). MGs that do not have a PCC are called isolated. In this operation mode, the MG must maintain the balance between the demand of the local loads and the active and reactive power generated in the MG, including the temporary transfer of power from/to the storage units (Olivares, Mehrizi-Sani, et al., 2014). In other words, any change in the demand must be balanced by the DERs (dispatchable and nondispatchable) and the ESSs. Isolated MGs are mainly in remote locations (e.g., remote communities or remote industrial sites) where an interconnection with the main grid is not feasible due to geographical, technical, and/or economic constraints; therefore MGs operate permanently in stand-alone mode (Olivares, Mehrizi-Sani, et al., 2014).

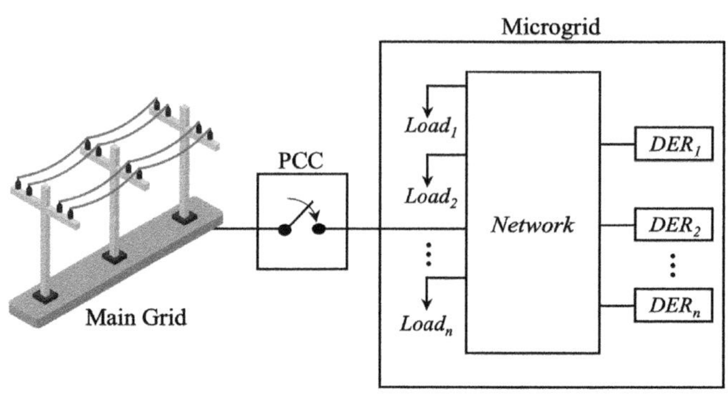

Figure 7.1 MG overview.

Control architectures: centralized, decentralized, distributed, and hybrid

The MG structure depicted in Fig. 7.1 can be operated in centralized, decentralized, distributed, or hybrid ways (Anvari-Moghaddam et al., 2021). The main goal of energy management is to determine generation setpoints for every dispatchable DER and manage controllable loads in the MG. Fig. 7.2 displays an overview of the different types of MG control architectures.

In a centralized control architecture, a main central controller makes decisions and transmits the most effective (or nearly optimal) operational settings to the local DER controllers. This central controller's primary goal is to minimize overall operating costs by ensuring critical loads remain intact even in adverse conditions (Anvari-Moghaddam et al., no date). The central controller relies on information from forecasting systems, such as load consumption, wind speed, solar radiation, and so forth (Espín-Sarzosa et al., 2020). This architecture is characterized by its straightforward design and satisfactory reliability when operating independently (Anvari-Moghaddam et al., 2021). However, this control approach may incur higher costs due to its reliance on extensive communication infrastructure. Practical

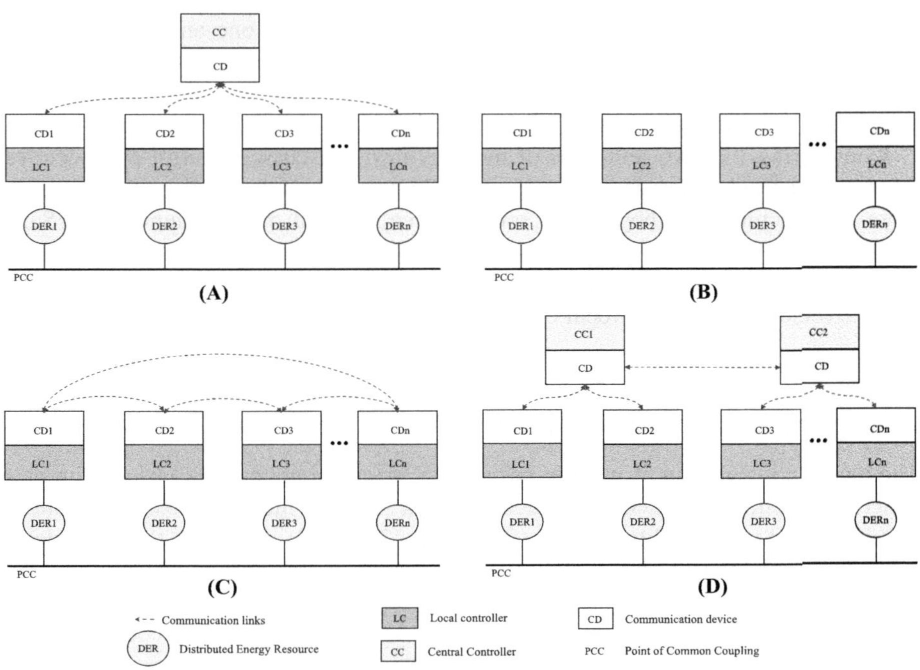

Figure 7.2 Different types of MG control architectures: (A) centralized, (B) decentralized, (C) distributed, and (D) hybrid.
Source: Adapted from Saleh, M., Esa, Y., El Hariri, M., & Mohamed, A. (2019). Impact of information and communication technology limitations on microgrid operation. *Energies, 12* (15) 2926, https://doi.org/10.3390/en12152926.

instances of this approach can be found in the literature (Olivares, Canizares, et al., 2014; Palma-Behnke et al., 2013).

A decentralized control architecture focuses on solving the energy management challenge while granting the utmost autonomy to DER units and loads (Olivares, Mehrizi-Sani, et al., 2014). Control decisions primarily occur at a local level, although this architecture can also incorporate a hierarchical control scheme. This control architecture relies heavily on the main grid when operating in on-grid mode. Consequently, an upsurge in uneconomical costs is anticipated. As a result, the MG utilizing a decentralized architecture, whether in on-grid or off-grid mode, is considered less practical and resilient (Anvari-Moghaddam et al., 2021). The distinctive features of the decentralized architecture have been extensively discussed in the literature using a multiagent system (MAS) approach. Broadly, a MAS system comprises numerous intelligent agents equipped with localized information, collaborating to attain global and local objectives (Olivares, Mehrizi-Sani, et al., 2014). For an in-depth examination of the benefits, limitations, control, operational hurdles, and communication requisites, refer to the analysis in Han et al. (2018).

Autonomous agents collaborate cooperatively in a distributed control architecture to fulfill the MG's goals (Espina et al., 2020). This control structure is often referred to as communication-based decentralized control (Sakurama & Miura, 2017), where the MG's agents communicate directly with one another, enhancing scalability by circumventing single-point failures. However, this architecture requires substantial processing capabilities at each local controller, leading to an anticipated rise in system costs (Saleh et al., 2019).

Considering the pros and cons of the previously discussed control architectures, a hybrid approach emerges as a viable solution to address the challenges inherent in these architectures, as they all, to some degree, depend on communication (Saleh et al., 2019). This hybrid control structure could entail multiple central controllers following a coordinated distributed topology. Furthermore, each central controller might oversee independent local controllers (Anvari-Moghaddam et al., 2021).

Configuration: Single and microgrid cluster

The massive integration of inverter-based RES is replacing conventional synchronous generators and, consequently, reducing power system inertia, potentially causing instability and vulnerability in MGs (Saha et al., 2023). To cope with these problems, instead of connecting a single MG to the main grid, interconnecting MGs as an MG cluster network has emerged as an appealing alternative because an MG cluster takes advantage of various complementary power sources and effectively coordinates the energy sharing/trading among the MGs and the main grid to improve the stability, reliability, and energy efficiency of the system (Zou et al., 2019).

Like the case of the single MG, energy management in the MG cluster is key to safe and reliable operation, whether grid-connected, stand-alone, or transition between the two modes of operation. Different control architectures for energy management in MG clusters are derived from the context of single MGs.

These architectures are centralized, decentralized, and distributed, as illustrated in Figs. 7.3–7.5, respectively.

The centralized control architecture comprises a central controller responsible for executing the energy management algorithms and transmitting control commands to the individual MGs. Its main advantage is facilitating coordination among several MGs because a single entity manages the control decisions. However, this centralized approach increases the risk of a single point of failure across the entire MG cluster system. In addition, this control architecture tends to require higher communication and computational resources, limiting its scalability (Zou et al., 2019).

In an MG cluster system with a decentralized control architecture, the controllers of the individual MGs operate without direct communication and rely on voltage and frequency signals (Bullich-Massagué et al., 2018). In this configuration, the local MG controllers operate autonomously, without exchanging information with external sources, reinforcing their independence and resilience against external disruptions or attacks. This feature makes reaching the global optimal operating point challenging (Du et al., 2018). Decentralized control systems can lead to a decrease in the performance of the MG cluster system due to the absence of collaboration between MGs (Kou et al., 2017).

Distributed control architecture involves communication between multiple controllers where collaboration is established to achieve a common goal. These controllers present a high level of adaptability, which allows for the integration of new MGs and facilitates plug-and-play functionality (Saha et al., 2023). A more detailed description of control architectures, operation objectives,

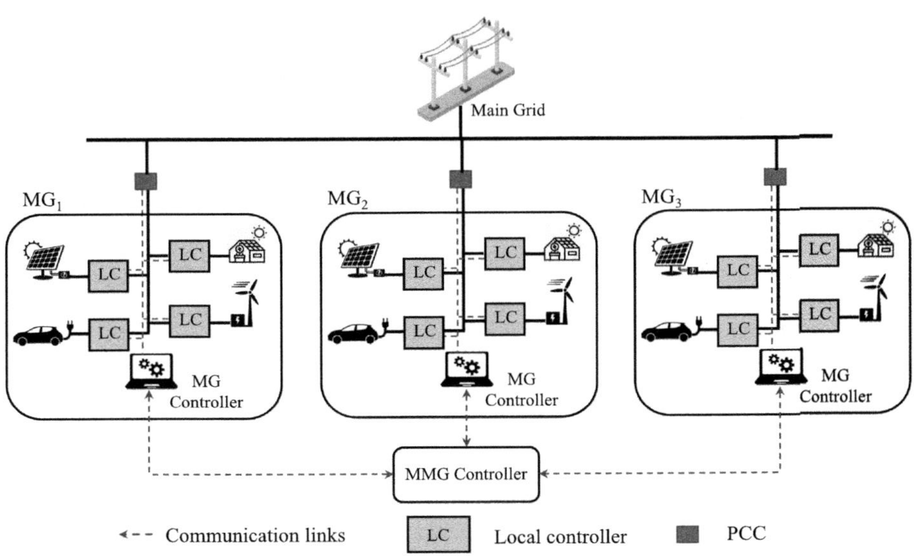

Figure 7.3 MG cluster centralized control architecture.

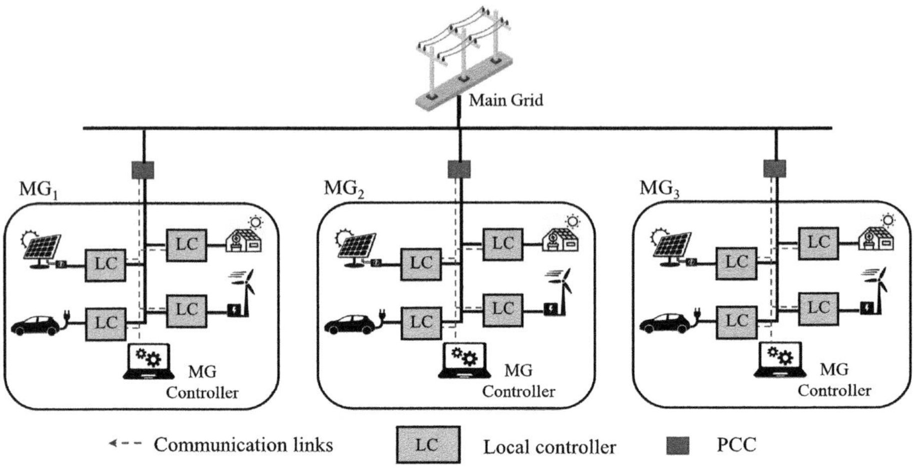

Figure 7.4 MG cluster decentralized control architecture.

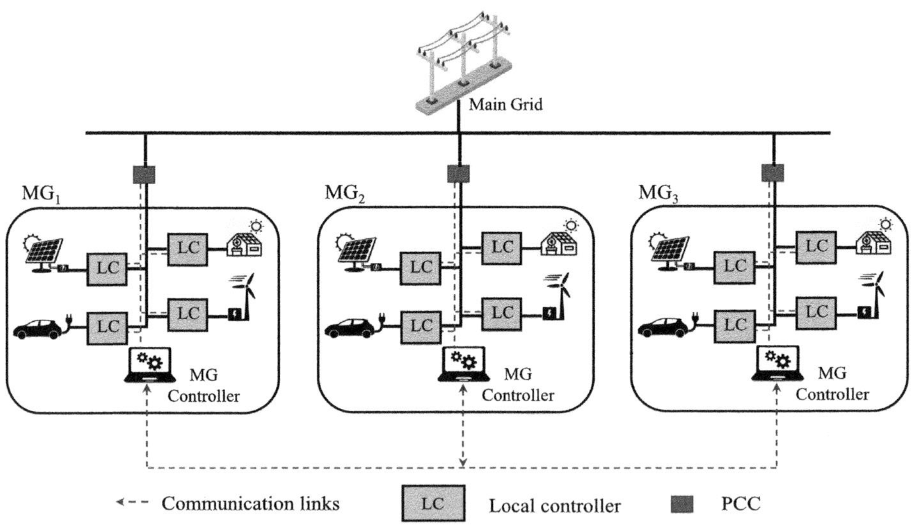

Figure 7.5 MG cluster distributed control architecture.

operational and technical constraints, and different topologies of MG cluster systems can be found in the literature (Bullich-Massagué et al., 2018; Saha et al., 2023; Zou et al., 2019).

It is worth noting that energy management and coordination among MGs and DSOs have become more challenging since various types of uncertainty arise from renewable resources, loads, and the electricity market, which are usually strongly correlated (Zou et al., 2019).

Microgrid legislation

Countries around the globe have taken different paths to foster the development of MGs. Usually, what is found is that the countries that have succeeded the most in promoving MG development have (i) promoted the development of specific MG legislation and normative and (ii) supported the development of demonstrative pilot projects. These projects allow stakeholders to verify capabilities and benefits, validate concepts, understand costs, and detect weaknesses of MGs.

Even though several norms and standards apply to MGs, grid code development varies widely from country to country. In several countries, there is no explicit mention of MGs, but instead, related or alternative concepts are used, for example, standalone power systems (SAPS) in Australia, and license exempt networks (LEN) in the United Kingdom, among others. These concepts do not necessarily coincide with all the attributes expected from an MG, which renders a direct comparison of MG legislation among different countries troublesome. However, some standards have been observed and followed more generally by countries around the globe, such as the IEEE 1547 series of standards, which has evolved to include the IEEE Std. 1547.4 specifically covering the design, operation, and integration of "Island Systems" (populated with distributed resources) with larger electric power systems, and the IEEE 2030 series of standards, focused on smart grid interoperability, with several dedicated specifically to the specification (IEEE Std. 2030.7) and testing (IEEE Std. 2030.8) of MG controllers, planning and design of MGs (IEEE Std. 2030.9), and extra low voltage DC MGs (IEEE Std. 2030.10).

In general, MGs are expected to operate mostly grid-connected (i.e., with normal operation connected to the distribution network), except in remote locations. On the one hand, in isolated MGs, to achieve sustainable isolated operation, an MG must meet high requirements in size and capacity of its resources to allow for continuous supply to loads or have significant demand flexibility (DF) by applying demand-side management (DSM) schemes. In either case, these MGs might not need to meet the same requirements of a distribution grid. On the other hand, in grid-connected mode, MGs are expected to follow the codes and rules that apply to the distribution systems to which they are connected without necessarily being involved in the rest of the system operation. Island mode is seldom activated, for example, in the event of a failure. Therefore MGs will be, in general, subject to rules and codes that apply to distribution grids; hence, rules and codes vary from country to country.

Below, we include a summary of selected countries' legislatures, rules, and codes that have successfully developed MGs and accompanying legislatures.

United States of America

The USA is one of the most prominent countries in MG deployment, with specific support for MG development at both federal and state levels. The United States Department of Energy (DOE), along with its MG Initiative, was able to fund demonstration projects through the Smart Grid Demonstration Program. Today, the

DOE's Office of Electricity manages the MG R&D program. It has several areas of action, including standardization, planning and design, control, and operation. The DOE keeps track of MGs in the United States: as of 2022, there were 687 MGs in the United States, with a total installed capacity of 4.357 MW, distributed as shown in Fig. 7.6.

According to the Smart Electric Power Alliance, 184 MG-related bills were either proposed or enacted across 24 states and Washington, D.C. Although many of the leading states (e.g., California) have had an interest in MGs due to extreme weather and climate issues that affect energy security, the recent increase in funding opportunities for MGs in the United States and the need to serve vulnerable populations and disadvantaged communities have also become drivers for the new legislature.

In the particular case of California, as a consequence of the environmental goals, the value of the power system's resilience (impacted by the 2000–2001 California electricity crisis, wildfires, and storms), as well as the evolution of generation technologies based on renewable energies and storage, there has been a boost to the development of MGs. This can be seen as materialized in its MG Roadmap, the financing of the Electric Program Investment Charge, and the "Microgrid Bill" (SB-1139). MGs in California emerged from the California Energy Commission's

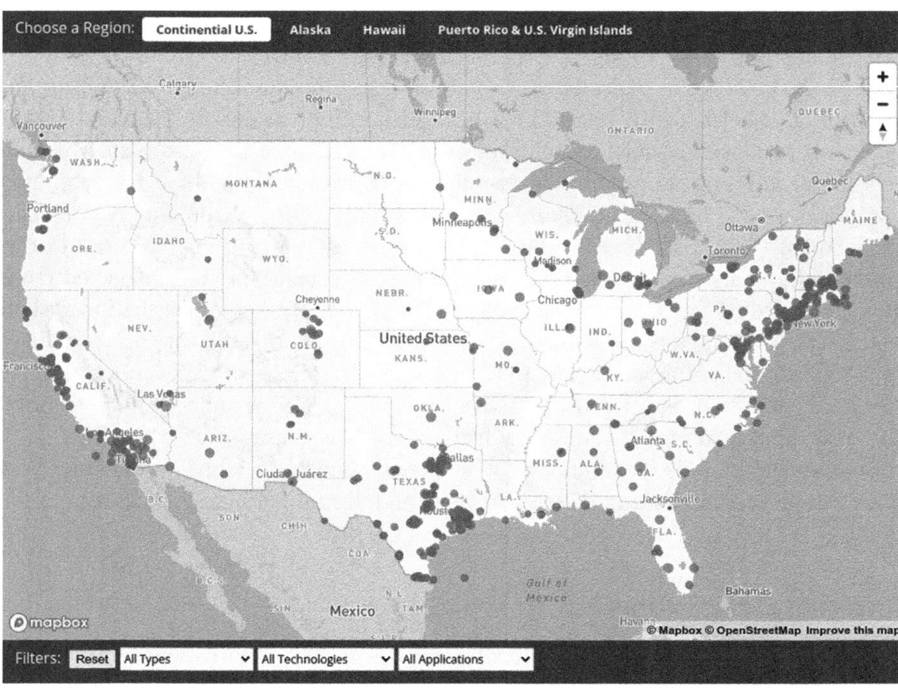

Figure 7.6 Microgrid installations in the United States.
Source: DOE (Corp, 1984). https://doe.icfwebservices.com/microgrid.

(CEC) Renewable Energy Secure Communities program. To support California's energy and greenhouse gas policies, CEC dedicated $45 million to MG demonstrations that generated successful, replicable prototypes for commercially viable MGs.

In 2018 the California Senate approved Senate Bill No. 1339 (SB-1339) regarding MGs. This law recognizes that MGs have a series of benefits: energy autonomy, resilience, and potential monetary benefits. A series of key issues for applying this technology in California are listed, such as the regulatory framework, technical challenges, and defining its commercial value and usefulness to meet energy policy goals. To encourage the use of MGs, the California Public Utility Commission (CPUC) is mandated to develop the following: a standard for interconnection between the MG and the general grid, a standard applicable to MGs (in accordance with other laws), and methods to reduce barriers to entrance. The law gives local public companies 180 days from the interconnection request to create a standard that regulates the interconnection between the MG and the general grid. The law only allows companies to have a differentiated MG rate if necessary. Additionally, cost transfer between customers without an MG and customers with an MG is prohibited.

Australia

Although there is no legal definition of an MG in Australia, there are different definitions in their National Electricity Law (NEL) or National Electricity Rules (NER) that fit specific cases of MGs (no date a):

- SAPS are MGs managed by the local distributor. These are subject to the national energy market registration requirements and the market and system procedures.
- Standalone MGs, managed by parties different from the local distributor, are not part of the interconnected electricity system and do not meet the definition of a regulated SAPS. Therefore they do not need to follow National Energy Laws and Standards. However, state and territory regulatory requirements may apply.
- Embedded MGs, managed by parties different from the local distributor, fall under the regulation of embedded network (i.e., private grid) and embedded generation (the Australian term for DG) that is operated in a coordinated manner. However, an embedded MG is distinguished from a traditional embedded network, as it can operate isolated from the main grid. As in the SAPS, they are subject to National Energy Laws and Regulations, predominantly through the framework of authorizations and exemptions administered by the Australian Energy Regulator (AER).

Starting August 1, 2022, distributed network service providers (DNSPs) in Australia can connect customers to a SAPS, which can be more economical, secure, and reliable than connecting to the grid. These become regulated SAPS. A regulated SAPS is then part of the national electricity system. It is part of a DNSP's network and must be regulated in the same way as if it were part of the connected network.

United Kingdom

As in Australia, there is no formal MG definition in the United Kingdom legislature. However, the regulatory agency, the Office of Gas and Electricity Markets

(OFGEM), recognizes an MG as "a network which distributes power without the need for an electricity distribution license." In this sense, there is a figure in the UK legislature called LEN, which are private networks, and "exempt/embedded distribution networks," which are not operated by the traditional DSOs.

According to these definitions, MGs can operate in the United Kingdom as long as they do within the framework provided for by the Electricity (Class Exemptions from the Requirement for a Licence) Order 2001 and appropriate provisions of the Electricity Act 1989, particularly Schedule 2ZA, which sets out the duties of distribution exemption holders and Schedule 2ZB which covers the duties of supply exemption holders (no date b).

The main regulatory challenge for MGs in the United Kingdom is related to energy measurement since current regulations do not allow clients located in a LEN to access better suppliers until the accuracy of the reports on monitoring and location of electrical consumption is improved.

Currently, the United Kingdom allows for pilot projects approved by the regulator through the "Energy Regulation Sandbox," where exceptions are established for implementing the test project. In August 2020, Emergent Energy applied for an Energy Regulation Sandbox for an MG implementation. Within the conditions imposed by the regulator, the maximum period was established for the duration of the project, being 2 years from the start date, along with 1 additional year as a transition period, in addition, to limiting the scope of the project to a maximum of 2000 customers. The proposal presented a meter reading based on the aggregation of on-site customer meters, where the total value that would have been calculated using a methodology provided for in the Balancing and Settlement Code is used. This allows each supplier to generate an invoice for the energy consumed without reaching a bilateral agreement with other suppliers. This action would reduce the overhead costs associated with supplying customers in the LENs, improving their access to the retail electricity market. The Sandbox was granted in January 2023, and as of 2024, the company is actively using its regulatory exemption in two projects.

Chile

Although there is no official definition of MGs in Chile, the current legal regime allows for two types of MGs. The first type is private systems ("Instalaciones de energía eléctrica de uso privado"), which operate under the same technical requirements as interconnected grids but are not subject to market rules or regulation. These have the limitation of always operating disconnected from an interconnected system.

The second type is a collective installation that falls under a new "residential generator" category under the Chilean net billing legislation (laws 20.571 and 21.118). In November 2018, the last update allowed for behind-the-meter installations, up to 300 kW, to inject surplus energy into the grid, as long as they are designed for self-consumption. Additionally, the generation equipment can be shared and treated as community generation, for which all community members can benefit from the injections of a single multiple-owner distributed generator. This last regime is suitable for some grid-connected MGs but is very limited in terms of consumer/prosumer freedom and access to tariffs, as the regulation at the distribution level does not allow for competitive energy retailing or aggregation. In terms of islanded operation, a grid-connected MG might be able to operate as long as there is an explicit agreement with the distribution company.

There is also a possibility of operating an MG as a distribution subsystem, as defined in the regulatory technical document RIC No. 16. A distribution subsystem is an internal distribution network owned by a private party, built within the limits of a property (e.g., building, shopping center, and condominium) and whose purpose would be to provide electrical energy to all its members, independently of the electricity distribution company in the area. This establishes the possibility that a group of facilities can purchase energy (e.g., through a Power Purchase Agreement (PPA)) at a single connection point. The RIC No. 16 establishes that the responsibility for the distribution subsystems' security, operation, and maintenance falls on those who constitute these systems, which allows for a usual MG aggregate operation of its resources. Finally, it establishes the isolation requirement under internal failures within a subsystem. In the event of grid failures, it must also be disconnected, enabling the islanded operation of an MG.

Local markets and mechanisms

Advances in information technologies and control systems have allowed for great developments to be carried out in smart networks, allowing systems to be monitored and controlled more effectively. These advantages have been quite useful for the transformation of the electric energy matrix and opened several ways to develop new market models that include electric energy production at the customer's side (e.g., generation at the distribution network) and mechanisms that incentivize the electric energy demand modification in accordance with the electric energy availability (Lampropoulos et al., 2013). In this context, DSM has become increasingly relevant for operating electrical energy distribution networks since it constitutes a group of actions designed to efficiently manage electrical energy consumption to reduce electricity supply costs, network charges, and general system charges, including taxes. To do this, the consumption patterns of the end user are modified (with reference to the usual consumption pattern) through incentives to change the times of use of electrical energy and the contractual parameters of the supply (contractual power and network connection parameters) and thus achieve a reduction in electricity rates and better use of energy generated in a distributed manner (Gellings, 1985).

DSM comprises different mechanisms such as energy efficiency, demand response (Dr), and DF (see Fig. 7.7). These mechanisms aim to (i) preserve the balance between supply and demand, (ii) improve the operational efficiency of the electrical system (generation, transmission, and distribution), (iii) reduce the price and volatility of electricity, and (iv) facilitate greater integration of renewable energies with variable behavior (Jabir et al., 2018). It should be noted that Dr and DF share similar characteristics. However, it is important to differentiate both mechanisms to achieve adequate implementation of them. The Dr mechanism brings together a set of auxiliary services, in which end users

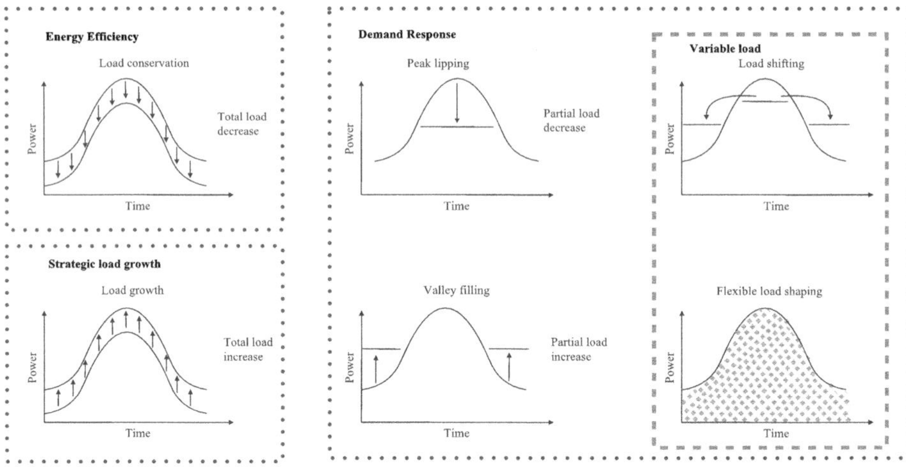

Figure 7.7 Actions typically included in DSM (Lampropoulos et al., 2013).

can adjust their energy consumption in response to a certain signal. This signal may contain information about the price of electricity, incentives to modify consumption, or reliability emergencies of the electrical system (Good, 2008). For its part, the DF mechanism extends the Dr mechanism by involving the user to continuously adjust their electricity consumption to its availability (following market price signals) and not occasionally (e.g., in emergency cases in the electrical grid) as occurs with the Dr mechanism (Corp, 1984). Table 7.1 summarizes the main differences between Dr and DF mechanisms.

Microgrid's participation in demand flexibility

One way to implement the DF mechanism is through MGs connected to the distribution network. Due to their special characteristics (for example, being able to operate connected or disconnected from a main network, taking advantage of local energy resources, and managing the use of electrical energy locally according to a usage policy), MGs facilitate the adaptation of the demand for electrical energy, in the node to which they are connected, to the needs of the system. These needs include its reduction, increase, or change in a specific period (Broberg et al., 2015), either to (i) adapt to the energy availability in the distribution network (energy exchange between different MGs for the use of resources), (ii) maximize the use of locally generated electrical energy, (iii) follow energy price signals at the connection point, (iv) provide complementary services to the distribution network (voltage and frequency support, demand reduction in peak hours, among others), and (v) present a deferrable charging behavior (the energy imported from the network is distributed over a time window) and/or controllable (the energy imported from the network is reduced or increased, as required, without compromising the local operating conditions of the MG) (Elghitani & Zhuang, 2018).

Table 7.1 Main differences between DR and DF mechanisms (Corp, 1984).

	Demand Response	**Demand Flexibility**
Financial model	Market-based model, involving direct bilateral agreements.	Customer-focused retail model, with tailored pricing structures.
Synchronization	Rarely applied: Typically reactive and reserved for critical power grid situations.	Continuous: Actively utilized to minimize expenses year-round.
Strategic business aim	Focused on managing the power grid, specifically the DSO.	Direct consumer involvement, influencing the DSO's operations.
Economic benefit	Payments made as part of incentive schemes.	Reduction in the consumer's bill through direct savings.

Despite this, the ability of MGs to implement DF mechanisms is directly related to the type of local energy sources being harnessed, the variability/volatility of local energy sources, the local pattern of energy use energy, the local contractual environment, the electrical energy distribution network (for example, payment for participation in complementary services), local energy management strategies, and the internal limitations of the MG (sensitive loads, infrastructure distribution location, conditions of the generation units, for example) (Razmara et al., 2017). To guarantee adequate interaction between MGs and the electricity distribution network, the figure of the aggregator agent has been defined in different countries (Sajjad et al., 2016). The aggregator agent is responsible for mediating between the DSO and the local MG operators to anticipate flexibility requirements and thus generate signals so that the MG locally modifies their behavior. It should be noted that the aggregator function can be done in a centralized, decentralized, or distributed manner. Furthermore, in the decentralized case, MG operators communicate directly with the distribution network operator (or receive flexibility requirement signals directly from it) (Tindemans et al., 2015). In addition, to the aggregator scheme used, it is also important to know whether the signals sent by the aggregator to the local MG operator will be incentive-based or rate-based. Depending on the type of signal used, the impacts of MGs' behavior on the energy market and the operation of the distribution system can be evaluated (Moghaddam et al., 2011), and operation strategies/policies can be defined for them.

Transactive energy and microgrid participation

Currently, the set made up of the implemented DF mechanism, the aggregator (or aggregation strategy), and the type of signal sent by the aggregator to the local MG operators is known as transactive energy (TE) (Council, 2015). As can be seen, the TE is strongly linked to how the electricity market is managed, considering that the end user is no longer a passive actor since MGs take advantage of energy resources. In fact, MGs are aimed at self-consumption, energy efficiency and the provision of services to the distribution system, such as dynamically balancing the supply and demand of electricity at the connection point of each of them. Consequently, MG users are expected to receive compensation in exchange for these services, becoming another player in the energy market. Although in the previous paradigm (consumers as passive market agents), the generation units were monitored as well as the load centers to guarantee the supply-demand balance, the paradigm behind the TE facilitates communication between all supply and demand points, creating an interoperability environment in which each point can exchange energy information and agree in real-time (or almost real-time) on the value of energy (Council, 2015). Based on those assessments, supply and demand points can execute energy transactions, maintaining the delicate balance required for a healthy electricity grid, encouraging dynamic demand-side energy activities based on economic incentives, and ensuring that economic signals are in line with the operational objectives to guarantee the reliability of the system (Guerrero et al., 2020).

It is important to note that interacting with the different supply and demand points makes it possible to reduce demand and avoid the need to build new infrastructure and expand the installed generation capacity, especially if the demand points are connection nodes of MGs. Because MGs are operated locally, the TE approach seeks to create incentives to ensure the generation and consumption of electrical energy (mainly in distribution networks) is collaborative, autonomous, and decentralized (Chen & Liu, 2017). According to what was stated in Chen and Liu (2017), this is achieved by considering a hierarchical structure composed of the following hierarchies:

1. Users and entities (which may or may not be MGs and their local operators), including a network operator.
2. Electrical network, including physical and technical limitations.
3. Regulatory and financial environment.
4. Computer and communication requirements.

Different proposed TE schemes containing the four aforementioned elements can be found in the specialized literature. For example, in Hildmann & Saffre, 2011), decision-making is based on energy price signals. In this case, the flexible loads (which can be MGs) negotiate with each other to modify their consumption pattern in case of high electricity price events. On the other hand, in Li et al. (2014), the authors analyze the trade-off between efficiency (understood as the success in flattening the demand curve at the connection point of flexible loads) and the use of the communications network (measured as the frequency of updating of energy prices and exchange of information between end users). From this study, it is concluded that the use of more elaborate negotiations between end users to define energy prices can lead to an overload of the communications network, a factor that significantly compromises its use in real cases.

Acknowledgment

This work was supported by SERC Chile FONDAP 1523A0006, CORFO technology development program 17PTECES-75830 Atacama Module Systems Technology, ATAMOSTEC, and by the Program "Estrategia de Transformación del Sector Energético Colombiano en el Horizonte de 2030" funded by Colciencias through the Grant 778 Ecosistema Científico under Contract FP44842−210-2018.

References

Anvari-Moghaddam, A., Abdi, H., Mohammadi-Ivatloo, B., & Hatziargyriou, N. (2021). *Microgrids: Advances in operation, control, and protection*. Springer International Publishing.

Broberg, R., Brännlund, A., Kazakauskas, L., & Persson, M. (2015).Vesterberg, an electricity market in transition—demand flexibility and preference heterogeneity.

Bullich-Massagué, E., Díaz-González, F., Aragüés-Peñalba, M., Girbau-Llistuella, F., Olivella-Rosell, P., & Sumper, A. (2018). Microgrid clustering architectures. *Applied Energy, 212*. Available from https://doi.org/10.1016/j.apenergy.2017.12.048, http://www.elsevier.com/inca/publications/store/4/0/5/8/9/1/index.htt.

Chen, S., & Liu, C. C. (2017). From demand response to transactive energy: State of the art. *Journal of Modern Power Systems and Clean Energy, 5*(1). Available from https://doi.org/10.1007/s40565-016-0256-x, http://www.springer.com/40565.

Council, T.G.A. (2015). GridWise transactive energy framework Version 1.0, GridWise. https://doi.org/10.1109/PEDG.2018.84478152015.

Corp, R. (1984). The customer behind the demand. *Electric Power Research Institute Journal*. 9(10). [Online]. Available: https://eprijournal.com/wp-content/uploads/2016/01/1984-Journal-No.-10.pdf.

Du, Y., Wang, Z., Liu, G., Chen, X., Yuan, H., Wei, Y., & Li, F. (2018). A cooperative game approach for coordinating multi-microgrid operation within distribution systems. *Applied Energy, 222*. Available from https://doi.org/10.1016/j.apenergy.2018.03.086, http://www.elsevier.com/inca/publications/store/4/0/5/8/9/1/index.htt.

Elghitani, F., & Zhuang, W. (2018). Aggregating a large number of residential appliances for demand response applications. *IEEE Transactions on Smart Grid, 9*(5). Available from https://doi.org/10.1109/TSG.2017.2679702.

Espina, E., Llanos, J., Burgos-Mellado, C., Cárdenas-Dobson, R., Martínez-Gómez, M., & Sáez, D. (2020). Distributed control strategies for microgrids: An overview. *IEEE Access, 8*. Available from https://doi.org/10.1109/ACCESS.2020.3032378, http://ieeexplore.ieee.org/xpl/RecentIssue.jsp?punumber = 6287639.

Espín-Sarzosa, D., Palma-Behnke, R., & Núñez-Mata, O. (2020). Energy management systems for microgrids: Main existing trends in centralized control architectures. *Energies, 13*(3). Available from https://doi.org/10.3390/en13030547, https://www.mdpi.com/1996-1073/13/3/547.

Gellings, C. W. (1985). The concept of demand-side management for electric utilities. *Proceedings of the IEEE., 73*(10). Available from https://doi.org/10.1109/PROC.1985.13318.

Good, D. E. (2008). Combined antithrombotic therapy. *CMAJ: Canadian Medical Association Journal, 178*(3), 327. Available from https://doi.org/10.1503/cmaj.1070122, http://www.cmaj.ca/.

Guerrero, J., Gebbran, D., Mhanna, S., Chapman, A. C., & Verbič, G. (2020). Towards a transactive energy system for integration of distributed energy resources: Home energy management, distributed optimal power flow, and peer-to-peer energy trading. *Renewable and Sustainable Energy Reviews, 132*. Available from https://doi.org/

10.1016/j.rser.2020.110000, https://www.journals.elsevier.com/renewable-and-sustainable-energy-reviews.

Han, Y., Zhang, K., Li, H., Coelho, E. A. A., & Guerrero, J. M. (2018). MAS-based distributed coordinated control and optimization in microgrid and microgrid clusters: A comprehensive overview. *IEEE Transactions on Power Electronics*, *33*(8). Available from https://doi.org/10.1109/TPEL.2017.2761438, http://ieeexplore.ieee.org/xpl/tocresult.jsp?isnumber = 4712525.

Hatziargyriou, N. (2014). *Microgrids: Architectures and control* (2014). John Wiley and Sons.

Islam, M. (2021). Control and optimization of grid-connected microgrids for tie-line smoothing.

Hildmann, H., & Saffre, F. (2011). Influence of variable supply and load flexibility on Demand-Side Management. *2011*. https://doi.org/10.1109/EEM.2011.5952980.

Jabir, H. J., Teh, J., Ishak, D., & Abunima, H. (2018). Impacts of demand-side management on electrical power systems: A review. *Energies*, *11*(5). Available from https://doi.org/10.3390/en11051050, http://www.mdpi.com/1996-1073/11/5/1050/pdf.

Khooban, M. H., Niknam, T., Shasadeghi, M., Dragicevic, T., & Blaabjerg, F. (2018). Load frequency control in microgrids based on a stochastic noninteger controller. *IEEE Transactions on Sustainable Energy*, *9*(2). Available from https://doi.org/10.1109/TSTE.2017.2763607.

Kou, P., Liang, D., & Gao, L. (2017). Distributed EMPC of multiple microgrids for coordinated stochastic energy management. *Applied Energy*, *185*. Available from https://doi.org/10.1016/j.apenergy.2016.09.092, http://www.elsevier.com/inca/publications/store/4/0/5/8/9/1/index.htt.

Lampropoulos, I., Kling, W.L., Ribeiro, P.F., & Van Den Berg, J. (2013). Netherlands history of demand side management and classification of demand response control schemes. *IEEE Power and Energy Society General Meeting* 10.1109/PESMG.2013.6672715 19449933.

Li, C., Cao, C., Cao, Y., Kuang, Y., Zeng, L., & Fang, B. (2014). A review of islanding detection methods for microgrid. *Renewable and Sustainable Energy Reviews*, *35*. Available from https://doi.org/10.1016/j.rser.2014.04.026.

Razmara, M., Bharati, G. R., Hanover, D., Shahbakhti, M., Paudyal, S., & Robinett, R. D. (2017). Building-to-grid predictive power flow control for demand response and demand flexibility programs. *Applied Energy*, *203*. Available from https://doi.org/10.1016/j.apenergy.2017.06.040.

Microgrids—large scale integration of micro-generation to low voltage grids. (2002).

Moghaddam, M. P., Abdollahi, A., & Rashidinejad, M. (2011). Flexible demand response programs modeling in competitive electricity markets. *Applied Energy*, *88*(9). Available from https://doi.org/10.1016/j.apenergy.2011.02.039, http://www.elsevier.com/inca/publications/store/4/0/5/8/9/1/index.htt.

Olivares, D. E., Canizares, C. A., & Kazerani, M. (2014). A centralized energy management system for isolated microgrids. *IEEE Transactions on Smart Grid*, *5*(4). Available from https://doi.org/10.1109/TSG.2013.2294187.

Olivares, D. E., Mehrizi-Sani, A., Etemadi, A. H., Cañizares, C. A., Iravani, R., Kazerani, M., Hajimiragha, A. H., Gomis-Bellmunt, O., Saeedifard, M., Palma-Behnke, R., Jiménez-Estévez, G. A., & Hatziargyriou, N. D. (2014). Trends in microgrid control. *IEEE Transactions on Smart Grid*, *5*(4). Available from https://doi.org/10.1109/TSG.2013.2295514.

Palma-Behnke, R., Benavides, C., Lanas, F., Severino, B., Reyes, L., Llanos, J., & Saez, D. (2013). A microgrid energy management system based on the rolling horizon strategy. *IEEE Transactions on Smart Grid*, *4*(2). Available from https://doi.org/10.1109/TSG.2012.2231440.

Saha, D., Bazmohammadi, N., Vasquez, J. C., & Guerrero, J. M. (2023). Multiple microgrids: A review of architectures and operation and control strategies. *Energies*, *16*(2). Available from https://doi.org/10.3390/en16020600, http://www.mdpi.com/journal/energies/.

Sajjad, I. A., Chicco, G., & Napoli, R. (2016). Definitions of demand flexibility for aggregate residential loads. *IEEE Transactions on Smart Grid*, *7*(6). Available from https://doi.org/10.1109/TSG.2016.2522961.

Sakurama, K., & Miura, M. (2017). Communication-based decentralized demand response for smart microgrids. *IEEE Transactions on Industrial Electronics*, *64*(6). Available from https://doi.org/10.1109/TIE.2016.2631133, http://ieeexplore.ieee.org/xpl/tocresult.jsp?isnumber = 5410131.

Saleh, M., Esa, Y., El Hariri, M., & Mohamed, A. (2019). Impact of information and communication technology limitations on microgrid operation. *Energies*, *12*(15), 2926. Available from https://doi.org/10.3390/en12152926.

Summary report: 2012 doe microgrid workshop. (2012).

Tindemans, S. H., Trovato, V., & Strbac, G. (2015). Decentralized control of thermostatic loads for flexible demand response. *IEEE Transactions on Control Systems Technology*, *23*(5). Available from https://doi.org/10.1109/TCST.2014.2381163.

Yang, L., & Hu, Z. (2021). Coordination of generators and energy storage to smooth power fluctuations for multi-area microgrid clusters: A robust decentralized approach. *IIEEE Access* (9 (). Available from https://doi.org/10.1109/ACCESS.2021.3052043, http://ieeexplore.ieee.org/xpl/RecentIssue.jsp?punumber = 6287639.

Zou, H., Mao, S., Wang, Y., Zhang, F., Chen, X., & Cheng, L. (2019). A survey of energy management in interconnected multi-microgrids. *IEEE Access*, *7*. Available from https://doi.org/10.1109/ACCESS.2019.2920008, http://ieeexplore.ieee.org/xpl/RecentIssue.jsp?punumber = 6287639.

Peer-to-peer and blockchain technologies in power systems

8

Edisson Villa-Ávila[1,2], *Paul Arévalo*[1,2] *and Esteban Albornoz*[2]
[1]Department of Electrical Engineering, University of Jaén, Jaén, Spain, [2]Faculty of Engineering, Department of Electrical, Electronics and Telecommunications Engineering (DEET), Universidad de Cuenca, Balzay Campus, Cuenca, Azuay, Ecuador

Introduction

In recent decades, the primary source of primary energy generation has been nonrenewable energy resources such as coal, natural gas, and oil (Kajaan et al., 2021). However, over time, these nonrenewable energy resources have become more costly and are proving challenging to meet the energy demand of a large population (Shu et al., 2023). Similarly, nonrenewable energy resources are environmentally unfriendly, with the energy generation process resulting in high carbon emissions. Consequently, numerous environmental organizations have advocated for the adoption of renewable energy sources (RES), such as solar, wind, tidal, biomass, and so forth. RES are environmentally friendly and are utilized to generate energy more affordably with lower transmission costs (Solanki et al., 2021). Furthermore, RES are also utilized to contribute to the main grid in order to meet the network load demand. The advancement in RES facilitates peer-to-peer (P2P) distributed energy trading, for example, within households and buildings (Hosseinnezhad et al., 2021). P2P energy trading is also referred to as trading among energy consumers and producers. Peers can exchange energy with each other without the involvement of any traditional energy distributor, such as the electrical grid (Yap et al., 2023). Innovations in the smart grid, such as distributed energy resources (DER) and microgrids, are poised to transform energy generation and consumption in two key aspects (Edmund et al., 2022). Firstly, the addition of an energy producer as a participant in the network assists in contributing energy to main grid storage and provides network decentralization. Secondly, there is a shift in utility for service providers from energy retailers, aimed at providing (leasing) transmission lines to the energy producer (Vasant et al., 2022). The transition from the traditional grid to the smart grid necessitates a reliable energy platform, mathematical modeling, distributed operations, and control algorithms to facilitate stable network functions, energy producer interaction, and intensive-based business models (Vasant et al., 2022). Outsourcing is a widespread practice across numerous industries and has been implemented in various fields such as cyber-physical systems, medicine, and engineering systems (Mololoth et al., 2023). The primary objective of outsourcing is to utilize crowd services and goods to achieve system goals. Outsourcing can also find applicability in the energy sector.

Towards Future Smart Power Systems with High Penetration of Renewables. DOI: https://doi.org/10.1016/B978-0-443-29871-4.00008-7

The energy outsourcing system proves beneficial in various applications such as battery charging and discharging, electric vehicle charging, and meeting real-time energy demand through RE resources like solar panels, wind, and other DER (Debe et al., 2021). These scenarios can be automated through the utilization of smart plugs, inverters, and digital meters with the involvement of an energy distributor and blockchain implementation (Tkachuk et al., 2023; Yap et al., 2023). The transformation in the continuous energy system, where energy producers are utilized for outsourcing, is accelerated in two ways:

- An outsourcing-based energy system that enhances network stability through real-time network management.
- A secure computing framework that supports millions of energy trading transactions (ETTs), such as from producer to consumer or from producer to utility.

Most existing studies rely on optimal power flow, which is used to estimate the optimal operating level for a utility to meet load demand while minimizing operational costs (Shu et al., 2023). Additionally, blockchain technology is also utilized by many researchers to provide data security in the energy system (Iskakova et al., 2020). However, both the energy system and the computing framework have several deficiencies. Firstly, the existing computing framework is not scalable enough to support millions of ETTs. Secondly, there is ambiguity regarding how trading between peers takes place. Lastly, there is uncertainty about how energy producers and subcontractors can adopt controllable loads and DER (Saxena et al., 2019).

Literature review

The scientific community has developed numerous optimization techniques specific to microgrids and their energy trading processes (Bukar et al., 2023; Debe et al., 2021; Edmund et al., 2022). Optimization, in this context, entails the selection of the most suitable option from among various alternatives (Alskaif et al., 2022; Mololoth et al., 2023; Yap et al., 2023). Furthermore, energy optimization is utilized in other fields of science, such as sustainable smart solutions. An optimization model is presented to enhance energy trading between two microgrids operating in isolated modes using a central controller (Das et al., 2023). The model meets the energy demand for a microgrid and reduces energy production costs. In Tkachuk et al. (2023), the authors introduced an incentive-based RE exchange technique to meet load demands with surplus energy, whereby energy is exchanged among multiple users simultaneously. An optimal electricity price for energy trading is derived using coalition game theory (Karumba et al., 2022). The model yields balanced revenues for small-scale energy producers and consumers. A cooperative mechanism for distributed energy generation and trading is introduced to enable multiple energy producers to exchange energy profitably (Edmund et al., 2022). Furthermore, various machine learning and blockchain techniques can be beneficial in addressing prediction and optimization challenges for energy trading in the smart grid (Okwuibe et al., 2020; Singh, et al., 2021). In Saxena et al.

(2021), a hybrid energy management system based on machine learning, fuzzy logic, and multiobjective optimization using linear programming is presented. The primary goal of this system is to minimize the operational cost of RE while maximizing energy generation. A vehicle-to-grid energy trading system based on edge computing and blockchain is presented (Debe et al., 2021). The efficient and secure trading system is a two-stage model composed of the Stackelberg leader-follower game and the reverse induction approach. The model is evaluated through numerical and theoretical analyses. Additionally, a blockchain-enabled energy trading system based on the Industrial Internet of Things known as FeneChain is introduced (Hosseinnezhad et al., 2021; Karumba et al., 2022). FeneChain is a secure energy trading platform based on Industry 4.0, which enhances and manages energy management in buildings. The following table summarizes the pros and cons of the state-of-the-art in energy trading in microgrids (Wittelsbürger, 2007).

Peer-to-peer and blockchain

Smart grids enable seamless control of energy generation and distribution, facilitating the entry of new players into electricity markets. Environmental concerns and the decreasing cost of REs are driving the integration of DER based on RE (Edmund et al., 2022). Advancements in sensor design and new communication protocols are paving the way for P2P energy trading mechanisms (Debe et al., 2021). These mechanisms allow electricity consumers with installed DER to trade their excess generated energy in the market and act as prosumers. The volatile and intermittent nature of RE-based DER generation requires grid connection for energy consumers to maintain reliability. Appropriate market mechanisms need to be designed to safely operate P2P energy trading in the local electricity market (Vasant et al., 2022).

Blockchain technology is presented as an efficient solution for implementing decentralized, reliable, and secure P2P energy trading. This technology is user-friendly and promising, enabling transparent communication among participants in the local market and empowering them to make decisions without the interference of a central authority. The key feature of blockchain is its ability to monitor all changes without altering any block, thus ensuring seamless transactions for cash, agreements, and properties. Additionally, information once stored in the blockchain cannot be altered (Bukar et al., 2023). The applications of blockchain are widely explored in areas such as load balancing, network operations, and various bidding strategies. However, further exploration is needed for implementing smart contracts using a blockchain mechanism in decentralized P2P markets (Das et al., 2023; Jamil et al., 2021; Mololoth et al., 2023).

Peer-to-peer

P2P energy markets can operate in a centralized manner (Solanki et al., 2021). However, in a centralized trading system, the security and privacy of consumer data are always a concern. Moreover, energy transactions among different peers are controlled by a central entity responsible for market management. This is reflected as an additional cost for peers due to the involvement of an extra entity, resulting in higher utility costs for consumers and fewer benefits for prosumers. This necessitates the transition from centralized systems to decentralized mechanisms. Decentralized P2P trading offers direct interaction between peers and incentivizes small-scale energy prosumers to trade their surplus generation. This allows participants in local energy markets to benefit more compared to trading via a centralized mechanism (Bukar et al., 2023). Therefore there is a need to design a mechanism for the secure operation of decentralized P2P trading in the local electricity market (see Fig. 8.1).

Blockchain

Smart contracts are an essential component of blockchain. They contain predefined rules and conditions to control the exchange of energy between market participants. The rules of the smart contract remain unaltered and ensure direct energy transactions in the local energy market (Hosseinnezhad et al., 2021). They ensure that the

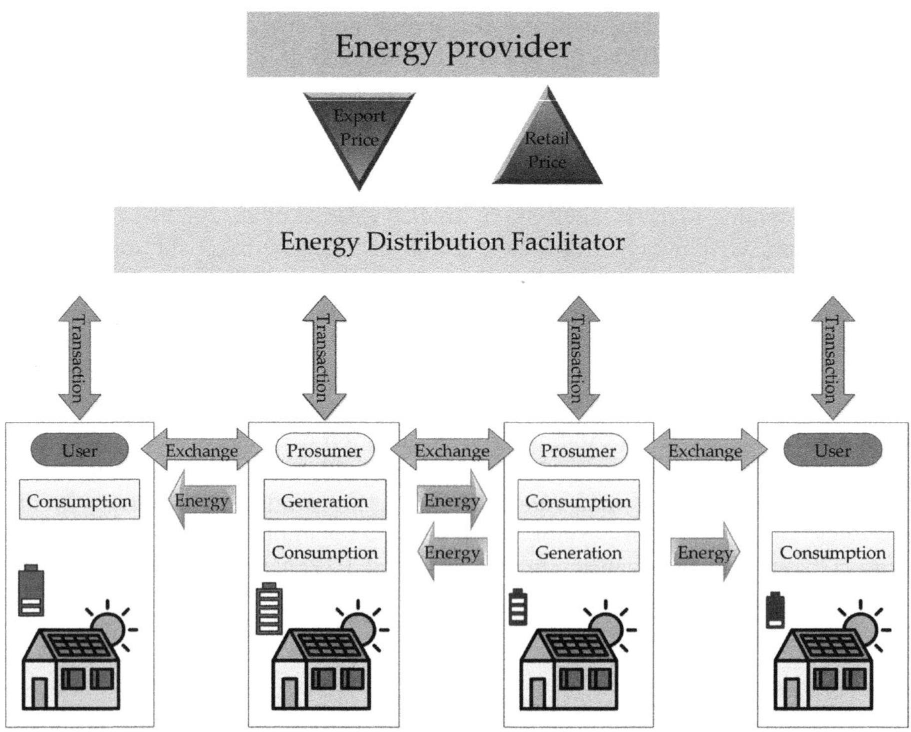

Figure 8.1 Peer-to-peer electricity market framework.

energy trading market is transparent for all participants. Smart contracts for P2P and P2G interaction are extensively explored in the literature. However, it is necessary to consider the electrical distance for the actual realization of network conditions. Therefore it is necessary to take into account the electrical distance to design an efficient platform for the decentralized P2P trading mechanism based on blockchain (Zaman & He, 2021).

Fig. 8.2 depicts the model of a smart P2P energy trading platform based on blockchain (Jamil et al., 2021). The smart energy trading enabled by smart contracts consists of two distinct modules, namely, real-time and forecasted energy trading based on preprocessed data and short-term energy prediction. Each node of the proposed platform is used to store and process energy trading data. In this study, we have considered solar energy generation (photovoltaic), dispatchable load (energy storage system), and flexible load (electric vehicle). The energy consumption data from these sources are analyzed using various machine learning approaches and data mining techniques to uncover useful temporal patterns and hidden knowledge in the data to meet future energy demand (Tkachuk et al., 2023). Preprocessed data are utilized for real-time and forecasted control and scheduling of DERs. Additionally, discovered temporal series features, such as hourly, daily, weekly, yearly, and seasonally, are used to predict short-term energy demand using machine learning models. Each transaction between nodes acting as prosumers and consumers is stored in the state database in the form of an energy trading transaction. Participants such as prosumers, consumers, and utility operators can interact with the system through the client application, which is used for secure energy trading (Karisma & Tehrani, 2022).

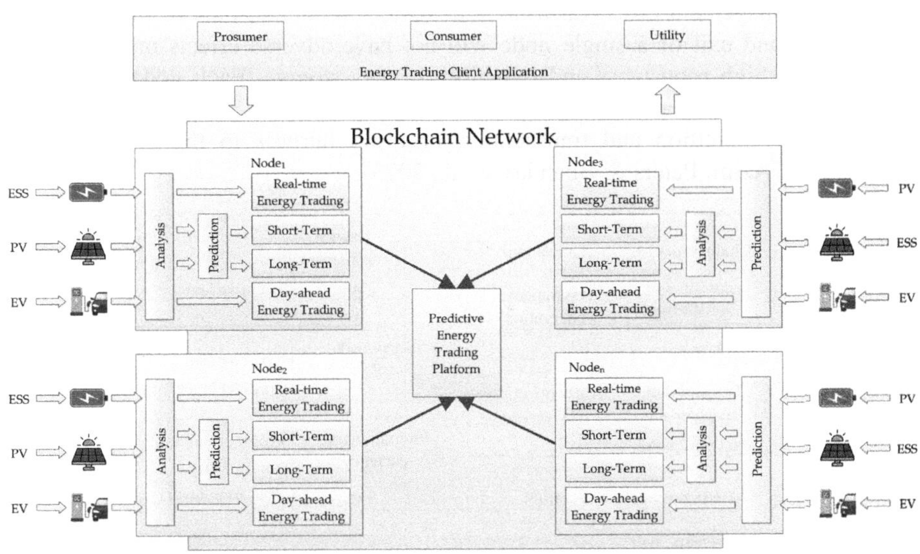

Figure 8.2 Blockchain network.

Blocks and blockchain

Blockchain technology is a system of decentralized and distributed ledgers. The structure of the blockchain system is divided into six components: the data layer, network layer, consensus layer, incentive layer, contract layer, and application layer (Shukla et al., 2020), as illustrated in Fig. 8.3.

1. The data layer is the bottom layer, which encapsulates basic data blocks and related data encryption technology. Each data block typically includes the block header and the block body (Edmund et al., 2022). The block header encapsulates the hash value of the previous block and connects to it. The block body contains multiple pairs of transaction information from the current block. After being encoded into a hash value, it is combined into a Merkle tree through certain connection methods; the binary Merkle tree is commonly used in blockchain and also serves as the unique identification of transactions in the block body.
2. The network layer mainly includes the system's connection mode (such as the distributed P2P network) (Pradhan & Singh, 2021), the data transmission protocol, and the verification mechanism. The P2P network is used to achieve decentralization in ethereum.
3. The consensus layer primarily consists of various consensus mechanisms to prove the computational power and workload of nodes. Currently, proof of work (POW) and proof of stake (POS) are widely used (Küster et al., 2023).
4. The incentive layer alleviates the computational power bottleneck of the blockchain through a certain incentive mechanism.
5. The contract layer is the central part of blockchain programmability. It includes the blockchain transaction protocol and the operation of block data.
6. The application layer is the occasion for blockchain application.

Each node in the blockchain has the following characteristics: Each node in the blockchain operates independently without unified management, which saves computation and management costs. Each node holds all transaction data of the system, so the addition and exit of a single node will not have adverse effects on the system, thus ensuring high robustness and reliability in data storage. Each node can participate in system maintenance (collective maintenance), enabling independent participation in fair competition and transactions, ensuring fairness to the greatest extent possible (Ali, Azim, Peters & Bhandari et al., 2023).

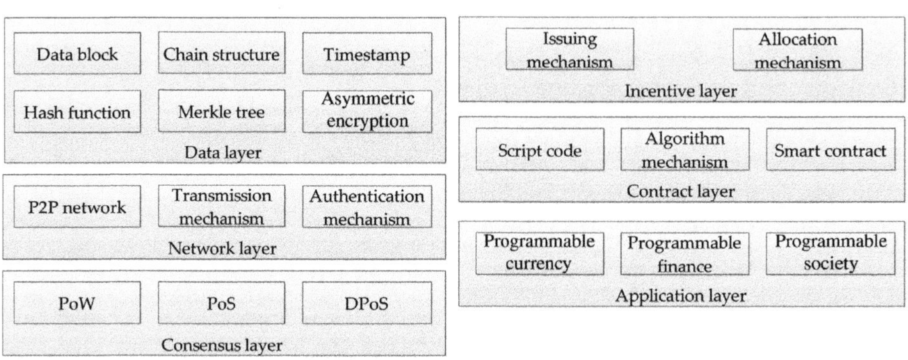

Figure 8.3 Blockchain infrastructure.

Current challenges in energy systems

Blockchain technology is emerging as a promising technology to address the challenges associated with a decentralized P2P system due to its versatile security features (Kajaan et al., 2021). In this process, the first step is to create user data blocks with a unique identity hash. This data block is then sent to smart contracts to enforce predefined rules. After the creation of the data block, the main smart contract validates these data blocks and adds them to the blockchain. Additionally, participants interact with P2P smart contracts until the market settlement time and then with P2G smart contracts. Since consumers and prosumers can interact both with their peers and with the grid, the proposed system evolves into a hybrid energy market (see Fig. 8.4) (Hosseinnezhad et al., 2021).

Shu et al. (2023) considered some assumptions for energy trading, such as:

- Only registered users are allowed to trade after validation.
- A market participant cannot bid as both a buyer and a seller simultaneously.

Centralization and vulnerabilities

In contrast to the fully decentralized P2P market scheme, the centralized P2P market scheme involves a coordinator who oversees the trading activities of the peers (Pradhan & Singh, 2021). This centralized scheme is more organized, with the coordinator managing trading activities within the community and the wider system, as depicted in Fig. 8.5. The main role of the coordinator is to facilitate communication among the entities involved in P2P electricity trading. Initially, the coordinator collects information from the peers and proposes energy export/import transactions among them. Afterward, the coordinator distributes the total P2P revenue and calculates the profits earned by each peer based on predetermined criteria, such as the selling price (Karisma &

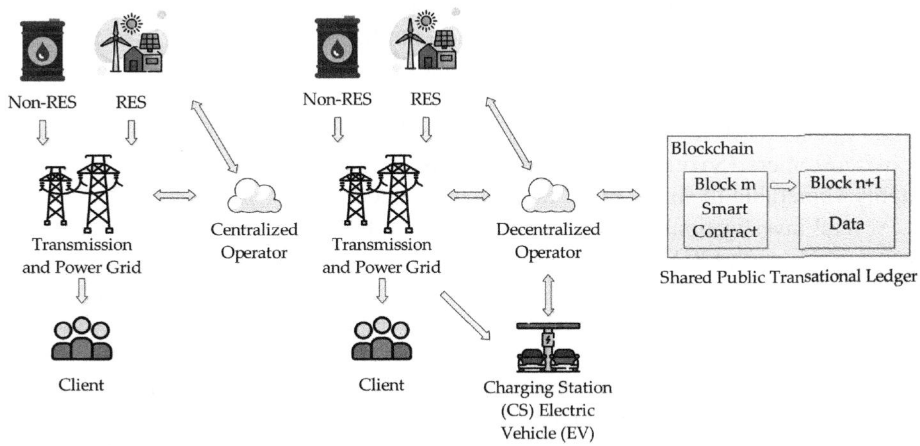

Figure 8.4 The distinctions between a traditional centralized operator and a decentralized operator leveraging blockchain technology.

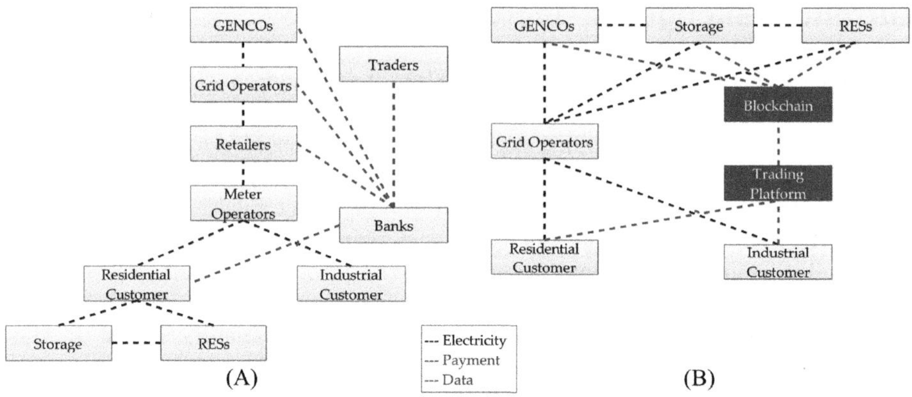

Figure 8.5 Structures of power markets: (A) Conventional wholesale power market and (B) power market utilizing blockchain technology.

Tehrani, 2022). A visual representation of the centralized P2P market scheme is depicted in Fig. 8.5. This market model can be implemented within groups of neighboring prosumers or communities that share common objectives and interests (Hosseinnezhad et al., 2021; Jamil et al., 2021), such as members of a community who are inclined to share RE but are situated in different regions. Centralized P2P markets have the potential to maximize the social welfare of the P2P community. Achieving this goal entails maximizing social welfare as the fitness function to be optimized, when establishing control decisions for the peers (Das et al., 2023). Additionally, in centralized market schemes, the power consumption and generation patterns of a peer are subject to uncertainty in the outcome. This uncertainty arises because, unlike in the complete P2P markets described in (De Diego et al., 2021), the coordinator directly controls the operational state of the peers (Karumba et al., 2020).

The primary drawback of centralized markets is that as the number of DER involved increases, the complexity of centralized management systems also increases (Edmund et al., 2022). Other issues associated with centralized markets include concerns regarding autonomy and privacy. Direct control of peer devices poses threats to privacy, security, and independence, which may be unacceptable to some customers (Saxena et al., 2021). Lastly, centralized P2P market schemes are highly vulnerable to single-point failures on the coordinator side.

Several investigations have explored centralized P2P electricity trading markets. For instance, Ali et al. (2023) introduced a community-oriented electricity market that promotes collaborative efforts among prosumers. Conejo and Baringo (2017) conducted research on the involvement of decentralized and centralized batteries in P2P electricity trading, presenting two centralized market schemes termed "Pool Hub" and "Flexi User." Kalakova et al. (2021) introduced an auction mechanism designed for sharing energy storage within a community. Their proposed system involves agents equipped with storage facilities and agents seeking to utilize the shared energy storage system. Zaman and He (2021) implemented a multiclass energy management system tailored for a community-based market. From the aforementioned studies, it becomes evident that centralized P2P market schemes hold potential for maximizing economic benefits, as global optimization can be conducted within such markets (Küster et al., 2023). Future research on centralized market schemes can progress along two avenues: one option involves enhancing the centralized P2P market scheme by integrating beneficial features like network constraint management and auxiliary services. The other avenue focuses on addressing the inherent shortcomings of centralized electricity trading related to privacy, autonomy, reliability, and scalability.

Applications of peer-to-peer and blockchain in energy systems

Peer-to-peer electricity trading platforms

P2P trading platforms play a pivotal role in implementing P2P market models. These platforms are typically categorized as either decentralized or centralized P2P. The centralized variant resembles eBay, with Elecbay being a typical example implemented in (Bukar et al., 2023). In recent years, considerable endeavors have been directed toward developing decentralized P2P platforms leveraging blockchain technology. These blockchain-based platforms have been proposed across various applications, employing related techniques to ensure privacy, high-level security, and methods aimed at minimizing chain length (Bukar et al., 2023). Furthermore, real-world projects have been undertaken in various industries to evaluate trading platforms integrated with blockchain technology. Despite substantial research and practical applications of blockchain-based decentralized P2P electricity trading, conducting an economic analysis is essential to ascertain the most suitable platform for each specific case (Shu et al., 2023). Additionally, while blockchain technology offers commendable features such as tamper-proof capabilities, avoidance of third-party involvement, transparency, and trustworthiness, it is imperative to scrutinize the drawbacks and costs associated with its implementation.

Peer-to-peer electricity networks

Government-owned or privately owned power lines can facilitate P2P electricity trading (Karumba et al., 2020). Private power line alternatives employing various control strategies have been acknowledged as beneficial in the literature. Nevertheless, they have yet to be implemented extensively. At present, government-owned power lines represent the most feasible option for P2P electricity trading, particularly in developing countries where the existing power transmission and distribution infrastructure is government-owned. In this scenario, the power line operates akin to a power pool, and P2P participants do not engage in direct physical energy trading in a P2P fashion. Certain technical considerations such as power line constraints and line losses have been identified as factors that can incentivize P2P participants to act in a manner conducive to the efficiency of the power lines. Further research could delve into practical aspects of line charges, including time-of-use costs, reinforcement avoidance, sunk cost recovery, and other relevant factors. Looking ahead, advancements in power routing devices and algorithms may eventually facilitate the physical exchange of electricity on a P2P basis.

Case studies implementing peer-to-peer and blockchain technologies

Various research works in the literature are available toward optimal power distribution, congestion management, and fault identification in a microgrid. A distributed control strategy was proposed for frequency control, congestion management, and optimal dispatch by Yap et al. (2023). The approach drives distributed generators in the microgrid toward optimal linear power flow. In Edmund et al. (2022), a greedy smallest cost path first algorithm is proposed, where the work employs integer linear programming for routing through the optimal path. Vasant et al. (2022) proposed a fault protection and localization method for DC microgrids. The proposed scheme is based on intelligent electronic devices to identify fault currents at the bus and isolate the segment to prevent collapse. A flexible services market for active management of distribution networks is proposed in Mololoth et al. (2023). More relevant works for fault identification in microgrids are available in the literature (Debe et al., 2021; Iskakova et al., 2020; Tkachuk et al., 2023). Contributions have been made in the domains of P2P energy trading in cyber-physical systems in the literature (Alskaif et al., 2022; Bukar et al., 2023; Saxena et al., 2019). However, to the best of our knowledge, no relevant work is available in the areas of optimal power dispatch and fault reporting in P2P microgrids using decentralized technologies. To address this gap, we propose a blockchain-based approach for optimal power dispatch and fault notification in P2P microgrids.

Challenges and future considerations

Security and privacy

To address the increasing uncertainty caused by the high proportion of RE, it is necessary to enhance the flexible adjustment capability of resources at every level of the electrical grid. In this context, the virtual power plant (VPP) can aggregate a variety of DERs and improve the efficiency of RE generation utilization, which will become a significant structural form of the electrical system in the future. Academics have achieved numerous advancements associated with VPP research (Huang et al., 2020). Huang et al. (2020) proposed an optimal management solution that provides the capability for the VPP to offer comprehensive services, both for production and demand management, by reducing the required reserve for balance. Another study demonstrated that by adjusting controllable loads and mitigating the impact of wind energy on the electrical system, the benefit of the VPP is improved. Furthermore, an operational strategy was designed for the VPP, constructing a model of optimal economic allocation to maximize the VPP's benefit in each period, and the VPP's energy output is obtained using a single-objective particle swarm optimization algorithm.

However, the existing VPP market trading platform has the following limitations. Firstly, the VPP is mainly composed of distributed energy sources, making it difficult to dispatch and operate with high efficiency. Secondly, there is currently a lack of an information security system for the VPP. Thirdly, centralization in VPP management can breed mistrust between distributed units and the control center (Sivianes et al., 2022). Hence, addressing the management of economic benefits, information privacy, and trust in administration arising from the organic integration of multiple energy sources has become a challenge to be resolved. The implementation of blockchain with features like decentralization, high security, and information transparency can enhance transaction efficiency, reduce programming complexity and operational costs, and improve trust among users stemming from the centralized management of VPP (Veerasamy et al., 2023). Additionally, blockchain and VPP are structurally compatible. VPP comprises various distributed energy sources, energy storage, flexible loads, among others, while the blockchain system consists of multiple distributed nodes (Yang et al., 2021). Therefore blockchain technology is highly suitable for building a VPP trading platform. Many scholars have been drawn to study the application of blockchain in VPP operation and scheduling. Yin et al. (2022) utilized blockchain's consensus mechanism to achieve distributed VPP scheduling and achieved optimal economic dispatch of VPP by combining it with the principle of equal increment. Kaif et al. (2024) proposed the design of a VPP market trading platform and a transaction mechanism based on a P2P network. Each node engages in bidding and transactions, with the platform displaying the final transaction outcomes. Furthermore, each blockchain node can modify data on the blockchain through a consensus mechanism, which aligns well with transactional requirements (Yu et al., 2024). The expected outcomes of the platform are as follows:

- Through the P2P transaction network, producers and consumers are paired to trade directly, which can reduce VPP operating costs and enhance transaction efficiency among users.
- Through direct control or price-driven management of distributed generation and flexible load in the VPP, generation and consumption are adjusted to achieve balance between energy supply and demand.
- We establish entry qualifications, allowing only qualified users to participate in the VPP transaction platform. Users on the consortium blockchain share key transaction information, such as their bidding power and price, which can ensure information transparency and prevent external malicious intrusion.

Limitations of blockchain technology

Furthermore, despite the advantages of blockchain technology, there are concerns regarding its limitations (Yap et al., 2023). One significant drawback is the excessive energy consumption associated with blockchains. Blockchains are known to consume substantial amounts of energy, and this energy consumption largely depends on the design choices made for the specific blockchain (Alskaif et al., 2022). For instance, Bitcoin networks operate on the POW algorithm, which is notorious for its high energy consumption. To address this issue, alternative consensus mechanisms such as POS and Proof of Authority can be employed, offering significantly faster performance and reduced energy consumption (Küster et al., 2023). However, there are concerns regarding processing capacity limitations for the energy industry in the medium and long term. The issue can be addressed by determining the current scope of blockchain applications for the energy sector. While security is one of the benefits of blockchain technology, privacy in private blockchains is a significant concern. Especially in the energy sector, maintaining the privacy of transactions is crucial (Iskakova et al., 2020). To tackle this problem, the use of public blockchain networks can be a promising solution.

Conclusion

This article provides an overview of the fundamental aspects of blockchain technology and its functionality in modern electrical systems, identifying the future of transactive energy systems through the use of blockchain technology. It is evident that blockchain holds significant potential in the energy sector, particularly in P2P virtual energy trading, electric vehicle charging, demand response, and VPPs. However, blockchain technology is still not fully mature and requires further research. Most blockchain projects are related to the industry, and there is limited research in this area. Nevertheless, there have been promising attempts to utilize blockchain technology in the energy sector. Many blockchain applications are still in their pilot stages, and governments are collaborating with technology companies to determine if blockchain can be used to address persistent issues such as traceability, identification, and trust. Furthermore, it is imperative to establish guidelines, laws, and regulations related to blockchain so that everyone can refer to them. It is worth noting that privacy, decentralized society and organization, technical complexity, connectivity and energy dependency, computation time, implementation cost, and ultimately secure integration are the long-term consequences of using blockchain technology.

Acknowledgments

Edisson Villa Ávila expresses his sincere gratitude for the opportunity to partially present the findings of his research, conducted as part of his doctoral studies in the Ph.D. program in Advances in Engineering of Sustainable Materials and Energies at the University of Jaen, Spain. Finally, the authors thank Universidad de Cuenca, Ecuador, for easing access to the facilities of the Micro-Grid Laboratory of the Centro Científico Tecnológico y de Investigación Balzay (CCTI-B), for allowing the use of its equipment, and for authorizing members of its staff the provision of technical support necessary to carry out the experiments described in this article.

The author (Paul Arévalo) thanks the Call for Grants for the Requalification of the Spanish University System for 2021–2023, Margarita Salas Grants for the training of young doctors awarded by the Ministry of Universities and financed by the European Union –Next Generation EU.

References

Ali, L., Azim, M. I., Peters, J., Bhandari, V., Menon, A., Tiwari, V., Green, J., & Muyeen, S. M. (2023). Application of a community battery-integrated microgrid in a blockchain-based local energy market accommodating P2P trading. *IEEE Access*, *11*. Available from https://doi.org/10.1109/access.2023.3260253.

Ali, L., Azim, M. I., Peters, J., Padmanaban, S., Bhandari, V., Menon, A., Green, J., Muyeen, S. M. (2023). Singapore local energy market development using blockchain enabled P2P trading, In *2023 IEEE 14th International Conference on Power Electronics and Drive Systems (PEDS)*, Montreal, QC, Canada, pp. 1–6. https://doi.org/10.1109/PEDS57185.2023.10246687.

Alskaif, T., Crespo-Vazquez, J. L., Sekuloski, M., Van Leeuwen, G., & Catalao, J. P. S. (2022). Blockchain-based fully peer-to-peer energy trading strategies for residential energy systems. *IEEE Transactions on Industrial Informatics*, *18*(1). Available from https://doi.org/10.1109/TII.2021.3077008, http://ieeexplore.ieee.org/xpl/RecentIssue.jsp?punumber = 9424.

Bukar, A. L., Hamza, M. F., Ayub, S., Abobaker, A. K., Modu, B., Mohseni, S., Brent, A. C., Ogbonnaya, C., Mustapha, K., & Idakwo, H. O. (2023). Peer-to-peer electricity trading: A systematic review on current developments and perspectives. *Renewable Energy Focus*, *44*. Available from https://doi.org/10.1016/j.ref.2023.01.008, https://www.sciencedirect.com/journal/renewable-energy-focus.

Conejo, A. J., & Baringo, L. (2017). *Power systems*. Springer Nature. Available from 10.1007/978-3-319-69407-8_1.

Das, A., Peu, S. D., Akanda, M. A. M., & Islam, A. R. M. T. (2023). Peer-to-peer energy trading pricing mechanisms: towards a comprehensive analysis of energy and network service pricing (NSP) mechanisms to get sustainable enviro-economical energy sector. *Energies*, *16*(5). Available from https://doi.org/10.3390/en16052198, http://www.mdpi.com/journal/energies/.

De Diego, S., Seco, I., Escalante, M., Vakakis, N., Alexandros, S., Drosou, A., Tzovaras, D., Votis, K., Thomas, T., Colet, A., & Paradell, P. (2021). A peer-to-peer verifiable and secure energy trading framework based on blockchain technology. In *2021 Fifth World Conference on Smart Trends in Systems Security and Sustainability* (WorldS4), London, United Kingdom, pp. 241–247. https://doi.org/10.1109/WorldS451998.2021.9514028.

Debe, M., Hasan, H. R., Salah, K., Yaqoob, I., & Jayaraman, R. (2021). Blockchain-based energy trading in electric vehicles using an auctioning and reputation scheme. *IEEE Access*, *9*. Available from https://doi.org/10.1109/ACCESS.2021.3133958, http://ieeexplore.ieee.org/xpl/RecentIssue.jsp?punumber = 6287639.

Edmund, E. C., Omitola, O., & Dada, J. O. (2022). A blockchain-based peer-to-peer energy trading platform for distributed energy resources, *In 2022 IEEE Nigeria 4th International Conference on Disruptive Technologies for Sustainable Development (NIGERCON)*, Lagos, Nigeria, pp. 1–5. https://doi.org/10.1109/NIGERCON54645.2022.9803115.

Hosseinnezhad, V., Hayes, B., O'Regan, B., & Siano, P. (2021). Practical insights to design a blockchain-based energy trading platform. *IEEE Access*, *9*. Available from https://doi.org/10.1109/ACCESS.2021.3127890, http://ieeexplore.ieee.org/xpl/RecentIssue.jsp?punumber = 6287639.

Huang, D., Zhang, C., Li, Q., Han, H., Huang, D., Li, T., & Wang, C. (2020). Consortium Blockchain-based Decentralized Energy Trading Mechanism for Virtual Power Plant, In *2020 IEEE 4th Conference on Energy Internet and Energy System Integration (EI2)*, Wuhan, China, pp. 3084–3089. https://doi.org/10.1109/EI250167.2020.9346653.

Iskakova, A., Kumar Nunna, H. S. V. S., & Siano, P. (2020). Ethereum blockchain-based peer-to-peer energy trading platform. In *2020 IEEE International Conference on Power and Energy (PECon)*. Penang, Malaysia, pp. 327–331. https://doi.org/10.1109/PECon48942.2020.9314591.

Jamil, F., Iqbal, N., Imran, S., Ahmad, D., & Kim. (2021). Peer-to-peer energy trading mechanism based on blockchain and machine learning for sustainable electrical power supply in smart grid. *IEEE Access, 9*. Available from https://doi.org/10.1109/ACCESS.2021.3060457, http://ieeexplore.ieee.org/xpl/RecentIssue.jsp?punumber = 6287639.

Kaif, A. M. A. D., Alam, K. S., & Das, S. K. (2024). Blockchain based sustainable energy transition of a virtual power plant: Conceptual framework design & experimental implementation. *Energy Reports*, *11*. Available from https://doi.org/10.1016/j.egyr.2023.11.061, https://www.sciencedirect.com/science/journal/23524847.

Karumba, S., Kanhere, S. S., Jurdak, R. (2020). A Relational Network Framework for Interoperability in Distributed Energy Trading. In *2020 IEEE International Conference on Blockchain and Cryptocurrency* (ICBC), Toronto, ON, Canada, pp. 1-3, https://doi.org/10.1109/ICBC48266.2020.9169462.

Kajaan, N. A. M., Salam, Z., & Radzi, R. Z. R. M. (2021). Review of market clearing method for blockchain-based P2P energy trading in Microgrid. In *2021 IEEE Conference on Energy Conversion (CENCON)*, Johor Bahru, Malaysia. (pp. 202–207). Available from https://doi.org/10.1109/CENCON51869.2021.9627254.

Kalakova, A., Zhanatbekov, A., Surash, A., Kumar Nunna, H.S. V.S., & Doolla, S. (2021). Blockchain-based decentralized transactive energy auction model with demand response. In 2021 IEEE Texas Power and Energy Conference (TPEC), College Station, TX, USA, pp. 1–6. https://doi.org/10.1109/TPEC51183.2021.9384958.

Karisma, K., & Tehrani, P. M. (2022). Legal challenges facing blockchain-based peer-topeer energy trading. In *2022 International Conference on Advanced Creative Networks and Intelligent Systems*, (pp. 1–7). Bandung, Indonesia: ICACNIS. Available from https://doi.org/10.1109/ICACNIS57039.2022.10055762.

Karumba, S., Kanhere, S.S., Jurdak, R. (2020). A Relational Network Framework for Interoperability in Distributed Energy Trading. In *2020 IEEE International Conference on Blockchain and Cryptocurrency (ICBC)*, Toronto, ON, Canada, pp. 1–3. https://doi.org/10.1109/ICBC48266.2020.9169462.

Küster, K. K., Gebbran, D., Aoki, A. R., Lambert-Torres, G., Navarro-Gevers, D., Impinisi, P. R., & Pinto, C. L. D. S. (2023). Adoption of local peer-to-peer energy markets: Technical and economical perspectives for utilities. *Energies*, *16*(5). Available from https://doi.org/10.3390/en16052364, http://www.mdpi.com/journal/energies/.

Mololoth, V. K., Ahlund, C., & Saguna, S. (2023). An energy trading framework using smart contracts. In *2023 IEEE Green Technologies Conference (GreenTech)*, Denver, CO, USA, pp. 214–218. https://doi.org/10.1109/GreenTech56823.2023.10173814.

Okwuibe, G. C., Zade, M., Tzscheutschler, P., Hamacher, T., & Wagner, U. (2020). A blockchain-based double-sided auction peer-to-peer electricity market framework. In *2020 IEEE Electric Power and Energy Conference (EPEC)*, Edmonton, AB, Canada, pp. 1–8. https://doi.org/10.1109/EPEC48502.2020.9320030.

Pradhan, N. R., & Singh, A. P. (2021). Performance analysis of a blockchain based peer-to-peer energy trading framework. In *2021 IEEE 4th International Conference on Computing, Power and Communication Technologies (GUCON)*, Kuala Lumpur, Malaysia, pp. 1–7. https://doi.org/10.1109/GUCON50781.2021.9573668.

Saxena, S., Farag, H., Brookson, A., Turesson, H., Kim, H. (2019). Design and field implementation of blockchain based renewable energy trading in residential communities.

In *2019 2nd International Conference on Smart Grid and Renewable Energy (SGRE)*, Doha, Qatar, pp. 1–6.https://doi.org/10.1109/SGRE46976.2019.9020672.

Shu, S., Wang, Z., Jiang, T., Zhang, Z., & Chen, Y. (2023). Research on the peer-to-peer market transaction mechanism of the future energy systems. In *2023 3rd Power System and Green Energy Conference (PSGEC)*, Shanghai, China, pp. 375–381. https://doi.org/10.1109/PSGEC58411.2023.10255982.

Shukla, D., Singh, S., Singh, S.P., Thakur, A.K., & Singh, S.P. (2020). Block-chain based energy trading in adn with its probable impact on aggregated load profile and available distribution capability. In *2nd International conference on smart power and internet energy systems*, SPIES 2020 (pp. 486–491). https://doi.org/10.1109/SPIES48661.2020.9242977, http://ieeexplore.ieee.org/xpl/mostRecentIssue.jsp?punumber = 9242535.

Singh, R., Singh, P., & Nandi, S. (2021). A blockchain-based approach for optimal energy dispatch and fault reporting in P2P microgrid. In *2021 IEEE Globecom Workshops (GC Wkshps)*, Madrid, Spain, pp. 1-6, https://doi.org/10.1109/GCWkshps52748.2021.9682176.

Sivianes, M., Zafra-Cabeza, A., & Bordons, C. (2022). Blockchain-based peer to peer energy trading using distributed model predictive control. In *2022 European Control Conference (ECC)*, London, United Kingdom, pp. 1832–1837. https://doi.org/10.23919/ECC55457.2022.9838195.

Solanki, B., Agarwal, A., Meena, R., Mahiya, N., Sharma, D., Kushwaha, P., Mathuria, P., & Bhakar, R. (2021). Blockchain-based decentralized hybrid P2P energy trading. In *2021 9th IEEE International Conference on Power Systems (ICPS)*, Kharagpur, India, pp. 1–5. https://doi.org/10.1109/ICPS52420.2021.9670159.

Tkachuk, R. V., Ilie, D., Robert, R., Kebande, V., & Tutschku, K. (2023). On the performance of consensus mechanisms in privacy-enabled decentralized peer-to-peer renewable energy marketplace. In *2023 26th Conference on Innovation in Clouds, Internet and Networks and Workshops*, (pp. 179–186). Paris, France: ICIN. Available from https://doi.org/10.1109/ICIN56760.2023.10073510.

Vasant, U. S., Shukla, D., Thakur, A. K., & Singh, S. P. (2022). Blockchain based transactive energy market using peer-to-peer(P2P) energy trading. In *2022 IEEE IAS Global Conference on Emerging Technologies (GlobConET)*, Arad, Romania, pp. 947-953. https://doi.org/10.1109/GlobConET53749.2022.9872378.

Veerasamy, V., Singh, S., Murshid, S., Nguyen, H.D., & Gooi, H.B. (2023). Blockchain-based microgrid frequency control and energy trading to incentivize EV aggregator. In *2023 IEEE 3rd international conference on sustainable energy and future electric transportation*, SeFet 2023, https://doi.org/10.1109/SeFeT57834.2023.10245092, http://ieeexplore.ieee.org/xpl/mostRecentIssue.jsp?punumber = 10244770.

Wittelsbürger, H. (2007). Users/hecto_000/AppData/Local/Mendeley Ltd./Mendeley Desktop/Downloaded/Wittelsb?rger—2007—La pol?tica energ?tica de Chile de la dependencia al desarrollo sostenible—el futuro es de las energ?as renovab.pdf. 4 2007.

Yang, Q., Wang, H., Wang, T., Zhang, S., Wu, X., & Wang, H. (2021). Blockchain-based decentralized energy management platform for residential distributed energy resources in a virtual power plant. *Applied Energy*, *294*, 117026. Available from https://doi.org/10.1016/j.apenergy.2021.117026.

Yap, K. Y., Chin, H. H., & Klemeš, J. J. (2023). Blockchain technology for distributed generation: A review of current development, challenges and future prospect. *Renewable and Sustainable Energy Reviews*, *175*. Available from https://doi.org/10.1016/j.rser.2023.113170, https://www.journals.elsevier.com/renewable-and-sustainable-energy-reviews.

Yin, S., Ai, Q., Li, J., Li, D., & Guo, Q. (2022). Trading mode design for a virtual power plant based on main-side consortium blockchains. *Applied Energy*, *325*. Available from https://doi.org/10.1016/j.apenergy.2022.119932, https://www.journals.elsevier.com/applied-energy.

Yu, S., Wei, Z., Sun, G., Zhou, Y., & Zang, H. (2024). A double auction mechanism for virtual power plants based on blockchain sharding consensus and privacy preservation. *Journal of Cleaner Production*, *436*, 140285. Available from https://doi.org/10.1016/j.jclepro.2023.140285.

Zaman, I., & He, M. (2021). A multilayered semi-permissioned blockchain based platform for peer to peer energy trading. In *IEEE green technologies conference*, 2021 (pp. 279–285). https://doi.org/10.1109/GreenTech48523.2021.00052, http://ieeexplore.ieee.org.

Storage economy and markets

9

Edisson Villa-Ávila[1,2], Paul Arévalo[1,2], Danny Ochoa-Correa[1], Michael Villa-Ávila[1], Juan Leonardo Espinoza[1] and Esteban Albornoz[1]
[1]Faculty of Engineering, Department of Electrical, Electronics and Telecommunications Engineering (DEET), Universidad de Cuenca, Balzay Campus, Cuenca, Azuay, Ecuador, [2]Department of Electrical Engineering, University of Jaén, Jaén, Spain

Introduction

In recent years, the increasing penetration of renewable electricity has been one of the reasons for the very low electricity prices in electricity markets (Abdon et al., 2017). The growing production of renewable energy has been accompanied by challenging situations for grid operators in terms of transmission and distribution loads, as well as monetary losses for conventional power generation plants due to production curtailment (Wang et al., 2023). This trend is expected to continue in the future as renewable electricity generation continues to grow (Xie et al., 2023). A possible remedy for the temporal mismatch between supply and demand is electricity storage (ES) (Lechón et al., 2023), which has led to increased research efforts in storage technologies over the past few years, as shown by several comprehensive technology reviews (Lechón et al., 2023). Since most new renewable energy sources are distributed and their success is at least partly due to the pursuit of higher levels of self-sufficiency (Newbery, 2023), interest in local storage solutions for districts, communities, and even individual households is growing (Chu et al., 2024). This contrasts with the existence of a large base of conventional large-scale pumped hydroelectric plants and has given rise to a growing market for stationary batteries and the development of new technologies, such as small-scale compressed air energy storage systems (Maghami & Mutambara, 2023).

The levelized cost of stored electricity (LCOES) is a fundamental economic parameter that evaluates both investment and operational costs over the lifespan of ES (Manandhar et al., 2023). It is essential to consider that the variability in the number of operational cycles can significantly influence the economic performance of storage (Manandhar et al., 2023). Additionally, each storage application is linked to a specific charge and discharge pattern, affecting its efficiency in different operational contexts. Regarding the environmental aspect, the use of electricity storage can contribute to reducing dependence on nonrenewable energy sources and mitigating associated carbon emissions (Cisterna et al., 2020). However, it is crucial to recognize that the implementation of these technologies may entail adverse environmental effects, depending on the context and the type of energy they replace. Cisterna et al. (2020) examined how

Towards Future Smart Power Systems with High Penetration of Renewables. DOI: https://doi.org/10.1016/B978-0-443-29871-4.00009-9

large-scale storage technologies affect carbon dioxide emissions in different renewable energy penetration scenarios. Abdon et al. (2017) evaluated the technoeconomic and environmental performance of stationary electricity storage, comparing technologies such as pumped hydro storage (PHS), isothermal and advanced adiabatic compressed air ES (AA-CAES), stationary lithium-ion batteries, and the power-to-gas-to-power (P2G2P) process. The rapid growth of intermittent renewable electricity has generated interest in the economic potential of ES. Kloess and Zach (2014) analyzed the economic performance of storage technologies for load leveling operation in the German and Austrian electricity markets. Kondziella et al. (2023) evaluated four technologies: PHS, AA-CAES, hydrogen storage (H_2), and methane storage (CH_4). There is potential seasonal use for H_2 and CH_4 if costs decrease. The results indicate low incentives for investment in storage, which could hinder the integration of renewable electricity.

The European Union is currently taking an effort to increase the share of renewable electricity (RES-E) in order to reduce carbon emissions in the energy sectors. The expected shares of fluctuating RES-E production will pose a major challenge for the electrical system, requiring greater flexibility from backup power plants to cover the residual load and higher transmission capacities to balance supply and demand over wider areas (McLaughlin et al., 2023). ES is considered a key component in addressing these challenges. García-Miguel et al. (2024) and Maghami et al. (2024) demonstrated the technical and economic benefits of using storage in isolated electrical systems. García-Miguel et al. (2024) showed the benefits of combining wind farms with ES. Palma et al. (2024) indicated that ES increases overall social welfare in the electricity market, and Maghami et al. (2024) found that ES leads to benefits in an electrical system with high wind penetration.

Cóndor (2021) and Gao et al. (2023) showed how electricity storage plants can help reduce transmission costs in an electrical system with wind generation. Benito & Arena (2020) found that storage plants are more viable than thermal power plants as a flexibility option up to a certain installed capacity. This aligns with the results found by Ölmez et al. (2024) and Zhu et al. (2024), who analyzed the optimal economic storage capacity for the German electrical system. Although the economic benefits of ES seem evident at the system level, its economic performance from an investor's perspective remains uncertain. Assareh and Ghafouri (2023) showed that the profitability of CAES remains negative despite the economic benefits it brings to the electrical system. Cetegen et al. (2024) and Saleh-Abadi et al. (2023) analyzed the current and future economic potential of CAES in the European market. Martinez and Iglesias (2024) conducted an economic assessment based on models of ES plants in different European countries according to historical price data. Altea and Yanagihara (2024) and Hadelu et al. (2024) showed that the injection of intermittent renewable energy has reduced the revenues of PHS plants in the German and Austrian electricity markets. In the economic evaluation, Le et al. (2024) analyzed these technologies in the German electricity market, noted for its rapid growth in intermittent RES-E. Hurta et al. (2022) examined the annual revenues of storage technologies in load leveling. Bulut and Özcan (2024) helped to understand the operation of ES and the differences between technologies, including the developing power-to-gas technologies. A focus on the Austrian–German market reveals the effects of increasing RES-E on the operation and revenues of storage.

Fig. 9.1 illustrates the traditional energy market, characterized by the use of conventional sources such as coal, oil, and natural gas for the production, distribution, and consumption of energy. Large utility companies control this market, generating and distributing electricity through a centralized electrical grid (Haji Bashi et al., 2022). Energy production is centralized, with power plants located far from end-users. Electricity is transmitted over long distances via high-voltage transmission lines and distributed to local communities via low-voltage distribution lines. Customers pay a fixed rate based on fuel costs, maintenance, operation, and regulatory requirements, although they often desire more control over their rates or the energy source (Liang et al., 2023). This market has been criticized for its negative environmental impact due to carbon emissions and its inefficiency in transmission and infrastructure construction (Watson et al., 2024).

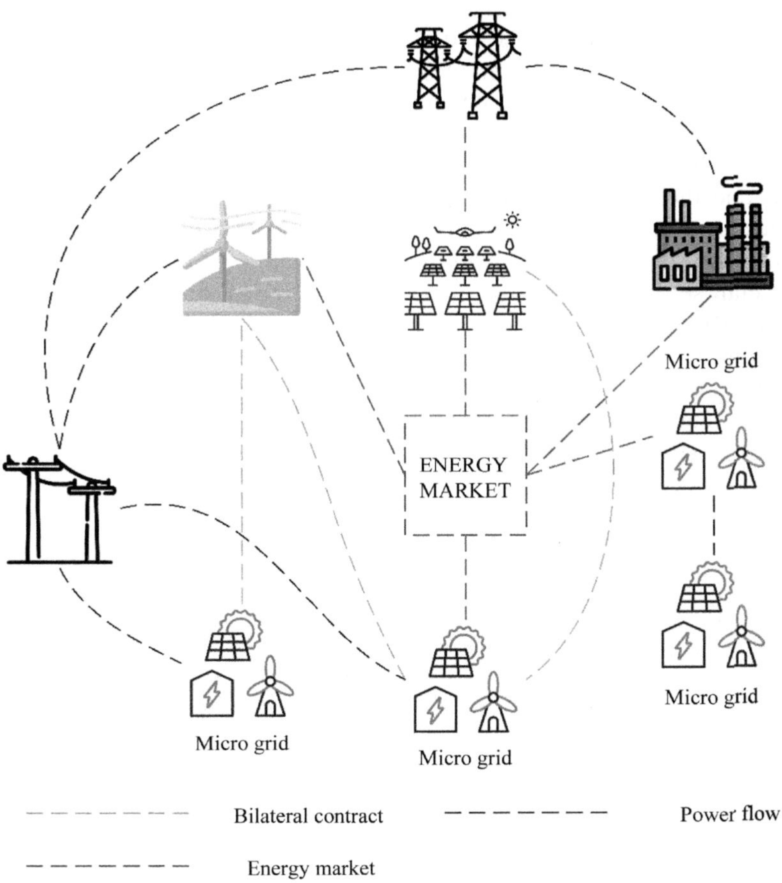

Figure 9.1 Block diagram of the energy market.

Literature review

In recent years, there has been significant interest in research on ES batteries in residential systems (Chreim et al., 2024; Al-Wreikat et al., 2022). This focus is due to their essential role in integrating renewable energies, balancing supply and demand, and allowing for active consumer participation in the electricity market. However, the high capital cost of these batteries remains a significant obstacle to their widespread adoption, even in the form of hybrid photovoltaic battery energy storage system (PV-BESS) installations (Hokmabad et al., 2024). The main research areas focus on optimal system sizing, technoeconomic analysis, and efficient battery management, considering different operational objectives (Cai et al., 2024). Bosu et al. (2023) and Jannesar Niri et al. (2024) addressed these topics from various perspectives. For instance, an overview of ES technologies has been provided, highlighting their importance in managing renewable energy fluctuations and addressing challenges such as frequency control and load management. Other studies have compared the advantages and disadvantages of ES batteries and pumped hydroelectric storage (Haji Bashi et al., 2022), emphasizing their role in improving the grid's capacity to handle intermittent renewable energy generation.

Chatzigeorgiou et al. (2024) examined the feasibility and sizing of PV systems, highlighting the widespread use of simulations for technoeconomic analysis and the growing popularity of PV-BESS systems. Energy management strategies have been proposed to improve the demand profile during peak hours, including the integration of photovoltaic systems with storage batteries in commercial settings (Kebir et al., 2023). In terms of future research, key areas identified include technoeconomic analysis (Lin & Bagnato, 2024), operational control (Yang et al., 2021), system sizing, and demand response (Chatzigeorgiou et al., 2024). The need to improve the economic viability of ES batteries is emphasized, as well as the development of effective demand-side management (DSM) strategies and operations in different application contexts (see Fig. 9.2) (Cha et al., 2023).

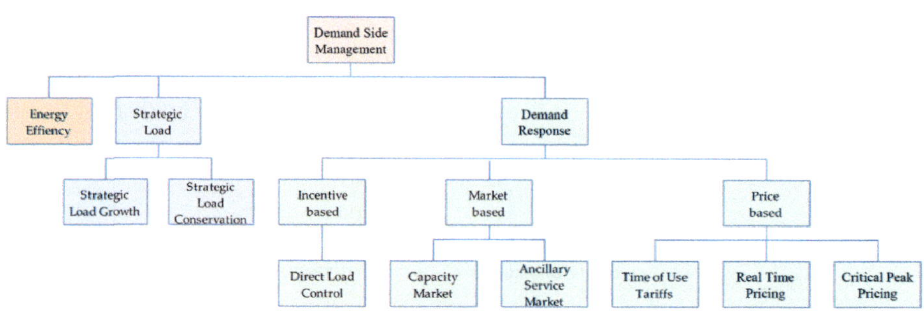

Figure 9.2 DSM techniques based on timing and impact of action. *DSM*, Demand-side management.

Storage technologies

This section presents the four ES technologies analyzed (Cho et al., 2015; Lee et al., 2023), for the integration of intermittent renewable energy into the European electricity market. In addition to PHS, three alternative options are analyzed (Altea & Yanagihara, 2024). Each technology has specific site requirements (Altea & Yanagihara, 2024). A brief description of their operation is provided, and their technical and economic parameters are analyzed, including storage components and the balance of plant (Khan et al., 2024). The storage capacity is decoupled from the discharge power for all technologies, except for stationary lithium-ion batteries and PHS (He et al., 2021).

Combined assessment methodology

Abdon et al. (2017) employed a combined assessment methodology to analyze the economic and environmental performance of electricity storage in various applications and system sizes. Scenarios are defined by considering different applications and sizes, then technologies are chosen, and technical parameters are assigned based on the literature (see Fig. 9.3). The levelized cost (EUR/kWh) and greenhouse gas emissions (CO_2 eq/kWh) are quantified to evaluate the economic and environmental performance of various technologies, including PHS (He et al., 2021), isothermal air energy storage and AA-CAES (Chen, 2023; Javaheri & Shafiei Ghazani, 2023), lithium-ion batteries, and P2G2P (Cruz-Soto et al., 2022;

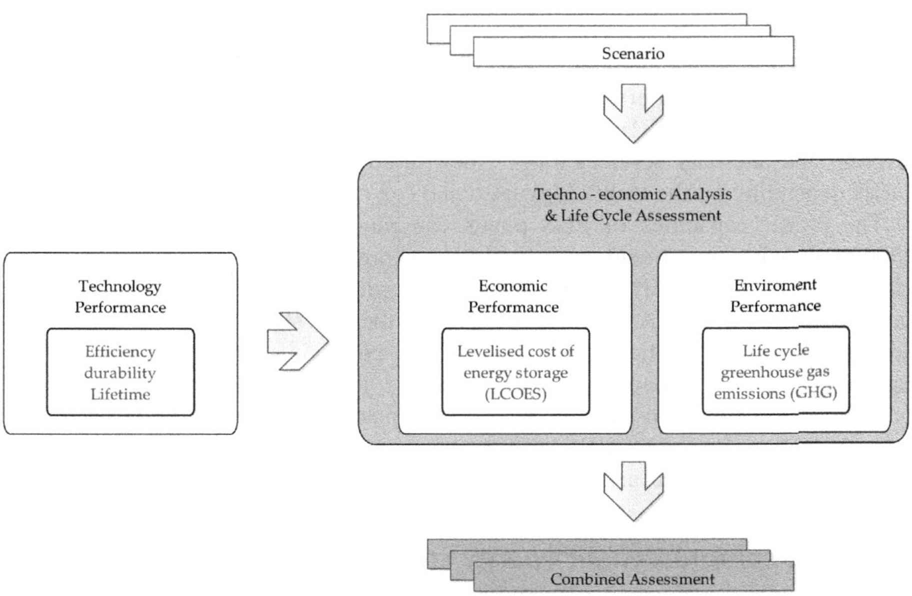

Figure 9.3 Methodology framework for the assessment of electricity storage technologies.

Park et al., 2024). Three generic time scales for stationary ES (short, medium, and long-term) (El-Sayed et al., 2023) are proposed, and two system scales are selected to characterize decentralized and centralized systems (Panda et al., 2023). Jamroen and Vongkoon (2023) and Williams et al. (2023) conducted a sensitivity analysis to understand the variability of the results, considering parameters such as lifespan, efficiency, costs, and prices of stored electricity. The aim is to understand the economic and environmental implications of various combinations of technology, application, and system scale.

Life cycle assessment

Niu et al. (2023a) used a life cycle assessment approach to evaluate the environmental impact of ES technologies. It focuses on climate change, measuring the carbon dioxide equivalent emissions per kWh of electricity supplied from storage. The functional unit is 1 kWh of electricity supplied from storage systems, covering from material production to final disposal. Life cycle inventory data comes from various sources, adjusted to ensure a consistent assessment. Sources of stored electricity include wind turbines, photovoltaic solar panels, and average grid electricity (Altea & Yanagihara, 2024).

Pumped hydro storage

PHS is a mature technology and the most common form of electricity storage, accounting for the majority of installed capacity worldwide. In Switzerland (Mearns & Sornette, 2023), for example, there are PHS plants ranging in size from small installations to large-scale projects under construction. Investment costs for PHS plants can vary significantly depending on local conditions such as topography and geology (Altea & Yanagihara, 2024). In general terms, PHS is widely used in Europe and has proven effective in integrating intermittent energy sources such as wind (Henke et al., 2022). These plants have the capability to provide both load leveling and ancillary services due to their rapid response. The efficiency of PHS varies depending on the technology used and can range between 65% and 87%.

The power capacities of PHS plants can range from a few hundred MW to 1500 MW, with investment costs typically expressed in euros per kW installed, ranging from 500 to 1500 €/kW. Specific investment costs related to ES capacity can range from 5 to 100 €/kWh. To estimate the investment costs of a new PHS project, a two-dimensional approach is considered, taking into account both power capacity and ES capacity. However, it is important to note that finding suitable sites for the construction of new PHS plants can be challenging, especially in regions like Germany and Austria, where suitable sites are often located in remote and mountainous areas (Henke et al., 2022).

Adiabatic compressed air energy storage

CAES plants harness the potential energy of compressed air to store electricity (Connolly et al., 2016). This process involves compressing air at high pressure and storing it in either pressurized containers or suitable geological formations such as salt caverns. When the compressed air is released, it powers turbines that produce electricity. There are two main types of CAES plants: diabatic and adiabatic (AA-CAES) (Kloess & Zach, 2014). Diabatic CAES plants are limited globally, while AA-CAES plants are in a more conceptual phase. Large-scale CAES plants rely on the availability of suitable geological formations for storing compressed air (Bennett et al., 2021).

Bennett et al. (2021) primarily focused on AA-CAES due to their higher efficiency and lack of need for fuels during discharge, unlike diabatic CAES, which require natural gas during this process. Although AA-CAES are still in the planning and demonstration phase, investment costs are estimated based on data found in CAES and thermal storage. Investment costs for CAES reported in the literature vary depending on storage capacity and power, ranging from 5 to 100 €/kWh and from 400 to 800 €/kW, respectively (Comodi et al., 2017). For estimating AA-CAES costs, both storage capacity-related and power-related components are considered. Capacity-related costs are set at 600 €/kW, in line with CAES costs found in the literature. Additionally, capacity-dependent storage costs are set at 70 €/kWh, reflecting the additional thermal storage requirements of AA-CAES compared to conventional CAES (Shabani et al., 2020).

Hydrogen storage

In hydrogen storage systems, electricity is converted into hydrogen through the process of electrolysis. This hydrogen can be stored in pressure vessels or in suitable geological formations such as salt caverns or rock, especially for large-scale central plants. To recover stored energy, hydrogen can be burned in thermal power plants or converted back into electricity through a fuel cell (Lin & Bagnato, 2024). In the context of the electric system, this would involve its use in single-cycle gas turbines (GT) or combined-cycle GT (CCGT), resulting in a storage cycle efficiency of 33%–40%, depending on component efficiencies reported by Li et al. (2024) and Martsinchyk et al. (2024). Since there are currently mostly demonstration plants for hydrogen storage, investment costs are estimated based on system components. According to Haroon Bukhari et al. (2023) and Yue et al. (2021), current electrolysis costs are around 1000 €/kW, with potential short- and medium-term reductions to around 500 €/kW. On the other hand, investment costs for thermal power plants vary depending on the technology applied, ranging from 300 €/kW for GT to 600 €/kW for CCGT plants in Europe (Peng et al., 2023). For hydrogen cavern storage, a cost of approximately 0.5 €/kWh is considered, according to Hassan et al. (2023) (see Table 9.1).

Renewable methane storage

Renewable methane storage (CH_4) offers an alternative for storing electricity through a process involving the conversion of electricity into hydrogen and then into methane by adding CO_2 (Abbas et al., 2023; De Maron et al., 2023). The produced methane can be injected into the natural gas grid or stored in specific facilities for this purpose.

Table 9.1 Efficiency and investment costs of hydrogen and methane storage components.

	Efficiency (%)	Investment costs (€/kW)	Sources
Hydrogen storage electrolyzer	75	500	Le et al. (2024); Lystbæk et al. (2023); Shirizadeh and Quirion (2023)
Compressor	89		Yang et al. (2023)
Thermal power plant (GT or CCGT)	50–60	300–600	Solovey et al. (2023); Mukherjee et al. (2023)
Total	33–40	800–1100	
Methane storage	75	500	Rodriguez et al. (2023); Bhandari et al. (2014)
Electrolyzer	80	500	Lystbæk et al. (2023); Shiva Kumar and Lim (2022); Niu et al. (2023b)
Compressor	97		Chen (2023)
Thermal power plant (GT or CCGT)	50–60	300–600	Abdon et al. (2017); Cruz-Soto et al. (2022); Quarton and Samsatli (2018)
Total	29–35	1300–1600	

CCGT, Combined-cycle gas turbines; GT, gas turbines.

This requires access to the natural gas grid near the production site. Reconversion to electricity can be achieved through conventional power plants fueled by natural gas, such as CCGT or GT (Aubin et al., 2023). This strategy leverages existing natural gas storage and distribution infrastructure, which can mitigate some challenges associated with the intermittency of renewable electricity. However, the storage efficiency of this process is relatively low, estimated between 27% and 32%, depending on the power plant efficiency (Wulf et al., 2020). Investment costs are calculated based on the components involved, with the methanation process being a factor of uncertainty due to the lack of available data. Therefore, investment costs are estimated to be similar to those of electrolysis (see Table 9.1).

Lithium-ion electrochemical battery

Lithium-ion battery technology has stood out due to its technical improvements compared to more traditional technologies. According to Cicconi and Kumar (2023) from IRENA in 2015, this technology represented the majority of installed capacity during that period. The lithium-ion battery is modular, and its available capacity ranges from several kWh up to 400 MWh in the market. In this analysis, the lithium titanate oxide (LTO) battery with nickel cobalt aluminum oxide (NCA) as the cathode material has been considered, known for its long lifespan, safety, and

Table 9.2 Storage technology specification for each application scenario.

	Short term	Medium term	Long term
1 MW	PHS LTO – NCA P2H + H_2 STORAGE + PEMFC	PHS I – CAES LTO – NCA P2H + H_2 STORAGE + PEMFC	PHS P2H + NG NETWORK + PEMFC
100 MW	PHS LTO – NCA P2M + NG NETWORK + CC	PHS AA-CAES LTO – NCA P2M + NG NETWORK + CC	PHS P2M + NG NETWORK + CC

LTO, Lithium titanate oxide; *NCA*, nickel cobalt aluminum oxide; *PHS*, pumped hydro storage.

performance, particularly suitable for stationary applications. Costs have been estimated considering both the upper and lower ranges, based on current and projected costs of the technology (see Table 9.2) (Okazaki et al., 2015).

Economic evaluation

In the evaluation, the perspective of an investor was adopted when analyzing the technologies. This involved operating plants with the aim of maximizing their own economic benefit, without necessarily prioritizing the improvement of the electrical system's efficiency or the energy market, unless contractual agreements stipulated otherwise. This approach aligned with typical framework conditions of liberalized energy markets, where ES plants are privately owned and offer their services in the open market (Amir et al., 2023).

Technological parameters

Alili and Mahmoudimehr (2023), Altea and Yanagihara (2024), and Assareh and Ghafouri (2023) presented a synthesis of the parameters used for the subsequent evaluation in Table 9.3 along with some of their key technical characteristics. Values were selected to reflect technological advancements, particularly for those technologies currently in experimental or demonstration phases. This approach provides a solid basis for comparison and analysis of the different ES technologies considered in the study, allowing for an accurate assessment of their performance and feasibility in various contexts.

Table 9.3 Storage parameters used in the analysis.

			Pumped hydro storage (PHS)	**Adiabatic compr. Air energy storage (AA-CAES)**	**Hydrogen storage (H_2)**	**Methane storage (CH_4)**
Efficiency	Charging	(%)	92	84	67	58
	Discharging	(%)	92	84	60	60
Investment costs						
Capacity specific		\$/kW	500	600	1000	1500
Energy specific		\$/kWh	30	70	0.5	0.15
Depreciation time		Years	25	20	20	20
O&M costs		\$/kW/year	4	20	20	20

Control reserve and intraday market

In some European countries (Cetegen et al., 2024; Green IRENA, 2020; Iribarren et al., 2023), control reserve has been a significant source of revenue for electricity storage systems, especially with the increasing intermittent production of renewable energy. However, the consolidation of control energy markets can lead to improved efficiency in control reserve utilization and, consequently, lower revenues for providers (Schwidtal et al., 2023). Demand and prices for control reserve vary among system operators in Europe, generating uncertainty in potential revenues for power plants. Although control reserve is not considered in this analysis, its provision could enhance economic performance, especially for highly flexible plants like PHS (Altea & Yanagihara, 2024). Intraday markets are expected to gain importance with the continuous integration of intermittent renewable energy, offering an alternative source of revenue for ES plants, thanks to price differences between base load and peak load periods in markets like EPEX SPOT in Germany (Wulf et al., 2020). Although traded volumes in these markets are still low compared to the day-ahead market, their relevance is increasing and expected to continue growing in the coming years.

Market-centric energy scenario

In a market-centric energy scenario (see Fig. 9.4), the goal of achieving zero carbon emissions is accomplished through the implementation of clean power stations, peak shaving, energy conservation, and efficiency improvement (Jamroen & Vongkoon, 2023). In this context, the energy side, the electricity grid side, and the

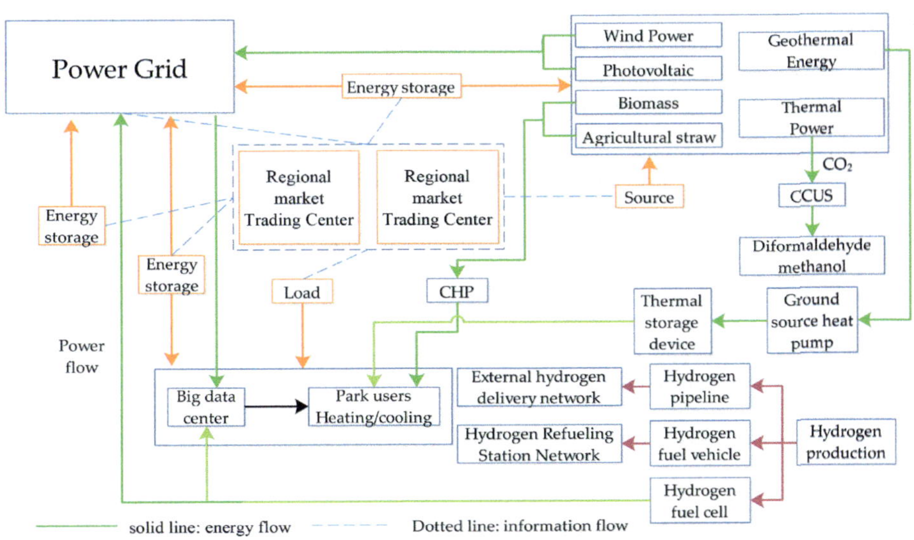

Figure 9.4 Design of market-centric energy scenarios for the industrial park.

load side share the responsibility of investing in, operating, and maintaining the ES station, jointly assuming the costs associated with zero carbon emissions (Fang et al., 2023).

Cost—benefit analysis

The cost of building an ES station is consistent across different scenarios, encompassing initial investment, operational and maintenance expenses, electricity procurement, and carbon-related costs (Yue et al., 2021; Zhu et al., 2024). This cost is directly related to the station's capacity and power. ES stations generate different benefits depending on the scenario. In the first scenario, they benefit from peak shaving, frequency modulation, ancillary services, and deferred device upgrades. In the second scenario, profitability is achieved through differential price arbitrage from peak to off-peak hours. In the third scenario, profitability comes from providing ancillary services and peak-to-off-peak price arbitrage. Hadelu et al. (2024) conducted a cost—benefit analysis, and estimates for each scenario are presented in Table 9.4.

Marketing measures

The application of ES has contributed to mitigating the volatility of renewable sources (Benavides et al., 2024), offering a technical approach to development with zero carbon emissions. However, in industrial parks (Fang et al., 2023), strategic planning and effective coordination of ES with sources, grids, and loads are essential to achieving zero carbon emissions goals. Several

Table 9.4 The cost and income of the three scenarios.

Scenarios	Power grid-centric scenario	Power user-centric scenario	Power market-centric scenario
Cost	One-time investment cost	One-time investment cost	One-time investment cost
	Operation and maintenance cost	Operation and maintenance cost	Operation and maintenance cost
	Electricity purchase cost	Electricity purchase cost	Electricity purchase cost
	Carbon emissions treatment cost	Carbon emissions treatment cost	Carbon emissions treatment cost
Income	Frequency regulation income	Peak and valley spread arbitrage	Frequency regulation income
	Benefits of peak reduction		Benefits of peak reduction
	Delay the benefits of equipment upgrades		Peak and valley spread arbitrage

Table 9.5 Source grid load storage coordination measures.

Subject	Synergistic measure	Synergistic effect	Benefits
Source	Power grid dispatching	Balance of supply and demand, abandon wind and light	Provide low-cost green electricity
	On-demand power supply	Balance of supply and demand, abandon wind and light	
	Smoothing output	Output adjustable	
Grid	Dispatching power	Balance of electricity supply and demand	Supply and demand balance and power quality
	Demand-side response	Dynamic grid power balance	
	Frequency modulation reserve	The system runs safety and reliably	
Load	Interruptible load	Lifting load elasticity	Green energy and elastic load
	Elastic load	Reduce electricity consumption	
	Cut peaks and fill valleys	The load moves with the source	
Storage	Power-side storage	Smooth output and energy storage	Supply and demand balance and power quality
	Grid-side energy storage	Frequency modulation, reserve, delay investment	
	Load-side energy storage	Peak-valley electricity price	

collaborative modes of grid-origin load and storage in industrial parks have been designed, including 4 collaborative subjects and 12 collaborative modes. These modes are classified according to their contribution to zero carbon emissions goals, maximizing the economic value of ES through specific collaboration measures (Fang et al., 2023). Table 9.5 shows collaboration measures and the synergistic effects of different entities. Among them, the electric grid as the main entity offers low-cost green energy and is responsible for grid regulation, energy supply according to demand, and smoothing output. Additionally, load-based synergy seeks to improve load elasticity and reduce fluctuation with energy supply, while synergy with ES aims to balance supply and demand, enhancing energy quality (Xie et al., 2024).

Analysis of trends and future perspectives

Energy storage

The analysis of trends and future perspectives highlights the crucial role that ESSs will play in the smart grid (Cha et al., 2023). The large-scale implementation of these systems will offer a variety of key applications:

- Energy arbitrage: It will enable buying electricity at low prices and storing it for use during periods of higher demand, generating income (Horesh et al., 2021).
- Frequency regulation: It will be vital for maintaining grid stability, especially with the increase in renewable energy sources.
- Peak load shaving: It will help reduce demand peaks, improving grid reliability and efficiency, and reducing costs for customers (Malka et al., 2023).
- Voltage support: It will provide services to keep voltage within safe ranges, ensuring proper operation of electrical equipment.
- Integration into virtual power plant systems: It will be an essential component for enhancing flexibility, reliability, and efficiency of the grid, facilitating the integration of renewable energies (Yu et al., 2024).
- Transient stability: It will play an important role in maintaining grid stability during abrupt changes in demand or supply, avoiding power outages or equipment damage (Kebir et al., 2023).

Future research directions for demand-side management implementation in the smart grid market

Several areas of research could further advance the use of DSM in energy markets:

- Advanced analytics and machine learning: Advanced analytics and machine learning techniques can be applied to extract insights from data generated by DSM programs. For example, predictive algorithms could anticipate changes in energy demand and adjust supply accordingly, improving grid efficiency and reducing the need for costly peak power plants (Kaif et al., 2024; Panda et al., 2023).
- Integration of renewable energy: DSM programs can help more effectively integrate renewable energy sources into the grid. Future research can explore how to manage fluctuations in supply and demand resulting from renewable energy sources and maximize their use (Chatzigeorgiou et al., 2024).
- Coordination of multiple DSM programs: Coordinating multiple DSM programs can enhance efficiency. Research can explore how to integrate and coordinate different programs to achieve optimal outcomes, as well as incentivize collaboration (Chatzigeorgiou et al., 2024).
- Policy and regulation: The design of policies and regulations can influence the adoption of DSM programs. Research can investigate how to design policies to encourage the use of technologies that enhance the effectiveness of DSM programs (Lee et al., 2023).
- Consumer behavior: Understanding how consumers respond to DSM programs is critical. Future research can explore behavioral interventions to promote more sustainable energy consumption (Cai et al., 2024).

Conclusions

This study addresses the evaluation of various electrical storage technologies for load leveling in the current market. It is highlighted that PHS plants exhibit remarkable efficiency and competitive costs, although their viability may be constrained by uncertainty in capital costs and site conditions. On the other hand, AA-CAES with enriched air is presented as a promising option for diurnal storage, although cost reductions are still needed for large-scale viability. Energy conversion technologies such as hydrogen (H_2) and methane (CH_4) are considered economically unviable due to their low storage efficiency, although they could find application as long-term storage plants to balance seasonal supply and demand, especially if integrated into existing natural gas infrastructure. In terms of LCOES, lithium-ion batteries are the most economical option for the short term, while PHS and AA-CAES present lower costs for larger systems. For small-scale seasonal storage, the electricity-to-gas-to-electricity conversion process (P2G2P) could be more cost-effective than PHS. Furthermore, it is highlighted that electricity storage from renewable sources can significantly reduce greenhouse gas emissions, making P2G2P competitive in environmental terms, especially in the short term. There is a need for more detailed analysis of the technoeconomic benefits and life cycle emissions of electrical storage technologies, as well as comparison with other alternatives such as grid expansion, to enhance the overall assessment of stationary electricity storage.

Acknowledgments

Edisson Villa Ávila expresses his sincere gratitude for the opportunity to partially present the findings of his research, conducted as part of his doctoral studies in the Ph.D. program in Advances in Engineering of Sustainable Materials and Energies at the University of Jaen, Spain. Finally, the authors thank Universidad de Cuenca, Ecuador, for easing access to the facilities of the Micro-Grid Laboratory of the Centro Científico Tecnológico y de Investigación Balzay (CCTI-B), for allowing the use of its equipment, and for authorizing members of its staff the provision of technical support necessary to carry out the experiments described in this article.

Paul Arévalo thanks the Call for Grants for the Requalification of the Spanish University System for 2021–23, Margarita Salas Grants for the training of young doctors awarded by the Ministry of Universities and financed by the European Union–Next Generation EU.

References

Abbas, Z., Waqas, M., & He, S. (2023). Environmental evaluation of electricity generation from landfill gas by using LEAP and IPCC model: A case study of Karachi. *Energy Systems*. Available from https://doi.org/10.1007/s12667-023-00574-3, https://www.springer.com/journal/12667.

Abdon, A., Zhang, X., Parra, D., Patel, M. K., Bauer, C., & Worlitschek, J. (2017). Techno-economic and environmental assessment of stationary electricity storage technologies for different time scales. *Energy*, *139*. Available from https://doi.org/10.1016/j.energy.2017.07.097, https://www.journals.elsevier.com/energy.

Alili, H., & Mahmoudimehr, J. (2023). Techno-economic assessment of integrating hydrogen energy storage technology with hybrid photovoltaic/pumped storage hydropower energy system. *Energy Conversion and Management*, *294*. Available from https://doi.org/10.1016/j.enconman.2023.117437, https://www.journals.elsevier.com/energy-conversio n-and-management.

Altea, C. d M., & Yanagihara, J. I. (2024). Energy, exergy and environmental impacts analyses of pumped hydro storage (PHS) and Hydrogen (H2) energy storage processes. *Journal of Energy Storage*, *76*109713. Available from https://doi.org/10.1016/j.est.2023.109713.

Al-Wreikat, Y., Attfield, E. K., & Sodré, J. R. (2022). Model for payback time of using retired electric vehicle batteries in residential energy storage systems. *Energy*, *259*. Available from https://doi.org/10.1016/j.energy.2022.124975, https://www.journals.elsevier.com/energy.

Amir, M., Deshmukh, R. G., Khalid, H. M., Said, Z., Raza, A., Muyeen, S. M., Nizami, A.-S., Elavarasan, R. M., Saidur, R., & Sopian, K. (2023). Energy storage technologies: An integrated survey of developments, global economical/environmental effects, optimal scheduling model, and sustainable adaption policies. *Journal of Energy Storage*, *72*108694. Available from https://doi.org/10.1016/j.est.2023.108694.

Assareh, E., & Ghafouri, A. (2023). An innovative compressed air energy storage (CAES) using hydrogen energy integrated with geothermal and solar energy technologies: A comprehensive techno-economic analysis—different climate areas- using artificial intelligent (AI). *International Journal of Hydrogen Energy*, *48*(34). Available from https://doi.org/10.1016/j.ijhydene.2022.11.233, http://www.journals.elsevier.com/international-journal-of-hydrogen-energy/.

Aubin, P., Wang, L., & Van herle, J. (2023). Evaporating water-cooled methanation reactor for solid-oxide stack-based power-to-methane systems: Design, experiment and modeling. *Chemical Engineering Journal*, *456*. Available from https://doi.org/10.1016/j.cej.2022.140256, http://www.elsevier.com/inca/publications/store/6/0/1/2/7/3/index.htt.

Benavides, D., Arévalo, P., Villa-Ávila, E., Aguado, J. A., & Jurado, F. (2024). Predictive power fluctuation mitigation in grid-connected PV systems with rapid response to EV charging stations. *Journal of Energy Storage*, *86*. Available from https://doi.org/10.1016/j.est.2024.111230, https://www.sciencedirect.com/science/journal/2352152X.

Benito, A. O,. & Arena, A. P. (2020). Distribuidos de Energía Eléctrica en Sistemas Energéticos Regionales: análisis con el modelo LEAP. *AJEA (Actas De Jornadas Y Eventos Académicos De UTN)*, (5). https://doi.org/10.33414/ajea.5.696.2020.

Bennett, J. A., Simpson, J. G., Qin, C., Fittro, R., Koenig, G. M., Clarens, A. F., & Loth, E. (2021). Techno-economic analysis of offshore isothermal compressed air energy storage in saline aquifers colocated with wind power. *Applied Energy*, *303*117587. Available from https://doi.org/10.1016/j.apenergy.2021.117587.

Bhandari, R., Trudewind, C. A., & Zapp, P. (2014). Life cycle assessment of hydrogen production via electrolysis—A review. *Journal of Cleaner Production*, *85*. Available from https://doi.org/10.1016/j.jclepro.2013.07.048, https://www.journals.elsevier.com/journal-of-cleaner-production.

Bosu, I., Mahmoud, H., & Hassan, H. (2023). Energy audit, techno-economic, and environmental assessment of integrating solar technologies for energy management in a university residential building: A case study. *Applied Energy*, *341*121141. Available from https://doi.org/10.1016/j.apenergy.2023.121141.

Bulut, M., & Özcan, E. (2024). How to build a state-of-the-art battery energy storage market? Challenges, opportunities, and future directions. *Journal of Energy Storage, 86.* Available from https://doi.org/10.1016/j.est.2024.111174, https://www.sciencedirect.com/science/journal/2352152X.

Cai, Q., Qing, J., Xu, Q., Shi, G., & Liang, Q. M. (2024). Techno-economic impact of electricity price mechanism and demand response on residential rooftop photovoltaic integration. *Renewable and Sustainable Energy Reviews, 189.* Available from https://doi.org/10.1016/j.rser.2023.113964, https://www.journals.elsevier.com/renewable-and-sustainable-energy-reviews.

Cetegen, S. A., Gundersen, T., & Barton, P. I. (2024). Evaluating economic feasibility of liquid air energy storage systems in US and European markets. *Energy, 300.* Available from https://doi.org/10.1016/j.energy.2024.131524, https://www.sciencedirect.com/science/journal/03605442.

Cha, M., Enshaei, H., Nguyen, H., & Jayasinghe, S. G. (2023). Towards a future electric ferry using optimisation-based power management strategy in fuel cell and battery vehicle application—A review. *Renewable and Sustainable Energy Reviews, 183*113470. Available from https://doi.org/10.1016/j.rser.2023.113470.

Chatzigeorgiou, N. G., Theocharides, S., Makrides, G., & Georghiou, G. E. (2024). A review on battery energy storage systems: Applications, developments, and research trends of hybrid installations in the end-user sector. *Journal of Energy Storage, 86.* Available from https://doi.org/10.1016/j.est.2024.111192, https://www.sciencedirect.com/science/journal/2352152X.

Chen, W. (2023). Optimal design and performance assessment of a proposed constant power operation mode for the constant volume discharging process of advanced adiabatic compressed air energy storage. *Renewable Energy., 221.* Available from https://doi.org/10.1016/j.renene.2023.1197282023.

Cho, J., Jeong, S., & Kim, Y. (2015). Commercial and research battery technologies for electrical energy storage applications. *Progress in Energy and Combustion Science, 48.* Available from https://doi.org/10.1016/j.pecs.2015.01.002, https://www.journals.elsevier.com/progress-in-energy-and-combustion-science.

Chreim, B., Esseghir, M., & Merghem-Boulahia, L. (2024). Recent sizing, placement, and management techniques for individual and shared battery energy storage systems in residential areas: A review. *Energy Reports, 11.* Available from https://doi.org/10.1016/j.egyr.2023.11.053, https://www.sciencedirect.com/science/journal/23524847.

Chu, W., Duić, N., & Wang, Q. (2024). Recent advances in energy storage and energy saving technologies: SDEWES special issue in 2022. *Energy Storage and Saving, 3*(1). Available from https://doi.org/10.1016/j.enss.2023.09.001, https://www.sciencedirect.com/science/journal/27726835.

Cicconi, P., & Kumar, P. (2023). Design approaches for Li-ion battery packs: A review. *Journal of Energy Storage, 73*109197. Available from https://doi.org/10.1016/j.est.2023.109197.

Cisterna, L., Améstica, L., & Piderit, M. (2020). Proyectos fotovoltaicos en generación distribuida ¿Rentabilidad privada o sustentabilidad ambiental? *Revista Politécnica, 45*(2), 2477–8990. Available from https://doi.org/10.33333/rp.vol45n2.03.

Comodi, G., Carducci, F., Sze, J. Y., Balamurugan, N., & Romagnoli, A. (2017). Storing energy for cooling demand management in tropical climates: A techno-economic comparison between different energy storage technologies. *Energy, 121.* Available from https://doi.org/10.1016/j.energy.2017.01.038, http://www.elsevier.com/inca/publications/store/4/8/3/.

Cóndor, H. (2021). Generación distribuida con energías renovables en Perú [http://purl.org/pe-repo/renati/type#tesis, Universidad de Piura]. https://hdl.handle.net/11042/4782.

Connolly, D., Lund, H., & Mathiesen, B. V. (2016). Smart energy Europe: The technical and economic impact of one potential 100% renewable energy scenario for the European Union. *Renewable and Sustainable Energy Reviews, 60*. Available from https://doi.org/10.1016/j.rser.2016.02.025.

Cruz-Soto, J. de la, Azkona-Bedia, I., Velazquez-Limon, N., & Romero-Castanon, T. (2022). A techno-economic study for a hydrogen storage system in a microgrid located in baja California, Mexico. Levelized cost of energy for power to gas to power scenarios. *International Journal of Hydrogen Energy*, *47*(70). Available from https://doi.org/10.1016/j.ijhydene.2022.03.026.

De Maron, J., Mafessanti, R., Gramazio, P., Orfei, E., Fasolini, A., & Basile, F. (2023). H2 production by methane oxy-reforming: Effect of catalyst pretreatment on the properties and activity of Rh-Ce0.5Zr0.5O2 synthetized by microemulsion. *Nanomaterials*, *13*(1). Available from https://doi.org/10.3390/nano13010053, http://www.mdpi.com/journal/nanomaterials.

El-Sayed, A. H. A., Khalil, A., & Yehia, M. (2023). Modeling alternative scenarios for Egypt 2050 energy mix based on LEAP analysis. *Energy*, *266*. Available from https://doi.org/10.1016/j.energy.2023.126615, https://www.journals.elsevier.com/energy.

Fang, Y., Fan, R., & Liu, Z. (2023). A study on the energy storage scenarios design and the business model analysis for a zero-carbon big data industrial park from the perspective of source-grid-load-storage collaboration. *Energy Reports*, *9*. Available from https://doi.org/10.1016/j.egyr.2023.05.202, http://www.journals.elsevier.com/energy-reports/.

Gao, Q., Yuan, R., Ertugrul, N., Ding, B., Hayward, J. A., & Li, Y. (2023). Analysis of energy variability and costs for offshore wind and hybrid power unit with equivalent energy storage system. *Applied Energy*, *342*. Available from https://doi.org/10.1016/j.apenergy.2023.121192, https://www.journals.elsevier.com/applied-energy.

García-Miguel, P. L. C., Zarilli, D., Alonso-Martinez, J., Plaza, M. G., & Gómez, S. A. (2024). Optimal operation and market integration of a hybrid farm with green hydrogen and energy storage: A Stochastic approach considering wind and electricity price uncertainties. *Sustainability (Switzerland)*, *16*(7). Available from https://doi.org/10.3390/su16072856, http://www.mdpi.com/journal/sustainability/.

Green IRENA (2020). Hydrogen.

Hadelu, L. M., Noorpoor, A., & Boyaghchi, F. A. (2024). Assessment and multiobjective dragonfly optimization of utilizing flash tank vapor injection cycle in a new geothermal assisted-pumped thermal energy storage system based on transcritical CO2 cycle. *Journal of Energy Storage*, *88*. Available from https://doi.org/10.1016/j.est.2024.111628, https://www.sciencedirect.com/science/journal/2352152X.

Haji Bashi, M., De Tommasi, L., & Lyons, P. (2022). Electricity market integration of utility-scale battery energy storage units in Ireland, the status and future regulatory frameworks. *Journal of Energy Storage*, *55*, 105442. Available from https://doi.org/10.1016/j.est.2022.105442.

Haroon Bukhari, M., Javed, A., Ali Abbas Kazmi, S., Ali, M., & Talib Chaudhary, M. (2023). Techno-economic feasibility analysis of hydrogen production by PtG concept and feeding it into a combined cycle power plant leading to sector coupling in future. *Energy Conversion and Management*, *282*. Available from https://doi.org/10.1016/j.enconman.2023.116814, https://www.journals.elsevier.com/energy-conversion-and-management.

Hassan, Q., Abdulrahman, I. S., Salman, H. M., Olapade, O. T., & Jaszczur, M. (2023). Techno-economic assessment of green hydrogen production by an off-grid photovoltaic energy system. *Energies*, *16*(2). Available from https://doi.org/10.3390/en16020744, http://www.mdpi.com/journal/energies/.

He, W., King, M., Luo, X., Dooner, M., Li, D., & Wang, J. (2021). Technologies and economics of electric energy storages in power systems: Review and perspective. *Advances in Applied Energy*, *4*, 100060. Available from https://doi.org/10.1016/j.adapen.2021.100060.

Henke, H. T. J., Gardumi, F., & Howells, M. (2022). The open source electricity Model Base for Europe—An engagement framework for open and transparent European energy modelling. *Energy*, *239*. Available from https://doi.org/10.1016/j.energy.2021.121973, https://www.journals.elsevier.com/energy.

Hokmabad, H. N., Husev, O., Kurnitski, J., & Belikov, J. (2024). Optimizing size and economic feasibility assessment of photovoltaic and energy storage setup in residential applications. *Sustainable Energy, Grids and Networks*, *38*, 101385. Available from https://doi.org/10.1016/j.segan.2024.101385.

Horesh, N., Quinn, C., Wang, H., Zane, R., Ferry, M., Tong, S., & Quinn, J. C. (2021). Driving to the future of energy storage: Techno-economic analysis of a novel method to recondition second life electric vehicle batteries. *Applied Energy*, *295*, 117007. Available from https://doi.org/10.1016/j.apenergy.2021.117007.

Hurta, A., Žilka, M., & Freiberg, F. (2022). Impact of the splitting of the German–Austrian electricity bidding zone on investment in a grid-scale battery energy storage system deployed for price arbitrage with gray and green power in Austrian and German day-ahead power markets. *Energy Reports*, *8*. Available from https://doi.org/10.1016/j.egyr.2022.09.045, http://www.journals.elsevier.com/energy-reports/.

Iribarren, A., Elizondo, D., Barrios, E. L., Ibaiondo, H., Sanchez-Ruiz, A., Arza, J., Sanchis, P., & Ursua, A. (2023). Dynamic modeling of a pressurized alkaline water electrolyzer: A multiphysics approach. *IEEE Transactions on Industry Applications*, *59*(3), 3741–3753. Available from https://ieeexplore.ieee.org/servlet/opac?punumber = 28, 10.1109/TIA.2023.3247405.

Jamroen, C., & Vongkoon, P. (2023). The role of state-of-charge management in optimal techno-economic battery energy storage sizing for off-grid residential photovoltaic systems. *Journal of Energy Storage*, *72*, 108246. Available from https://doi.org/10.1016/j.est.2023.108246.

Jannesar Niri, A., Poelzer, G. A., Zhang, S. E., Rosenkranz, J., Pettersson, M., & Ghorbani, Y. (2024). Sustainability challenges throughout the electric vehicle battery value chain. *Renewable and Sustainable Energy Reviews*, *191*. Available from https://doi.org/10.1016/j.rser.2023.114176, https://www.sciencedirect.com/science/journal/13640321.

Javaheri, M., & Shafiei Ghazani, A. (2023). Energy and exergy analysis of a novel advanced adiabatic compressed air energy storage hybridized with reverse osmosis system. *Journal of Energy Storage*, *73*, 109250. Available from https://doi.org/10.1016/j.est.2023.109250.

Kaif, A. M. A. D., Alam, K. S., & Das, S. K. (2024). Blockchain based sustainable energy transition of a virtual power plant: Conceptual framework design & experimental implementation. *Energy Reports*, *11*. Available from https://doi.org/10.1016/j.egyr.2023.11.061, https://www.sciencedirect.com/science/journal/23524847.

Kebir, N., Leonard, A., Downey, M., Jones, B., Rabie, K., Bhagavathy, S. M., & Hirmer, S. A. (2023). Second-life battery systems for affordable energy access in Kenyan primary schools. *Scientific Reports*, *13*(1). Available from https://doi.org/10.1038/s41598-023-28377-7, https://www.nature.com/srep/.

Khan, M. K., Raza, M., Shahbaz, M., Farooq, U., & Akram, M. U. (2024). Recent advancement in energy storage technologies and their applications. *Journal of Energy Storage*, *92*, 112112. Available from https://doi.org/10.1016/j.est.2024.112112.

Kloess, M., & Zach, K. (2014). Bulk electricity storage technologies for load-leveling operation—An economic assessment for the Austrian and German power market. *International Journal of Electrical Power & Energy Systems, 59*. Available from https://doi.org/10.1016/j.ijepes.2014.02.002.

Kondziella, H., Specht, K., Lerch, P., Scheller, F., & Bruckner, T. (2023). The techno-economic potential of large-scale hydrogen storage in Germany for a climate-neutral energy system. *Renewable and Sustainable Energy Reviews, 182*. Available from https://doi.org/10.1016/j.rser.2023.113430, https://www.journals.elsevier.com/renewable-and-sustainable-energy-reviews.

Le, S. T., Nguyen, T. N., Bui, D. K., Teodosio, B., & Ngo, T. D. (2024). Comparative life cycle assessment of renewable energy storage systems for net-zero buildings with varying self-sufficient ratios. *Energy, 290*. Available from https://doi.org/10.1016/j.energy.2023.130041, https://www.sciencedirect.com/science/journal/03605442.

Lechón, Y., Lago, C., Herrera, I., Gamarra, A. R., & Pérula, A. (2023). Carbon benefits of different energy storage alternative end uses. Application to the Spanish case. *Renewable and Sustainable Energy Reviews, 171*. Available from https://doi.org/10.1016/j.rser.2022.112985, https://www.journals.elsevier.com/renewable-and-sustainable-energy-reviews.

Lee, J. Y., Ramasamy, A. K., Ong, K. H., Verayiah, R., Mokhlis, H., & Marsadek, M. (2023). Energy storage systems: A review of its progress and outlook, potential benefits, barriers and solutions within the Malaysian distribution network. *Journal of Energy Storage, 72*. Available from https://doi.org/10.1016/j.est.2023.108360, http://www.journals.elsevier.com/journal-of-energy-storage/.

Li, J., Wang, C., Zhai, G., Li, Q., Lim, S. , H., Abdoli, S., Kook, S., Yeoh, G. H., & Chan, Q. N. (2024). Evaluating the techno-economic feasibility of hydrogen-fuelled reciprocating engines for renewable base-load power generation. *Energy Conversion and Management, 311*, 118515. Available from https://doi.org/10.1016/j.enconman.2024.118515.

Liang, Y., Galiano, J. C., & Zhou, H. (2023). The environmental impact of stock market capitalization and energy transition: Natural resource dynamics and international trade. *Utilities Policy, 82*. Available from https://doi.org/10.1016/j.jup.2023.101517, http://www.elsevier.com/inca/publications/store/3/0/4/7/8/index.htt.

Lin, A., & Bagnato, G. (2024). Revolutionising energy storage: The latest breakthrough in liquid organic hydrogen carriers. *International Journal of Hydrogen Energy, 63*. Available from https://doi.org/10.1016/j.ijhydene.2024.03.146, https://www.sciencedirect.com/science/journal/03603199.

Lystbæk, N., Gregersen, M., & Shaker, H. R. (2023). Review of energy portfolio optimization in energy markets considering flexibility of power-to-X. *Sustainability (Switzerland), 15*(5). Available from https://doi.org/10.3390/su15054422, http://www.mdpi.com/journal/sustainability/.

Maghami, M. R., & Mutambara, A. G. O. (2023). Challenges associated with hybrid energy systems: An artificial intelligence solution. *Energy Reports, 9*. Available from https://doi.org/10.1016/j.egyr.2022.11.195, http://www.journals.elsevier.com/energy-reports/.

Maghami, M. R., Pasupuleti, J., & Ekanayake, J. (2024). Energy storage and demand response as hybrid mitigation technique for photovoltaic grid connection: Challenges and future trends. *Journal of Energy Storage, 88*. Available from https://doi.org/10.1016/j.est.2024.111680, https://www.sciencedirect.com/science/journal/2352152X.

Malka, L., Bidaj, F., Kuriqi, A., Jaku, A., Roçi, R., & Gebremedhin, A. (2023). Energy system analysis with a focus on future energy demand projections: The case of Norway. *Energy, 272*127107. Available from https://doi.org/10.1016/j.energy.2023.127107.

Manandhar, P., Rafiq, H., & Rodriguez-Ubinas, E. (2023). Current status, challenges, and prospects of data-driven urban energy modeling: A review of machine learning methods. *Energy Reports*, *9*. Available from https://doi.org/10.1016/j.egyr.2023.01.094, http://www.journals.elsevier.com/energy-reports/.

Martinez, A., & Iglesias, G. (2024). Levelized cost of energy to evaluate the economic viability of floating offshore wind in the European Atlantic and Mediterranean. *e-Prime—Advances in Electrical Engineering, Electronics and Energy*, *8*100562. Available from https://doi.org/10.1016/j.prime.2024.100562.

Martsinchyk, K., Martsinchyk, A., Łazor, M., Shuhayeu, P., Kupecki, J., Niemczyk, A., Błesznowski, M., & Milewski, J. (2024). Feasibility study and techno-economic assessment of power-to-gas (P2G) technology based on solid oxide electrolysis (SOE). *Journal of Environmental Management*, *354*. Available from https://doi.org/10.1016/j.jenvman.2024.120425, https://www.sciencedirect.com/science/journal/03014797.

McLaughlin, H., Littlefield, A. A., Menefee, M., Kinzer, A., Hull, T., Sovacool, B. K., Bazilian, M. D., Kim, J., & Griffiths, S. (2023). Carbon capture utilization and storage in review: Sociotechnical implications for a carbon reliant world. *Renewable and Sustainable Energy Reviews*, *177*113215. Available from https://doi.org/10.1016/j.rser.2023.113215.

Mearns, E., & Sornette, D. (2023). Are 2050 energy transition plans viable? A detailed analysis of projected Swiss electricity supply and demand in 2050. *Energy Policy*, *175*113347. Available from https://doi.org/10.1016/j.enpol.2022.113347.

Mukherjee, S. S., Meshram, H. A., Rakshit, D., & Saha, B. B. (2023). A comparative study of sensible energy storage and hydrogen energy storage apropos to a concentrated solar thermal power plant. *Journal of Energy Storage*, *61*. Available from https://doi.org/10.1016/j.est.2023.106629, http://www.journals.elsevier.com/journal-of-energy-storage/.

Newbery, D. M. (2023). High renewable electricity penetration: Marginal curtailment and market failure under "subsidy-free" entry. *Energy Economics*, *126*. Available from https://doi.org/10.1016/j.eneco.2023.107011, http://www.elsevier.com/inca/publications/store/3/0/4/1/3/.

Niu, D., Fang, J., Yau, W., & Goetz, S. M. (2023a). Comprehensive evaluation of energy storage systems for inertia emulation and frequency regulation improvement. *Energy Reports*, *9*. Available from https://doi.org/10.1016/j.egyr.2023.01.110, http://www.journals.elsevier.com/energy-reports/.

Niu, M., Li, X., Sun, C., Xiu, X., Wang, Y., Hu, M., & Dong, H. (2023b). Operation optimization of wind/battery storage/alkaline electrolyzer system considering dynamic hydrogen production efficiency. *Energies*, *16*(17). Available from https://doi.org/10.3390/en16176132, http://www.mdpi.com/journal/energies/.

Okazaki, T., Shirai, Y., & Nakamura, T. (2015). Concept study of wind power utilizing direct thermal energy conversion and thermal energy storage. *Renewable Energy*, *83*. Available from https://doi.org/10.1016/j.renene.2015.04.027, http://www.journals.elsevier.com/renewable-and-sustainable-energy-reviews/.

Ölmez, M. E., Ari, I., & Tuzkaya, G. (2024). A comprehensive review of the impacts of energy storage on power markets. *Journal of Energy Storage*, *91*. Available from https://doi.org/10.1016/j.est.2024.111935, https://www.sciencedirect.com/science/journal/2352152X.

Palma, G., Guiducci, L., Stentati, M., Rizzo, A., & Paoletti, S. (2024). Reinforcement learning for energy community management: A European-scale study. *Energies*, *17*(5). Available from https://doi.org/10.3390/en17051249, http://www.mdpi.com/journal/energies/.

Panda, S., Mohanty, S., Rout, P. K., Sahu, B. K., Parida, S. M., Samanta, I. S., Bajaj, M., Piecha, M., Blazek, V., & Prokop, L. (2023). A comprehensive review on demand side management and market design for renewable energy support and integration. *Energy Reports, 10*. Available from https://doi.org/10.1016/j.egyr.2023.09.049, http://www.journals.elsevier.com/energy-reports/.

Park, J., Kang, S., Kim, S., Kim, H., Cho, H. S., & Lee, J. H. (2024). Enhancing the economic viability and reliability of renewables based electricity supply through power-to-gas-to-power with green hydrogen. *Energy Conversion and Management, 310*. Available from https://doi.org/10.1016/j.enconman.2024.118485, https://www.sciencedirect.com/science/journal/01968904.

Peng, Y., Ma, X., Wang, Y., Li, M., Gao, F., Zhou, K., & Aemixay, V. (2023). Energy performance assessment of photovoltaic greenhouses in summer based on coupled optical-electrical-thermal models and plant growth requirements. *Energy Conversion and Management, 287*117086. Available from https://doi.org/10.1016/j.enconman.2023.117086.

Quarton, C. J., & Samsatli, S. (2018). Power-to-gas for injection into the gas grid: What can we learn from real-life projects, economic assessments and systems modelling? *Renewable and Sustainable Energy Reviews, 98*. Available from https://doi.org/10.1016/j.rser.2018.09.007, https://www.journals.elsevier.com/renewable-and-sustainable-energy-reviews.

Rodriguez, E. J. R., Monzer, D., & Bouallou, C. (2023). Study of interest for integrating wind energy with a power-to-methane system based on the Molten Carbonate Electrolyzer. *Journal of Energy Storage, 63*. Available from https://doi.org/10.1016/j.est.2023.107021, http://www.journals.elsevier.com/journal-of-energy-storage/.

Saleh-Abadi, M., Rostami, M., & Farajollahi, A. (2023). 4-E analysis of a hybrid integrated mechanical/chemical/electrochemical energy storage process based on the CAES, amine-based CO2 capture, SOEC, and CO2 electroreduction cell. *Journal of Energy Storage, 72*. Available from https://doi.org/10.1016/j.est.2023.108278, http://www.journals.elsevier.com/journal-of-energy-storage/.

Schwidtal, J. M., Piccini, P., Troncia, M., Chitchyan, R., Montakhabi, M., Francis, C., Gorbatcheva, A., Capper, T., Mustafa, M. A., Andoni, M., Robu, V., Bahloul, M., Scott, I. J., Mbavarira, T., España, J. M., & Kiesling, L. (2023). Emerging business models in local energy markets: A systematic review of peer-to-peer, community self-consumption, and transactive energy models. *Renewable and Sustainable Energy Reviews, 179*113273. Available from https://doi.org/10.1016/j.rser.2023.113273.

Shabani, M., Dahlquist, E., Wallin, F., & Yan, J. (2020). Techno-economic comparison of optimal design of renewable-battery storage and renewable micro pumped hydro storage power supply systems: A case study in Sweden. *Applied Energy, 279*115830. Available from https://doi.org/10.1016/j.apenergy.2020.115830.

Shirizadeh, B., & Quirion, P. (2023). Long-term optimization of the hydrogen-electricity nexus in France: Green, blue, or pink hydrogen? *Energy Policy, 181*. Available from https://doi.org/10.1016/j.enpol.2023.113702, http://www.journals.elsevier.com/energy-policy/.

Shiva Kumar, S., & Lim, H. (2022). An overview of water electrolysis technologies for green hydrogen production. *Energy Reports, 8*. Available from https://doi.org/10.1016/j.egyr.2022.10.127, http://www.journals.elsevier.com/energy-reports/.

Solovey, V., Zipunnikov, M., & Rusanov, R. (2023). Increasing the manoeuvrability of power units of the thermal power plants due to applying the hydrogen-oxygen systems. *Institute of Physics, Ukraine JPhys Energy, 5*(1). Available from https://doi.org/10.1088/2515-7655/aca9ff, opscience.iop.org/journal/2515-7655.

Wang, S., Li, F., Zhang, G., & Yin, C. (2023). Analysis of energy storage demand for peak shaving and frequency regulation of power systems with high penetration of renewable energy. *Energy, 267*126586. Available from https://doi.org/10.1016/j.energy.2022.126586.

Watson, N. E., Huebner, G. M., Fell, M. J., & Shipworth, D. (2024). Consumer preferences for business models with multiple electricity suppliers: Online choice experiments in the United Kingdom. *Energy Research and Social Science, 109*. Available from https://doi.org/10.1016/j.erss.2023.103403, https://www.sciencedirect.com/science/journal/22146296.

Williams, B., Gallardo, P., Bishop, D., & Chase, G. (2023). Impacts of electric vehicle policy on the New Zealand energy system: A retro-analysis. *Energy Reports, 9*. Available from https://doi.org/10.1016/j.egyr.2023.02.080, http://www.journals.elsevier.com/energy-reports/.

Wulf, C., Zapp, P., & Schreiber, A. (2020). Review of power-to-X demonstration projects in Europe. *Frontiers in Energy Research, 8*. Available from https://doi.org/10.3389/fenrg.2020.00191, http://www.frontiersin.org/Energy_Research/.

Xie, H., Ahmad, T. R., Zhang, D., Hwang Goh, H., & Wu, T. (2024). Community-based virtual power plants' technology and circular economy models in the energy sector: A techno-economy study. *Renewable and Sustainable Energy Reviews, 192*114189. Available from https://doi.org/10.1016/j.rser.2023.114189.

Xie, Y., Wu, X., Hou, Z., Li, Z., Luo, J., Lüddeke, C. T., Huang, L., Wu, L., & Liao, J. (2023). Gleaning insights from German energy transition and large-scale underground energy storage for China's carbon neutrality. *International Journal of Mining Science and Technology, 33*(5). Available from https://doi.org/10.1016/j.ijmst.2023.04.001, http://www.elsevier.com/wps/find/journaldescription.cws_home/727915/description#description.

Yang, Q., Wang, H., Wang, T., Zhang, S., Wu, X., & Wang, H. (2021). Blockchain-based decentralized energy management platform for residential distributed energy resources in a virtual power plant. *Applied Energy, 294*117026. Available from https://doi.org/10.1016/j.apenergy.2021.117026.

Yang, R., Kweon, H., & Kim, K. (2023). Preliminary study for the commercialization of a electrochemical hydrogen compressor. *Energies, 16*(7). Available from https://doi.org/10.3390/en16073128, http://www.mdpi.com/journal/energies/.

Yu, S., Wei, Z., Sun, G., Zhou, Y., & Zang, H. (2024). A double auction mechanism for virtual power plants based on blockchain sharding consensus and privacy preservation. *Journal of Cleaner Production, 436*140285. Available from https://doi.org/10.1016/j.jclepro.2023.140285.

Yue, M., Lambert, H., Pahon, E., Roche, R., Jemei, S., & Hissel, D. (2021). Hydrogen energy systems: A critical review of technologies, applications, trends and challenges. *Renewable and Sustainable Energy Reviews, 146*111180. Available from https://doi.org/10.1016/j.rser.2021.111180.

Zhu, Y., Wu, S., Li, J., Jia, Q., Zhang, T., Zhang, X., Han, D., & Tan, Y. (2024). Towards a carbon-neutral community: Integrated renewable energy systems (IRES)—sources, storage, optimization, challenges, strategies and opportunities. *Journal of Energy Storage, 83*110663. Available from https://doi.org/10.1016/j.est.2024.110663.

Part 3

Real projects and case studies

Renewables for isolated and rural areas, the case of Ecuador

10

Paul Arévalo[1,2], *Darío Benavides*[1,3] *and Danny Ochoa-Correa*[2]
[1]Department of Electrical Engineering, University of Jaén, Jaén, Spain, [2]Faculty of Engineering, Department of Electrical, Electronics and Telecommunications Engineering (DEET), Universidad de Cuenca, Balzay Campus, Cuenca, Azuay, Ecuador, [3]Department of Electrical Engineering, University of Málaga, Malaga, Spain

Introduction

In the pursuit of comprehensive and equitable development, rural electrification stands out as a fundamental driver of progress, impacting elements such as per capita energy consumption, educational opportunities, economic prosperity, and overall social well-being (Singh et al., 2020). As the exploration of electrification in the context of Ecuador unfolds, inspiration is drawn from global efforts and lessons learned. The imperative to electrify rural areas transcends being solely a technical challenge; it acts as a catalyst to address socioeconomic disparities, echoing sentiments found in studies examining the impact of rural electrification on lighting, study habits, energy expenditure, and income in diverse regions. Experiences from countries like Ethiopia, where electrification played a pivotal role in achieving development goals, resonate with the transformative potential of rural electrification (Singh et al., 2020). Acknowledging the interconnected nature of global challenges, including climate change, the shift toward renewable energy sources emerges as a beacon of hope. However, understanding the nuanced impact of renewable energy consumption on the environment, society, and the economy becomes paramount (Farghali et al., 2023). This involves assessing the potential for job creation, economic benefits, and environmental improvements stemming from the integration of solar photovoltaic, battery storage, wind, hydropower, and bioenergy.

Navigating the complexities of rural electrification, lessons drawn from World Bank projects implementing solar home systems offer valuable insights. Examining 16 projects spanning from 2000 to 2020 in remote rural areas of developing countries identifies internal and external factors influencing the success of these initiatives (Shyu, 2023). This exploration sheds light on financial, technical, and implementation barriers, as well as sociopolitical and cultural dynamics shaping the outcome of electrification projects. In the context of these global narratives, the focus sharpens on Ecuador. The motivation to investigate rural electrification in Ecuador stems not only from the pressing need to bridge energy disparities but also from the potential for this effort to catalyze socioeconomic development. By analyzing the impact on lighting usage, educational opportunities, and income

Towards Future Smart Power Systems with High Penetration of Renewables. DOI: https://doi.org/10.1016/B978-0-443-29871-4.00010-5

generation, the goal is to contribute valuable insights to shape effective electrification policies and initiatives. This chapter aligns with the broader vision of sustainable development, aiming to achieve universal access to electricity by 2030 (Shyu, 2023). Embarking on this research journey, the objective is not only to understand challenges but also to pave the way for informed strategies that illuminate the path to a brighter, more electrified future for rural Ecuador.

Rural electrification has been the subject of extensive studies in various regions and countries, revealing valuable experiences and innovative solutions to address associated challenges. South Africa, in its postapartheid effort, stands out with the National Electrification Program, led by Eskom, which electrified 300,000 households annually since 1994, generating positive impacts on employment, especially for women (Dinkelman, 2011). In Bangladesh, the feasibility of a hybridized energy system (PV/Wind/Diesel) for a remote village was explored. The study emphasized economic efficiency and environmental benefits, including a 62% reduction in CO_2 emissions compared to traditional kerosene use (Soumya et al., 2018). A comprehensive literature review focused on off-grid electrification systems, categorizing them by technology, simulation methods, feasibility analyses, case studies, and policy considerations. This review highlighted the growing importance of off-grid solutions and underscored the need to consider local contexts (Mandelli et al., 2016). In Uttarakhand, India, an integrated renewable energy system (IRES) for rural electrification was proposed, considering multiple scenarios. This study highlighted the importance of optimizing reliability, cost, and customer interruptions (Kanase-Patil et al., 2010). Saudi Arabia examined the potential for PV-diesel hybrid systems, revealing a 27% PV penetration in an optimized hybrid system, with an emphasis on reducing diesel operational hours and achieving fuel savings (Shaahid & El-Amin, 2009). Colombia addressed rural electrification challenges through the evaluation of PV panels, wind turbines, and diesel generators in a stand-alone hybrid system. This study underscored the low carbon footprint and economic barriers to full-scale implementation (Mamaghani et al., 2016).

A study in Pakistan explored the feasibility of a grid-tied hybrid microgrid system in Kallar Kahar, highlighting the generation of over 50 MW and emphasizing the importance of developing sustainable energy systems (Ahmad et al., 2018). Other studies addressed the feasibility of hybrid systems in Nigeria (Rajbongshi et al., 2017), highlighting the economic and environmental superiority over standalone diesel systems, as well as hybrid systems for rural electrification in areas where grid expansion is unfeasible (Olatomiwa et al., 2015). Additionally, independent and hybrid solar generation systems for off-grid electrification were analyzed. These studies emphasized the efficiency of hybrid systems in addressing solar radiation variability and their viability to provide a sustainable and economically viable energy source in rural areas (Akikur et al., 2013). In a more generalized review, literature related to rural electrification and photovoltaic solar technologies was grouped into experiences from rural electrification programs and technoeconomic aspects. The conclusion highlighted the ongoing relevance of decentralized photovoltaic systems at both local and global levels (Chaurey & Kandpal, 2010).

Rural electrification in Ecuador has been the subject of various studies addressing the sustainability of programs. The literature examines several relevant research efforts with the aim of evaluating the sustainability of existing programs and highlighting research gaps. The study (Feron et al., 2016) focuses on the challenges faced by electrification programs, especially those targeting indigenous communities in the Amazon basin. The lack of institutional stability is identified as a significant obstacle, with frequent disruptive changes in electrification policies. While there is consensus on providing access to energy for all, a lack of funding allocation for the operation and maintenance of off-grid photovoltaic solutions is evident, compromising their sustainability. The work (Lillo et al., 2021) delves into the comparison of management models in renewable energy projects, analyzing cases in Peru, Ecuador, and Bolivia. It emphasizes the importance of strengthening community participation and proposes models based on microenterprises and cooperatives as the most comprehensive in terms of technical, economic, social, organizational, and environmental sustainability. However, this research highlights the lack of comparisons with similar projects in other countries and insufficient attention to trends and future challenges in rural electrification. The study (Galleguillos-Pozo et al., 2022) addresses the complexity of estimating energy demand in rural electrification projects and proposes a fuzzy mixed-integer linear programming model to incorporate uncertainty. While emphasizing the importance of balancing costs and user satisfaction, the research lacks comparisons with experiences in other countries and a detailed analysis of trends and future challenges in this context. Finally, the Yantsa Ii Etsari project in the Amazon region of Ecuador (presented in "Sustainability of renewable energy projects in the Amazonian region," 2018) illustrates a successful case of solar electrification. The collaboration between the electric company and academia underscores the importance of applied research to improve sustainability. However, there is a lack of comparisons with similar projects and limited attention to the assessment of long-term impacts.

Collectively, these studies highlight significant gaps in the literature on rural electrification in Ecuador. There is a need for comparative studies with international experiences, analysis of trends and future challenges, as well as comprehensive evaluations of environmental, social, and cultural impacts. These gaps need to be addressed to inform more effective and sustainable policies in the field of rural electrification in Ecuador. Therefore, this chapter addresses all identified gaps by conducting an in-depth study on rural and isolated electrification in Ecuador. The remainder of the chapter is organized as follows: Section "Current state of studies related to electrical power generation in rural and isolated areas in Ecuador" explores the current state of studies related to electrical power generation in rural and isolated areas in Ecuador. In Section "Contextualizing the energy landscape in remote and rural areas of Ecuador, a global and local perspective", we provide a comprehensive contextualization of the energy landscape in remote and rural areas of Ecuador, offering a global and local perspective. Section "Analysis of trends in the energy sector (2011–21): production, importation, exportation, consumption, and losses" offers an in-depth analysis of trends in the energy sector from 2011 to 2021, encompassing production, importation, exportation, consumption, and losses. In Section "Availability of renewable resources in Ecuador", we delve into the availability of renewable resources in Ecuador, while Section "Renewable energy projects in rural areas" sheds light on existing and upcoming renewable energy projects in rural areas. Specific challenges and their corresponding solutions for Ecuador in electrification and renewable sources are addressed in Section "Specific challenges and solutions for ecuador in electrification and renewable sources". The social and economic impact of the transition to renewable energies in Ecuador is discussed in detail in Section "Social and economic impact of the transition to renewable energies in Ecuador". Finally, in Section "Future perspectives and recommendations", we offer insights into future perspectives and provide recommendations based on the findings and analysis presented throughout the chapter.

Current state of studies related to electrical power generation in rural and isolated areas in ecuador

The results in WoS using the keyword "rural electrification" and refining results with "Ecuador" amount to 27, and they are divided and classified, as shown in Fig. 10.1. It is noteworthy that no specific results are found before 2016. Among the most important studies is, for instance, the study by Feron et al. (2016). Comparative studies across Peru, Ecuador, and Bolivia stress the necessity of strengthening community involvement, proposing models like microenterprises and cooperatives for holistic sustainability (Lillo et al., 2021). However, these models lack sufficient international comparisons and foresight into emerging challenges. The estimation of energy demand in rural electrification projects is addressed through a proposed linear

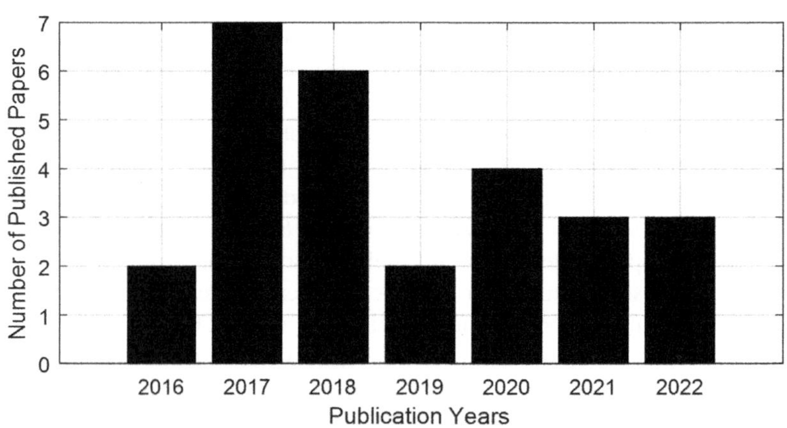

Figure 10.1 Publication trends of papers on rural and remote electrification in Ecuador (2016–22). Examining paper trends on rural and remote electrification in Ecuador from 2016 to 2022.

programming model; however, this study lacks comprehensive international comparisons and detailed future trend analyses (Galleguillos-Pozo et al., 2022).

In the Amazon region of Ecuador, the Yantsa Ii Etsari project exemplifies successful solar electrification, emphasizing collaborative efforts between electric companies and academia ("Sustainability of renewable energy projects in the Amazonian region," 2018). Nevertheless, the study reveals a need for broader project comparisons and a more in-depth assessment of long-term impacts. A broader examination of rural electrification efforts in Andean countries reveals a literature gap in the comparative analysis of project design processes, calling for a more nuanced understanding of decision-making implications (López-González et al., 2017). A technical-economic analysis in southern Ecuador assesses an off-grid renewable hybrid system's optimization, showcasing the feasibility of providing electricity through various energy sources (Cano et al., 2020). Finally, a cross-country evaluation of off-grid electrification in Chile, Ecuador, and Peru exposes challenges in sustainability dimensions, emphasizing the role of strong and decentralized institutions for success (Feron et al., 2017). In conclusion, the existing literature highlights the multidimensional challenges in rural electrification, advocating for comprehensive analyses, increased international comparisons, and holistic approaches to ensure sustainability.

The research on the South American Electric Energy System emphasizes the transition from fossil fuel-based energy systems to renewable energies, with a specific focus on the Ecuadorian case (Icaza et al., 2022). The study presents a flexible analysis using Energy Plan, proposing a 100% renewable electricity generation system for Ecuador by 2050. The suggested energy matrix includes 20 GW of installed power, with hydro (6.02 GW), solar PV (5.7 GW), and wind (5.61 GW) as major contributors. The average production cost is estimated to be approximately 18 US cents per MWh. The transition is expected to positively impact the economy, production levels, and citizens' quality of life. The research concludes by recommending stronger policies for a successful transition, acknowledging potential challenges related to historical dependence on oil. In another study focused on Cuenca, Ecuador, the research analyzes the city's electricity generation system through renewable energies, considering the growing demand and various projects (Icaza & Borge-Diez, 2019). Two power generation systems are proposed to meet the demand, emphasizing the need for reviewing land use ordinances and conducting prospective analyses for the city's growth. The article also discusses the potential of biomass production from solid waste and suggests a review of relevant aspects of IRES for Cuenca.

Rural electrification is addressed in a study designing an autonomous hybrid system for a national park in Ecuador (Lata-Garcia et al., 2018). The research combines fluvial and solar energies to meet electricity demand, aiming to reduce environmental impact and conserve endemic fauna and flora. The proposed hybrid system includes a hydrokinetic turbine and photovoltaic panels, with a backup subsystem consisting of a diesel generator. The simulation results show an optimized system fulfilling energy demand while minimizing diesel generator usage.

Another research focuses on off-grid hybrid power systems for the Cerrito de los Morreños community in Ecuador (Hidalgo-Leon et al., 2022). Four configurations, including diesel generators, solar photovoltaic systems, and battery storage, are analyzed for technical, economic, and environmental feasibility. The study emphasizes the positive impact of energy efficiency measures on the performance and cost-effectiveness of off-grid power systems. The configurations with energy efficiency initiatives are found to have lower costs and reduced CO_2 emissions.

In the Amazon Region of Ecuador, the technoeconomic assessment of microgrids, including photovoltaic arrays, a microhydro turbine, and diesel generation, is presented (Clairand et al., 2022). The study evaluates two scenarios, one considering diesel generation and another analyzing a carbon-free microgrid. The results suggest that the PV/hydro/diesel microgrid configuration is the most economically viable, providing new electrical alternatives for isolated communities. The study includes a sensitivity analysis based on projected diesel prices. The research on rural and indigenous electrification in Ecuador's Amazon Region addresses the critical issue of limited electricity access (Domenech et al., 2022). The proposed tool for designing standalone electrification systems based on solar energy considers economic, technical, and social aspects. The study emphasizes the feasibility of electrification, providing higher supplies than international benchmarks and economically competitive costs ranging from 2500 to 3000 USD per consumer. The tool is considered adaptable to various geographic contexts, supporting electrification policy-making in remote communities. The research on rural electrification in the Amazon region investigates the challenges of integrating indigenous visions of Buen Vivir and energy into the design of off-grid electrification projects (Palomares et al., 2016). The study explores the electrification priorities and aspirations of Kichwa, Siona, and Achuar communities in the Ecuadorian Amazon. The paper argues that recognizing the diversity of these visions provides an opportunity for new rural electrification logics that challenge the dominant development paradigm. The conclusion emphasizes the need for electrification projects to align with indigenous values, promote social transformation, and view energy beyond a commodity or poverty alleviation tool.

In another study, the focus shifts to the hill Curiquinga of Quingeo, Ecuador, where a solar–wind system is modeled and analyzed (Icaza, Cordova et al., 2018). The numerical model, constructed with 18 equations and coded in MATLAB®, aims to optimize resources and designs of hybrid systems for the hill Curiquinga. The research concludes that the model, validated with experimental data, provides accurate and reliable results. It suggests the potential implementation of the model by local authorities to harness the area's significant tourist potential. The design of a photovoltaic solar system for the GAD's Parish facilities of Ecuador is presented in another article (Daniel, Cabrera et al., 2018). The system is intended to provide electric power independently, especially during unforeseen cuts or when the local GAD requires it. The research discusses the significance of promoting such initiatives at the national level, aligning with the National Plan a Lifetime and the Constitution of the Republic of Ecuador of 2009, recognizing electric power services as a basic service. The introduction of electricity in rural areas as a means to achieve Buen Vivir and stimulate production development is explored in a study focused on Taday and Rivera in Ecuador (Vicuña et al., 2017). Interviews with various stakeholders highlight the positive impacts of electricity on health, education, women's participation in the labor force, reduction of social exclusion, and local economic activity. The study emphasizes the role of local agencies and government policies in stimulating rural development through the productive use of electricity. The importance of providing universal access to electricity in rural and remote areas is addressed in a comprehensive review focusing on the Amazon region of Ecuador (Apolo Cristhian, 2017). The research evaluates the use of Renewable Energy Sources for electricity supply, considering the challenges of implementing transmission and distribution lines in remote areas. The study presents a thorough analysis of photovoltaic home systems and centralized photovoltaic systems (CPVS), comparing their net present cost through microgrid optimization software. The CPVS is recommended as the best option for supplying energy to the rural community of Yuwints in the Ecuadorian Amazon.

Contextualizing the energy landscape in remote and rural areas of ecuador, a global and local perspective

Rural electrification has undergone a significant transformation over time, initially focusing on urban and industrial areas. According to data from the World Bank (Access to electricity in rural areas (% of rural population) | Data, 2023), by the mid-20th century, over 50% of the global population in rural areas lacked access to electricity, directly impacting quality of life and economic development. In Africa, over 50% of the rural population still lacks access to electricity, according to the International Energy Agency (IEA), with Congo having the lowest access at only 1% in 2022 (Access to electricity—SDG7: data & projections—analysis—IEA, 2023). The Sustainable Energy for All initiative in Africa (Sustainable Energy for All, 2023) works to address this challenge, aiming to provide universal access to modern energy services. In China, through the National Energy Administration (National Energy Administration, 2023), electricity has been brought to millions of rural households, improving living conditions. India, through the Saubhagya Program (Saubhagya, 2020), strives to electrify all households, benefiting a significant proportion of its rural population.

Latin America has improved rural electrification, reducing the rate of the population without access to electricity in rural areas from 46.1% in 2000 to 16.7% in 2022, according to the Economic Commission for Latin America and the Caribbean (Statistics & indicators—CEPALSTAT Databases and Statistical, 2022). However, challenges persist in remote areas, and solar energy projects are being implemented to close these gaps. Europe has largely achieved rural electrification, with a total of people with access to electricity reaching 99.79% by 2022. However, the focus now is on more sustainable solutions. The International Renewable Energy Agency (IRENA) (IRENA, 2022a,b) emphasizes the importance of microgrids and renewable energies for achieving a sustainable supply. In North America, most rural areas have access to electricity and are focused on making the grid more efficient and sustainable. The U.S. Energy Information Administration (Homepage—U.S. Energy Information Administration EIA, 2024) provides data on the integration of renewable energies in rural areas, with 99.4% of the population electrified, similar to China. Despite advancements, global challenges persist. Insufficient funding and the need for sustainable solutions are critical areas, according to the International Telecommunication Union (ITU) (International Telecommunication Union (ITU) Strategic plan for 2024–2027, 2023). The transition to renewable energies, supported by data from the International Food Policy Research Institute (IFPRI) (Home, IFPRI, 2023), is crucial for equitable and sustainable access to electricity worldwide. In summary, numerous projects aim to improve the quality of life for many people. However, as of 2023, 84.5% of the rural population still lacks access to electricity, according to the World Bank (Access to electricity in rural areas (% of rural population) | Data, 2023).

Fig. 10.2 illustrates the evolution of rural electrification in four countries. It is noteworthy that Germany has maintained 100% access to rural electricity since 1995. Regarding Colombia and Ecuador, they follow a similar trajectory, achieving full access in 2021. In the case of Ghana, there has been significant growth, starting from 7% in 1995 and reaching 74% in 2021, according to World Bank data. However, it is essential to acknowledge that these figures may not entirely reflect the reality. According to information from the Ministry of Energy and Mines (MERNNR) in Ecuador, a coverage of 97.46% is projected to be achieved by 2024, enhancing accessibility in both rural and urban areas. Furthermore, the Insular region already attained a coverage of 99.34% in 2019 (Loachamin et al., 2020).

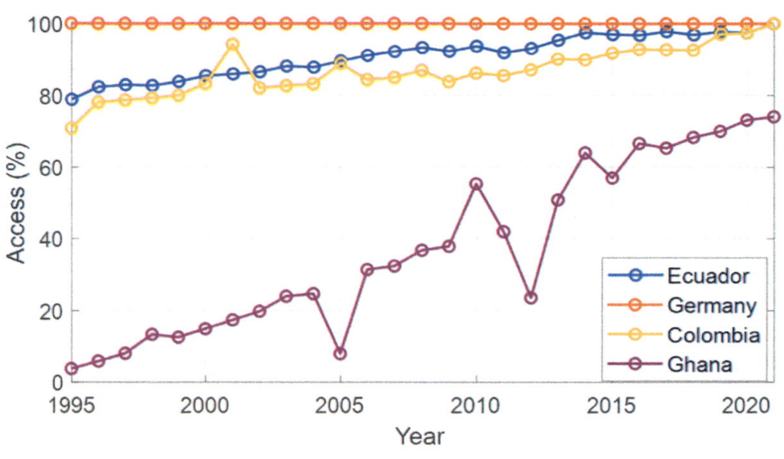

Figure 10.2 Rural electrification rates (% of rural population). Percentage of rural population with access to electricity, a key indicator of rural electrification rates.

Analysis of trends in the energy sector (2011–21): production, importation, exportation, consumption, and losses

In the realm of Ecuador's energy landscape, the dynamics of electricity supply and demand have been integral in shaping the nation's power infrastructure. The following data encapsulates a comprehensive analysis of the offering and requisition of electricity spanning the years 2011 to 2021. From the robust production figures to intricate details such as imports, exports, losses, and consumption patterns, this insightful examination provides a nuanced understanding of Ecuador's journey in the electricity sector. Each category, meticulously outlined with corresponding variations and percentages, sheds light on the intricate interplay between production capabilities, external dependencies, internal consumption, and the ever-evolving demands of a burgeoning nation. The units are measured in GWh, offering a granular perspective on the transformational decade in Ecuador's electrical landscape (Lasso & Xavier, 2022).

- **Production**. Referring to the gross energy generated, it has steadily increased since 2011, reaching its peak in 2021 at 32,207 gigawatt-hours (GWh).
- **Importation**. Representing the influx of energy from Peru and Colombia, despite notable fluctuations, there was a significant increase of 45.1% recorded in 2021.
- **Exports**. Destined for Peru and Colombia, exports exhibited notable variations, with a noteworthy decrease of 60.9% in 2021 compared to the previous year.
- **Total supply**. Encompassing energy generated not available for public service and energy delivered to large consumers in subtransmission, it increased by 6.3% in 2021, contributing to an average annual variation of 3.9%.
- **Own consumption**. In generation for public service, own consumption experienced a modest increase of 2.1% in 2021 compared to the previous year.
- **Losses**. Encompassing technical and nontechnical losses in transmission and distribution, losses increased by 5.2% in 2021.
- **Total energy consumption**. Encompassing own consumption and adjustments, total energy consumption saw an increase of 6.1% in 2021.
- **Total energy demand**. Reflecting consumption and losses, total energy demand increased by 6.3% in 2021 compared to 2020.

In Ecuador's pursuit of a sustainable and resilient energy landscape, the last decade has seen a notable transformation in the utilization of renewable energy sources, particularly in remote and rural areas. Examining the annual generation figures from 2011 to 2021 provides a comprehensive overview of the dynamics within various renewable energy categories, as shown in Fig. 10.3 (Lasso & Xavier, 2022).

- **Hydropower**. The ascendancy of hydropower has been a consistent theme, reaching a zenith of 25,575 GWh in 2021. With an average annual increase of 8.67%, hydropower stands as a cornerstone in Ecuador's renewable energy portfolio.
- **Thermal power**. In contrast, thermal power generation faced a decline, marking 6118 GWh in 2021. This decline, averaging 3.92% annually, signifies a shift toward cleaner and more sustainable energy alternatives.

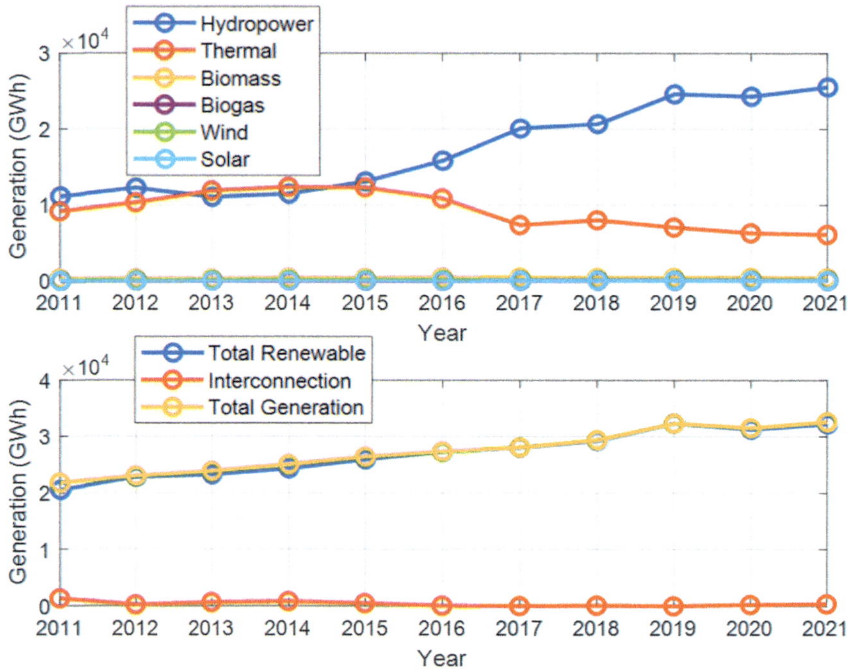

Figure 10.3 Renewable energy generation (2011–21). Decade in review: renewable energy generation trends (2011–21).

- **Renewable subcategories, biomass**. Biomass generation experienced a 12.61% decrease in 2021, totaling 373 GWh.
- **Biogas**. Despite its nascent stage, biogas contributed 42 GWh in 2021, with a decline of 4.23%.
- **Wind**. Wind energy, although witnessing a reduction of 19.57%, reached 62 GWh in 2021.
- **Solar**. The solar energy sector displayed remarkable growth, contributing 37 GWh in 2021 and witnessing a substantial increase of 90.65%.
- **Overall renewable generation**. The aggregate renewable energy generation experienced a 12.24% decrease in 2021, reaching a total of 514 GWh. This fluctuation underscores the dynamic nature of renewable energy production influenced by varying weather conditions and evolving technologies.
- **Interconnection and total generation**. Energy exchange demonstrated a noteworthy 45.06% increase in 2021, emphasizing the strategic importance of interconnected energy systems. The overall total generation saw a 3.40% rise, reflecting Ecuador's commitment to diversify its energy portfolio.

The data on effective power by the system in Ecuador from 2011 to 2021 provides insights into the country's diversified energy landscape, highlighting the contributions of interconnected systems and isolated subsystems, with a particular focus on the unique challenges and successes observed in the Galápagos Islands, as shown in Fig. 10.4 (Lasso & Xavier, 2022).

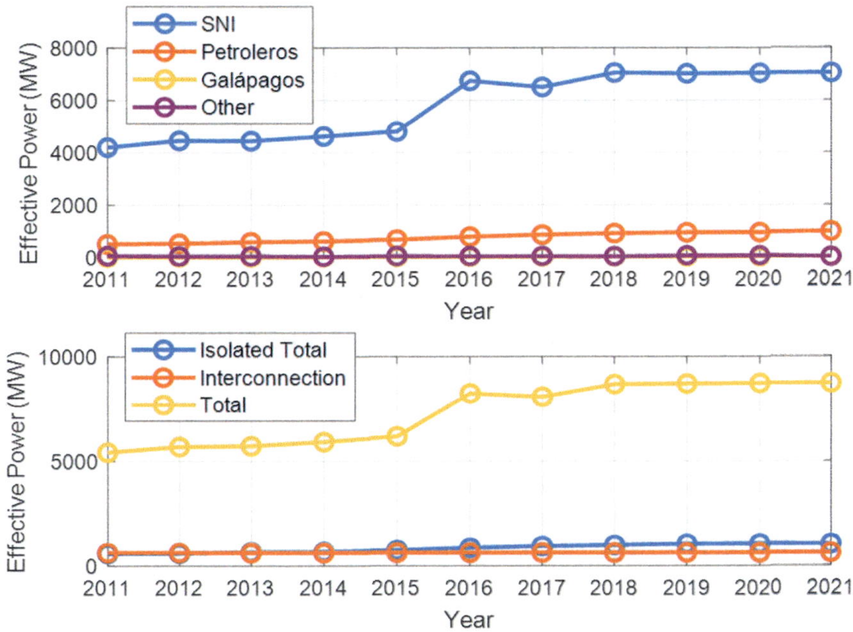

Figure 10.4 Effective Power by the System in Ecuador (2011–21). Empowering Ecuador: a concise glimpse at Effective Power by System from 2011 to 2021.

- **National Interconnected System (SNI)**. The SNI exhibits a consistent growth pattern, with an average annual increase of 5.3% in effective power. The SNI's effective power rose from 4201 MW in 2011 to 7048 MW in 2021, reflecting Ecuador's commitment to expanding its national grid and meeting growing energy demands.
- **Isolated Systems—Petroleros**. The Petroleros subsystem, within isolated systems, experienced a steady growth of 6.8% on average annually. Effective power in the Petroleros subsystem increased from 517 MW in 2011 to 1000 MW in 2021, emphasizing the role of isolated systems in contributing to Ecuador's overall energy capacity.
- **Isolated Systems—Galápagos**. The Galápagos Islands, functioning as an isolated subsystem, demonstrated a remarkable average annual growth of 10.7% in effective power. Despite its comparatively smaller contribution, this growth underscores the successful integration of sustainable energy practices in the Galápagos region.
- **Isolated Systems—Other**. The "other" isolated systems experienced a notable decrease of 65.8% in effective power on average annually. Contextual factors specific to these subsystems likely influenced this decline.
- **Total isolated systems**. The aggregate effective power from isolated systems saw a 5.9% average annual increase, contributing to a total of 1052 MW in 2021. This growth highlights the resilience and adaptability of isolated systems within Ecuador's energy infrastructure.
- **Interconnection**. The interconnection between the SNI and isolated systems remained constant at 635 MW throughout the entire period, showcasing a balanced approach to energy distribution between centralized and distributed sources.

- **Overall national effective power**. The total national effective power experienced a 4.9% average annual increase, reaching 8736 MW in 2021. This growth signifies Ecuador's ongoing efforts to enhance its energy capacity and infrastructure.
- **Galápagos Islands—A Unique Perspective**. The consistent growth in effective power for the Galápagos Islands reflects Ecuador's dedication to sustainable development in this ecologically sensitive region. The unique challenges of meeting energy demand while preserving the fragile ecosystem of the Galápagos Islands highlight the success of Ecuador's approach.

Availability of renewable resources in Ecuador

Ecuador holds substantial untapped renewable energy potential across various sources, as outlined in key energy atlases and resource inventories:

- **Wind energy**. According to the "Wind Atlas of Ecuador for Electricity Generation," (Wind Atlas of Ecuador: For Electrical Generation Purposes. ARIAE, 2023), the gross wind–electric potential in the country is 1691 MW, considering areas with an average annual wind speed exceeding 7 m/s. This potential could generate a mean energy output of 2869 GWh. The short-term feasible potential is estimated at an installable power of 884 MW and an annual mean energy of 1518 GWh. Notably, this potential could increase by including areas with lower average annual wind speeds (between 5 and 6 m/s).
- **Solar energy**. The "Solar Atlas of Ecuador for Electricity Generation" (Solar Atlas of Ecuador for Electrical Generation Purposes ARIAE, 2023) indicates an average direct insolation value of 2543 Wh/m^2/day, diffuse insolation of 2032 Wh/m^2/day, and global insolation of 4575 Wh/m^2/day across continental Ecuador. These values signify a significant solar energy resource that remains largely untapped.
- **Biomass**. The "Bioenergy Atlas of Ecuador" (Bionergetic Atlas of the Republic of Ecuador. ARIAE, 2023) estimates a biomass resource of 18.4 million tons/year, encompassing agricultural, livestock, and forestry residues. This biomass could potentially yield an estimated energy potential of 230,959 TJ/year, equivalent to 12,700 GWh/year. Utilizing just 50% of these residues could theoretically generate around 500 MW of firm electricity throughout the year.
- **Geothermal energy**. Regarding geothermal potential, CELEC EP is actively studying various sites. Based on the "Energy Resources Inventory of Ecuador for Electricity Generation" (2015) (Inventory of energy resources of Ecuador. ARIAE, 2023), the maximum potentials for certain sites are:
- Chachimbiro: Estimated potential of 178 MWe
- Tufiño-Chiles-Cerro Negro (binational): 330 MWe
- Chalupas: 283 MWe
- Chacana-Cachiyacu: 83 MWe
- Jamanco: 26 MWe

These geothermal resources present an opportunity for sustainable electricity generation. To provide a comprehensive visual overview, the energy source maps are attached at the end of the Master Electricity Plan (Electricity Master Plan, 2023). These maps delineate the diverse renewable energy potentials across the country, highlighting key regions for harnessing these valuable resources.

Renewable energy projects in rural areas

International overview

Renewable energy initiatives have become crucial drivers of sustainable development and improved living conditions, with numerous success stories at both national and international levels. India's Saubhagya Program, launched with the goal of electrifying all households, especially in rural areas, has successfully illuminated millions of homes, contributing significantly to India's rural electrification (Saubhagya, 2020). Similarly, China's unwavering commitment to rural electrification has resulted in remarkable success, providing electricity to millions of rural households and transforming communities (National Energy Administration, 2023). Moreover, the African Renewable Energy Initiative has played a pivotal role in accelerating the deployment of renewable energy projects across Africa, contributing significantly to rural electrification (Dinkelman, 2011). The Global Himalayan Expedition is another noteworthy initiative that has implemented solar microgrids in remote Himalayan villages, providing sustainable energy access and positively impacting the lives of isolated communities (Travel for Impact, 2023). Lighting Africa has successfully brought affordable and clean lighting solutions to rural communities across multiple African countries, contributing to improved education and livelihoods (Homepage—Lighting Global, 2024). In Bangladesh, the Renewable Energy for Rural Livelihood project has been instrumental in providing renewable energy solutions to rural communities, promoting sustainable livelihoods (Soumya et al., 2018). Additionally, SolarAid's initiatives in Malawi have showcased the transformative impact of solar lighting on education, health, and economic activities in rural areas, underscoring the potential of solar energy for rural development (SolarAid, 2023). These projects serve as inspiring examples of the successful implementation of renewable energy solutions, fostering sustainable development, economic empowerment, and an improved quality of life in rural regions.

National overview

In Ecuador, the Yantsa Ii Etsari Project in the Amazon region stands out as a successful solar electrification initiative, showcasing effective collaboration between electric companies and academia and emphasizing the potential of renewable energy in rural development ("Sustainability of renewable energy projects in the Amazonian region," 2018). In Ecuador, significant improvements in electrical service coverage have been achieved through strategic investments and projects focused on expanding and enhancing distribution systems. Emphasis has been placed on electrifying rural and marginalized urban areas, resulting in positive impacts on approximately 974 thousand households and around 4 million people. As of 2021, a substantial total investment of USD 431 million has been recorded in rural electrification projects. The ongoing Expansion and Improvement of Distribution Plan aims to achieve an impressive coverage of 97.99% by 2027 (2023).

An exemplary initiative in rural electrification is the Rural and Marginal Urban Electrification Fund (FERUM). FERUM's objective is to execute electrical projects that significantly improve living conditions in historically marginalized rural and urban areas. The focus is on promoting equity and creating favorable conditions for development, encompassing aspects such as education, health, communication, tourism, and agroindustry—all powered by electricity. In terms of investments, FERUM allocated USD 259 million from 1999 to 2006 (over 8 years), and this increased substantially to USD 440 million for the 2007–14 period, representing a doubling of investment in the latter years. The tangible results of these efforts are evident in the electrical service coverage, which stood at an impressive 96.77% by December 2013. The established target for December 2014 is 97.04%, positioning Ecuador as one of the countries with the highest electrical service coverage in the region (2009). Furthermore, in various rural areas of Ecuador, renewable energy projects have been implemented with the purpose of reducing environmental impact and preserving native fauna and flora. These projects involve the installation of renewable energy systems, such as solar panels and hydrokinetic turbines (More than 400 inhabitants of Mindo, 2021; Photovoltaic solar panels as electrification solutions in isolated communities in the Amazon, 2016).

One of the conducted studies focused on analyzing the technical, economic, and environmental feasibility of off-grid hybrid systems in the Cerrito de los Morreños community. These hybrid systems encompass diesel generators, photovoltaic systems, and energy storage through batteries, aiming to provide a solution that optimizes efficiency and reduces environmental impact (Juan et al., 2016). Additionally, a specific project has been developed for electrification in the GAD Parish. This project centered on the design of a photovoltaic system that ensures the electrical supply to the GAD Parish facilities. The main objective is to guarantee the continuity of the electrical service, especially during unplanned outages or situations requiring power supply in rural areas (Daniel, Cabrera et al., 2018).

The Electrification Project in Ecuador, known as MEER (Implementation of Energy Sources with Renewable Resources in Isolated Areas), is currently one of the most successful initiatives. It is spearheaded by the Ministry of Energy and Non-Renewable Natural Resources (MERNNR). Under the Subsecretariat of Generation and Transmission of Electric Energy, the project focuses on installing photovoltaic generation systems for individual solutions and mini networks to expand electrical coverage in isolated areas of the country. This project, funded through a nonreimbursable agreement from the Global Environment Facility managed by the Inter-American Development Bank (IDB), aims to supply electrical energy to households in remote areas of the Amazon Region and certain parts of the Coast, where access to conventional electrical services has been challenging.

MEER includes the implementation of individual photovoltaic systems of 350 Wp and systems for modular mini networks of 2.5 kWp, complete with storage and distribution networks. The service is specifically directed toward communities in the Amazon Region and the Coast, where geographical inaccessibility and socioeconomic conditions have hindered conventional electrification. With a 36-month execution period and a total budget of USD 4,219,090.00, the project aims to increase the installed capacity of renewable energy generation by 0.019%. It is expected to benefit approximately 832 families with access to electrical energy, contributing to the sustainable development of these communities. The implementation of suitable and sustainable systems aims to ensure the dignified delivery of electrical services in the long term, utilizing individual systems of 350 Wp and modular mini networks of 2.5 kWp. The necessary resources for project execution include transportation of materials to communities and labor for system implementation. The project's focus is on leveraging available renewable resources and avoiding contaminating thermal generation in these areas, marking a significant advancement toward sustainable electrification in Ecuador (Loachamin et al., 2020).

Projects in the Galápagos Islands

Under the "Renewable Energy in the Galápagos Islands" project, the expansion of the Isabela Hybrid Plant is underway. With a funding of 4.4 million Euros from Germany's KfW development bank, the goal is to increase the capacity of the photovoltaic solar plant and the energy storage system. The project encompasses the hiring of a company for engineering, procurement, construction, and commissioning, including the interconnection with the existing plant. Consultancy services are also sought for technical specifications, bidding documents, project oversight, and monitoring. The objective is to achieve energy transition in the Galápagos Islands, with 85% of energy from renewable sources by 2030 and a comprehensive electric transportation plan (Isabela Hybrid System—Phase II, 2022).

Annual investments (USD) in Rural and Marginal Urban Electrification:

- 2018: USD 89.07 million
- 2019: USD 40.07 million
- 2020: USD 25.93 million
- 2021: USD 18.84 million
- **Total: USD 173.91 million**

The Energy Transition Plan in the Galápagos Islands, named Evolution Energy, aims to reduce greenhouse gas emissions, mitigate the fiscal impact of subsidies, empower users, and ensure long-term sustainability. It proposes an expansion of photovoltaic and wind turbine capacity, along with active demand optimization measures to achieve 85% renewable energy by 2030. The plan has the potential to reduce 122,915.55 ton CO_2 between 2025 and 2030, with an estimated investment of USD 147 million. Emphasis is placed on women's participation in the sector, and there is the possibility of obtaining financing resources through carbon reduction certificates in the relevant markets (Isabela Hybrid System—Phase II, 2022).

Green hydrogen in Ecuador

Ecuador, as a signatory of the Paris Agreement, is committed to limiting global warming below 2°C and, preferably, 1.5°C. The country's National Determined Contributions (NDC) outline specific actions and targets to reduce greenhouse gas emissions. Ecuador's NDC aims to unconditionally reduce emissions in energy, industrial processes, waste, and agriculture by 9% or conditionally by 20.9% with international cooperation by 2025. Additionally, a 4% unconditional reduction or a 20% reduction with international cooperation is set for the land use sector. Coordinated efforts between public and private stakeholders, facilitated by effective governance, are crucial for achieving the outlined hydrogen capacity goals: 1 GW by 2030, 3 GW by 2040, and 6 GW by 2050 (Green Hydrogen roadmap in Ecuador, 2023).

In recent years, Ecuador has made substantial investments in renewable energy sources such as hydroelectric, wind, and solar, resulting in a 60% share of renewable energy in installed capacity for electricity generation. The National Energy Efficiency Plan 2016–35 aims to reduce fossil fuel usage, especially in the transportation sector. One noteworthy project is the Demonstration Center for the Production of Green Hydrogen and Methanol in the province of Manabí. This project focuses on producing green hydrogen and transforming it into green methanol, contributing to the decarbonization of the existing industrial sectors like fats and oils industries and maritime fishing vessels. The proposed electrolyzer of 200 MW would yield 22,000 tons of hydrogen annually, facilitating the installation of a green methanol production plant with a maximum capacity of 165,000 tons per year (Green Hydrogen roadmap in Ecuador, 2023).

Cost projections for hydrogen production until 2050 anticipate a decrease, particularly for solar photovoltaic and wind energy technologies. Despite the current higher costs of solar production, significant future reductions are expected. By 2030, an average levelized cost of hydrogen (LCOH) between 2.36 and 3.26 USD/kg is expected. Costs for 2040 are estimated to be around 1.71 and 1.66 USD/kg for wind and solar energy, respectively. By 2050, further cost reductions are anticipated, enabling Ecuador to produce hydrogen from wind and solar energy with LCOH of 1.27 and 0.89 USD/kg, respectively. Although geothermal and biomass residual energy incur higher costs, they provide an opportunity for Ecuador to leverage its potential in these areas (Green hydrogen roadmap in Ecuador, 2023).

Specific challenges and solutions for ecuador in electrification and renewable sources

The adoption of photovoltaic energy in Ecuador faces numerous challenges that hinder its full potential. This paper explores key obstacles and proposes strategic solutions to promote the widespread use of renewable energy in rural and isolate areas, aiming to contribute to a sustainable and diversified energy landscape (Reinoso & Geovanna, 2023).

State control and energy policy obstacle

State control across all energy aspects, coupled with a lack of a transparent energy policy, impedes private sector involvement. This limits comprehensive support and planning for unconventional renewable energies (UREs). Encouraging decentralization and enhancing private sector participation in the electricity market could provide a solution, fostering increased diversification and promotion of URE.

Hydropower favoritism obstacle

The prioritization and oversupply of electricity from hydropower, supported by political favoritism, have marginalized URE. This results in a reduced development of these renewable sources. A solution lies in developing a comprehensive plan and establishing a clear long-term energy policy to prioritize URE projects with specific goals.

Corruption and lack of transparency obstacle

Cases of corruption associated with mega hydropower plant construction have created distrust and deterred potential investors. This acts as a disincentive for participating in public procurement processes. A solution involves improving the technical evaluation of projects, steering clear of unsupported targets, and thereby increasing the viability and success of renewable generation projects.

Fossil fuel subsidies obstacle

Government policies providing subsidies to fossil fuels create a barrier, affecting the competitiveness of photovoltaic energy, particularly at a small scale. To address this, promoting the diversification of generation sources, especially harnessing solar and wind potential, would reduce risk and improve the attractiveness of photovoltaic energy adoption.

Uncertainty and nonrenewable dependency obstacle

A lack of clarity in future energy policy direction and dependence on nonrenewable sources create uncertainty and hinder the consolidation of renewable sources. A potential solution involves reviewing and adjusting subsidy policies to enhance URE competitiveness, eliminating cost distortions and encouraging photovoltaic energy adoption.

Lack of transparency in generation costs obstacle

Opacity in energy generation costs and a lack of mechanisms to reduce the kWh price hinder the competitiveness of photovoltaic solar energy. Incentivizing distributed generation can empower users, reduce dependence on large corporations, and increase citizen participation, contributing to the energy transition.

Political and institutional obstacles

A statist view of the electricity sector, coupled with a lack of a clear energy policy, politicization of energy planning, and insufficient interministerial coordination, hampers the promotion of photovoltaic energy. Addressing this involves establishing long-term energy policies, depoliticizing energy planning, and enhancing interministerial coordination for a smoother energy transition.

Regulatory obstacles

Bureaucracy in project development processes, a lack of separation between electricity service commercialization and distribution, and delays in project authorizations impede private sector participation. Simplifying bureaucratic processes, considering commercialization and distribution separation, and expediting authorizations for new projects can address these regulatory challenges.

Economic obstacles

A lack of tax incentives, difficulties accessing credit, and unattractive interest rates hinder investment in photovoltaic energy. To overcome these economic barriers, reducing tariffs and VAT, improving tax incentives, facilitating access to credit, and negotiating competitive interest rates are potential solutions.

Technical obstacles

Insufficient technical progress at the national level, a shortage of expert professionals, reliance on imported technical personnel, and limited renewable energy training constrain photovoltaic technology development. Solutions include investing in local research and training, encouraging specialization in renewable energy, and establishing quality control and local certification mechanisms.

Sociocultural obstacles

General unawareness of the benefits of photovoltaic energy, lack of available information, and a deeply rooted culture favoring hydropower complicate social acceptance. Addressing sociocultural barriers involves implementing energy education campaigns, incorporating renewable energy topics into national education, and highlighting the social and economic benefits of photovoltaic energy. In summary, these challenges and proposed solutions offer a roadmap for overcoming barriers and advancing the widespread adoption of photovoltaic energy in Ecuador.

Social and economic impact of the transition to renewable energies in ecuador

The social and economic impact of Ecuador's transition to renewable energy is poised for acceleration through proposed reforms. Facilitated by a $500 million credit approved by the IDB, Ecuador aims to instigate a fair energy transition encouraging both public and private investment. The program outlines support for policies promoting the decarbonization of the energy sector, emphasizing URE sources, alternative generation methods, energy efficiency measures, demand management, regional integration, and electric mobility. Additionally, the plan targets the energy transition in the Galápagos Islands, universal electricity access, and gender and disability gap reduction in the electricity sector. Anticipated outcomes of the program include increased renewable energy participation, the establishment of new electric vehicle charging stations, and improved access to electricity in rural and marginalized urban areas. The population, benefitting from expanded electricity services, envisions a sustainable, just, inclusive, and clean transition, leading to a reduction in greenhouse gas emissions (Ecuador will promote its energy transition & investments in the sector with support from the IDB, 2023).

Notably, the program plans to provide electricity services to 80,358 homes nationwide, with 25,645 in rural and marginalized urban areas. The private sector is also expected to thrive, especially in renewable energy generation, distributed generation, electric mobility, and alternative energy sources, generating employment opportunities and economic activity. Women and people with disabilities stand to gain from new job opportunities and access to technical training, violence prevention, and community engagement. The $500 million IDB credit combines $450 million from the IDB and $50 million from the Korea Infrastructure Development Co-financing Facility. The IDB segment features a 19-year repayment term with a 7-year grace period and an interest rate based on SOFR. The $50 million segment from the Korea Infrastructure Development Co-financing Facility has a 15-year repayment term with a 3-year grace period and a 2.5% annual interest rate (Ecuador will promote its energy transition & investments in the sector with support from the IDB, 2023; Leonardo & Aldás, 2019).

In assessing Ecuador's progress toward efficiency and sustainability, the paper uses indicators such as final energy intensity and levelized cost of electricity (LCOE). Final energy intensity shows a decrease, reflecting increased energy efficiency, particularly in the scenario with deeper intensity reduction. The LCOE analysis underscores Ecuador's abundant potential for renewable energy, suggesting a shift to a 100% renewable matrix, especially in scenarios emphasizing renewable energy sources. The National Energy Plan (PEN) 2050 further analyzes scenarios for a shift in energy and production matrices. With the ambition to transform from a raw material exporter to one of high-value goods and services, the PEN envisions changes in the energy sector, with scenarios ranging from Business as Usual to more ambitious alternatives. As Ecuador endeavors to develop its first National Energy Plan with support from the IDB and the French Development Agency, the focus is on internalizing scenario modeling capacities. This ensures future planning updates can be carried out independently by the MERNNR (Leonardo & Aldás, 2019, 2022; Ecuador will promote its energy transition & investments in the sector with support from the IDB, 2023).

Future perspectives and recommendations

In Ecuador, ongoing projects are paving the way for the implementation of renewable energies. Germany, a global leader in energy transition, is fostering collaboration with Ecuador to ensure the successful execution of clean energy initiatives. By 2030, Germans aim for at least 80% of electricity consumption to come from renewable sources, laying the groundwork to become a climate-neutral industrial nation by 2045. The Ecuadorian–German Chamber conducted a forum titled "Renewable Energies for Self-Consumption (Reducing Fossil Fuels) in Touristic and Conservation Areas in Ecuador." This event aimed to present plans and exchange viable experiences in the field. Academics, officials from the German and Ecuadorian public sectors, an international and national panel of entrepreneurs, along with representation from the Deputy Ministry of Energy, discussed implementation alternatives for these energies. The focus was on establishing an accelerated plan for Ecuador's energy matrix transition, minimizing environmental impact, and ensuring sustainability in touristic and conservation areas (This is what the future looks like for Ecuador in renewable energies—MINERGÍA Magazine, 2023).

Until 2022, Ecuador's electrical sector's energy matrix was primarily composed of hydroelectric power, followed by biomass (2.66%), wind energy (0.98%), solar energy (0.53%), and biogas (0.15%). The prospects for energy transition are promising, with URE projects like photovoltaic solar, wind, geothermal, and biomass already underway. A roadmap for the production and use of green hydrogen by 2030 has also been proposed. Given Ecuador's geographical features, especially in touristic and conservation areas (representing over 13.78% of the territory), the country holds significant potential for clean energy applications. The government, through the MERNNR, is actively working on a Universal Access policy and programs for Electricity by 2030. This involves rural and marginal urban electrification, with targets aiming for 97.99% national coverage by 2027. The ultimate goal is to structure programs and projects that achieve 100% coverage by 2030, aligning with the United Nations' Sustainable Development Goal No.7. Efforts are underway to identify families without access in rural areas, with a focus on electrifying households in unregulated housing developments in rural and marginal urban areas. The IDB has contributed $123.5 million in financing over the past decade, but increasing demand poses a challenge to achieving universal access (Electricity Master Plan, 2023).

In rural areas without electricity, two types of projects are being considered: connecting houses to the national distribution system and providing individual or clustered houses in remote areas with independent or centralized energy generation solutions. The government's current initiative, supported by the IDB, seeks to empower families in vulnerable areas by providing normalized electrical services. However, challenges include the identification of families without access and the need to maintain investment pace with population growth and rural economic development. The role of women in using electricity for productive purposes is crucial for the sustainability of specific initiatives. The IDB pledges to continue supporting Ecuador's efforts through technical assistance and financing to achieve the coveted goal of universal access by 2030. Additionally, a significant focus is on addressing challenges related to informal connections to the electrical grid, which lead to commercial losses and affect service quality in marginalized urban areas (Galapagos Islands Energy Transition Plan, 2023).

The Galápagos Islands present a unique opportunity for energy transition through the Evolutionary Energy Transition Plan. This plan aims to significantly reduce fossil fuel use in the short (2026) and medium term (2030) and achieve zero consumption in the long term (2050). The plan focuses on incorporating URE sources, developing energy efficiency and demand management programs, implementing energy storage systems, and ensuring 85% of demand is met by clean sources by 2030. The plan also emphasizes the evaluation of an integrated public transportation plan to reduce diesel demand, especially in major urban centers. It proposes assessing the development and implementation of a comprehensive maritime transport plan to make diesel use more efficient. The success of the energy transition process is contingent upon the successful execution of the first action, as it facilitates fundamental changes in available transportation fuels, which account for around 69% of the Archipelago's liquid fuel consumption (Galapagos Islands Energy Transition Plan, 2023).

References

Access to electricity in rural areas (% of rural population) | data (2023). https://data.worldbank.org/indicator/EG.ELC.ACCS.RU.ZS.

Access to electricity—SDG7: data and projections—analysis—IEA (2023). https://www.iea.org/reports/sdg7-data-and-projections/access-to-electricity.

Ahmad, J., Imran, M., Khalid, A., Iqbal, W., Ashraf, S. R., Adnan, M., Ali, S. F., & Siddique Khokhar, K. (2018). Techno economic analysis of a wind-photovoltaic-biomass hybrid renewable energy system for rural electrification: A case study of Kallar Kahar. *Energy.*, *148*. Available from https://doi.org/10.1016/J.ENERGY.2018.01.133.

Akikur, R. K., Saidur, R., Ping, H. W., & Ullah, K. R. (2013). Comparative study of stand-alone and hybrid solar energy systems suitable for off-grid rural electrification: A review. *Renewable and Sustainable Energy Reviews.*, *27*. Available from https://doi.org/10.1016/J.RSER.2013.06.043.

Apolo, C. (2017). Exploration of renewable sources for isolated systems. Doctoral Thesis, Polytechnic University of Leiria. https://iconline.ipleiria.pt/handle/10400.8/2832.

Bionergetic Atlas of the Republic of Ecuador. ARIAE (2023). https://www.ariae.org/servicio-documental/atlas-bionergetico-de-la-republica-del-ecuador.

Cano, A., Arévalo, P., & Jurado, F. (2020). Energy analysis and techno-economic assessment of a hybrid PV/HKT/BAT system using biomass gasifier: Cuenca-Ecuador case study. *Energy.*, *202*. Available from https://doi.org/10.1016/J.ENERGY.2020.117727.

Chaurey, A., & Kandpal, T. C. (2010). Assessment and evaluation of PV based decentralized rural electrification: An overview. *Renewable and Sustainable Energy Reviews.*, *14*(8). Available from https://doi.org/10.1016/J.RSER.2010.04.005.

Clairand, J. M., Serrano-Guerrero, X., González-Zumba, A., & Escrivá-Escrivá, G. (2022). Techno-economic assessment of renewable energy-based microgrids in the amazon remote communities in Ecuador. *Energy Technology.*, *10*(2). Available from https://doi.org/10.1002/ENTE.202100746.

Dinkelman, T. (2011). The effects of rural electrification on employment: New evidence from South Africa. *American Economic Review.*, *101*(3078). Available from https://doi.org/10.1257/aer.101.7.3078.

Domenech, B., Ferrer-Martí, L., García, F., Hidalgo, G., Pastor, R., & Ponsich, A. (2022). Optimizing PV microgrid Isolated electrification projects—A case study in Ecuador. *Mathematics.*, *10*(8). Available from https://doi.org/10.3390/MATH10081226.

Ecuador will promote its energy transition and investments in the sector with support from the IDB. (2023). https://bit.ly/3U25kFB.

Electricity Master Plan. (2023). https://www.recursosyenergia.gob.ec/plan-maestro-de-electricidad/.

Farghali, M., Osman, A. I., Chen, Z., Abdelhaleem, A., Ihara, I., Mohamed, I. M. A., Yap, P. S., & Rooney, D. W. (2023). Social, environmental, and economic consequences of integrating renewable energies in the electricity sector: A review. *Environmental Chemistry Letters.*, *21*(3). Available from https://doi.org/10.1007/S10311-023-01587-1.

Feron, S., Cordero, R. R., & Labbe, F. (2017). Rural electrification efforts based on off-grid photovoltaic systems in the andean region: Comparative assessment of their sustainability. *Sustainability.*, *9*(10). Available from https://doi.org/10.3390/SU9101825.

Feron, S., Heinrichs, H., & Cordero, R. R. (2016). Are the rural electrification efforts in the ecuadorian Amazon sustainable. *Sustainability.*, *8*(5). Available from https://doi.org/10.3390/SU8050443.

Galapagos Islands Energy Transition Plan. (2023). http://bit.ly/3RXHmsA.

Galleguillos-Pozo, R., Domenech, B., Ferrer-Martí, L., & Pastor, R. (2022). Balancing cost and demand in electricity access projects: Case studies in Ecuador, Mexico and Peru. *Mathematics*, *10*(12). Available from https://doi.org/10.3390/MATH10121995.

Green Hydrogen roadmap in Ecuador. (2023). https://bit.ly/48R6tUO.

Hidalgo-Leon, R., Amoroso, F., Urquizo, J., Villavicencio, V., Torres, M., Singh, P., & Soriano, G. (2022). Feasibility study for off-grid hybrid power systems considering an energy efficiency initiative for an Island in Ecuador. *Energies*, *15*(5). Available from https://doi.org/10.3390/EN15051776.

Home, IFPRI: International Food Policy Research Institute. (2023). https://www.ifpri.org/.

Homepage—Lighting Global. (2024). https://www.lightingglobal.org/.

Homepage—U.S. Energy Information Administration (EIA). (2024). https://www.eia.gov/.

Icaza, D., & Borge-Diez, D. (2019). Potential sources of renewable energy for the energy supply in the city of cuenca-ecuador with towards a smart grid. In *8th International conference on renewable energy research and applications* (ICRERA) (pp. 603–610). Brasov, Romania. Available from https://doi.org/10.1109/ICRERA47325.2019.8997114.

Icaza, D., Borge-Diez, D., & Galindo, S. P. (2022). Analysis and proposal of energy planning and renewable energy plans in South America: Case study of Ecuador. *Renewable Energy*, *182*. Available from https://doi.org/10.1016/J.RENENE.2021.09.126.

Daniel, I., Cabrera, J.B., & Arias, P. (2018). *Solar energy supply for the rural Parish GAD's of Ecuador*. Santiago de Cali, Colombia: IEEE ANDESCON. Available from https://doi.org/10.1109/ANDESCON.2018.8564659.

Icaza, D., Cordova, F., & Toledo, F. (2018). System of electrical generation by wind and solar sources in the archaeological surroundings of the Hill Curiquinga of Quingeo-Ecuador. In *6th IEEE international conference on smart grid(icSmartGrids)*. Nagasaki, Japan: IEEE. Available from https://doi.org/10.1109/ISGWCP.2018.8634462.

International Telecommunication Union (ITU) Strategic plan for 2024−2027 (2023). https://www.itu.int/es/Pages/default.aspx#/es.

Inventory of energy resources of Ecuador. ARIAE (2023). https://www.ariae.org/servicio-documental/inventario-de-recursos-energeticos-del-ecuador.

IRENA (2022a). https://www.irena.org/How-we-work/Europe.

IRENA (2022b). Scenarios for the energy transition: Experience and good practices in America. http://www.irena.org.

Isabela Hybrid System—Phase II. (2022). https://www.recursosyenergia.gob.ec/sistema-hibrido-isabela-fase-ii/.

Juan, P., Emérita, D., Benitez, A., Neira, A., & Aveiga, G. (2016). Evaluation of rural power grids systems of the cerrito de los Morreños Island in the Gulf of Guayaquil, Ecuador. In *14th LACCEI international multi-conference for engineering, education, and technology*, 20−22 July 2016. San Jose, Costa Rica: Unpublished content, 7. Available from https://doi.org/10.18687/LACCEI2016.1.1.276.

Kanase-Patil, A. B., Saini, R. P., & Sharma, M. P. (2010). Integrated renewable energy systems for off grid rural electrification of remote area. *Renewable Energy*, *35*(6). Available from https://doi.org/10.1016/J.RENENE.2009.10.005.

Lasso, G., & Xavier, V. (2022). Ministry of energy and mines. National Energy Balance. https://www.recursosyenergia.gob.ec/5900-2/.

Lata-Garcia, J., Jurado-Melguizo, F., Sanchez-Sainz, H., Reyes-Lopez, C., & Fernandez-Ramirez, L. (2018). Optimal sizing hydrokinetic-photovoltaic system for electricity generation in a protected wildlife area of Ecuador. *Turkish Journal of Electrical Engineering and Computer Sciences*, *26*(2). Available from https://doi.org/10.3906/elk-1706-23.

Leonardo, J., & Aldás, H. (2019). Diversification of the energy matrix in Ecuador: energy, socio-economic and environmental indicators. Master's Thesis, Flacso Ecuador, Unpublished content. http://hdl.handle.net/10469/15640.

Lillo, P., Ferrer-Martí, L., & Juanpera, M. (2021). Strengthening the sustainability of rural electrification projects: Renewable energy, management models and energy transitions in Peru, Ecuador and Bolivia. *Energy Research & Social Science*, *80*. Available from https://doi.org/10.1016/J.ERSS.2021.102222.

Loachamin, C., Vinicio, E., Delgado, V., & Oswaldo, M. (2020). Project information sheet 2020. MERNNR-Ministry of Energy and Non-Renewable Natural Resources, Unpublished content, https://bit.ly/3UkmJKb.

López-González, A., Domenech, B., Gómez-Hernández, D., & Ferrer-Martí, L. (2017). Renewable microgrid projects for autonomous small-scale electrification in Andean countries. *Renewable and Sustainable Energy Reviews*, *79*. Available from https://doi.org/10.1016/J.RSER.2017.05.203.

Mamaghani, A. H., Escandon, S. A. A., Najafi, B., Shirazi, A., & Rinaldi, F. (2016). Techno-economic feasibility of photovoltaic, wind, diesel and hybrid electrification systems for off-grid rural electrification in Colombia. *Renewable Energy*, *97*. Available from https://doi.org/10.1016/J.RENENE.2016.05.086.

Mandelli, S., Barbieri, J., Mereu, R., & Colombo, E. (2016). Off-grid systems for rural electrification in developing countries: Definitions, classification and a comprehensive literature review. *Renewable and Sustainable Energy Reviews*, *58*. Available from https://doi.org/10.1016/J.RSER.2015.12.338.

More than 400 inhabitants of Mindo, Ecuador, use renewable energy in their homes—PV magazine Latin America. (2021). https://bit.ly/3O4gpT2.

National Energy Administration (2023). https://www.nea.gov.cn/.

Olatomiwa, L., Mekhilef, S., Huda, A. S. N., & Ohunakin, O. S. (2015). Economic evaluation of hybrid energy systems for rural electrification in six geo-political zones of Nigeria. *Renewable Energy*, *83*. Available from https://doi.org/10.1016/J.RENENE.2015.04.057.

Ten Palomares, M., Aristizabal, A. B., Palomares, M. T., & Aristizabal, A. B. (2016). Visions of rural electrification in the Ecuadorian Amazon: Disputing hegemonic logics. *Letras Verdes, Revista Latinoamericana de Estudios Socioambientales*, *20*(20). Available from https://doi.org/10.17141/LETRASVERDES.20.2016.2181.

Photovoltaic solar panels as electrification solutions in isolated communities in the Amazon (2016). https://bit.ly/3vCCYYB.

Rajbongshi, R., Borgohain, D., & Mahapatra, S. (2017). Optimization of PV-biomass-diesel and grid base hybrid energy systems for rural electrification by using HOMER. *Energy*, *126*. Available from https://doi.org/10.1016/J.ENERGY.2017.03.056.

Reinoso, R., & Geovanna, P. (2023). Obstacles to the energy transition in Ecuador: the case of electricity generation from photovoltaic sources. Thesis (Master in Climate Change, Sustainability and Development). Simón Bolívar Andean University, Ecuador Headquarters. Environment and Sustainability Area. http://hdl.handle.net/10644/9311.

Saubhagya. (2020). Government of India, Ministry of Power. https://powermin.gov.in/en/content/saubhagya.

Shaahid, S. M., & El-Amin, I. (2009). Techno-economic evaluation of off-grid hybrid photovoltaic–diesel–battery power systems for rural electrification in Saudi Arabia—A way forward for sustainable development. *Renewable and Sustainable Energy Reviews*, *13* (3). Available from https://doi.org/10.1016/J.RSER.2007.11.017.

Shyu, C. W. (2023). Lessons from the World Bank's solar home system-based rural electrification projects (2000–2020): Policy implications for meeting Sustainable Development Goal 7 by 2030. *Energy Reports*, *9*. Available from https://doi.org/10.1016/J.EGYR.2023.01.128.

Singh, P., Khan, B., Haes Alhelou, H., & Prakash Mahela, O. (2020). Impressions of remote area electrification on social and economic indicators. *AIMS Energy*, *8*(1045). Available from https://doi.org/10.3934/energy.2020.6.1045.

Solar Atlas of Ecuador for Electrical Generation Purposes ARIAE (2023). https://www.ariae.org/servicio-documental/atlas-solar-del-ecuador-con-fines-de-generacion-electrica.

SolarAid. (2023). Combatting poverty and climate change. https://solar-aid.org/.

Soumya, M., Das Barun, K., & Hoque, N. (2018). Optimum sizing of a stand-alone hybrid energy system for rural electrification in Bangladesh. *Journal of Cleaner Production*, *200*. Available from https://doi.org/10.1016/J.JCLEPRO.2018.07.257.

Statistics and indicators—CEPALSTAT Databases and Statistical (2022). Publications https://statistics.cepal.org/portal/cepalstat/dashboard.html?indicator_id = 4178&area_id = 409&lang = es.

Sustainable Energy for All. (2023). https://www.seforall.org/.

Sustainability of renewable energy projects in the Amazonian Region. In *Sustainable energy mix in fragile environments*. Springer (2018). Available from https://doi.org/10.1007/978-3-319-69399-6_7.

This is what the future looks like for Ecuador in renewable energies—MINERGÍA Magazine. (2023). https://bit.ly/47E1mWU.

Travel for Impact. (2023). https://www.ghe.co.in/.

Vicuña, D., Rosa, M., Pizarro, J., Escribano, P., & Javier, E. (2017). Electrification, rural development and Buen Vivir. An analysis from the experience of rural territories of Taday and Rivera (Ecuador). *Cuadernos Geográficos*, *56*(2), 306—327.

Wind Atlas of Ecuador: For Electrical Generation Purposes. ARIAE. (2023). https://www.ariae.org/servicio-documental/atlas-eolico-del-ecuador-con-fines-de-generacion-electrica.

Advances in power smoothing techniques in renewable microgrids: a comprehensive review of the state of the art

11

Paul Arévalo[1,2], *Darío Benavides*[1,3], *Danny Ochoa-Correa*[2] *and Edisson Villa-Ávila*[1,2]

[1]Department of Electrical Engineering, University of Jaén, Jaén, Spain, [2]Faculty of Engineering, Department of Electrical, Electronics and Telecommunications Engineering (DEET), Universidad de Cuenca, Balzay Campus, Cuenca, Azuay, Ecuador, [3]Department of Electrical Engineering, University of Málaga, Malaga, Spain

Context and motivation

The transition toward a more sustainable and diversified energy matrix has driven the growth of renewable energies, such as wind, solar, and hydrokinetic, in modern electrical systems (Barra et al., 2021). However, the intermittent and variable nature of these sources presents significant challenges for the stability and reliability of energy systems (Zhou et al., 2013). Variability in energy generation can cause fluctuations in frequency and energy quality in electrical grids, underscoring the urgent need to develop effective power smoothing strategies. Recent research has shown that power smoothing techniques play a crucial role in mitigating power fluctuations, thus ensuring the operational stability of electrical systems. A variety of approaches have been developed to address this challenge, with a growing emphasis on the use of high-power energy storage systems (ESSs) (Barra et al., 2021).

In the realm of grid-connected residential microgrids, which integrate renewable generation and storage capacity, energy management strategies based on fuzzy logic have been designed to smooth the network power profile (Arcos-Aviles et al., 2019). Additionally, the utilization of kinetic energy from wind turbines for grid connections has been explored as a promising strategy for smoothing wind power (Wang et al., 2018). Kinetic ESSs, such as flywheel storage systems, have emerged as an attractive option for enhancing stability and energy quality in power systems and microgrids (Zhou et al., 2013). By leveraging advances in energy storage technologies and control strategies, it is possible to address challenges associated with renewable generation variability, thus advancing toward a more sustainable and reliable energy future (Jarnut et al., 2017; Mansouri et al., 2020; McCarthy et al., 2016). All of this, together with the increasing adoption of renewable energies such

Towards Future Smart Power Systems with High Penetration of Renewables. DOI: https://doi.org/10.1016/B978-0-443-29871-4.00011-7

as solar and wind, has posed significant challenges to the stability of electrical systems due to the inherent variability of these energy sources. To address this variability and ensure the reliability of electrical systems, further research into effective power smoothing strategies, especially in isolated and renewable-based systems, is required (Zhou et al., 2013). The study of energy storage technologies has become a key area of research in this field. Electrochemical supercapacitors, for example, have proven to be rapidly rechargeable energy storage devices used in a variety of applications, from pulse technology to electrical energy storage (Volfkovich, 2021). Research has also demonstrated the potential of porous media in improving the performance of photovoltaic (PV) thermal systems (Sangtarash et al., 2024).

However, the proper selection and comparison of these technologies are key aspects that require comprehensive and global assessment to determine their effectiveness and viability in different environments (Barra et al., 2021; Sebastián, 2021; Wang et al., 2018). Additionally, the study of island systems presents particular challenges in terms of energy stability and reliability due to their geographic isolation and dependence on external energy sources. Therefore, research on power smoothing techniques adapted to these specific conditions is essential to ensure a successful transition to more sustainable and autonomous electrical systems (Jarnut et al., 2017; Mansouri et al., 2020; Tripathy et al., 2017). In this context, this review study aims to provide a comprehensive and comparative overview of the different power smoothing techniques available globally, with a specific focus on their application in island systems. By analyzing and comparing the results of previous research, we hope to identify the most promising technologies and best practices for improving the stability and reliability of renewable energy-based electrical systems, thus contributing to a more effective transition towards a sustainable and reliable energy future.

Literature review

The existing literature provides a comprehensive insight into power smoothing strategies applied to various renewable energy sources. For instance, studies such as Barra et al. (2021) offer a thorough review of high-power ESSs specifically used for wind power smoothing, highlighting the increase in publications on this topic in recent years and the need for further research to address wind energy fluctuations. Furthermore, research like that presented in Arcos-Aviles et al. (2019) explores fuzzy logic-based energy management strategies to smooth the power profile of grid-connected residential microgrids, emphasizing fuzzy logic-based control design and its effectiveness in improving the power profile of residential microgrids. The utilization of kinetic energy from wind turbines for grid connection has also been investigated, as described in Wang et al. (2018), where emerging technologies for frequency regulation and low-voltage ride-through capability utilizing kinetic energy stored in wind turbines are analyzed, emphasizing applications of this energy for wind smoothing.

Likewise, the use of flywheel-based ESSs has been extensively researched, as presented in Zhou et al. (2013), where the structures and applications of these systems in power systems and microgrids are examined, highlighting their role in improving energy quality, supporting renewable energy integration, and enhancing system stability. Studies like Zhou et al. (2013) offer an overview of energy storage technologies used in marine current energy systems, emphasizing the importance of high ESSs for long-term energy fluctuations smoothing due to tidal phenomena. Within the same context, research on electrochemical supercapacitors, as presented in Volfkovich (2021), provides a detailed review of these rapidly rechargeable energy storage devices, highlighting their application in electrical energy storage and peak load mitigation in electrical grids.

The presented works address various crucial aspects in renewable energy integration, energy storage, and energy system stability. Beginning with the analysis of short-term solar variability in Rajasthan, India, the importance of understanding and managing this variability for efficient large-scale solar energy integration into grids is highlighted. Utilization of ground-measured solar irradiation data allows for quantification of the impact of this variability and assessment of the effectiveness of spatial smoothing techniques in reducing fluctuations, which is fundamental for successful solar energy implementation in the region (Menadi et al., 2024). Furthermore, deeper dives into energy storage technology and applications recognize their vital role in energy system operation to mitigate renewable energy source intermittency and enhance system stability. From biomass and gas storage to more advanced technologies like zinc-bromine flow batteries, performance characteristics are examined, and future development trends, challenges, and opportunities in energy storage are identified (Tripathy et al., 2017).

Isolated microgrids with wind and diesel power are also under study, with a focus on dynamic simulation to evaluate system quality and stability. The review highlights different operation modes of these microgrids as well as the role of various energy storage technologies in improving their quality and stability. Dynamic simulation examples are provided to illustrate microgrid behavior under wind speed variations and load consumption (Sebastián, 2021). On the other hand, the selection of energy storage technologies for prosumer microgrids is also examined, emphasizing the importance of location and technical characteristics of storage devices to optimize renewable energy integration and increase locally generated energy self-consumption. Comparison of different technologies, such as lithium batteries and vanadium flow batteries, provides a comprehensive view of available options and their potential applications in prosumer microgrids (Jarnut et al., 2017).

Some authors have conducted review studies discussing the importance of ESSs in improving system quality and stability, as well as providing ancillary services for grid support (Mansouri et al., 2020). Additionally, the potential of flutter-based energy harvesting devices in fluid streams is explored, highlighting research approaches and potential applications of this emerging technology. Key challenges, such as low power density and structural fatigue issues, requiring further research to overcome, are identified (McCarthy et al., 2016). The review on battery storage utilization in the value chain of energy systems underscores the increasing importance of this system in addressing congestion and imbalances in conventional electrical grids. An exhaustive analysis of various battery technologies, from well-known lithium-ion and lead-acid batteries to emerging technologies like vanadium flow batteries, is conducted. Operational principles, control strategies, and technical characteristics of each technology are highlighted, as well as their application in ancillary services for microgrids and large-scale energy systems (Ratshitanga et al., 2024).

Research gaps

Despite the exhaustive research provided by scholars, several criteria remain to be studied regarding the advancement in power smoothing techniques, some of which are enumerated below:

- Despite the diversity of energy storage technologies mentioned in the literature (Barra et al., 2021; Volfkovich, 2021; Zhou et al., 2013), such as flywheel storage systems, electrochemical supercapacitors, and vanadium flow batteries, there is still a lack of comprehensive comparative evaluation of these technologies in terms of effectiveness and specific applicability for smoothing power generated by renewable sources. This gap can help guide the appropriate selection of storage technologies for different power smoothing applications.
- Current research highlights the need for strategies tailored to the particular conditions of island systems, such as their geographic isolation and dependence on external energy sources. However, there is a gap in the development and evaluation of specific power smoothing strategies for these systems, considering the integration of multiple renewable energy sources and storage systems, and partial research on this topic is presented in the literature (Benavides et al., 2024; Villa-Ávila et al., 2024; Zhou et al., 2013).
- Despite the variety of energy storage technologies currently available and studied in the literature (Gao et al., 2024; Wu et al., 2024; Zhou et al., 2013), there is a lack of research on how to integrate these technologies synergistically to maximize power smoothing efficiency, especially in microgrid systems or island environments. This could explore the potential of combining different technologies to enhance energy supply stability and reliability.
- A significant gap lies in the optimization of ESSs for particular applications, such as residential microgrids or marine current energy systems presented in the literature (Arcos-Aviles et al., 2019; Jarnut et al., 2017; Wang et al., 2018). Further research is required to understand how to design and operate these systems optimally, considering the variability of renewable generation and specific environmental characteristics.
- Although some emerging technologies are mentioned, such as flutter-based energy harvesting devices (McCarthy et al., 2016; Ratshitanga et al., 2024), deeper research is needed on these technologies. This includes their technical feasibility, implementation challenges, and potential applications in power smoothing, which could contribute to the development of innovative and effective solutions to address renewable generation variability.

Contributions and chapter organization

To address the identified research gaps in the field of power smoothing techniques, several key contributions are proposed. Firstly, a comprehensive comparative evaluation of various energy storage technologies mentioned in the literature, such as flywheel storage systems, electrochemical supercapacitors, and vanadium flow batteries, is suggested to better understand their effectiveness and specific applicability for smoothing power generated by renewable sources. Additionally, the development of specific power smoothing strategies adapted to the unique conditions of island systems is proposed, considering the integration of multiple renewable energy sources and storage systems. Furthermore, investigating how to synergistically integrate different energy storage technologies to maximize power smoothing efficiency, especially in microgrid systems or island environments, is suggested. Research is also proposed to optimize ESSs for particular applications, such as residential microgrids or marine current energy systems. Finally, deeper research into emerging technologies, such as flutter-based energy harvesting devices, is suggested to explore their technical feasibility, implementation challenges, and potential applications in power smoothing. The subsequent sections of this chapter are structured as follows: Section "Contextualization and state of the art" provides a thorough contextualization and state of the art, offering a comprehensive overview of the current landscape. Following this, Section "Power smoothing techniques" delves into the intricacies of power smoothing techniques, detailing various methodologies and their applications. Section "Current challenges and future directions" addresses the contemporary challenges and outlines future directions, presenting an analysis of the obstacles ahead and potential avenues for advancement. Lastly, Section "Conclusion" encapsulates our findings and deductions in a concise conclusion.

Contextualization and state of the art

Definition of power smoothing

"Power smoothing" refers to the action of reducing fluctuations in energy generation, especially in systems incorporating renewable energy sources. The goal is to maintain a more stable and uniform power output, which is essential for preserving the stability and reliability of electrical systems, particularly in situations where energy generation is intermittent and variable. This process relies on various techniques and strategies designed to smooth variations in generated power. It involves the use of high-power ESSs, as well as the application of adaptive control techniques and the integration of different storage technologies. The implementation of these techniques allows for the optimization of renewable energy utilization and reduces dependence on conventional energy sources, contributing to a more effective transition toward a sustainable and reliable energy future (Barra et al., 2021; Jarnut et al., 2017; Mansouri et al., 2020; Sebastián, 2021; Wang et al., 2018). In Fig. 11.1, the methodology used in this chapter is depicted.

History and evolution of power smoothing technique

Indeed, the history of power smoothing in renewable energy is marked by a series of technological advancements and contributions from various researchers over time. Although there are no exact data on its inception or who specifically is

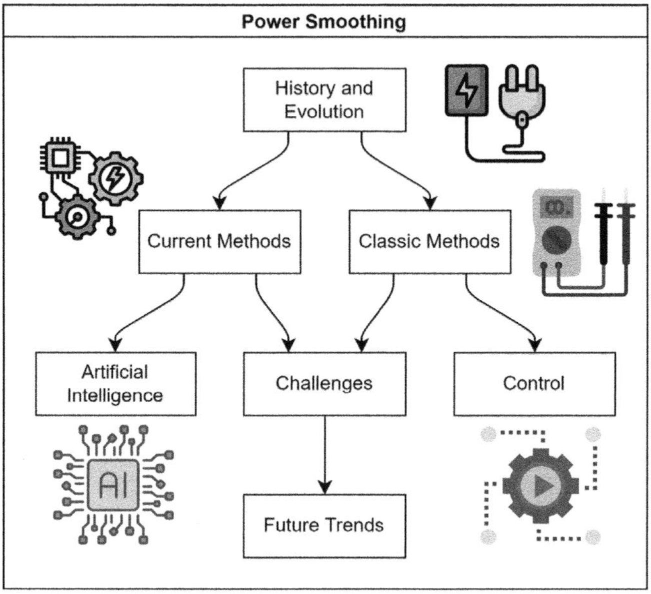

Figure 11.1 Proposed methodology overview. Theoretical framework.

credited with inventing this technique, it can be said that the earliest works in the field of power smoothing in renewable energy likely date back to the 1970s and 1980s when the first commercial applications of wind and solar technologies began to emerge. Authors and pioneers in the field of renewable energy, such as William Heronemus (Manwell, et al., 2001; Prof Dr & Goetzberger, 2023), may have contributed to laying the groundwork for research in this field. As research in renewable energy progressed and challenges associated with variability in energy generation were identified, other researchers and scientists began to explore solutions to mitigate these fluctuations. Researchers such as Daniel Kirschen, Goran Strbac, and Jianzhong Wu have made significant contributions to the development of control techniques and ESSs for power smoothing in electrical systems. In recent decades, the field of power smoothing has experienced significant growth, with a wide range of research and technological developments conducted by scholars, engineers, and scientists worldwide. Scientific and technical literature on this topic has increased considerably, addressing aspects such as control algorithms, ESSs, renewable integration into the grid, and specific applications in island systems (McCarthy et al., 2016).

In the current energy landscape, power smoothing techniques have become crucial due to the growing dependence on renewable sources and the urgent need to maintain stability in electrical grids. With the expansion of wind and solar energies, whose generation is inherently variable and intermittent, significant challenges arise to ensure a constant and reliable electricity supply. In the case of wind systems, fluctuations in wind speed can result in drastic variations in power production. Recent research has shown that integrating technologies such as lithium-ion batteries and blade angle regulation can mitigate these fluctuations, improving supply quality and reducing the associated economic impacts (Galli et al., 2024). Similarly, in PV solar systems, the use of advanced ESSs alongside control algorithms can smooth power fluctuations, ensuring grid stability even under variable solar irradiance conditions. The combination of storage technologies, such as lithium-ion batteries and supercapacitors, has been shown to significantly reduce power oscillations while extending system lifespan (Yan et al., 2024).

Furthermore, the development of predictive control techniques based on neural networks has opened up new possibilities for reducing energy fluctuations and optimizing battery state of charge, outperforming conventional power smoothing methods (Cano et al., 2024). In microgrids, innovative control strategies, such as the use of multiple hydrogen storage tanks and hierarchical control systems, can significantly improve stability and responsiveness to load demand, enabling efficient autonomous operations (Pelosi et al., 2024).

It is also crucial to consider the evaluation of storage requirements for power smoothing applications in PV plants. Detailed methodologies for determining the optimal size of storage systems, considering technical and economic aspects, can result in significant reductions in required storage volume (Amorim et al., 2024). Similarly, the implementation of sensorless controllers and the use of multistage flying capacitor inverters are technological advancements that allow for the optimization of energy extraction and smoothing of power injection into the grid, increasing reliability and reducing system costs (Khlifi et al., 2024). In an increasingly renewable energy-focused world, power smoothing techniques are essential to ensure effective integration, improve grid stability, and offer significant economic and environmental benefits. Their relevance in the current context is undeniable, as they address critical challenges on the path toward a sustainable energy future (Abdelghany, Al-Durra, et al., 2024).

Currently, when performing a search with the term "power smoothing" in Web of Science, the results shown in Fig. 11.2 are obtained and Fig. 11.3 illustrates the quantity of publications per country, delineating the primary keywords preceding a deeper exploration of the topic.

Power smoothing techniques

A comprehensive classification of various power smoothing techniques encompasses a wide range of methods aimed at stabilizing energy generation and consumption in renewable energy systems. In this chapter, these techniques will be classified under certain indices that are important within the field of electrical microgrids specifically. These techniques fall into different categories, including control strategies, objective function optimization methods, and energy conversion

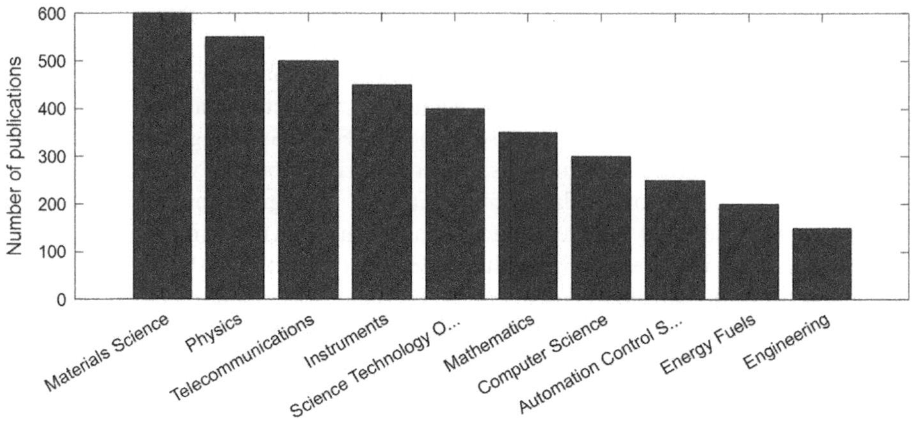

Figure 11.2 Analysis of publication distribution by area.

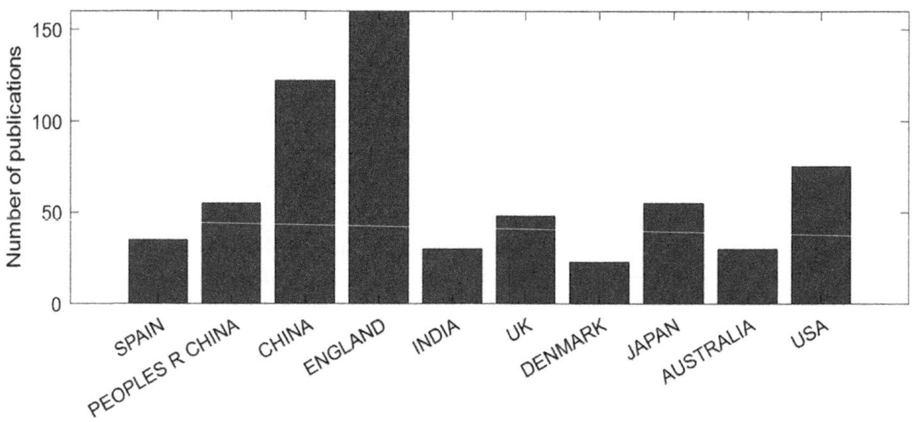

Figure 11.3 Analysis of publication distribution by country.

approaches, among others. To better understand what the power smoothing technique entails, Fig. 11.4 displays the PV output power without applying any power smoothing method and applying the moving average method. It is noticeable that the reduction in peaks is significant despite it being a basic technique.

Adaptive control

Adaptive control emerges as an essential component in the landscape of power smoothing, offering adjustable dynamics to balance energy supply with demand in real-time. Strategies such as adaptive control, discussed in Wu et al. (2024), allow for dynamic allocation of power among different energy storage devices, such as batteries and supercapacitors, thus optimizing energy flow according to real-time

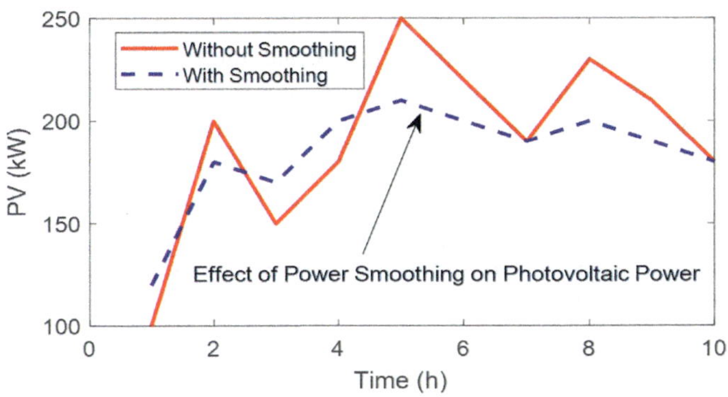

Figure 11.4 Effect of power smoothing on PV output power.

conditions. This type of control is based on learning from past errors, continuously adjusting the system based on the amount of wasted or surplus energy, which may be related to the intermittency of renewable sources. Various research works explore adaptive approaches to enhance system efficiency, ranging from the use of adaptive affine projection algorithms in superconducting energy storage units to adaptive fuzzy logic control techniques in hybrid microgrid energy management.

For instance, Hasanien (2014) presented an adaptive control scheme for superconducting energy storage units aimed at smoothing the power output of wind farms. This technique, based on the Set-membership Affine Projection Algorithm, allows for faster adjustment and lower computational complexity, demonstrating its effectiveness in improving grid stability. On the other hand, Ambia and Al-Durra (2015) introduced an innovative adaptive power smoothing control technique for hybrid microgrids, incorporating an adaptive fuzzy logic control algorithm to efficiently manage energy during transitions between grid-connected and islanded modes, demonstrating its ability to maintain energy balance even under extreme conditions. Similarly, Meng et al. (2018) proposed an adaptive control design for autonomous operation of multiple ESSs in power smoothing applications, using an adaptive cutoff frequency method to optimize the operation of distributed systems autonomously and without the need for external communication. Jamsheed and Iqbal (2022) investigated the use of a battery-based ESS to improve the stability of wind energy systems through a smart control approach based on neural identification and simultaneous control, highlighting the controller's adaptability to changing operating conditions.

Predictive control

Predictive control, using advanced algorithms such as neural networks, is employed to anticipate energy fluctuations and proactively adjust energy production. This approach is crucial for managing the variable generation of renewable sources such as PV systems, especially with the increasing demand for fast charging stations for electric vehicles. To effectively manage power fluctuations in wind-solar microgrids, a cascaded predictive control is proposed, integrating multiple hydrogen storage tanks. This method allows the microgrid to operate autonomously without connection to the electrical grid. An innovative approach combines neural predictive control and exponential moving average to smooth power, optimize battery charging, and mitigate ramp rates. This system outperforms conventional methods.

Feedback output control in PV systems, using a multilevel flying capacitor inverter, ensures maximum power and corrects power factor. To smooth wind power, a dynamic approach based on dynamic waves, using a hybrid ESS, is proposed. This method addresses uncertainty in wind energy through robust predictive control. Additionally, frequency control mechanisms monitor and regulate grid frequency.

To delve deeper into these techniques, the most significant studies have been analyzed from the perspective of relevance in the Web of Science. For example, in the study presented in Khalid and Savkin (2010), a model-based predictive control approach is proposed to smooth the output of wind energy using controlled battery storage. Similarly, Di Cairano et al. (2013) described a model predictive control (MPC) approach to smooth motor transitions in a hybrid powertrain, using battery storage. Furthermore, Khalid and Savkin (2014) focused on minimizing distributed battery storage capacity in a wind farm using predictive control and wind energy prediction. Meanwhile, Liu et al. (2017a, 2017b) presented an operation model of a hybrid ESS based on wavelet dynamic decomposition and predictive control to smooth wind energy output. Other studies, such as Meng et al. (2017), proposed a flexible control strategy for maximum power point tracking in wind turbines, considering both conversion efficiency and power smoothing. On the other hand, Bo and Johansen (2017) described a power variation smoothing method using batteries in a marine power plant, with nonlinear predictive control. Similarly, Lei et al. (2018) presented an MPC-based energy storage control method to smooth PV energy output. Nasiri et al. (2019) discussed predictive control of a CHB STATCOM in a wind farm application to smooth active and reactive power. In the field of PV energy, Han et al. (2020) proposed a dual regulation control method for a distributed ESS in smoothing PV power fluctuations. Meanwhile, Shaltout et al. (2020) presented an economic MPC control approach to smooth wind power and mitigate tower loading.

For wind energy, Wu et al. (2021) described a stochastic predictive control approach for wind energy smoothing using ESSs. On the other hand, Park and Zadeh (2021) proposed a predictive control approach for onboard dc energy hybrid systems in ships to improve voltage regulation and power smoothing. Additionally, Syed and Khalid (2021) described a neural network-based predictive control approach for PV power smoothing with battery energy storage. Meanwhile, Syed and Khalid (2023) presented an IMPC control strategy for stable dispatch of solar-wind renewable energy with battery energy storage and hydrogen electrolysis. In the transportation field, Lian et al. (2023) proposed a real-time energy management strategy for hybrid fuel cell electric buses using short-term power smoothing prediction. Finally, Abdelghany, Mariani, et al. (2024) described a unified control platform for integrating wind-hydrogen systems into the electrical grid using predictive control and mixed-logical dynamic modeling, while Lin, Cao, et al. (2024) proposed an MPC and WMA control approach for energy management and HESS capacity allocation in wind farms.

Objective function optimization

On the other hand, objective function optimization methods focus on maximizing energy efficiency and minimizing fluctuations through mathematical modeling and algorithmic optimization. These techniques involve formulating optimization problems to determine the optimal size and operation of ESSs. By defining objective functions that prioritize factors such as energy quality, system stability, and profitability, these methods allow for the design of robust and efficient power smoothing solutions.

Specifically, objective function optimization methods aim to maximize energy efficiency and minimize fluctuations through the use of mathematical modeling and algorithmic optimization techniques. This approach stands as a cornerstone in the quest for solutions that enable more effective management of ESSs, addressing crucial aspects such as energy quality, system stability, and profitability. In a recent study on wave energy conversion for generator sizing optimization and hybrid system integration, the importance of smoothing power variations in wave energy converters to optimize generator size and associated converters is highlighted. This study reveals the need to understand the impact of mechanical ESSs on the performance of wave energy converters, leading to the proposal of a method for selecting the optimal generator capacity based on different wave climates.

Furthermore, technology and applications of ESSs have been thoroughly reviewed, recognizing their crucial role in grid operation to smooth the intermittency of renewable energy generation and improve system stability. This analysis covers a variety of ESS technologies, from biomass storage to gas storage, highlighting their historical evolution, performance characteristics, and advanced materials. Additionally, applications of these technologies in short, medium, and long-term power systems are identified, ranging from energy generation to distribution and final consumption.

On the other hand, the use of battery-based ESSs has emerged as a key solution to address instability challenges in conventional power grids. This study details the global landscape of integrating battery technologies into renewable energy production systems and utility power grids. A comparative analysis of various battery technologies, such as lithium, lead-acid, sodium-sulfur, and vanadium flow batteries, is conducted, highlighting their application in ancillary services and control techniques that improve the performance of microgrids and conventional power grids. These studies are focused on renewable energy microgrids, within a world with similar studies that are widely studied in the literature review.

Power smoothing techniques using energy storage and energy conversion systems

Lastly, energy conversion techniques play a crucial role in transforming energy from variable renewable sources into stable electricity compatible with the grid. Hybrid optimization methods combine different energy conversion technologies to maximize energy capture and utilization in variable environmental conditions. Additionally, innovative approaches such as using electric vehicle batteries for power smoothing leverage the existing infrastructure to store and distribute energy efficiently, reducing dependence on traditional grid resources. To delve into the details of these power smoothing techniques, energy conversion techniques play an essential role in the process of transforming variable renewable sources into a form of electricity stable and compatible with the power grid. These methodologies not only address the capture and efficient utilization of energy in changing environmental conditions but also explore new innovative ways to improve energy system stability and efficiency.

A key strategy in this regard is hybrid optimization, which merges various energy conversion technologies to maximize the utilization of energy resources in variable environments. This approach, supported by research such as that conducted by Menadi et al. (2024) and Pelosi et al. (2024), aims to combine, for example, wind and solar energy with storage systems such as electric vehicle batteries. The latter point is particularly relevant given the increase in electric vehicle adoption and associated charging infrastructure. As mentioned in Villa-Ávila et al. (2024), the strategic use of these batteries not only contributes to smoothing energy fluctuations but also optimizes system efficiency by innovatively leveraging the existing infrastructure. A notable example is the power smoothing method presented in the study titled "Innovative Power Smoothing Technique for Enhancing Renewable Integration in Insular Power Systems Using Electric Vehicle Charging Stations." This approach proposes an improved "ramp-rate" technique that incorporates fuzzy logic to control the ramp rate in real time, using electric vehicle batteries to optimize self-consumption and reduce power fluctuations. Experimental validation of this technique demonstrates a significant 14% reduction in energy generation variability, resulting in a more stable and reliable power supply profile.

Additionally, the importance of real-time implementation of hybrid approaches is highlighted, as discussed in Menadi et al. (2024), where the implementation of a hybrid ESC control approach to maximize PV power extracted under partial shading conditions is analyzed. This study demonstrates how the combination of methods such as super twisting control and modified extreme control can improve the efficiency of solar energy generation systems by optimizing maximum power point tracking, even under adverse conditions. Then, the integration of hybrid ESSs, as described in Pelosi et al. (2024), offers a promising solution to smooth power oscillations in energy production from sources such as wave energy. By combining technologies such as lithium-ion batteries and supercapacitors, the stochastic nature of wave energy generation is mitigated, thus improving the quality and stability of the energy delivered to the power grid.

Lastly, hydrogen plays a crucial role in power smoothing in PV generation systems, as described in several studies. For example, in Takahashi et al. (2019), a novel power smoothing control method for a PV generation system using a water electrolyzer is proposed. This method distributes the long and short-term components of PV power fluctuations, supplying the long-term components to the power grid and the short-term components to the electrolyzer, which converts energy into hydrogen for fuel cell vehicles. On the other hand, Kong et al. (2021) focuses on the technoeconomic analysis of hydrogen energy for power smoothing in renewable energies, optimizing the capacity of electrolyzers and fuel cells to maximize system benefits. Additionally, Huang et al. (2021) presented a wind–hydrogen– superconducting magnetic energy storage (SMES) hybrid system to improve power smoothing and system reliability, using an alkaline electrolyzer and SMES to provide a cost-effective solution. On the other hand, Mariani et al. (2022) proposed a model-based predictive control to operate hydrogen-based ESSs in wind farms, allowing for smooth power injections and avoiding penalties for contracted power deviations. Likewise, Rufer (2023) described a new experimental power smoothing method using supercapacitors and hydrogen for PV/HKT hybrid systems, demonstrating a significant reduction in power fluctuations and an improvement in hydrogen production. Finally, Behera and Saikia (2023) presented a hybrid ESS including a hydrogen/bromide redox flow battery and a supercapacitor, controlled by an adaptive energy management system to enhance energy and power support in PV systems. Table 11.1 provides a brief summary of the different existing power smoothing methods classified according to the methodology of the algorithm used in each case.

Current challenges and future directions

The current challenges of power smoothing span a wide range of areas and technologies. These details are shown in Table 11.2. Among these challenges are the significant costs associated with battery maintenance and replacement, as well as the storage of chemical fuels, which represent a major economic obstacle to the implementation and maintenance of ESSs. The integration of hybrid systems with the existing infrastructure presents technical and compatibility difficulties that must be overcome to ensure smooth operation. Another crucial challenge is user acceptance, as users often show resistance to adopting new technologies without clear incentives. Concerns about reliability and associated costs also play a significant role in users' perception of these technologies.

The complexity in algorithm control is another prominent challenge, with high computational capacity requirements and large volumes of data needed to train advanced algorithms. Additionally, uncertainty in prediction models and prediction accuracy represent a constant challenge in the energy smoothing process. Numerous proposals and technical and operational improvements are being made to address these challenges, from intelligent control of battery state of charge to the development of

Table 11.1 Classification of the most important power smoothing methods, according to the methodology used.

Method	Description	References
Autoadaptive control	Uses electric vehicle batteries to smooth energy fluctuations with fuzzy logic.	Villa-Ávila et al. (2024)
	Autoadaptive approach to allocate power between grid, battery, and supercapacitor.	Wu et al. (2024)
Predictive control	Innovative method using hybrid energy storage to smooth energy fluctuations.	Benavides et al. (2024), Mansouri et al. (2020), Ratshitanga et al. (2024), Abdelghany, Al-Durra, et al. (2024), Cano et al. (2024), Khlifi et al. (2024), Guo et al. (2020), Arévalo et al. (2023), Benavides et al. (2023), Khalid and Savkin (2010), Abdelghany, Mariani, et al. (2024), Lin, Cao, et al. (2024), Takahashi et al. (2019), Kong et al. (2021), Huang et al. (2021), Mariani et al. (2022), Rufer (2023), Behera and Saikia (2023)
	Reduction of fluctuations in renewable sources through predictive control with neural networks.	Cano et al. (2024), Guo et al. (2020), Arévalo et al. (2023), Benavides et al. (2023), Reihani et al. (2016), Nempu et al. (2021), Hadi et al. (2021), Cao, Wu, Zhang and Li (2022), Cao, Wu, Zhang, Li, and Zhang (2022)
	Combines techniques to maximize energy under partial shading conditions.	Menadi et al. (2024)
	Hybrid storage combined with wave energy converter to smooth power oscillations.	Pelosi et al. (2024)
Objective function (optimization)	Proposes a method to optimize marine wind and wave generator sizing.	Gao et al. (2024)
	Energy storage to smooth fluctuations in renewable energy generation.	Tripathy et al. (2017), Ratshitanga et al. (2024), Benavides et al. (2022), Shim et al. (2019), Anwar et al. (2017), Zhang et al. (2017)
	Uses multiple storage technologies to provide greater flexibility in energy management.	Tripathy et al. (2017), Kong et al. (2021), Behera and Saikia (2023)

(*Continued*)

Table 11.1 (Continued)

Method	Description	References
	Short-term energy storage to mitigate energy fluctuations in short periods.	Tripathy et al. (2017)
	battery energy storage system (BESS) efficiency in Maui, Hawaii, through nonlinear programming and real-time control.	Reihani et al. (2016)
	Power variability smoothing with Li-Ion battery.	Galli et al. (2024)
	Algorithm to reduce necessary storage capacity in photovoltaic systems.	Yan et al. (2024)
	Improves energy quality in hybrid microgrids.	Nempu et al. (2021)
	Peaks reduction in machining processes without productivity loss.	Hadi et al. (2021)
	Balances power smoothing and frequency regulation.	Cao, Wu, Zhang and Li (2022)
	Addresses uncertain power fluctuations and frequency deviation.	Cao, Wu, Zhang, Li, and Zhang (2022)
	Enhances stability in renewable systems.	Benavides et al. (2022)
	Optimizes size and operability of supercapacitors.	Arévalo et al. (2023)
	Enhances battery lifespan.	Benavides et al. (2023)
Dynamic simulation	Dynamic simulation of renewable energy generation systems to understand and optimize energy quality and stability.	Sebastián (2021), Mansouri et al. (2020)
Adaptive frequency control	Maintains grid frequency within acceptable limits by adjusting instantaneous power generation.	Mansouri et al. (2020), Ratshitanga et al. (2024)
Power reduction	Controlled reduction of active power generated by renewable energy facilities to maintain grid stability.	Mansouri et al. (2020)
Restrictive control	Controlled injection of reactive power into the grid to maintain voltage quality within acceptable limits.	Mansouri et al. (2020), Ratshitanga et al. (2024)
	Sensorless control of photovoltaic systems with a flywheel inverter.	Khlifi et al. (2024)
Frequency control	Integration of wind turbine pitch regulation with a battery to avoid ramp-up violations.	Galli et al. (2024)

Table 11.2 Current challenges of the power smoothing method for electrical systems including renewable sources and microgrids.

Current Challenges	References	Description
Costs	Cunha and Pereira (2024), Fan et al. (2024), Liu and Peng (2024), Syed et al. (2024), Yu et al. (2024), Zhang et al. (2024)	High costs due to battery maintenance and replacement, and chemical fuel storage. Economic challenges in implementing and maintaining energy storage systems.
Integration with Infrastructure	Hou et al. (2024), Li, Wang, et al. (2024), Li, Ding, et al. (2024), Shi et al. (2024), Syed et al. (2024), Wang, Zhou, et al. (2024), Ye et al. (2024), Yu et al. (2024), Zhang et al. (2024)	Difficulty in integrating hybrid systems with existing infrastructure. Technical and compatibility challenges.
User Acceptance	Liu and Peng (2024), Shi et al. (2024), Yu et al. (2024), Zhang et al. (2024)	User resistance to new technologies without clear incentives. Concerns about reliability and costs.
Control Complexity and Algorithms	Li, Wang, et al. (2024), Li, Ding, et al. (2024), Wang, Zhou, et al. (2024), Ye et al. (2024)	High computational complexity and large data volumes needed for training advanced algorithms.
Inaccurate Models and Prediction of Errors	Li, Wang, et al. (2024), Li, Ding, et al. (2024)	Uncertainty in wind predictions and variability in generation models.
System Security and Stability	Hou et al. (2024), Wang, Zhou, et al. (2024)	Need for real-time monitoring and proper parameter adjustment to ensure stability.
Selection of K-means Clusters	Hu et al. (2024)	Optimization using Pearson correlation coefficient and greedy algorithm.
Spectral Comparison between Locations	Bel and Bandi (2024)	Systematic decomposition of solar irradiance time series needed.

(*Continued*)

Table 11.2 (Continued)

Current Challenges	References	Description
Technical and Operational Proposals and Improvements	Deng et al. (2024), Jena and Ray (2024), Lei et al. (2024), Li, Ding, et al. (2024), Li, Wang, et al. (2024), Liang et al. (2024), Lin, Lin, et al. (2024), Wang, Cui, et al. (2024)	Various proposals and improvements related to energy smoothing.
Prediction of Lithium-ion Battery Health	Fauzi et al. (2024)	SGEformer model improves prediction accuracy, setting a new standard.
Battery State of Charge Control	Lin, Cao, et al. (2024), Syed and Khalid (2023)	Intelligent control to extend battery life and stabilize wind power.
Real-time Energy Management Strategy	Lian et al. (2023), Mariani et al. (2022)	Real-time strategy for electric buses and energy storage system operation in wind farms.
Power Smoothing Method for PV Systems	Behera and Saikia (2023), Takahashi et al. (2019)	Control for PV systems to smooth fluctuations and experimental method with supercapacitors and hydrogen.
Techno-economic Analysis of Hydrogen Energy for Renewable Energy Smoothing	Kong et al. (2021), Rufer (2023)	Economic analysis of electrolyzer optimization and gas reinforcements for renewable energy smoothing.
Cooperative Control of Wind-hydrogen Hybrid Systems	Arevalo, Cano, et al. (2023), Huang et al. (2021)	Cooperative strategy to enhance resilience and smoothing in wind-hydrogen systems. Experimental method to reduce fluctuations with supercapacitors and hydrogen.
Integration of Renewable Energy Prediction Systems into Power Smoothing Controllers	Khalid and Savkin (2010), Syed and Khalid (2021), Wu et al. (2021)	Development of controllers using energy prediction systems to smooth renewable energy output.
Optimization of Energy Management Strategies for Hybrid Electric Vehicles	Di Cairano et al. (2013)	Proposal of energy management strategies to enhance efficiency in hybrid vehicles.

(*Continued*)

Table 11.2 (Continued)

Current Challenges	References	Description
Battery Storage Capacity Minimization for Cost Reduction	Khalid and Savkin (2014), Benavides et al. (2022)	Development of schemes to minimize battery storage capacity in distributed configurations.
Optimal Coordination of Energy Storage Systems for Wind Power Smoothing	Liu et al. (2017a, 2017b), Wu et al. (2021)	Development of operation models to smooth wind power output.
Conversion Efficiency and Power Smoothing Balance in Large Inertia Wind Turbines	Meng et al. (2017)	Methods proposed to balance conversion efficiency and power smoothing in wind turbines.
Fatigue Mitigation in Wind Turbine Towers through Economic Wind Energy Control	Shaltout et al. (2020)	Economic control methods proposed to reduce fatigue load on wind turbine towers.
Uncertainty Management and Energy Storage Capacity Planning in Wind Energy Systems	Wu et al. (2021)	Methods developed to manage uncertainty and plan energy storage capacity in wind systems.
Voltage Regulation and System Stability Improvement in Hybrid Energy Systems on Vessels	Bo and Johansen (2017), Park and Zadeh (2021)	Methods proposed to improve voltage regulation and stability in hybrid energy systems on vessels.
Utilization of Neural Networks for Solar Energy Fluctuations Smoothing with Energy Storage Systems	Bo and Johansen (2017), Syed and Khalid (2021)	Development of controllers based on neural networks to smooth solar energy fluctuations with energy storage.
Optimal BESS Configuration with Frequency Constraints	Cao, Wu, Zhang, Li, and Zhang (2022), Cao, Wu, Zhang and Li (2022)	Proposal of optimal configuration for battery energy storage systems (BESS) to maintain power fluctuations and frequency deviation within predefined limits, improving system stability.

(Continued)

Table 11.2 (Continued)

Current Challenges	References	Description
Experimental Study of Power Smoothing Methods	Arevalo, Benavides, et al. (2023), Benavides et al. (2022, 2023)	Evaluation of power smoothing algorithms for renewable systems, combining solar, wind, and hydrokinetic energy with hybrid storage systems, showing reduction in energy fluctuations and increased self-consumption.
Adaptive Control for Power Smoothing in Wind Farms	Hasanien (2014)	Adaptive control scheme proposed for energy storage units (SMES) to smooth power output from wind farms, improving grid stability through fuzzy learning algorithms and adaptive proportional-integral control.
Adaptive Control Design for Autonomous ESS Operation	Ambia and Al-Durra (2015)	Proposal for adaptive control method for energy storage systems (ESS) operating autonomously and without external communication, using an adaptive high-pass filter to smooth power and maintain state of charge within safe limits.
Improvement of Dynamic Stability in Wind Systems	Meng et al. (2018)	Examination of how energy storage systems (BESS) can enhance stability of wind systems through intelligent control based on neural identification and simultaneous control.

real-time energy management strategies for electric vehicles and wind farms. Furthermore, innovative approaches are being explored, such as the use of neural networks to smooth solar energy fluctuations and the design of optimal configuration schemes for ESSs with frequency constraints.

After a comprehensive analysis of the literature, several future research directions can be identified, guiding technological development in the field of power smoothing. Below is a concise summary of potential future trends in power smoothing methods for renewable systems. Additionally, Table 11.3 provides an overview of these future trends in a more concise format.

Energy storage systems

The development of efficient ESSs emerges as a key trend in power smoothing. Advancing storage technologies to store excess energy during high generation periods and release it during low generation intervals stabilizes the grid (Barra et al., 2021; Benavides et al., 2024; Menadi et al., 2024; Pelosi et al., 2024; Ratshitanga et al., 2024; Tripathy et al., 2017; Wu et al., 2024; Zhou et al., 2013).

Control and energy management

Effective control and management of energy resources are crucial for ensuring grid stability. Sophisticated control algorithms dynamically adjusting energy flow, optimizing

Table 11.3 Key trends and future challenges of power smoothing techniques in renewable systems.

Future challenge	References
Development of adaptive strategies for renewable energy storage system control	Benavides et al. (2024), Khlifi et al. (2024), Shim, Kim, and Hur (2019), Guo et al. (2020), Nempu et al. (2021)
Optimization of integrated hybrid renewable energy systems	Anwar et al. (2017), Guo et al. (2020)
Advancement of predictive control techniques	Benavides et al. (2024), Hadi et al. (2021)
Investigation of load management strategies for high penetration renewable energy networks	Reihani et al. (2016), Zhang et al. (2017)
Optimization of energy storage system size and placement	Reihani et al. (2016), Zhang et al. (2017)
Research on robust methods for wind power smoothing	Guo et al. (2020)
Development of power smoothing strategies using superconductors for AC-DC hybrid microgrids	Nempu et al. (2021)

energy utilization, and responding to grid disturbances are needed (Abdelghany, Al-Durra, et al., 2024; Arcos-Aviles et al., 2019; Benavides et al., 2024; Villa-Ávila et al., 2024; Wang et al., 2018; Wu et al., 2024; Yan et al., 2024).

Integration of renewable technologies

Seamless integration of various renewable energy sources is essential for achieving a reliable and sustainable energy mix. Optimizing the integration of solar, wind, and other renewables to maximize system efficiency is emphasized (Barra et al., 2021; Benavides et al., 2024; Galli et al., 2024; Mansouri et al., 2020; Menadi et al., 2024; Pelosi et al., 2024; Sebastián, 2021; Villa-Ávila et al., 2024; Wang et al., 2018; Zhou et al., 2013).

Optimization of photovoltaic systems

Optimizing PV systems plays a critical role in improving their performance and reliability. Enhancing the design and operation of PV systems to maximize energy yield and mitigate grid disturbances is a focus (Galli et al., 2024; Jarnut et al., 2017; Menadi et al., 2024; Yan et al., 2024).

Advanced predictive control

Predictive control techniques based on neural networks are gaining prominence in improving the stability and efficiency of renewable energy systems. The potential of predictive control algorithms to anticipate grid conditions and optimize energy management accordingly is highlighted (Cano et al., 2024).

Microgrid and vehicle-to-grid applications

Microgrids and vehicle-to-grid (V2G) systems offer innovative solutions for decentralized energy management. The applications of microgrids and V2G technologies in optimizing energy utilization and grid stability are explored (Abdelghany, Al-Durra, et al., 2024; Arcos-Aviles et al., 2019; Benavides et al., 2024; Mansouri et al., 2020; Prof Dr & Goetzberger, 2023; Villa-Ávila et al., 2024; Wu et al., 2024).

Robust wind power smoothing

Robust methodologies utilizing hybrid ESSs are essential for mitigating wind generation variability. Developing robust control strategies to ensure reliable and consistent integration of wind energy is emphasized (Guo et al., 2020).

Power smoothing in microgrids with superconductors

Utilizing superconductors in AC-DC hybrid microgrids presents a promising pathway for power smoothing. Exploring the potential of superconductor technologies in improving grid stability and efficiency is discussed (Nempu et al., 2021).

Conclusion

The transition toward a more sustainable and diversified energy matrix has increased the demand for power smoothing techniques due to the intermittent and variable nature of renewable energies. The need to maintain stable and uniform energy generation to ensure the operational stability of electrical systems is becoming increasingly crucial. Over the decades, the field of power smoothing has experienced significant growth, with contributions from various researchers and technological advancements. From early work in the 1970s and 1980s to recent developments in adaptive and predictive control, the historical evolution highlights continuous progress in this field. This chapter provided a comprehensive review of these techniques applied to renewable energies. The following conclusions are drawn:

Although there is a wide variety of power smoothing techniques, ranging from adaptive and predictive control strategies to objective function optimization methods and energy conversion approaches, this diversity reflects the complexity of the challenge and the need for innovative and adaptive solutions. Power smoothing techniques are applied in a variety of contexts, from wind and solar systems to microgrids and transportation applications. Each application has its own challenges and specific requirements, underscoring the importance of adapting power smoothing techniques to the conditions and needs of each environment.

Technological advancements, such as the development of adaptive and predictive control algorithms, advanced ESSs, and optimization techniques, are continually driving the field of power smoothing forward. These advancements are enabling more efficient and effective solutions to address the variability of renewable energy generation. Objective function optimization methods focus on maximizing energy efficiency and minimizing fluctuations through mathematical modeling and algorithmic optimization. This demonstrates a comprehensive approach to addressing the challenges associated with renewable energy generation variability.

Various energy storage technologies, such as electric vehicle batteries, hydrogen storage, and hybrid storage systems, are highlighted. These technologies offer innovative solutions to address power fluctuations and improve electrical system stability. Hybrid approaches, which combine different energy conversion technologies, are critical for maximizing energy capture and utilization in variable environmental conditions. These approaches allow for the optimal use of available energy resources and improve overall energy efficiency.

Further research is needed to optimize ESSs for particular applications, such as residential microgrids or energy systems based on marine currents. Understanding how to design and operate these systems optimally is crucial to ensuring energy supply stability and reliability. Additionally, it is essential to continue exploring emerging technologies, such as aerodynamic flutter-based energy harvesting devices, to drive the development of power smoothing and improve overall energy efficiency.

Acknowledgments

The authors thank Universidad de Cuenca (UCUENCA), Ecuador, for easing access to the facilities of the Micro-Grid Laboratory, Faculty of Engineering, where this research was conducted.

References

Abdelghany, M. B., Al-Durra, A., Zeineldin, H., & Hu, J. (2024). Integration of cascaded coordinated rolling horizon control for output power smoothing in islanded wind–solar microgrid with multiple hydrogen storage tanks. *Energy, 291.* Available from https://doi.org/10.1016/j.energy.2024.130442, https://www.sciencedirect.com/science/journal/03605442.

Abdelghany, M. B., Mariani, V., Liuzza, D., Natale, O. R., & Glielmo, L. (2024). A unified control platform and architecture for the integration of wind-hydrogen systems into the grid. *IEEE Transactions on Automation Science and Engineering*, 1545–5955. Available from https://doi.org/10.1109/tase.2023.3292029.

Ambia, M. N., & Al-Durra, A. (2015). Adaptive power smoothing control in grid-connected and islanding modes of hybrid micro-grid energy management. *Journal of Renewable and Sustainable Energy*, *7*(3). Available from https://doi.org/10.1063/1.4921370, Article 033104.

Amorim, W. C. S., Cupertino, A. F., Pereira, H. A., & Mendes, V. F. (2024). On sizing of battery energy storage systems for PV plants power smoothing. *Electric Power Systems Research*, *229*, 110114. Available from https://doi.org/10.1016/j.epsr.2024.110114.

Anwar, M. B., El Moursi, M. S., & Xiao, W. (2017). Novel power smoothing and generation scheduling strategies for a hybrid wind and marine current turbine system. *IEEE Transactions on Power Systems*, *32*(2). Available from https://doi.org/10.1109/TPWRS.2016.2591723.

Arcos-Aviles, D., Guinjoan, F., Pascual, J., Marroyo, L., Sanchis, P., Gordillo, R., Ayala, P., & Marietta, M. P. (2019). *A review of fuzzy-based residential grid-connected microgrid energy management strategies for grid power profile smoothingenergy. Environment, and sustainability*. Ecuador: Springer Nature. Available from springer.com/series/15901, 10.1007/978-981-13-3284-5_8.

Arevalo, P., Benavides, D., Tostado-Veliz, M., Aguado, J. A., & Jurado, F. (2023). Smart monitoring method for photovoltaic systems and failure control based on power smoothing techniques. *Renewable Energy*, *205*, 366–383. Available from https://doi.org/10.1016/j.renene.2023.01.059.

Arévalo, P., Benavides, D., Tostado-Véliz, M., Aguado, J. A., & Jurado, F. (2023). Smart monitoring method for photovoltaic systems and failure control based on power smoothing techniques. *Renewable Energy*, *205*. Available from https://doi.org/10.1016/j.renene.2023.01.059, http://www.journals.elsevier.com/renewable-and-sustainable-energy-reviews/.

Arevalo, P., Cano, A., & Jurado, F. (2023). A novel experimental method of power smoothing using supercapacitors and hydrogen for hybrid system PV/HKT. *Journal of Energy Storage*, *73*. Available from https://doi.org/10.1016/j.est.2023.108819, Article 108819.

Barra, P. H. A., de Carvalho, W. C., Menezes, T. S., Fernandes, R. A. S., & Coury, D. V. (2021). A review on wind power smoothing using high-power energy storage systems. *Renewable and Sustainable Energy Reviews*, *137*, 110455. Available from https://doi.org/10.1016/j.rser.2020.110455.

Behera, M. K., & Saikia, L. C. (2023). A Novel Resilient Control of Grid-Integrated Solar PV-Hybrid Energy Storage Microgrid for Power Smoothing and Pulse Power Load Accommodation. *Ieee Transactions on Power Electronics*, *38*(3), 3965–3980. Available from https://doi.org/10.1109/tpel.2022.3217144.

Bel, G., & Bandi, M. M. (2024). Spectral analysis of solar-irradiance fluctuations. *Physical Review Applied*, *21*(3). Available from https://doi.org/10.1103/PhysRevApplied.21.034019, Article 034019.

Benavides, D., Arevalo, P., Aguado, J. A., & Jurado, F. (2023). Experimental validation of a novel power smoothing method for on-grid photovoltaic systems using supercapacitors. *International Journal of Electrical Power & Energy Systems*, *149*. Available from https://doi.org/10.1016/j.ijepes.2023.109050, Article 109050.

Benavides, D., Arévalo, P., Tostado-Véliz, M., Vera, D., Escamez, A., Aguado, J. A., & Jurado, F. (2022). An experimental study of power smoothing methods to reduce renewable sources fluctuations using supercapacitors and lithium-ion batteries. *Batteries*, *8* (11), 228. Available from https://doi.org/10.3390/batteries8110228.

Benavides, D., Arévalo, P., Villa-Ávila, E., Aguado, J. A., & Jurado, F. (2024). Predictive power fluctuation mitigation in grid-connected PV systems with rapid response to EV charging stations. *Journal of Energy Storage*, *86*. Available from https://doi.org/10.1016/j.est.2024.111230, https://www.sciencedirect.com/science/journal/2352152X.

Bo, T. I., & Johansen, T. A. (2017). Battery Power Smoothing Control in a Marine Electric Power Plant Using Nonlinear Model Predictive Control. *Ieee Transactions on Control Systems Technology*, *25*(4), 1449–1456. Available from https://doi.org/10.1109/tcst.2016.2601301.

Cano, A., Arévalo, P., & Jurado, F. (2024). Neural network predictive control in renewable systems (HKT-PV) for delivered power smoothing. *Journal of Energy Storage*, *87*, 111332. Available from https://doi.org/10.1016/j.est.2024.111332.

Cao, Y., Wu, Q., Zhang, H., & Li, C. (2022). Optimal sizing of hybrid energy storage system considering power smoothing and transient frequency regulation. *International Journal of Electrical Power and Energy Systems*, *142*. Available from https://doi.org/10.1016/j.ijepes.2022.108227, https://www.journals.elsevier.com/international-journal-of-electrical-power-and-energy-systems.

Cao, Y., Wu, Q., Zhang, H., Li, C., & Zhang, X. (2022). Chance-Constrained Optimal Configuration of BESS Considering Uncertain Power Fluctuation and Frequency Deviation Under Contingency. *Ieee Transactions on Sustainable Energy*, *13*(4), 2291–2303. Available from https://doi.org/10.1109/tste.2022.3192087.

Cunha, J. L. R. N., & Pereira, C. M. N. A. (2024). A hybrid model based on STL with simple exponential smoothing and ARMA for wind forecast in a Brazilian nuclear power plant site. *Nuclear Engineering and Design*, *421*. Available from https://doi.org/10.1016/j.nucengdes.2024.113026, Article 113026.

Deng, L., Shen, W., Xu, K., & Zhang, X. (2024). An Adaptive Modeling Method for the Prognostics of Lithium-Ion Batteries on Capacity Degradation and Regeneration. *Energies*, *17*(7). Available from https://doi.org/10.3390/en17071679, Article 1679.

Di Cairano, S., Liang, W., Kolmanovsky, I. V., Kuang, M. L., & Phillips, A. M. (2013). Power Smoothing Energy Management and Its Application to a Series Hybrid Powertrain. *Ieee Transactions on Control Systems Technology*, *21*(6), 2091–2103. Available from https://doi.org/10.1109/tcst.2012.2218656.

Fan, Y., Li, L., & Sun, S. (2024). Powerful and accurate detection of temporal gene expression patterns from multi-sample multi-stage single-cell transcriptomics data with TDEseq. *Genome Biology*, *25*(1). Available from https://doi.org/10.1186/s13059-024-03237-3, Article 96.

Fauzi, M. R., Yudistira, N., & Mahmudy, W. F. (2024). State-of-Health Prediction of Lithium-Ion Batteries Using Exponential Smoothing Transformer With Seasonal and Growth Embedding. *Ieee Access*, *12*, 14659–14670. Available from https://doi.org/10.1109/access.2024.3357736.

Galli, C., Superchi, F., Papi, F., Ferrara, G., & Bianchini, A. (2024). Innovative power smoothing techniques for wind turbines using batteries and adaptive pitch regulation. *Journal of Energy Storage*, *84*, 110964. Available from https://doi.org/10.1016/j.est.2024.110964.

Gao, Q., Ertugrul, N., Ding, B., Negnevitsky, M., & Soong, W. L. (2024). Analysis of wave energy conversion for optimal generator sizing and hybrid system integration. *IEEE Transactions on Sustainable Energy*, *15*(1). Available from https://doi.org/10.1109/TSTE.2023.3318010, https://ieeexplore.ieee.org/servlet/opac?punumber = 5165391.

Guo, T., Liu, Y., Zhao, J., Zhu, Y., & Liu, J. (2020). A dynamic wavelet-based robust wind power smoothing approach using hybrid energy storage system. *International Journal of Electrical Power and Energy Systems*, *116*. Available from https://doi.org/10.1016/j.ijepes.2019.105579, https://www.journals.elsevier.com/international-journal-of-electrical-power-and-energy-systems.

Hadi, M. A., Brillinger, M., Wuwer, M., Schmid, J., Trabesinger, S., Jäger, M., & Haas, F. (2021). Sustainable peak power smoothing and energy-efficient machining process thorough analysis of high-frequency data. *Journal of Cleaner Production*, *318*, 128548. Available from https://doi.org/10.1016/j.jclepro.2021.128548.

Han, X., Liu, X., & Wang, H. (2020). Dual-regulating feedback optimization control of distributed energy storage system in power smoothing scenariox based on KF-MPC. *IEEE Access*, *8*. Available from https://doi.org/10.1109/ACCESS.2020.3025334, http://ieeexplore.ieee.org/xpl/RecentIssue.jsp?punumber = 6287639.

Hasanien, H. M. (2014). A Set-Membership Affine Projection Algorithm-Based Adaptive-Controlled SMES Units for Wind Farms Output Power Smoothing. *Ieee Transactions on Sustainable Energy*, *5*(4), 1226–1233. Available from https://doi.org/10.1109/tste.2014.2340471.

Hou, E., Xu, Y., Tang, J., & Wang, Z. (2024). Strategy of Flywheel-Battery Hybrid Energy Storage Based on Optimized Variational Mode Decomposition for Wind Power Suppression. *Electronics*, *13*(7). Available from https://doi.org/10.3390/electronics13071362, Article 1362.

Hu, S., Yang, J., Wang, Y., Chen, C., Nan, J., Zhao, Y., & Bi, Y. (2024). Feature Extraction Approach for Distributed Wind Power Generation Based on Power System Flexibility Planning Analysis. *Electronics*, *13*(5). Available from https://doi.org/10.3390/electronics13050966, Article 966.

Huang, C., Zong, Y., You, S., Traholt, C., Zheng, Z., & Xie, Q. (2021). Cooperative control of wind-hydrogen-SMES hybrid systems for fault-ride-through improvement and power smoothing. *IEEE Transactions on Applied Superconductivity*, *31*(8). Available from https://doi.org/10.1109/TASC.2021.3103729, https://ieeexplore.ieee.org/servlet/opac?punumber = 77.

Jamsheed, F., & Iqbal, S. J. (2022). A neuro-adaptive control scheme to improve dynamic stability of wind power system using battery energy storage. In *11th IFAC symposium on Control of Power and Energy Systems* (*CPES*). Electr Network.

Jarnut, M., Wermiński, S., & Waśkowicz, B. (2017). Comparative analysis of selected energy storage technologies for prosumer-owned microgrids. *Renewable and Sustainable*

Energy Reviews, *74*. Available from https://doi.org/10.1016/j.rser.2017.02.084, https://www.journals.elsevier.com/renewable-and-sustainable-energy-reviews.

Jena, C. J., & Ray, P. K. (2024). Power allocation scheme for grid interactive microgrid with hybrid energy storage system using model predictive control. *Journal of Energy Storage*, *81*. Available from https://doi.org/10.1016/j.est.2023.110401, Article 110401.

Khalid, M., & Savkin, A. V. (2010). A model predictive control approach to the problem of wind power smoothing with controlled battery storage. *Renewable Energy*, *35*(7), 1520–1526. Available from https://doi.org/10.1016/j.renene.2009.11.030.

Khalid, M., & Savkin, A. V. (2014). Minimization and control of battery energy storage for wind power smoothing: Aggregated, distributed and semi-distributed storage. *Renewable Energy*, *64*. Available from https://doi.org/10.1016/j.renene.2013.09.043.

Khlifi, Y. E., Magri, A. E., Lajouad, R., Bouattane, O., & Giri, F. (2024). Output-feedback control of a grid-connected photovoltaic system based on a multilevel flying-capacitor inverter with power smoothing capability. *ISA—Instrumentation, Systems, and Automation Society, Morocco ISA Transactions*, *147*. Available from https://doi.org/10.1016/j.isatra.2024.02.007, https://www.sciencedirect.com/science/journal/00190578.

Kong, L., Li, L., Cai, G., Liu, C., Ma, P., Bian, Y., & Ma, T. (2021). Techno-economic analysis of hydrogen energy for renewable energy power smoothing. *International Journal of Hydrogen Energy*, *46*(3), 2847–2861. Available from https://doi.org/10.1016/j.ijhydene.2020.07.231.

Lei, C., Bu, S., Wang, Q., Chen, Q., Yang, L., & Chi, Y. (2024). Look-Ahead Rolling Economic Dispatch Approach for Wind-Thermal-Bundled Power System Considering Dynamic Ramping and Flexible Load Transfer Strategy. *Ieee Transactions on Power Systems*, *39*(1), 186–202. Available from https://doi.org/10.1109/tpwrs.2023.3238035.

Lei, M., Yang, Z., Wang, Y., Xu, H., Meng, L., Vasquez, J. C., & Guerrero, J. M. (2018). An MPC-Based ESS control method for PV power smoothing applications. *IEEE Transactions on Power Electronics*, *33*(3). Available from https://doi.org/10.1109/TPEL.2017.2694448, http://ieeexplore.ieee.org/xpl/tocresult.jsp?isnumber = 4712525.

Li, G., Ding, C., Zhao, N., Wei, J., Guo, Y., Meng, C., & Zhu, R. (2024). Research on a novel photovoltaic power forecasting model based on parallel long and short-term time series network. *Energy*, *293*. Available from https://doi.org/10.1016/j.energy.2024.130621, Article 130621.

Li, J., Wang, X., & Liu, B. (2024). Event-Based Deep Reinforcement Learning for Smoothing Ramp Events in Combined Wind-Storage Energy Systems. *Ieee Transactions on Industrial Informatics*, *20*(5), 7871–7882. Available from https://doi.org/10.1109/tii.2024.3353859.

Lian, J., Han, P., Li, L., & Sun, X. (2023). Real-time energy management strategy for fuel cell plug-in hybrid electric bus using short-term power smoothing prediction and distance adaptive state-of-charge consumption. *Energy Technology*, *11*(8). Available from https://doi.org/10.1002/ente.202300099, http://onlinelibrary.wiley.com/journal/10.1002/(ISSN)2194-4296.

Liang, T., He, W., Ahmad, A., Li, Y., & Ding, Y. (2024). Integration of liquid air energy storage with wind power - A dynamic study. *Applied Thermal Engineering*, *242*. Available from https://doi.org/10.1016/j.applthermaleng.2024.122415, Article 122415.

Lin, L., Cao, Y., Kong, X., Lin, Y., Jia, Y., & Zhang, Z. (2024). Hybrid energy storage system control and capacity allocation considering battery state of charge self-recovery and capacity attenuation in wind farm. *Journal of Energy Storage*, *75*, 109693. Available from https://doi.org/10.1016/j.est.2023.109693.

Lin, X., Lin, X., Zhong, W., Cong, F., & Zhou, Y. (2024). Cross-level steam load smoothing and optimization in industrial parks using data-driven approaches. *Energy and Ai, 16*. Available from https://doi.org/10.1016/j.egyai.2024.100344, Article 100344.

Liu, Y., & Peng, M. (2024). Research on peak load shifting for hybrid energy system with wind power and energy storage based on situation awareness. *Journal of Energy Storage, 82*. Available from https://doi.org/10.1016/j.est.2024.110472, Article 110472.

Liu, Y., Zhu, Y., Xu, W., Gou, J., Xiang, Y., & Liu, J. (2017a). Dynamic wavelet decomposition based multi-objective operation model for HESS enabling wind power output smoothing. *Energy Procedia* [Proceedings of the 9th international conference on applied energy]. 9th International Conference on Applied Energy (ICAE), Cardiff, ENGLAND.

Liu, Y., Zhu, Y., Xu, W., Gou, J., Xiang, Y., & Liu, J. (2017b). Dynamic wavelet decomposition based multi-objective operation model for HESS enabling wind power output smoothing. *Energy Procedia, 142*, 1462–1467. Available from https://doi.org/10.1016/j.egypro.2017.12.593, http://www.sciencedirect.com/science/journal/18766102.

Mansouri, N., Lashab, A., Guerrero, J. M., & Cherif, A. (2020). Photovoltaic power plants in electrical distribution networks: A review on their impact and solutions. *IET Renewable Power Generation, 14*(12). Available from https://doi.org/10.1049/iet-rpg.2019.1172, http://www.theiet.org/.

Manwell, J. F., Rogers, A. L., & McGowan, J. G. (2001). Status of offshore wind energy in the United States. In *Proceedings of the IEEE power engineering society transmission and distribution conference*. Institute of Electrical and Electronics Engineers Inc. Available from https://doi.org/10.1109/pess.2001.969971.

Mariani, V., Zenith, F., & Glielmo, L. (2022). Operating hydrogen-based energy storage systems in wind farms for smooth power injection: A penalty fees aware model predictive control. *Energies, 15*(17). Available from https://doi.org/10.3390/en15176307, http://www.mdpi.com/journal/energies/.

McCarthy, J. M., Watkins, S., Deivasigamani, A., & John, S. J. (2016). Fluttering energy harvesters in the wind: A review. *Journal of Sound and Vibration, 361*. Available from https://doi.org/10.1016/j.jsv.2015.09.043.

Menadi, A., Boukahil, F. Z., & Betka, A. (2024). Real-time implementation of a hybrid ESC approach for maximising the extracted photovoltaic power under partial shading conditions. *Power Electronics and Drives, 9*(1). Available from https://doi.org/10.2478/pead-2024-0013.

Meng, H., Yang, T., Liu, J. Z., & Lin, Z. (2017). A flexible maximum power point tracking control strategy considering both conversion efficiency and power fluctuation for large-inertiawind turbines. *Energies, 10*(7). Available from https://doi.org/10.3390/en10070939, http://www.mdpi.com/journal/energies/.

Meng, L., Dragicevic, T., & Guerrero, J. M. (2018). Adaptive Control Design for Autonomous Operation of Multiple Energy Storage Systems in Power Smoothing Applications. *Ieee Transactions on Industrial Electronics, 65*(8), 6612–6624. Available from https://doi.org/10.1109/tie.2017.2756584.

Nasiri, M. R., Farhangi, S., & Rodriguez, J. (2019). Model predictive control of a multilevel CHB STATCOM in wind farm application using diophantine equations. *IEEE Transactions on Industrial Electronics, 66*(2). Available from https://doi.org/10.1109/TIE.2018.2833055, http://ieeexplore.ieee.org/xpl/tocresult.jsp?isnumber = 5410131.

Nempu, P. B., Sabhahit, J. N., Gaonkar, D. N., & Rao, V. S. (2021). Novel Power smoothing technique for a hybrid AC-DC microgrid operating with multiple alternative energy sources. *Advances in Electrical and Computer Engineering, 21*(2). Available from https://doi.org/10.4316/AECE.2021.02011, http://www.aece.ro.

Park, D., & Zadeh, M. (2021). Modeling and predictive control of shipboard hybrid DC power systems. *IEEE Transactions on Transportation Electrification*, *7*(2). Available from https://doi.org/10.1109/TTE.2020.3027184, https://www.ieee.org/membership-catalog/productdetail/showProductDetailPage.html?product = PER473-ELE&utm_source = Mainsite_CSE&utm_medium = CSE_Promotion&utm_campaign = Catalog_Promotion-PER473.

Pelosi, D., Gallorini, F., Alessandri, G., & Barelli, L. (2024). A Hybrid energy storage system integrated with a wave energy converter: Data-driven stochastic power management for output power smoothing. *Energies*, *17*(5). Available from https://doi.org/10.3390/en17051167, http://www.mdpi.com/journal/energies/.

Prof Dr, A., & Goetzberger. (2023). Founder of the Fraunhofer Institute for solar Power Systems, deceased Solar Industry loses highly valued Scientist and visionary Mastermind. *Vakuum in Forschung Und Praxis*, *35*.

Ratshitanga, M., Ayeleso, A., Krishnamurthy, S., Rose, G., Aminou Moussavou, A. A., & Adonis, M. (2024). Battery storage use in the value chain of power systems. *Energies*, *17*(4). Available from https://doi.org/10.3390/en17040921, http://www.mdpi.com/journal/energies/.

Reihani, E., Motalleb, M., Ghorbani, R., & Saad Saoud, L. (2016). Load peak shaving and power smoothing of a distribution grid with high renewable energy penetration. *Renewable Energy*, *86*. Available from https://doi.org/10.1016/j.renene.2015.09.050, http://www.journals.elsevier.com/renewable-and-sustainable-energy-reviews/.

Rufer, A. (2023). Increasing the energy efficiency of gas boosters for hydrogen storage and for refueling stations. *Energies*, *16*(4). Available from https://doi.org/10.3390/en16041763, http://www.mdpi.com/journal/energies/.

Sangtarash, A., Maadi, S. R., Taylor, R. A., Arabkoohsar, A., Wongwises, S., Sheremet, M., Mahian, O., & Sahin, A. Z. (2024). A comprehensive investigation of porous media's effects on the performance of photovoltaic thermal systems. *Applied Thermal Engineering*, *245*, 122766. Available from https://doi.org/10.1016/j.applthermaleng.2024.122766.

Sebastián, R. (2021). Review on dynamic simulation of wind diesel isolated microgrids. *Energies*, *14*(7), 1812. Available from https://doi.org/10.3390/en14071812.

Shaltout, M. L., Alhneaish, M. M., & Metwalli, S. M. (2020). An economic model predictive control approach for wind power smoothing and tower load mitigation. *Journal of Dynamic Systems, Measurement and Control, Transactions of the ASME*, *142*(6). Available from https://doi.org/10.1115/1.4046278, http://dynamicsystems.asmedigitalcollection.asme.org/journal.aspx?journalid = 117.

Shi, S., Wang, P., Zheng, Z., & Zhang, S. (2024). Two-Layer Optimization Strategy of Electric Vehicle and Air Conditioning Load Considering the Benefit of Peak-to-Valley Smoothing. *Sustainability*, *16*(8). Available from https://doi.org/10.3390/su16083207, Article 3207.

Shim, J., Kim, H., & Hur, K. (2019). Incorporating state-of-charge balancing into the control of energy storage systems for smoothing renewable intermittency. *Energies*, *12*(7). Available from https://doi.org/10.3390/en12071190, https://www.mdpi.com/1996-1073/12/7.

Syed, M. A., & Khalid, M. (2021). Neural network predictive control for smoothing of solar power fluctuations with battery energy storage. *Journal of Energy Storage*, *42*. Available from https://doi.org/10.1016/j.est.2021.103014, http://www.journals.elsevier.com/journal-of-energy-storage/.

Syed, M. A., & Khalid, M. (2023). An Intelligent Model Predictive Control Strategy for Stable Solar-Wind Renewable Power Dispatch Coupled with Hydrogen Electrolyzer and Battery Energy Storage. *International Journal of Energy Research*, *2023*. Available from https://doi.org/10.1155/2023/4531054, Article 4531054.

Syed, M. A., Siddiqui, O., Kazerani, M., & Khalid, M. (2024). Analysis and Modeling of Direct Ammonia Fuel Cells for Solar and Wind Power Leveling in Smart Grid Applications. *Ieee Access*, *12*, 46512–46523. Available from https://doi.org/10.1109/access.2024.3376513.

Takahashi, A., Goto, A., Machida, Y., & Funabiki, S. (2019). A power smoothing control method for a photovoltaic generation system using a water electrolyzer and its filtering characteristics. *Electrical Engineering in Japan*, *206*(2). Available from https://doi.org/10.1002/eej.23191.

Tripathy, S. K., Mitra, I., Heinemann, D., Giridhar, G., & Gomathinayagam, S. (2017). Impact assessment of short-term variability of solar radiation in Rajasthan using SRRA data. *Renewable and Sustainable Energy Reviews*, *78*. Available from https://doi.org/10.1016/j.rser.2017.05.014, https://www.journals.elsevier.com/renewable-and-sustainable-energy-reviews.

Villa-Ávila, E., Arévalo, P., Ochoa-Correa, D., Iñiguez-Morán, V., & Jurado, F. (2024). Innovative power smoothing technique for enhancing renewable integration in insular power systems using electric vehicle charging stations. *Applied Sciences*, *14*(1), 375. Available from https://doi.org/10.3390/app14010375.

Volfkovich, Y. M. (2021). Electrochemical Supercapacitors (A review). *Journal of Electrochemistry*, *57*(4). Available from https://doi.org/10.1134/S1023193521040108, https://www.springer.com/journal/11175.

Wang, D., Gao, X., Meng, K., Qiu, J., Lai, L. L., & Gao, S. (2018). Utilisation of kinetic energy from wind turbine for grid connections: A review paper. *IET Renewable Power Generation*, *12*(6). Available from https://doi.org/10.1049/iet-rpg.2017.0590, http://www.theiet.org/.

Wang, X., Cui, J., Ren, B., Liu, Y., & Huang, Y. (2024). Integrated energy system scheduling optimization considering vertical-axis wind turbines and thermal inertia in oilfield management areas. *Frontiers in Energy Research*, *12*. Available from https://doi.org/10.3389/fenrg.2024.1340580, Article 1340580.

Wang, X., Zhou, J., Qin, B., & Guo, L. (2024). Coordinated Power Smoothing Control Strategy of Multi-Wind Turbines and Energy Storage Systems in Wind Farm Based on MADRL. *Ieee Transactions on Sustainable Energy*, *15*(1), 368–380. Available from https://doi.org/10.1109/tste.2023.3287871.

Wu, C., Gao, S., Liu, Y., Han, H., & Jiang, S. (2021). Wind Power Smoothing With Energy Storage System: A Stochastic Model Predictive Control Approach. *Ieee Access*, *9*, 37534–37541. Available from https://doi.org/10.1109/access.2021.3063740.

Wu, X., Wu, Y., Tang, Z., & Kerekes, T. (2024). An adaptive power smoothing approach based on artificial potential field for PV plant with hybrid energy storage system. *Solar Energy*, *270*, 112377. Available from https://doi.org/10.1016/j.solener.2024.112377.

Yan, H. W., Liang, G., Beniwal, N., Rodriguez, E., Farivar, G. G., & Pou, J. (2024). Flexible power point tracking aided power ramp rate control for photovoltaic systems with small energy storage capacity. *IEEE Transactions on Power Electronics*, *39*(2). Available from https://doi.org/10.1109/TPEL.2023.3333534, https://ieeexplore.ieee.org/xpl/mostRecentIssue.jsp?punumber = 63.

Ye, G., Yang, J., Xia, F., Shao, F., Xu, J., Yang, Z., & Zheng, Z. (2024). Hybrid energy storage system control strategy to smooth power fluctuations in microgrids containing photovoltaics. *International Journal of Low-Carbon Technologies*, *19*, 676–682. Available from https://doi.org/10.1093/ijlct/ctae027.

Yu, S., Liu, X., Li, R., Yin, Z., Chen, Z., Han, F., & Li, S. (2024). Demand side management full season optimal operation potential analysis for coupled hybrid photovoltaic/thermal, heat pump, and thermal energy storage systems. *Journal of Energy Storage, 80*. Available from https://doi.org/10.1016/j.est.2023.110375, Article 110375.

Zhang, F., Meng, K., Xu, Z., Dong, Z., Zhang, L., Wan, C., & Liang, J. (2017). Battery ESS Planning for wind smoothing via variable-interval reference modulation and self-adaptive SOC control strategy. *IEEE Transactions on Sustainable Energy, 8*(2). Available from https://doi.org/10.1109/TSTE.2016.2615638.

Zhang, H., Yang, J., Zhang, B., Tang, Y., Du, W., & Wen, B. (2024). Enhancing generalized spectral clustering with embedding Laplacian graph regularization [; Early Access]. *Caai Transactions on Intelligence Technology*. Available from https://doi.org/10.1049/cit2.12308.

Zhou, Z., Benbouzid, M., Frédéric Charpentier, J., Scuiller, F., & Tang, T. (2013). A review of energy storage technologies for marine current energy systems. *Renewable and Sustainable Energy Reviews, 18*. Available from https://doi.org/10.1016/j.rser.2012.10.006.

Real energy communities worldwide

12

Carlos Cruz
Department of Electronics, University of Alcalá, Alcalá de Henares, Madrid, Spain

The reader of this chapter will:

- Explore the main challenges facing EC implementation
- Learn about the main benefits of ECs for future energy systems
- Learn about real EC models currently implemented worldwide.

Introduction

Nowadays, energy consumers and utilities are innovating in a constant evolution aimed at achieving the transition to low-carbon energies (Gui & MacGill, 2018). Recent record rises in energy costs and heavy reliance on energy imports, especially natural gas, and oil, have increased vulnerability to shocks that potentially increase prices. One way for countries to secure and optimize energy supplies is to develop better cooperation in the integration of energy networks. In this respect, government policies and incentives can play an important role in fostering innovative initiatives.

The emerging concept of an energy community (EC) refers to the organization and management of energy production, distribution, and consumption at the local or community level. EC development is also directly linked to self-consumption in terms of encouraging energy efficiency, citizen participation in energy-related decision-making, and, most especially, energy generation using renewable energy sources (e.g., wind, solar, and biomass). ECs install solar panels, wind turbines, or other generation capabilities on their properties, as well as energy storage systems (e.g., batteries), to more effectively harness energy and ensure a steady supply. Locally generated energy is shared among community members through an internal distribution network based on metering and innovative smart-grid technologies. By optimizing demand management and reallocation capabilities, ECs promote energy efficiency in both generation and consumption. A suitable framework includes democratic decision-making regarding investments, regulations, and policies to reduce carbon emissions and energy costs while promoting sustainability.

Obstacles to the development of ECs are sociotechnical (digitalization, decarbonization, and decentralization), geographical (governance and social configuration), and technical (regulation, financing, and infrastructure). Objectives and constraints depending on regulation may result in ECs having different bureaucratic and

Towards Future Smart Power Systems with High Penetration of Renewables. DOI: https://doi.org/10.1016/B978-0-443-29871-4.00012-9

organizational structures, and legislation may impose constraints on interactions between EC members and, consequently, on technical design. Within the European Union (EU), for instance, each member state is free to enact its own specific legislation provided it complies with specific directives.

EC models, which vary in size and scope (ranging from small local energy cooperatives to larger projects at the neighborhood, municipal, regional, and even national level), aim to reduce energy dependence and promote the transition to a more sustainable energy model. The main models are based on legal persons developing their own renewable energy projects and consumers cooperating to produce their own renewable energy supply and to manage its consumption.

A fundamental aspect of ECs is the active participation of consumers in decision-making and energy management. Participation in organized energy production markets requires a minimum number of partners. However, in the case of renewable ECs, members must be located closed to the renewable energy facilities. Here, the radius of action for small communities is extended to ensure an adequate number of members in line with the objectives established for real implementation.

The focus is therefore on decentralizing energy production and energy management, increasing resilience, and reducing dependence on nonrenewable energy sources. As stated in Fig. 12.1, successful EC implementation generates benefits for participants, especially

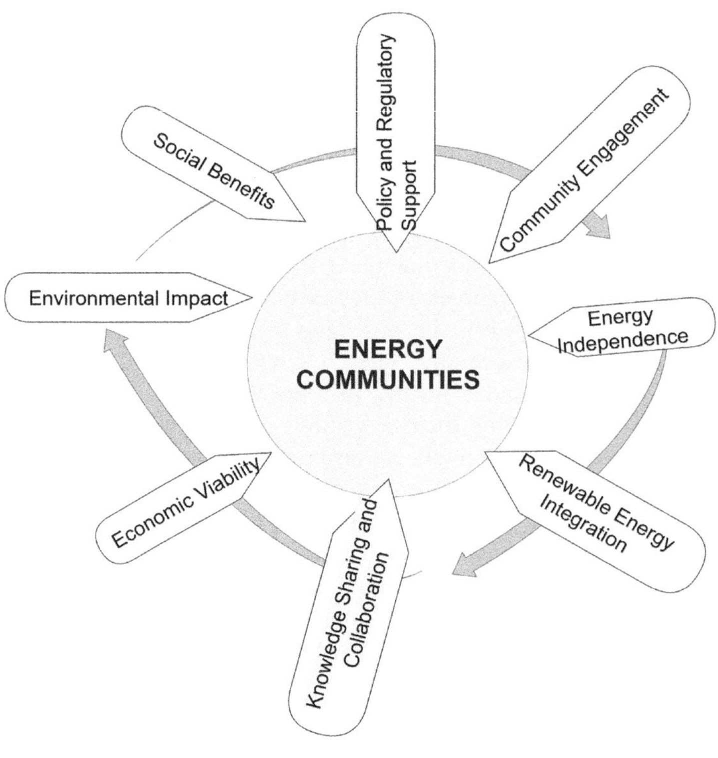

Figure 12.1 Benefits of energy communities implementation.

in the form of energy cost savings and new revenue opportunities from the sale of surplus energy production. The benefits are evaluated through various metrics, including: (1) the level of involvement and participation from community members (community engagement); (2) the degree dependence on centralized energy sources (energy independence); (3) the proportion of renewable energy sources integrated (renewable resources); (4) reducing greenhouse gas emissions capabilities (environmental impact); (5) social benefits of ECs (social energy); (6) the presence of supportive policies, regulations, and incentives (policy and regulatory support); (7) the deployment of technological innovation (Information and Communication Technology); and (8) fostering collaboration and replication of successful models elsewhere (knowledge sharing and collaboration).

The chapter considers EC functioning and the role of ECs in the transition to low-carbon energy sources. The current scenario in the five continents is briefly described, illustrated with examples of real ECs. The discussion covers the potential of ECs and the challenges facing them and highlights the need for suitable incentives, resources, and acceptance for implementation. The limited synergies among ECs, aggregators, and grid services are also highlighted, as well as the need for business models and regulatory frameworks adapted to ECs. Finally, this review suggests that ECs will be an important component of future energy systems and can set an example for sustainable development worldwide.

Real-world energy community examples

Since the physical environment, the social dimension, and the governmental perspective affect the deployment of ECs, implementation results vary depending on the regulations established by each country. Reviewed and analyzed below are current EC developments in various regions and cities on the five continents, aimed at improving understanding of the concepts, components, and technical needs of real ECs in particular contexts as governing by existing regulations.

Europe

Directives focused on electricity energy have been introduced for citizen and renewable sources 944/EU (BOE, 2019) or 2001/EU (BOE, 2018), which can adopt a multiple legal entity form through regulations, as well as providing flexibility services through demand response, storage, or electricity sale. EU recognizes and promotes the role of ECs in directives based on electricity market needs and focusing on promotion of renewable sources use. These directives set out the general principles and rights, as well as obligations of member states to facilitate their development by offering economic and/or environmental benefits according to:

- Reducing dependence on fossil fuels and the conventional electricity grid
- Encouraging investment in renewable energy and local job creation
- Reducing greenhouse gas emissions and carbon footprints
- Driving innovation in clean energy technologies and sustainable business models.

In order to ensure greater visibility between possible partners, a novel European Commission initiative based on a list created by the EU-funded COMETS project is the creation of an interactive map of ECs (see Fig. 12.2), with a website providing information on the number of available ECs and other data of interest (e.g., type of community, energy production, etc.). Currently, the largest number of ECs are in Germany (approximately 1800), Denmark (700), and the Netherlands (500). Spain is also experiencing an increase in EC numbers (currently 33), thanks to the EU-funded Recovery, Transformation, and Resilience Plan.

Inspirational examples in Europe (Tarpani, 2022; Kubli, 2023)

- Samsø (Clean-energy, 2022), Denmark: An emblematic example of a local EC with a 100% self-sufficient use in renewable energy by installing wind turbines and solar systems.
- Feldheim (Ecoinventos, 2008), Germany: A small and self-sufficient town in renewable energy by creating its own electricity grid. The community provides a wind farm, a biogas plant, and solar systems able to generate clean energy.
- Güssing (Energeticafutura, 2013), Austria: A hub of innovation in a role model for the energy transition. This city has achieved energy independence through the production biomass energy and other green sources.
- Totnes (Energylocal, 2020), United Kingdom: It has adopted the concept of "transition" to foster local resilience and energy sustainability. The community has implemented numerous projects, such as installing solar panels on public buildings and promoting energy efficiency in households.

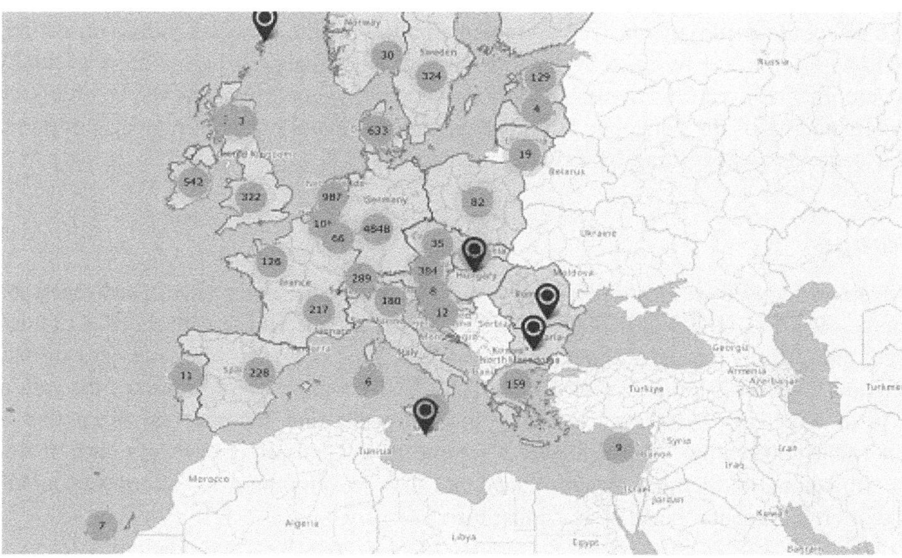

Figure 12.2 EC Map. Example of interactive map of energy communities in Europe. Map lines delineate study areas and do not necessarily depict accepted national boundaries. *Source*: Data from http:/energy-communities-repository.ec.europa.eu.

- Som Energía (SomenergiaCoop, 2010), Spain: An energy cooperative dedicated to the promotion of renewable energy and citizen participation. Different projects have been launched (e.g., shared solar installations, community wind farms), followed by up to 60,000 members. Som Energía was created for promoting sustainable development actions with active participation. Its main activities are focused on the electricity commercialization and the generation of renewable sources.
- REScoop (Rescoop, 2020), EU countries: A European federation of cooperatives bringing together more than 1500 EC in 15 countries. REScoop promotes collaboration between EC and facilitates the exchange of knowledge and resources to drive the transition to renewable energy in Europe.
- BeauVent (Beauvent, 2000), Belgium: An energy cooperative organization up to 5000 participants. Its activities are focused on the generation of renewable energy, including the electricity sale to customers with photovoltaic panels.
- Courant d'Air (Courantdair, 2009), Belgium: A cooperative society with activities focused on renewable generation (e.g., wind and solar). The cooperative carries open and share subscription, promoting awareness and improving efficiency in the green energy field.
- Ecopower (Ecopower, 1992), Belgium: A company with activities centered on generation and renewable supply. The project is followed by more than 56,000 participants helping citizens to a deep energy renovations in their households.
- Marstal Fjernvarme (Solarmarstal, 1962), Denmark: A not-for-profit client-owned company with up to 1600 members. The main activities follow up the development of a district heating framework based on renewable energies distribution, generation, or supply, by approximately 2200 customers. The project tries to demonstrate an innovative, cost-effective large-scale renewable energy system.
- Svalin Shared Housing Complex (The-energy-collective, 2020), Denmark: A collective energy project based on renewable electricity generation and consumption and community-based energy sharing of 20 households in Roskilde. The Svalin project aims to lead to local self-sufficiency, sharing economy, and collectively consuming 100% renewable, avoiding traditional intermediaries. The infrastructure was implemented to household solar panels or heat pumps.
- Enercoop (Energetic coop, 2005), France: A cooperative provider in the form of a cooperative "social enterprise" buying energy from producers. 11 regional renewable EC are developed with up to 70,000 members acting as a supplier of 100% renewable electricity. The main objective is to meet positive environmental and socioeconomic goals offering only renewable sources.
- Bioenergiedorf Jühnde eG (Bioenergiedorf, 2015), Germany: A cooperative organization with renewable electricity generation activities. Jühnde is the first village producing heat and electricity using renewable, becoming the first self-sufficient place producing renewable sources with the consumer involvement.
- Elektrizitätswerke Schönau eG (Ews, 2015), Germany: A cooperative consisting to a total of 7300 members providing energy services. The cooperative organization was leader electricity supply to the EC by selling renewable energy to local electricity customers.
- Sprakebüll Village eG (co2, 1998): A company focused on buying heat from a privately energy plant and distributing through the heating network to the consumers. A total number of 247 members focus their activities on the generation of renewable electricity.
- Amelander Energie Coöperatie (Amelandenergie, 2009), Netherlands: A cooperative with up to 290 members with activities focused on the generation of renewable energy. The company aims energy users to be supported by sustainable electricity.

- Duurzaam Ameland (Duurzaamameland, 2007), Netherlands: A public–private association formed by 9 members with activities focused on the generation and supply of renewable energy. The project is a cooperation example of involving utilities and distribution system operators as main actors of a real EC.
- Spółdzielnia Nasza Energia (Blue-fifty, 2014), Poland: An association focused on energy supply and distribution forecasting. It is the first large energy cooperative started in this country consisting of 300 Members. The cooperative was create in response to the high prices of electricity.
- Żywiecka Energia Przyszłości (Klasterzywiec, 2014), Poland: An association formed with the aim of generating renewable electricity for the supply and distribution of energy. The association is a public–private cooperation network focused on Dr solutions and electricity production.
- Słupsk (H2020, 2018), Poland: It aims to remove energy poverty by becoming a city with zero air pollution. This pilot project of H2020 is formed by 200 members and started with the main objective to increase energy efficiency by renovating households.
- Bostadsrättsföreningen Lyckansberg (HSB, 2018), Sweden: An association with activities focused on renewable electricity generation (e.g., solar plant). The organization consists of 85 members aiming to generate electricity for common purposes (e.g., transmission or energy storage).
- Edinburgh community solar limited (Solar, 2013), UK: A company operating with solar systems of 24 Edinburgh City Council buildings. It is the largest type of EC in this city. The company has up to 550 members, and the main objective is to help kickstart low-carbon initiatives in buildings.
- Energy4All (Energy4all, 2002), UK: A company made up of 27 cooperatives and provide up to 16978 members. The objectives aim to develop cooperatively and community-owned renewable energy.
- Isle of Eigg (Isleofeigg, 2008), UK: A cooperative founded by 96 local resident members. The island is one of the first communities in the world to have put in place an off-grid electricity system powered by wind, hydro and solar power (Eigg was not connected to the UK electricity grid).
- Viure (viuredelaire, 2017), Spain: A project aiming to install a wind turbine of shared ownership, which allows for the generation of net and green electricity, making solidarity between possible partners.
- Smart Åland energy project (Smartenergysystems, 2023), Finland: A project aiming to develop and build a social-scale energy system based on renewable energy. The overall goal here is to test and demonstrate a fully renewable energy system that is both, technically and economically viable in Åland City.
- Tiko energy business (Tiko, 2021), EU countries: It provides more than 40,000 installations across Europe, being one of the growing leaders in the energy management market. The company's goal is to create effective, flexible, and easy-to-use solutions that make energy decentralized, shared, and sustainable use.
- Feldheim Energy project (Feldheim, 2023), Germany: A project in a Brandenburg village supplied with 100% renewable energy. Households and businesses receive electricity and heat from their own wind and photovoltaic power plants. Biogas plants are connected to a local heat and power grid in this village.
- Schoonschip energie cooperatie residential neighborhood (Schoonschipamsterdam, 2022), Netherlands: A floating residential neighborhood of 47 dwellings located in Amsterdam cooperating for sustainable purposes by generating renewable sources for self-consumption.

- Green energy cooperative (Zez, 2022), Croatia: A green energy cooperative that raise funds for energy saving kits for citizens at risk of fuel poverty. EON Solar cloud Energy solution is another provider of smart grid solutions, which tries to expand its portfolio of digital solutions for the transition to green energy.
- Tibber energy management platform (Tibber, 2022), Norway: It provides a complete overview of the electricity consumption, making solutions easier to reducing household's electricity consumption. The platform does not execute any profit from electricity price.
- Open Energy management platform (Openenergi, 2023), UK: It offers an automated optimization service optimizing storage assets in all energy markets.
- Smartly Energy solution provider (Smartly, 2022), Norway: It develops simple and efficient energy management solutions for companies and residential businesses (e.g., metering, control of energy consumption).
- Entelios energy solution provider (Entelios, 2019), Norway: It provides energy, knowledge, and clean technology that empowers businesses, organizations, and the public sector to take the lead in the transition to a sustainable and renewable future in Norway.
- Quartierstrom energy project (Quartier, 2022), Switzerland: A local electricity market that was built in the pilot region of Walenstadt. The project enables the participation up to 40 households to buy and sell locally the produced solar electricity in their neighborhood. The project was finished at the end of July 2020.
- Flexens energy service provider (Flexens, 2021), Northern Europe: A partner for communities, asset owners, investors, and commercial and industrial customers seeking renewable solutions and arranging project finance to enable companies to operate with a 100% renewable energy mix.
- Lumenza energy management platform provider (Lumenaza, 2018), Germany: It enables green utilities to start, grow, and evolve their business by connecting producers and consumers of distributed green energy.
- Lemene energy project (Elcogen, 2020), Finland: It produces renewable energy for customers' needs. It is considered a local, expert, and innovative energy production company that produce district heating with biobased fuels acting as an energy system operator.
- Collective energy (Coen, 2020), Greece: A nonprofit citizen EC founded in Athens. A collective energy is considered to explore multiple paths leading to the transformation of the existing energy model, on the basis of social, environmental, economic, and political reform.
- NCEI cooperative (NCEI, 2022), Ireland: A cooperative supporting citizens, cooperatives, and cities to drive a citizen-led energy transition.
- CommonEn/Electra Energy (Electraenergy, 2021), Greece: A cooperative where members own as part of the collective solar farm needed to cover the annual electricity needs. For instance, a solar farm has been developed for covering the electricity needs of 35 local families.
- Rising.eco REC facilitator (Rising, 2020), Hungary: A project with the aim of becoming self-sufficient in electricity supply. The project helps organizing into EC becoming an independent energy producer, with renewable energy sources.
- EPV CEC facilitator (Enr, 2017), France: A project where all the volunteers and employees of the structures and networks associated with "Énergies citoyennes" have been working together for many years for the reappropriation of energy by citizens, involving them as much as possible in the energy and social transition.
- Som serveis energetics CEC (Cleanwatts, 2017) and Cleanwatts REC facilitator, Spain: It offers individual services aimed at individuals and member companies. The following activities stand out: collective actions promoted by the cooperative with participation; collective financing of projects; and development of self-production and/or sales projects promoted as local EC.

- CER energy city hall Magliano Alpi REC facilitator (Cermaglianoalpi, 2020), Italy: A project focused on building rehabilitation and EC implementation for value creation and post-pandemic revitalization aiming an innovative energy management models.
- EmasP S.coop Urroz REC (Ceurroz, 2018), Spain: A local energy community promoted by the town council of Urroz Villa. They have started a project that, through the self-consumption of renewable energy, can achieve a saving of 54% in the overall bill and contribute to the care of the planet by avoiding CO_2.
- EmasP S.coop Lasierra REC (Lasierra, 2021), Spain: A Consortium of Mendi Haran administrative board that advices and offers technical energy support. This community takes the experience of the community for the management of water, without any kind of profit or benefit, and transfers it to the production and management of zero kilometer renewable energy, for the self-consumption of the components of the EC.
- MegaWattPuur cooperative (MegaWatt, 2020), Belgium: A cooperative and social enterprise and Wase citizens' association focused on investing renewable energy and energy saving measures.
- Druifkracht cooperative (Druifkracht, 2020), Belgium: A citizens' cooperative (CVSO) and a nonprofit association. It is based on seven principles of openness, democracy, participation, independence, knowledge sharing, cooperation, and energy commitment.
- Noordlicht cv cooperative (Noordlicht, 2019), Belgium: A cooperative enterprise that aims to bring citizens together to invest in renewable energy projects, energy saving, and rational energy consumption, thus contributing to the reduction of greenhouse gases.
- Energy Revolt s.c. cooperative (Energyrevolt, 2018), Turkey: It provides a number of associated benefits for a region of Turkey. It aims to boost the local economy by ensuring that profits stay in the area, encouraging visitors and raising the profile of the local area.
- UrStrom eG (Islandinstitute, 2015), Germany: An initiative of the citizens of Mainz aiming to contribute energy transition supported up to 730 members. The project currently operates 18 photovoltaic systems, producing clean solar energy for around 300 households. In addition, they offer 100% green electricity for private and commercial customers, an increasing proportion of which comes from community energy systems.

The analyzed EC projects cover a wide variety of EU energy structures, including energy generation and energy sharing. Many are successful EC examples of stable systems that integrate the goals of using renewable energy sources and reducing emissions (Koirala, 2016), reflecting a growing tendency in the EU of focusing on both environmental and economic benefits (Caramizaru & Uihlein, 2020). Renewable installations and district heating distribution systems and networks are increasingly popular technologies deployed by ECs. Although the overall share of citizens and ECs as investors in renewable energy is currently insufficient, the future potential is undoubtedly great.

America

In the United States, most incentives available for renewable energy production are based on tax reductions, with investors required to meet specific conditions. Several EC financing models have been designed for the creation of wind and photovoltaic production plants. Some features of EC implementation in the United States are summarized as follows:

- Citizens have sought ways to collectively invest in renewable sources by financing EC models that maximize federal and state incentives.
- Many utilities are governed by their own customers, as approximately a quarter of the US population owns their own energy company (as an electric cooperative or public company).
- Smart net metering (a billing mechanism that compensates customers of energy utilities for the electricity they sell to their utility) allows for virtual net metering in some states.

Currently, approximately 900 electric cooperatives serve a million people in 47 US states and around 15% of meters correspond to consumers (NRECA, 2023), while cooperatives or state-level associations in 38 of those 47 states channel interactions with public regulatory authorities and state legislatures (Electric, 2023). Electric cooperatives have traditionally promoted energy efficiency in the interest of cost savings, and nowadays, consumption efficiency is maximized by shifting peak demand and achieving electricity supply targets while maintaining positive consumer relationships.

Fig. 12.3 depicts electric cooperatives with solar energy capacity currently installed across the United States. Around 96% of cooperatives operate some form of energy efficiency program, 70% offer financial incentives to promote energy efficiency, and 73% are planning to expand existing energy efficiency programs in

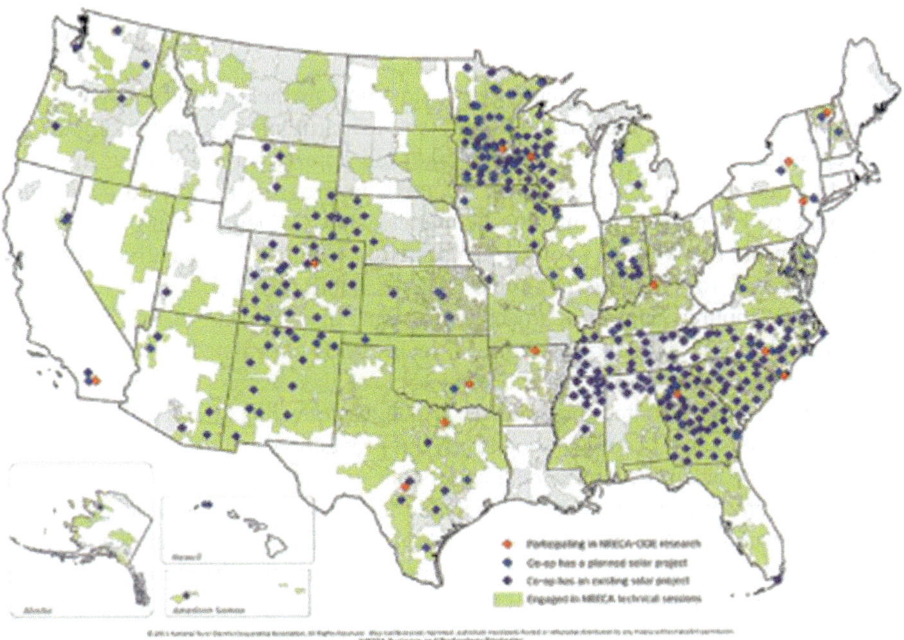

Figure 12.3 Electric cooperatives. Map of electric cooperatives with a solar energy installation underway or in the pipeline in United States. Map lines delineate study areas and do not necessarily depict accepted national boundaries.
Source: Data from https://www.seia.org/.

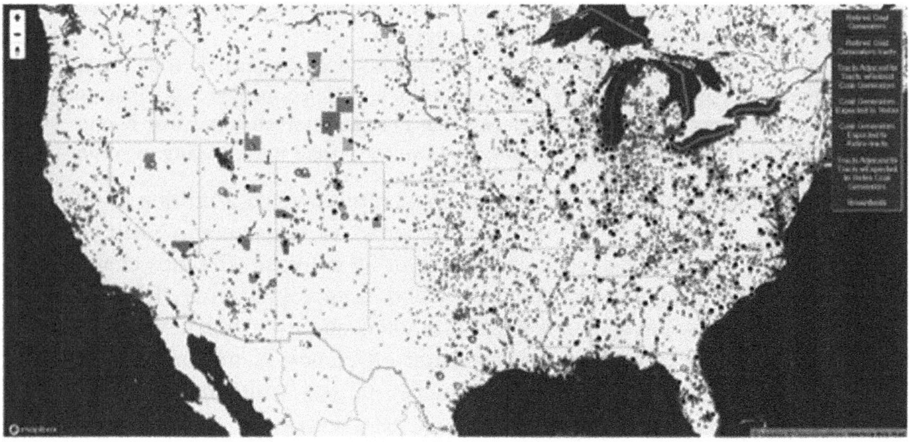

Figure 12.4 Energy communities. Example of Energy communities in United States—sneak peek.
Source: Data from https://www.seia.org/research-resources/energy-communities-map-sneak-peek.

the upcoming 2 years. Furthermore, the Solar Energy Industries Association is creating resources to identify areas for ECs, including an interactive map as shown in Fig. 12.4 (NRECA, 2023).

EC projects and directives that support start-ups are helping reduce greenhouse gas emissions and are motivating US citizens to participate in neighborhood ECs with renewable energy capabilities. A market-led approach to providing incentives to utilities and start-ups appears to be effective in progressing toward achieving a wide range of EC objectives in some US states. The US Department of Energy has published an interactive map showing census tracts eligible for incentives according to specific goals, for example, coalmine closures (DOE, 2023).

Canada exemplifies another approach, with forms of ownership of ECs regulated at the municipal (45%) and cooperative (33%) levels. Most distributed utilities are owned by cooperatives, especially in Ontario province. Another feature of Canada is that a wide variety of energy sources predominate in supplies covering all the country.

In central and south American countries, the production and consumption of local energy is clearly identified as a resilience strategy, with self-consumption and EC deployment conceived as encouraging dynamic investment, social cohesion, and environmental awareness. Newer governments are also leading a positive energy transition discourse. Mexico, however, as yet has no specific legislation in place regarding EC implementation, as authorization from the Energy Regulatory Commission is lacking.

Inspirational examples in America

- Fox Islands wind (Islandinstitute, 2015), United States: A project providing 3 general electric turbines of 1.5 MW located on Vinalhaven island (Maine). All the produced electricity is sold to the utility that it is supplied itself.

- Local governments (Celar, 2016), United States: They strive to create ECs by investing in future-focused clean energy solutions for their communities. As an example, NYSERDA's new Clean EC Leadership Round provides resources and funding helping for a successful EC implementation.
- Celar (Celar, 2016), Argentina: An energy cooperative that distributes electricity with photovoltaic panels by promoting user participation through smart grids and a distributed energy generation. The cooperative is implemented in the Armstrong city (Santa Fe). The cooperative and other entities have launched another renewable energy project by creating EC with a photovoltaic plant in the city's industrial park. The EC is performed by installing photovoltaic systems and wind turbines on the rooftops of residential households, and public spaces as well.
- Potencialidad (Revolusolar, 2021), Argentina: It offers different initiatives of ECs for modeling extension to other regions such as Córdoba city by promoting the potential of a real cooperation between partners.
- Photovoltaic Community (Revolusolar, 2021), Brazil: It is created in the Babilônia favela of Rio de Janeiro to promote a sustainable development of low-income communities through solar energy. The project was successfully executed through a collective volunteer work and involving community leaders. In addition, the cooperative provides information and educational tools to the local population about the benefits, social, economic, and environmental aspects of solar energy use.
- Comuna Energética (Gub, 2015), Chile: A successful initiative created from the Swiss model Ba, which was created to promote the energy development. The initiative supports municipalities by developing local energy strategies, with EC visions and action plans in renewable and energy efficiency fields.
- The Energy Communes program, Uruguay (2018): A participatory tool where civil society raises problems associated with energy and seeks sustainable solutions in conjunction with the national energy consortium. The initiative is also presented in "Atlántida" and "Solís de Mataojo" and has reached the municipalities of "Nueva Helvecia" and "La Paloma." The goal here is to promote a local development and capacity building in energy issues. More specifically, the latest EC pilots currently provide a local energy strategy.

Asia

Asian countries have made significant progress in EC deployment, despite the lack of viable frameworks for EC coordination and clean energy production and consumption. The complexities in facilitating energy transmission require the creation of a regional-level energy network, which would have political and economic implications (Tuerk et al., 2022).

China is the world's largest producer and consumer of electricity and accounts for around 50% of Asia's consumption and production. Consumption outruns production, so China is a net importer of electricity. As an important followup to its previous commitment of peak carbon emissions by 2030, China announced plans in 2020 to be carbon-neutral by 2060. Currently, however, a decentralized energy industry and real EC projects are still pending development, as many challenges have to be overcome, including the ongoing influence of the existing policies, the lack of overall energy planning, and the need for basic technology and innovative equipment. Local governments are active in issuing policies to foster efficient development of EC-based decentralized energy. For instance, the International EC-China, with 40 representatives from governments and multinational companies (Openenergi, 2023), aims to accelerate energy transition by showcasing energy-related activities focused on partners.

Japan is another major energy consumer in Asia and, since 2007, has been a fully self-sufficient electricity producer due to declining consumption combined with increased efficiency. EC pilots inspired by movements in Denmark and Germany began to appear in Japan in the early 2000s, mainly motivated and organized by committed activists. However, after the Fukushima nuclear disaster of 2011, broader sectors joined the renewable energy movement, explicitly adopting the position that ECs were key to Japan's economic growth strategy (DeWit, 2018). ECs emerging as companies are by now a key part of the Japanese energy sector, and smart grids have introduced revolutionary modifications in existing energy management systems. EC implementation is focused on government-led, community-oriented, and business-driven models (Liu et al., 2022)., with government constraints somewhat undermined by progress in EC development. As an example, a model initially deployed in the European Union has been implemented in Japan after suitable adaptation, offering solutions related to membership, nondiscriminatory treatment, barriers, support schemes, grid management, and connection (Sokołowski & Visvizi, 2023).

India has pledged to reach net-zero emissions by 2070, but there is a 20-year gap between that pledge and scientific consensus on the need to globally reach net-zero emissions by 2050. India, however, is still developing its economy, and energy consumption is projected to grow by almost 70% by 2040. Many challenges remain to be overcome, including energy reliability, as rural communities experience recurrent and prolonged energy outages that hinder essential services such as health and education. Nonetheless, deployment of clean energy and EC implementation will be vital to controlling emissions in this country, currently the third largest energy consumer in the world. Other EC projects developed in Asia (Neurasia, 2023; Wongbumru & Dewancker, 2014; Yilmaz & Li, 2018).

Inspirational examples in Asia

- Japanese government deployed pilot cities of EC projects by different characteristics in four cities (Wongbumru & Dewancker, 2014): (1) Yokohama city, a residential and business area; (2) Toyota city, industry and living area; (3) Kansai Science city, Kyoto prefecture, education and research area; and (4) Kitakyushu city, an industrial and household area.
- Tolo 1 Wind Farm (Namienergy, 2020), Indonesia: One of the largest projects based on ECs in the country with a electricity capacity of 72 MW. The project was developed by Equis Energy, a Singapore-based company that specializes in renewable energy projects in the Asia-Pacific region.
- Bau Bang Solar Power Plant (Namienergy, 2020), Vietnam: A company with a capacity of 450 MW being one of the largest power plants in South East Asia. The project was developed by Singapore-based Sunseap Group and has helped to significantly increase Vietnam's renewable energy capacity.
- Green Innovations for Tomorrow Corp, Philippines (Reurasia, 2014): A biomass energy plant deploying agricultural waste to fuel a steam turbine.

- Ohisama Shimpo Energy (Eco-business, 2010), Japan: An EC solar photovoltaic project. Iida city installed solar PVs on the rooftops of 38 public facilities including kindergartens and community centers.
- Asia Super Grid, Pacific Power Grid (UNESCAP, 2014), multiple Asian countries: It maximizes the use of renewable energy resources to deliver safe and inexpensive electricity. GRENATEC (2010) is another Asian energy infrastructure nation formed by Australia, China, Japan, Korea, and different Chinese cities (e.g., Taipei).
- Cooperation organization initiative based on energy interconnection (GEIDCO), China, efforts of hundreds of companies, institutions, and individuals undertaking the research on international interconnection projects based on ECs (Yilmaz & Li, 2018).
- Asia Pacific Power Grid company (2011), multiple Asian countries: A pilot example of renewable energy resources use to deliver safe electricity around different regions (see Fig. 12.5).
- NEA Super Grid (Unescap, 2014), multiple Asian countries, formed by Korea Electric Power Corporation, China, Japan, Korea, Mongolia and Russia. NEA Interconnection would be an alternative to contribute to the government policy of the around these countries.
- Gobitec and Asian Super Grid (Eirenewable, 2020), multiple Asian countries: It produces energy from renewable energy sources in the Gobi desert focusing their activities by supplying the energy produced to regions with a high electricity demand.

Africa

In Africa, energy access remains unequal (a million people lack access to electricity in a context of a rapidly growing population), and access levels are particularly low in rural areas, as shown in Fig. 12.6 (Fabini et al., 2014). In the 20 years from 1990 to 2010, access increased only modestly, from 76% to 83%. The few real ECs that

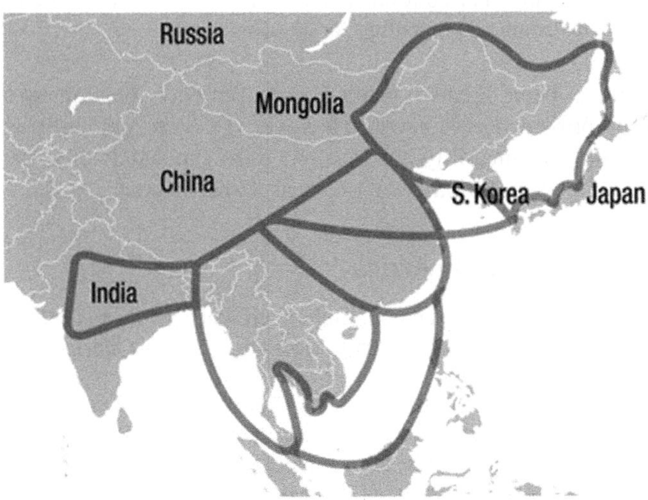

Figure 12.5 Interconnections. Example of realized interconnections between China, Mongolia, Russia and other neighboring countries. Map lines delineate study areas and do not necessarily depict accepted national boundaries.
Source: Data from https://www.renewable-ei.org/en/asg/.

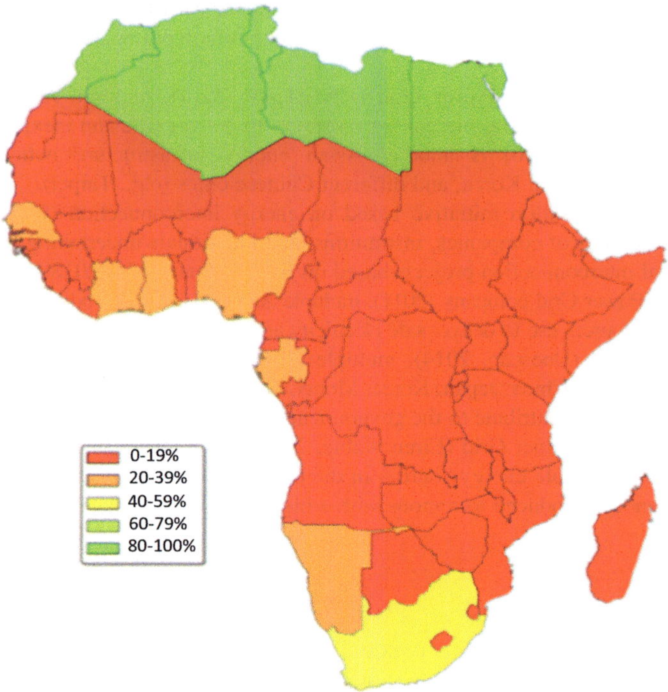

Figure 12.6 Rural map. Example of rural electrification rates in Africa. Map lines delineate study areas and do not necessarily depict accepted national boundaries.
Source: Data from Mapping induced residential demand for electricity in KenyalProceedings of the fifth ACM symposium on computing for development.

are operative have been created primarily as projects meeting specific poverty reduction and improved health needs. However, ECs are gradually emerging as a means for ensuring sustainability and resilience for thousands of households. Other EC projects developed in Africa (Ambole et al., 2021; Hoeck, 2022).

Inspirational examples in Africa

- Malawi Renewable energy acceleration program presents actions to achieve the renewable energy use in Malawi city by an universal access to renewable electricity (Dauenhauer & Frame, 2016).
- Renewable energy minigrids in Zambia's solar sector has attracted global attention of solar production by low costs of electricity procured. Power generation in Zambia accounts for nearly 90% of its total installed solar generation capacity (Muhoza & Johnson, 2018).
- Plantation-based biofuel schemes based on real EC are considered a real option to cover the lack of accessibility to a modern energy supply in Mpumalanga District-South Africa (Röder et al., 2017).
- Ethiopia Village biogas and solar energy scheme aims to determine efficient options for the production and use of biogas and solar energy for small-scale electricity production (Tucho & Nonhebel, 2017).

- Community electricity minigrids address the biggest gap in Uganda's electrification program by accessing to the electricity needs (Eder et al., 2015).
- Redavia Rental Solar Power rents preassembled solar photovoltaic systems to local operators as electricity minigrid deployment in Tanzania (Odarno et al., 2017).
- Lucingweni (Gauteng Province) investigates the feasibility of a hybrid minigrid as a solution for rural development in South Africa. Community electricity minigrids have the greatest potential to meet household energy needs in a cost-effective manner (Azimoh et al., 2016).
- Community electricity minigrids project relies heavily on coal-based electricity imports, limiting the country's ability to achieve promised sustainability goals in Namibia (Tsumkwe Village). The project highlights legislation and the neglect of educational outreach among the EC (Azimoh et al., 2017).
- Rwanda Solar powered microgrids and smart metering proposes an operator model where solar energy and smart metering systems increase the reliability of electricity and energy supply to different consumers (Dieu Niyonteze et al., 2019).
- South Africa household energy models directive aims at the development and the implementation of energy efficiency practices by providing practical guidelines for EC operation (Masekoameng et al., 2005).
- Off-grid schemes Jasper 96 MW (Sudáfrica), Gomoa Onyandze 20 MW (Ghana), or Soroti 10 MW (Uganda) are examples of large photovoltaic solar parks aiming to generate electricity that is fed directly into the national grid (Monyei et al., 2018).
- SSA Sustainable energy for grid connected rural EC is a decentralized renewable technology in the electrification of the rural population (Hubble & Ustun, 2016).

Oceania

EC implementation is primarily focused on economic benefits that could be produced for the community. In New Zealand, for instance, the national government plays an important role in the growth of investment in ECs. However, a proper framework is lacking, highlighting the challenges facing successful EC development. Pilot projects with a strategy for monitoring and evaluating impacts aim to identify risks and barriers to EC initiatives (Berka, 2020). Consortia accounting for at least 50% of energy projects are the most common form of EC creation. Partnerships also play a very important role, confirming the experience of many countries where local actors collaborate with each other (Hoicka, 2018).

Inspirational examples in Oceania

- Community energy project (2019) offers services to create ECs arising from diverse needs and available resources of a EC. The instruments can range from solar panels on the roof of a school to a small wind farm on the outskirts of town or a micro-hydro system owned by two nearby municipalities. Fig. 12.7 shows some of the community energy clusters for which information is available.
- King (Tasmania), Norfolk (Australia), or Rotuna Islands (Fijis) are examples of type of diesel power plant that currently cover more than 50% of renewable supply for ECs (Neves, 2014).

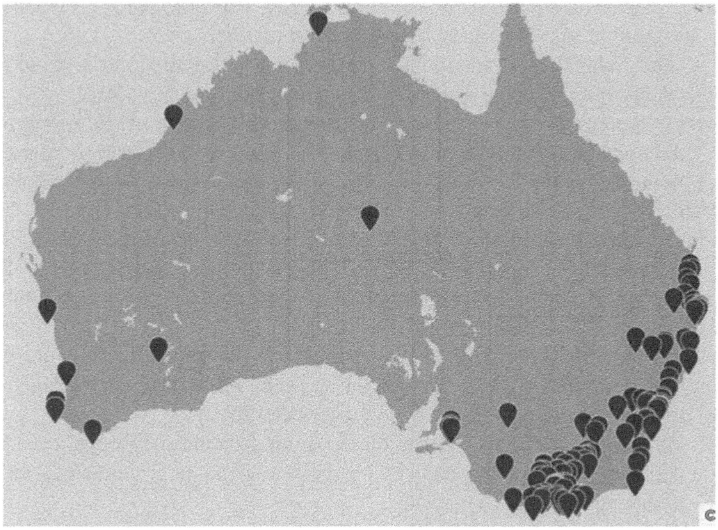

Figure 12.7 Organizations. Records of energy communities organizations of the Australian community energy database. Map lines delineate study areas and do not necessarily depict accepted national boundaries.
Source: Data from https://cpagency.org.au/resources-2/map.

Discussion

ECs have been demonstrated internationally as a key element in promoting a range of benefits, including sustainability, energy security, self-sufficiency, cost savings, and energy independence. For instance, use of renewable energy and applying a fair cost-allocation criterion based on energy coo peration can reduce prosumer and consumer bills by 20%–30% compared to the use of dispatchable electricity sources. Furthermore, price-based demand response systems deployed in an EC can additionally reduce community costs by up to 50% (Volpato et al., 2022). Effective participation in ECs based on collective purchasing and cooperative operation make citizens more energy aware and so contribute to improved energy efficiency and reduced emissions. ECs have also the potential to provide supply flexibility as well as supply security and, furthermore, contribute to renewable energy penetration and energy efficiency targets as well as climate change objectives (Brugger, 2021). Participation and collaboration in ECs can help defer new investments in power lines, reduce system peaks, and distribute loads more evenly throughout the day.

While the rise of individualized prosumers and the development of microgrids are widely discussed in energy policy circles and beyond, institutions remain slow in formulating new legislation focused on EC regulation, which is currently insufficient in many parts of the world. The use of indicators to quantify the social impact of ECs has also been vaguely explored. Although ECs function differently in a wide variety of locations, most of the barriers to effective EC implementation are based on social acceptance, motivations to participate, actor and network dynamics, and preliminary exploratory studies. Experience shows that real ECs commonly spread their investments over time, depending on annual savings, the evolution of local financial opportunities, and community interests (Gjorgievski et al., 2021). A favorable institutional environment can influence EC success at the individual and community levels, with sociocultural norms considered to be beneficial for EC emergence (Lode et al., 2022). Willingness to participate in ECs cannot be only based on evaluation of value-for-money and the economic point of view. Clearly, consumer participation is essential to achieve multiple objectives, including decarbonization of the energy system (Iazzolino et al., 2022).

The availability of energy generation and storage facilities will allow ECs to contribute flexibility as needed to today's energy system and market. Cooperatively optimizing supply and demand in ECs, while keeping overall maintenance and operating costs low and ensuring high supply security, will contribute to ultimately achieving climate policy objectives set for the coming years. The organization of ECs, as local exchanges for interactions between members, will have direct and indirect impacts on different actors. Finally, it is crucial that costs and benefits are shared equitably among stakeholders.

Conclusions

The world faces increasing pressure to accelerate decarbonization and mitigate the effects of the energy crisis on households and businesses. EC initiatives are receiving attention as an effective vehicle toward more equitable and resilient energy systems. Advances in digitalization are also expanding the potential of ECs to act as a central pillar in the transition to clean energy. The deployment of policy support and local buy-ins for EC-based infrastructure developments is key to accelerating this transition by encouraging citizens to be active participants in energy systems.

In this chapter covering the concept of ECs and their role in the transition to clean energy, several examples were provided of real ECs on different continents. In the European Union, for instance, energy directives have been introduced aimed at fostering ECs and renewable energy sources, while in the United States, electric cooperatives promote energy efficiency and renewable energy use. The examples of real ECs come almost exclusively from industrialized countries; in less industrialized countries, progress is slow mainly due to limited investment. The main challenges to be overcome are summarized as the development of suitable frameworks, incentives, and resources for EC implementation, the fostering of EC acceptance, and the evaluation of current practices in collaborative energy systems. Synergies among ECs, aggregators, and grid services are as yet limited in terms of meeting energy transition needs, and business models and regulatory frameworks have not yet been globally adapted to incoming ECs. Concerning renewable energy development, compared to photovoltaic deployment, certain renewable technologies (e.g., hydro and biomass) are sparingly used. A quantitative evaluation using empirical data from pilot projects is recommended to obtain valuable insights into EC needs.

An analysis based on the current state of ECs would increase the knowledge of the different actors involved, enhance the impact, and further the development of smart energy management systems. Growing concern for the environment and a renewed focus on universal energy access are the main drivers of further progress in ECs for both developed and developing countries. EC and local energy projects clearly have benefits around the world in terms of renewable energy deployment and efficiency, reduced costs, and accessible and reliable supply.

ECs are increasingly being adopted and modified, thanks to new energy policies and development and application of specific technologies, and environmental, technical, socioeconomic, and institutional issues are increasingly being considered. ECs will undoubtedly be an important component of future energy systems, as they can transform current energy systems and set an inspiring example for sustainable development around the world.

References

Ambole, A., Koranteng, K., Njoroge, P., & Luhangala, D. L. (2021). A review of energy communities in sub-Saharan Africa as a transition pathway to energy democracy. *Sustainability*, *13*(4), 2128.

Amelandenergie. (2009). https://www.amelandenergie.nl/.

Azimoh, C. L., Klintenberg, P., Mbohwa, C., & Wallin, F. (2017). Replicability and scalability of mini-grid solution to rural electrification programs in sub-Saharan Africa. *Renewable energy*, *106*, 222–231.

Azimoh, C. L., Klintenberg, P., Wallin, F., Karlsson, B., & Mbohwa, C. (2016). Electricity for development: Mini-grid solution for rural electrification in South Africa. *Energy Conversion and Management*, *110*, 268–277.

Beauvent (2000). https://www.beauvent.be.

Berka, A. (2020). Explaining inclusivity in energy transitions: Local and community energy in Aotearoa New Zealand. *Environmental Innovation and Societal Transitions*, *34*, 165–182.

Bioenergiedorf. (2015). http://www.bioenergiedorf.de/en/home.html/.

Blue-fifty. (2014). https://blue-fifty.com/pl/rozwoj-projektu/spoldzielnia-nasza-energia/.

BOE. (2018). https://www.boe.es/doue/2018/328/L00082-00209.pdf.

BOE. (2019). https://www.boe.es/doue/2019/158/L00125-00199.pdf.

Brugger, H. (2021). Energy Efficiency Vision 2050: How will new societal trends influence future energy demand in the European countries? *Energy Policy*, *152*, 112216.

Caramizaru, A., & Uihlein, A. (2020). *Energy communities: an overview of energy and social innovation* ((Vol. 30083).). Luxembourg: Publications Office of the European Union.

Celar. (2016). https://www.celar.com.ar/index.php/features/la-cooperativa/.

Cermaglianoalpi. (2020). https://cermaglianoalpi.it/.

Ceurroz. (2018). https://comunidadesenergeticas.org/urroz/.

Clean-energy. (2022). https://clean-energy-islands.ec.europa.eu/news/profile-samso-island-inspiring-energy-communities-around-world/.

Cleanwatts. (2017). https://www.cleanwatts.energy/.

co2. (1998). http://co2mmunity.eu/.

Coen. (2020). https://coen.coop/en/.

Courantdair. (2009). https://www.courantdair.be/wp/.

Dauenhauer, P. M., & Frame, D. F. (2016). Critical review of the Malawi community energy model. *IEEE PES PowerAfrica*, 78–82.

DeWit, A. (2018). Japanese smart communities as industrial policy. In *Sustainable Cities and Communities Design Handbook*, (pp. 421–452). Butterworth-Heinemann.

Dieu Niyonteze, J. D., Zou, F., Osarumwense Asemota, G. N., & Bimenyimana, S. (2019). *Solar-powered mini-grids and smart metering systems, the solution to Rwanda energy crisis*, . *Journal of Physics: Conference Series*, (1st, p. 012002)(Vol. 1311, p. 012002). IOP Publishing.

DOE. (2023). https://www.energy.gov/.

Druifkracht. (2020). https://www.druifkracht.be/.

Duurzaamameland. (2007). https://www.duurzaamameland.nl/over-ons/.

Eco-business. (2010). https://www.eco-business.com/opinion/southeast-asia-en-route-to-pan-asian-energy-network/.

Ecoinventos. (2008). https://ecoinventos.com/.

Ecopower. (1992). https://www.ecopower.be/.

Eder, J. M., Mutsaerts, C. F., & Sriwannawit, P. (2015). Mini-grids and renewable energy in rural Africa: How diffusion theory explains adoption of electricity in Uganda. *Energy Research & Social Science*, *5*, 45–54.

Eirenewable. (2020). https://www.renewable-ei.org/en/images/pdf/20140124/Gobitec_and_ASG_report_ENG_BOOK.pdf/.

Elcogen. (2020). https://elcogen.com/casestudies/lemene-fuel-cell-system-to-be-commercialised-in-finland/.

Electraenergy. (2021). http://electraenergy.coop/commonen-energy-community/.

Electric. (2023). https://www.electric.coop/.

Energetic coop. (2005). http://www.enercoop.fr/.

Energeticafutura. (2013). https://energeticafutura.com/.

Energy4all. (2002). https://energy4all.co.uk/.

Energylocal. (2020). https://energylocal.org.uk/.

Energyrevolt. (2018). http://energyrevolt.lu/.

Enr. (2017). https://www.enr-citoyennes.fr/.

Entelios. (2019). https://www.entelios.com/en/.

Ews. (2015). https://www.ews-schoenau.de/.

Fabini, D. H., Baridó, D. P. D. L., Omu, A., & Taneja, J. (2014). *Mapping induced residential demand for electricity in Kenya. Proceedings Symposium on Computing for Development* (pp. 43–52).

Feldheim. (2023) https://nef-feldheim.info/the-energy-self-sufficient-village/?lang = en.

Flexens. (2021) https://flexens.com/.

Gjorgievski, V. Z., Cundeva, S., & Georghiou, G. E. (2021). Social arrangements, technical designs and impacts of energy communities: A review. *Renewable Energy*, *169*, 1138–1156.

Gub. (2015). https://www.gub.uy/.

Gui, E. M., & MacGill, I. (2018). Typology of future clean energy communities: An exploratory structure, opportunities, and challenges. *Energy research & social science*, *35*, 94–107.

H2020. (2018) https://www.score-h2020.eu/pilot-projects/slupsk/

Hoeck, I. (2022). Challenges for off-grid electrification in rural areas. Assessment of the situation in Namibia using the examples of Gam and Tsumkwe. *Energy, Ecology and Environment*, *7*(5), 508–522.

Hoicka, C. (2018). From tip to toes: Mapping community energy models in Canada and New Zealand. *Energy Policy*, *121*, 162–174.

HSB. (2018). https://www.hsb.se/sydost/brf/lyckansberg/miljo/solceller/

Hubble, A. H., & Ustun, T. S. (2016). *Scaling renewable energy based microgrids in underserved communities: Latin America, South Asia, and Sub-Saharan Africa* (pp. 134–138). IEEE.

Islandinstitute. (2015). https://www.islandinstitute.org/ii-solution/fox-islands-wind/.

Isleofeigg. (2008). http://isleofeigg.org/eigg-electric/.

Iazzolino, G., Sorrentino, N., Menniti, D., Pinnarelli, A., De Carolis, M., & Mendicino, L. (2022). Energy communities and key features emerged from business models review. *Energy Policy*, *165*, 112929.

Klasterzywiec. (2014). http://klasterzywiec.pl/.

Koirala, B. (2016). Energetic communities for community energy: A review of key issues and trends shaping integrated community energy systems. *Renewable and Sustainable Energy Reviews*, *56*, 722–727.

Lasierra. (2021). https://comunidadesenergeticas.org/lasierra/.

Liu, J., Zhang, X., Lin, J., & Li, Y. (2022). Beyond government-led or community-based: Exploring the governance structure and operating models for reconstructing China's hollowed villages. *Journal of Rural Studies*, *93*, 273–286.

Lode, M.L., Te Boveldt, G., Coosemans, T., & Camargo, L.R. (2022). A transition perspective on Energy Communities: A systematic literature review and research agenda. *Renewable and Sustainable Energy Reviews*, *163*, 112479.

Lumenaza. (2018). https://www.lumenaza.de/en/.

MegaWatt. (2020). https://www.megawattpuur.be/.

Masekoameng, K. E., Simalenga, T. E., & Saidi, T. (2005). Household energy needs and utilization patterns in the Giyani rural communities of Limpopo Province, South Africa.. *Journal of Energy in Southern Africa*, *16*(3), 4–9.

Monyei, C. G., Adewumi, A., & Jenkins, K. E. H. (2018). Energy (in)justice in off-grid rural electrification policy: South Africa in focus. *Energy Research & Social Science*, *2018* (44), 152–171.

Muhoza, C., & Johnson, O. W. (2018). Exploring household energy transitions in rural Zambia from the user perspective. *Energy Policy*, *121*, 25–34.

Namienergy. (2020). https://namienergy.com/du-an/bau-bang-industry-zone/.

NCEI. (2022). https://www.ncei.ie/.

Neurasia. (2023). https://reurasia.com/the-benefits-of-renewable-energy-for-the-environment-and-local-communities-in-southeast-asia/.

Neves, D. (2014). Design and implementation of hybrid renewable energy systems on micro-communities: A review on case studies. *Renewable and Sustainable Energy Reviews*, *31*, 935–946.

Noordlicht. (2019). https://www.noordlicht.be/.

NRECA. (2023). https://www.nrecainternational.coop/.

Odarno, L., Sawe, E., Swai, M., Katyega, M.J., & Lee, A. (2017). Accelerating mini-grid deployment in Sub-Saharan Africa: Lessons from Tanzania.

Openenergi. (2023). https://openenergi.com/.

Quartier. (2022). https://quartier-strom.ch/index.php/en/homepage/.

Rescoop. (2020). https://www.rescoop.eu/.

Revolusolar. (2021). https://revolusolar.org.br/sobre-nos/.

Rising. (2020). https://rising.eco/.

Reurasia. (2020). https://reurasia.com/.

Röder, M., Stolz, N., & Thornley, P. (2017). Sweet energy–Bioenergy integration pathways for sugarcane residues. A case study of Nkomazi, District of Mpumalanga, South Africa. *Renewable energy*, *113*, 1302–1310.

Schoonschipamsterdam. (2022). https://schoonschipamsterdam.org/.
Smartenergysystems. (2023). https://www.eranet-smartenergysystems.eu/ll/120/Smart-Energy-land.html.
Smartly. (2022). https://www.smartly.no/.
Sokołowski, M. M., & Visvizi, A. (2023). *Exploring the Smart cities-Energy Communities Nexus.*. Routledge Handbook of energy Communities and Smart Cities..
Solar. (2013). https://www.edinburghsolar.coop/projects/how-the-co-op-works/.
Solarmarstal. (1962). https://www.solarmarstal.dk/.
SomenergiaCoop. (2010). https://www.somenergia.coop/.
The-energy-collective. (2020). http://the-energy-collective-project.com/.
Tibber. (2022). https://tibber.com/no/.
Tiko. (2021). https://tiko.energy/.
Tucho, G. T., & Nonhebel, S. (2017). Alternative energy supply system to a rural village in Ethiopia. *Energy, Sustainability and Society*, *7*, 1–14.
Tuerk, A., Neumann, C., Gubina, A. F., Medved, T., Singh, P. K., Nathan, H. S. K., & Xu, X. (2022). *Overview of international approaches for local energy systems and energy communities*. Slovenia: Compile: Luce.
UNESCAP. (2014). https://www.unescap.org/sites/default/d8files/event-documents/3-4_Hoseok%20Kim.pdf/.
Viuredelaire. (2017). https://www.viuredelaire.cat/.
Volpato, G., Carraro, G., Cont, M., Danieli, P., Rech, S., & Lazzaretto, A. (2022). General guidelines for the optimal economic aggregation of prosumers in energy communities. *Energy*, *258*, 124800.
Wongbumru, T., & Dewancker, B. (2014). Smart communities for future development: Lessons from Japan. *International Journal of Building, Urban, Interior and Landscape Technology (BUILT)*, *3*, 69–76.
Yilmaz, S., & Li, X. (2018). Energy socialization: The Northeast Asia energy grid and the emergence of regional energy cooperation framework. *Energy Strategy Reviews*, *22*, 279–289.
Zez. (2022). https://www.zez.coop/en/.

Measures for renewable integration in developing countries: the case of Jordan

13

Ra'ed Nahar Myyas[1,2,3], *Mohamad Al-Widyan*[3] *and Alaa Zetawi*[1]
[1]Department of Electrical Engineering, University of Jaén, Linares, Spain, [2]Jordan Design & Development Bureau (JODDB), R&D Department, Amman, Jordan, [3]Amman Arab University, Department College of Engineering, Mubis, Amman, Jordan

Introduction

Renewable energy, like solar and wind power, is now the cheapest option for making electricity in most places. Over the past decade, the cost of setting up big solar panels (photovoltaics [PV]) has dropped by 85%, while other types of renewable energy, like wind, have also become much cheaper. This means that in almost every country, when they want to build new power plants, they are choosing renewables more often. With fossil fuel prices going up, it is even more likely that we will see more investment in renewable energy because it is cheaper and better for the environment (World Energy Transitions Outlook 2022, 2022).

At the same time, achieving the goal of limiting global warming to 1.5°C is becoming increasingly difficult, and without urgent and significant action, it may become impossible to attain. This latest edition of the World Energy Transitions Outlook outlines how we can address both challenges by accelerating the transition to renewable energy, expanding its use across various sectors. The economic case for embracing renewables is becoming stronger, and the benefits will be widespread. However, achieving this requires clear plans and strategies. The time for action is now. The next decade leading up to 2030 is crucial in ensuring that the target of limiting global warming to 1.5°C by 2050 remains achievable (World Energy Transitions Outlook 2022, 2022). Additionally, the global increase in population density and energy consumption for basic needs and improved living standards has heightened the demand for energy, prompting many developed countries to seek alternative energy solutions. Renewable energies are considered the most viable solution to meet energy needs while reducing reliance on traditional energy sources and lowering greenhouse gas emissions (Zittis et al., n.d.).

Global perspectives on renewable energy integration

Energy is not just about electricity; it is a fundamental driver of economic growth and a cornerstone of sustainable development. Despite progress in using

Towards Future Smart Power Systems with High Penetration of Renewables. DOI: https://doi.org/10.1016/B978-0-443-29871-4.00013-0

renewable energy sources, the global energy challenge still stems from the need for diversity. While renewable energy usage has improved, it has not reached the necessary levels yet. Jordan stands out as having significant potential to tap into renewable energy. With resources like hydropower, wind, solar energy, and biomass, along with vast desert areas ideal for renewable energy projects, Jordan is primed for development in this area. Therefore, the government has established a thorough, long-term energy strategy to guarantee a steady supply of energy. Increasing the usage and growth of clean energy sources in the power sector is the objective of this strategy, which seeks to achieve the goal of obtaining 50% of the nation's electricity from renewable sources by 2030 (Zittis et al., n.d.). The potential for increasing disparity in energy distribution is high when relying on renewable energy sources, as they offer a sustainable way to ensure continuous energy supply while also reducing greenhouse gas emissions. For this reason, a lot of people are working on distributed energy resource designs that use renewable power sources. These sources can include things like solar PV, wind turbines, diesel engines, biomass gasifiers, battery banks, fuel cells, and more. These resources are particularly important in rural areas where they can be used to power essential loads (Generated energy surges by 8% last year, consumed energy by 6.6%, 2023).

Since 2015, about 14% of Jordanian houses have installed solar water heating systems (SWHS) (Dehghan et al., 2021). The cumulative count of SWHSs rose to 16,000 at the end of 2018, reached 24,000 in 2020, and further escalated to 30,000 by the end of 2021, which is attributed to an annual growth rate averaging 20%. The upward trajectory would not have been possible without initiatives such as those funded by the Jordan Renewable Energy and Energy Efficiency Fund (JREEEF).In 2019, 2200 solar water heaters were installed, and in 2020, that number jumped to 3000. Nevertheless, additional expansion has been hindered by the sluggish implementation of rules mandating the installation of SWHS in bigger residential complexes and commercial buildings (Muheisen, 2023). Although practically every Jordanian now has access to electricity, several areas are also installing off-grid renewable energy technologies, such as standalone units. The use of such systems is more common in rural areas, border checkpoints, telecommunications stations, desert camps, and communities (Hamed & Bressler, 2019). Solar PV systems installed independently have shown to be a practical and economical way to power these regions. In support of renewable energy adoption in agriculture and water management, the Renewable Energy and Energy Efficiency Program, funded by the EU, installed approximately 106 solar pumps in the Highlands and 214 in the Jordan Valley (Ministry of Energy and Mineral Resources MEMR, 2020.; Muheisen, 2023). These systems can provide more than 1000 kW of power when used together. Another sign of increasing interest in these systems is the recent green light for the first 9 MW standalone project (no feed-in). These projects, which require distribution company authorization, are expected to be overseen by the Energy and Minerals Regulatory Commission (EMRC) soon (MEMR, 2020; Muheisen, 2023).

Energy sector in Jordan

Jordan is a Middle Eastern nation that shares boundaries with both the Red Sea and the Gulf of Aqaba at its southernmost point. Its geographical position is between 29.1830° N latitude and 36.2677° E longitude. Just 60 km via Palestine separates you from the Mediterranean. The overall area of the land is 89,320 km^2 (34,487 mi^2), and the length of the shoreline is 26 km (16.2 mi). This land area makes up around 112% of Maine's total area. Consequently, Jordan ranks as the 112th largest nation globally and the 21st smallest in Asia. Urban areas accommodate a notably high percentage of 92% of the population, with Amman housing over one-third of the populace (Latlong, n.d.). Fig. 13.1 displays the average daily sunshine hours for each month in the region of Amman, Jordan. It is important to note that the number of hours of sunshine represents the time when the sun is visible without any obstruction caused by clouds, fog, or mountains. As shown in Table 2.1, July has the highest number of daily sunshine hours, with an average of 12 hours/day (worlddata, n.d.).

Challenges arising from rapid population increase, expanding industrial and commercial sectors, expensive imported energy sources, and elevated greenhouse gas emissions are causing the Jordanian government to scramble. So, making sure people can get their hands on affordable, sustainable energy has become a major issue. Renewable energy projects, including those that use waste, biomass, solar, and wind power, have been prioritized by the government as a means to overcome these challenges. Ranked 66th in the world for energy transition by the World Economic Forum, Jordan fell sharply to 82nd place. Overall, the country's score dropped to 49.5% in 2019, despite attaining 53%. The energy system efficiency

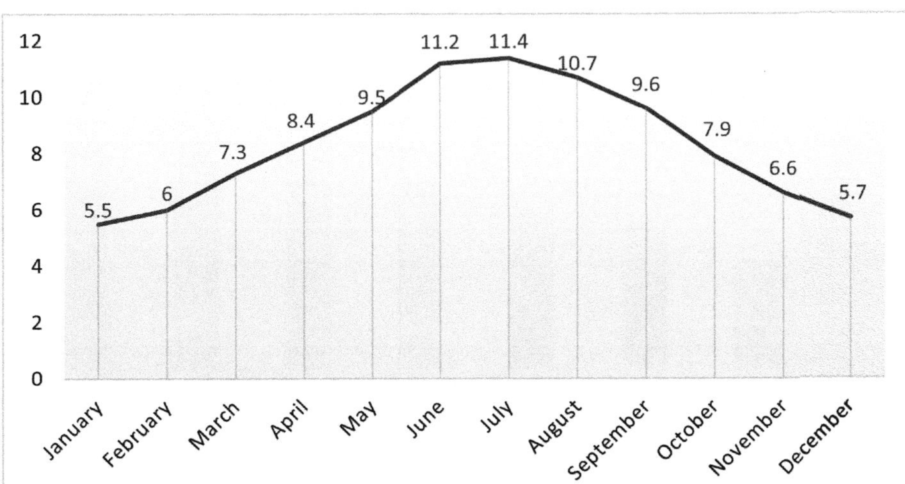

Figure 13.1 Average daily sunshine hours for each month in Amman.
Source: From worlddata (n.d.). http://www.worlddata.info.

score for Jordan dropped from 56% to 46%, which is noteworthy, even if it had improved slightly from 50% to 53% before the default. Furthermore, despite a 0.4% decline in the labor force participation rate during the fourth quarter of 2020, the unemployment rate surged to 24.7% from 9.3% in the first quarter (Myyas & Almajali, 2022). Fig. 13.2 provided graph illustrates Jordan's total energy supply (TES) from 1990 to 2019, categorized by source.

In 2023, a slowing economy will result in a 1.3% increase in global energy demand. Despite decarbonization goals, coal use will increase somewhat to make up for gas supply shortages. The energy transition will be delayed by more frequent extreme weather occurrences that will compel many nations to revert to using fossil fuels. Asia is poised to spearhead a global surge in renewable energy consumption, with an anticipated 11% increase, even as investment in the sector is projected to decline. Concurrently, certain countries may reverse their plans to gradually phase out nuclear power usage in response to the energy crisis.

Energy source and consumption patterns

With increases of 5.5% in 2019 and 7.4% in 2020, Jordan's peak load was noticeably higher than in 2019. Energy usage increased by around 3.1% per year on average between 2010 and 2020 (Latlong, n.d.). Renewable energy sources, which are alternative, clean, and dependable, are becoming increasingly important in addressing the increasing energy demands, as the ability to generate electricity reached 5424 MW by 2020. In 2020, traditional sources accounted for 72% of the local available capacity, while renewables made for about 28%. It is worth mentioning

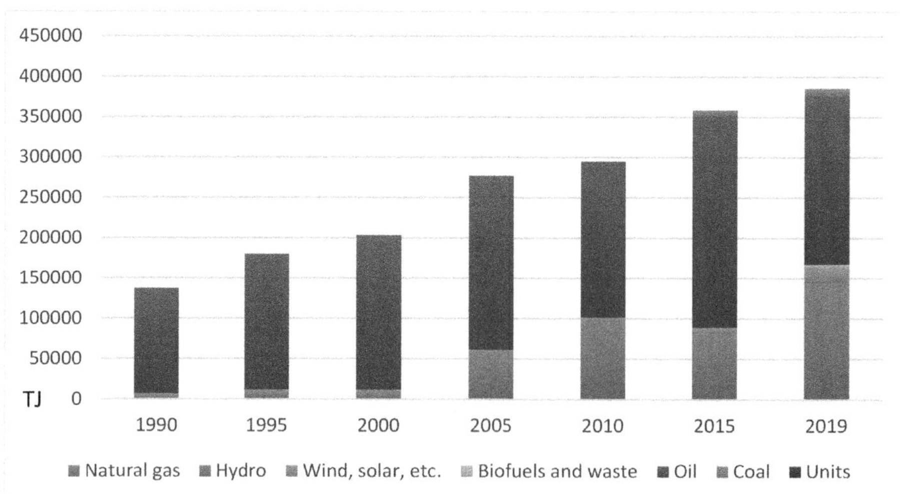

Figure 13.2 Jordan 1990–2019 total energy supply (TES).
Source: From International Energy Agency (n.d.). http://www.iea.org. (International Energy Agency n.d.)

that in 2020, just 8% of the potential capacity of renewable energy systems was utilized for power generation. Consequently, the government of Jordan is looking into ways to sell power to nearby nations. the data shows a 4.3% increase in per capita electricity usage compared to the previous year, with consumption reaching 1821 kW/hour in 2023, up from 1746 kW/hour in the previous year. Residential electricity consumption made up 47.9% of the total, while industrial usage accounted for 21.2%, and commercial consumption stood at 14.7%. Additionally, 14.2% of electricity consumption was attributed to water pumping, and street lighting contributed 2% to the overall usage. Furthermore, the data indicated a 13.3% energy loss, with electricity being supplied to 99.9% of the population Fig. 13.3 shows main power consumption in 2023 (Latlong, n.d.)

Policy and regulatory framework in Jordan

Fossil fuels constitute 52% of the nation's overall energy consumption, with natural gas closely trailing at 38%. The average per capita consumption stands at 0.7 toe/cap, which includes 1850 kWh of electricity. Households are charged \$0.100 kWh^{-1} for electricity, while businesses pay slightly more at \$0.123 kWh^{-1}. These rates encompass all associated fees, including taxes, distribution expenses, and energy charges. For homes and businesses, respectively, the average global price of energy during that time was \$0.139 kWh^{-1} (Global Petrol Prices, 2022).

The amount of energy that Jordan imported in 2021 was 2.8 billion dinars, the amount of energy that was used up was 8166 tons of oil equivalent, and the amount of electricity and natural gas that was consumed per capita was 1.746 kWh. Power generation also saw a decrease, with the contribution falling to 73% from 80% in

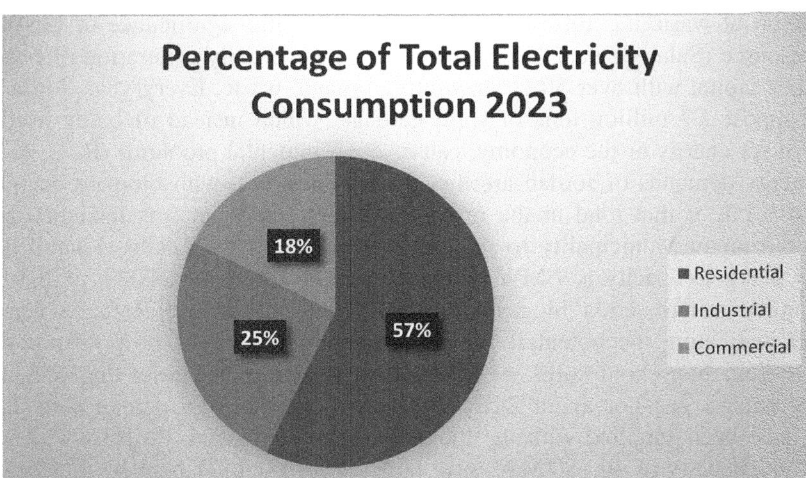

Figure 13.3 Percentage of total electricity consumption, 2023.
Source: From Latlong (n.d.). http://www.latlong.net.

2020. In 2021, 28% of the electricity came from local sources, which was an increase from 22% the year before. Imports of crude oil totaled 13 million barrels from September to December of 2021, while output from the Hamza oil field was 105,000 barrels (Jordan News, 2022).

Wind generation in Jordan

If renewable energy is to become more resilient and sustainable, the incorporation of wind turbines is now essential. Strategic investments in wind turbine projects have allowed Jordan to make substantial progress in this sector. In 2014, the nation began wind farm projects with modest capacities of approximately 1.4 MW. With the aim of boosting power generation in Jordan by 4%, advancements were evident in the establishment of four wind power plants by 2016: Ibrahimya, Hofa, Ma'an, and Tafilah. In early 2020, two wind farms with a combined capacity of 145 MW were commissioned, as per an agreement reached in 2016. It followed that by 2021, Jordan's total installed wind generating capacity had risen to 622 MW, and projections suggested that the nation could produce 1595–2076 GWh of energy annually. According to both (Jordan News, 2022; Ministry of Energy and Mineral Resources MEMR, 2020), this would have supplied roughly 11% of the nation's electricity needs that year. About 15.3% of the system-peak demand in 2021 came from wind-generated units, according to National Electric Power Company (NEPCO)'s annual report.

Biomass energy in Jordan

Jordan primarily relies on municipal solid waste (MSW) as its main bioenergy resource. The country inaugurated its initial MSW-based pilot plant near Amman in 2000, boasting a capacity of 3.5 MW. With the population of Jordan on the rise, the generation of waste has followed suit, emphasizing the significance of MSW as a vital resource (Salah et al., 2023). Estimates suggest a daily generation of one kilogram per capita, with over 50% constituting organic waste. Every year, Jordan produces almost 2.7 million tons of solid garbage, which instead of being used as a resource for energy or the economy, causes environmental problems (R. E. A, n.d.). The energy demands of Jordan are met by other sources, with biomass accounting for just 0.1% of that total at the moment. A biogas project was launched by the Greater Amman Municipality to recover methane from the Ghabawi landfill. The project's target capacity is 7 MW, which should be reached by 2023. at 2019, The government sanctioned a biogas pilot facility at the Zaatari Refugee Camp in Mafraq, indicating the potential for investment in generating power from biogas derived from household solid waste deposits. Evidence suggests that 656 metric tons of natural gas—or about 24% of Jordan's main energy demands in 2012—could have been supplied with the biogas that was produced. Projects with waste-to-energy capacity of 40–50 MW were targeted for establishment by the Jordanian government by 2020 (Abu-Rumman et al., 2020; IGA, n.d.).

Jordan is home to a variety of biomass resources, such as agricultural dung, olive groves, and industrial byproducts. Home HVAC systems are the intended beneficiary of these funds from the government. The olive oil harvesting process produces around 19,000 metric tons (MT) of Jift each year, which could be used as fuel or feed (Stoyanov et al., 2021). Even if these materials are plentiful, we still need to figure out how to make the most of our biomass resources to generate renewable energy.

Geothermal projects in Jordan

Jordan's energy sector is progressively exploring geothermal energy as a promising resource. The nation possesses geothermal reservoirs, particularly hot mineral water sources found in springs and wells located along the Rift Valley. Notably, the northeastern region and the Dead Sea escarpment exhibit significant temperature variations, with gradients reaching up to 50°C per kilometer of depth (°C/km). Despite these resources, the temperature remains below the threshold required for electricity production (below 100°C). Nonetheless, hot water is currently utilized for recreational purposes in numerous hotels and resorts across Jordan (R. E. A, n. d.). Several public facilities in Jordan have adopted small-scale indirect geothermal systems to provide space heating and cooling Instances comprise the National Center for Research, the American University of Madaba, and the development building of the Higher Council of Science and Technology. According to data from the International Geothermal Association (Abu-Rumman et al., 2020), Jordan's current geothermal energy utilization capacity stands at 153.3 MW thermal (MWTH), constituting merely 0.42% of the nation's total geothermal energy potential. Geothermal energy offers potential for diverse applications like aquaculture, horticultural heating, and usage in the food industry. Nevertheless, further investigation is crucial to assess its viability and encourage its adoption in Jordan.

Hydropower projects in Jordan

Jordan's investment in hydropower faces challenges due to the scarcity of suitable water bodies necessary for constructing large-scale hydroelectric plants. Currently, only two operational hydropower facilities exist: the King Talal Dam, with a capacity of 4 MW, and the Aqaba Power Station, with a capacity of 6 MW (R. E. A, n.d.). The hydroelectric plant at Aqaba utilizes water from the Gulf of Aqaba to provide cooling for a thermal power plant. Together, these installations generate approximately 60 GWh/year, representing about 0.4% of Jordan's total power production in 2018 (R. E. A, n.d.; Stoyanov et al., 2021). Moreover, the considerable difference in elevation between the Red Sea and the Dead Sea presents a significant opportunity for hydropower development. Estimates indicate that this could result in an electrical output capacity ranging from 400 to 800 MW. Several studies conducted in Jordan, such as those by Stoyanov et al. (2021), Al-Addous et al. (2020), and Al-Masri et al. (2019), have investigated the feasibility of establishing hydropower facilities in the region, highlighting promising investment opportunities. Al-Addous et al. (2020) assessed the viability of microhydro power generation in Jordan, focusing on its cost-effectiveness and sustainability as a solution for off-grid and rural areas.

Policy and regulatory framework in Jordan

With the hope of increasing the proportion of renewable energy in its overall energy use, Jordan has enacted a number of laws and regulations. Energy Efficiency and Conservation Law, Net Metering, the National Energy Strategy, and Feed-in Tariff Program are all part of this (R. E. A, n.d.). Nevertheless, there are still obstacles to overcome in order for these policies to be effectively implemented and utilized. One area of special concern is the regulation of energy storage across all stages of the energy life cycle, from production to transmission, distribution, and consumption. The development of battery storage infrastructure, which is essential for improving grid stability and reliability, is impeded by the lack of comprehensive laws. To ensure the most efficient use of resources and the long-term viability of the environment, new policies are needed to promote the use of electric vehicles and handle problems with recycling and decommissioning of solar power plants.

To promote the usage of solar energy for water heating in several sectors, including healthcare facilities, schools, and corporations, Jordan has initiated awareness programs and performed energy audits (Hamed & Bressler, 2019). A growing proportion of residential and medium-sized commercial buildings in Jordan are integrating solar energy into their energy plans by installing grid-connected photovoltaic systems. Financial institutions, in collaboration with the Ministry of Energy and Mineral Resources (MEMR), have joined forces to provide financial assistance covering 30% of the total installation expenses for solar water heating and PV systems in households.

Based on the EMRC study (Ministry of Energy and Mineral Resources MEMR, 2020), Jordan has approximately 506 registered installation businesses, out of which 20 are actively involved in the renewables sector. In addition, nine energy audit-service companies have obtained permits to operate in the country. The proliferation of enterprises in the renewables industry has resulted in heightened competition, ultimately advantageous for both consumers and businesses. The variability in solar radiation has important implications for the development and implementation of solar energy projects in Jordan, as certain places are more suitable for large-scale solar projects than others. Therefore, it is crucial for policymakers and investors to have a thorough understanding of solar radiation patterns in various regions and provinces. This knowledge will enable them to make well-informed judgments on the most suitable location and design for solar energy projects.

According to the MEMR (2013), the transportation sector accounts for 51% of Jordan's overall energy consumption. Biodiesel, bioethanol, biogas, and algae fuels are sustainable alternatives to conventional transportation fuels, and the government should investigate them. Yet, the distribution, creation, and implementation of alternative fuels must be in sync with the aims and policies of the local government. The growth of Jordan's biofuel industry is being fueled by the country's efforts to reduce reliance on imported fossil fuels and address the impacts of climate change

(Al-Kurdi, 2021). The MEMR has released guidelines outlining the criteria for issuing permits to facilities engaged in the production of biofuel. These rules cover both industrial fuel and biodiesel manufacturing processes. Investors are required to meet some broad criteria; this involves conducting an assessment of environmental impact, feasibility, and network impact before obtaining clearance. The results will have importance for developing policies and improving services, with the goal of offering better assistance to the public in dealing with and recovering from complex difficulties that impact their lifestyles and health situations.

Challenges facing the renewable energy sector in Jordan

The net-metering system and JREEEF are two programs that the Jordanian government has implemented to encourage small-scale investments in renewable energy. Because of their narrow focus and lack of financial incentives, these rules have come under fire from renewable energy industry professionals (Kiwan & Al-Gharibeh, 2020; R. E. A, n.d.). A lack of easily available finance options and the high initial expenses of installing renewable energy systems at the household level are major obstacles to the success of small-scale initiatives in Jordan. In addition, the difficulties in building small-scale renewable energy projects have been exacerbated by the recent adoption of new electricity rates in Spring 2022. The economic feasibility of such undertakings is undermined since these tariffs do not provide system owners with enough assistance. As a result, it is becoming more and more challenging to encourage participation in these programs, especially for low-income households.

Major problems caused by variations in energy demand were brought to light in 2021 by the widespread power outages that occurred across the nation. There is a lack of flexibility and capacity in the existing electrical system to integrate many power generation sources. A renewable energy control center must be set up in order to improve the ability of government institutions to implement these technologies and to enable thorough monitoring, reporting, and analysis in the energy sector (Alrwashdeh, 2022).Through the implementation of the Green Corridor project and the introduction of innovative grid designs, Jordan is actively addressing technological challenges. A total investment of $172 million was allocated for the Green Corridor project in the southern region. As part of this initiative, a new 400/132 kV substation was constructed in Ma'an, along with the 132/33 kV substations in Qatrana and at the airport. By 2019, the project was completed, with the Ma'an, Qatrana, and Queen Alia International Airport substations interconnected via double-circuit overhead lines. One of the primary objectives of the Green Corridor project is to alleviate the strain on transmission lines by facilitating the transfer of power from the southern region to the central loads in the country's interior. Additionally, in 2019, Jordan launched a 23 MW/12.6 MWh lithium-ion solar battery project, recognizing the importance of storage in grid management (Updated Submission of Jordan's 1st Nationally Determined Contribution [NDC], 2021).

As local power demand and renewable energy generation both surge, Jordan aims to expand its presence in the energy export market by linking to transmission lines in neighboring nations to export its renewable energy output. Encouraging the transportation sector to transition to electric vehicles is one strategy to mitigate the gradual increase in electricity demand. However, Jordan is concerned about the potential adverse impacts of climate change, such as hotter temperatures, reduced rainfall, and accelerated evaporation rates, which could lead to higher energy costs and shortages. Climate variability can also disrupt service reliability, damage infrastructure, and exacerbate these challenges. Peak and average power consumption in Jordan fell sharply from 2019 levels in 2020 as a result of the COVID-19 pandemic's influence on the country's power sector. The primary causes of this drop were a decrease in industrial activity and an increase in the connection of residential renewable energy systems to the distribution grid. In order to guarantee energy security, the National Control Center placed restrictions on the deployment of renewable energy sources (Alasali, n.d.; Prabheesh et al., 2020). As a consequence, the grid's reliability and stability were further compromised, and all renewable wheeling projects that were linked to it were disconnected. Research conducted by the Association for Energy, Water, and Environment (EDAMA) underscored the negative impacts of the pandemic on companies operating in the renewable energy sector. According to the survey, 37% of these companies reported a decline in revenues, while 32% were unable to generate any income due to the COVID-19 lockdown measures implemented in 2020 (EDAMA, n.d.). Furthermore, several scheduled renewable energy initiatives encountered setbacks, affecting electricity production costs. Despite these challenges, the Jordanian government emphasizes the importance of integrating clean energy and environmental objectives into recovery programs to achieve more efficient outcomes. Investing in renewable energy resources can bolster socioeconomic resilience and accelerate economic revitalization. Jordan is actively shifting towards a renewable energy-oriented framework to align with its economic development objectives across multiple sectors. However, the advancement of the renewable energy sector faces several challenges, broadly categorized as Challenges in Jordan.

- **Grid infrastructure**: Jordan's grid infrastructure may not be adequately developed to handle the integration of large-scale renewable energy projects, leading to challenges in grid stability and reliability.
- **Energy storage:** The limited capacity for energy storage presents a notable obstacle to the integration of renewable energy in Jordan, as it restricts the capacity to store surplus energy generated during peak production for later use when renewable energy generation is low.
- **Investment and financing**: Securing long-term investments and financing for renewable energy projects can pose challenges due to perceived risks, including political and economic instability, as well as uncertainties regarding project financial viability. Sufficient financial resources are crucial for the advancement and execution of renewable energy initiatives in Jordan. Despite attracting investments in the renewable energy sector, ongoing financial backing is vital to broaden renewable energy adoption and overcome financial challenges. Access to financing mechanisms, such as grants, loans, and investment incentives, can facilitate the development of renewable energy projects and ensure their economic feasibility. Furthermore, innovative financing models and collaborations between the public and private sectors can mobilize additional financial resources and expedite the transition to renewable energy (Khawaja et al., 2021).

- **Counterparty risks**: Renewable energy projects in Jordan rely on agreements with entities like the NEPCO for power purchase, but these entities may face financial challenges or delays in payments, posing risks to project developers.
- **Regulatory framework**: Despite efforts to improve the regulatory framework for renewable energy, challenges remain in terms of bureaucracy, permitting processes, and regulatory uncertainty, which can hinder project development and investment.
- **Electricity tariffs and subsidies**: The structure of electricity tariffs and subsidies in Jordan can impact the economic viability of renewable energy projects, especially when renewable energy producers are not fairly remunerated for their role in supporting the grid.
- **Intermittency and forecasting**: Solar and wind, as renewable energy sources, exhibit intermittency, presenting grid operators with challenges in forecasting and balancing supply and demand, thereby potentially resulting in grid instability.
- **Competition with fossil fuels**: The fossil fuel industry in Jordan may resist the expansion of renewable energy due to vested interests and economic dependencies, creating barriers to renewable energy sector growth.
- **Possible roadblocks** to speeding up Jordan's transition to renewable energy sources also include the fact that the general people may be unaware of and unconvinced by the advantages of renewable energy sources.

Impact indicators

When evaluating the success and results of renewable energy projects in Jordan, impact indicators are crucial. To measure progress toward sustainability goals including reducing carbon emissions, increasing energy efficiency, and encouraging the adoption of renewable energy sources, these metrics are used. Important metrics include the share of energy that comes from renewable sources, changes in carbon emissions over time, trends in investments in renewable energy projects, and the pace of implementation of renewable energy solutions. The impact of renewable energy policies and projects may be assessed, problem areas can be identified, and informed decisions can be made to advance Jordan's sustainable energy environment by keeping an eye on these indicators.

Environment indicator

Carbon emissions must be closely monitored, and efforts like the Global Carbon Stocks Program must be actively pursued if we are to fight climate change and join the worldwide movement toward sustainable economies. A decrease in greenhouse gas emissions and home air pollution was an unintended consequence of the COVID-19 epidemic. This was because energy usage was lowered. The implementation of statewide curfews in Jordan led to a precipitous drop-in industrial activity, which in turn reduced energy demand and emissions of greenhouse gases. The Ministry of Environment has been working on the Green Growth National Action Plan 2021–2025 since the latter half of 2018, with the goal of putting the plan's proposals into action. In order to help Jordan achieve its goal of sustainable and ecologically friendly economic development and to make the country more resilient to unexpected shocks like the COVID-19 pandemic, this plan provides a thorough framework for implementing green growth policies and initiatives across different sectors. The Eastern Mediterranean and Middle East region is seeing a concerning trend of rapidly increasing greenhouse gas emissions, which are now higher than emissions from EU member states, according to recent research. This area has warmed up at a faster rate than other heavily populated regions in the past several years, making a big contribution to global warming. Temperature increases are expected to continue, although this trend might be slowed with international cooperation to cut emissions of greenhouse gases. Jordan highlighted energy, transportation, water, health, and agriculture as critical areas for reducing emissions at the 2019 United Nations Framework Convention on Climate Change's 25th Conference of the Parties in Madrid. Since energy-related activities are responsible for 64% of Jordan's overall national emissions, the government has pledged to reduce greenhouse gas emissions by 31% by 2030. Consequently, it is reasonable to put an emphasis on reducing emissions in the power generating sector, where renewable energy projects such as solar PV and wind farms are crucial in achieving this goal.

New goals for reducing greenhouse gas emissions through a variety of programs, policies, laws, and foreign assistance are outlined in the NDC. By putting an emphasis on renewable energy sources and improving energy efficiency, we can significantly reduce emissions, according to forecasts, and bring overall greenhouse gas emissions down by 2040. The NDC has set a goal of increasing energy efficiency by 9% and increasing the proportion of renewable energy in electricity production to 35% by 2030. Furthermore, there is hope for substantial reductions in greenhouse gas emissions (Fig. 13.4) through the implementation of concentrated solar power plants. To evaluate the potential effects on the local ecosystem and population, comprehensive environmental evaluations must precede the green light of any renewable energy project.

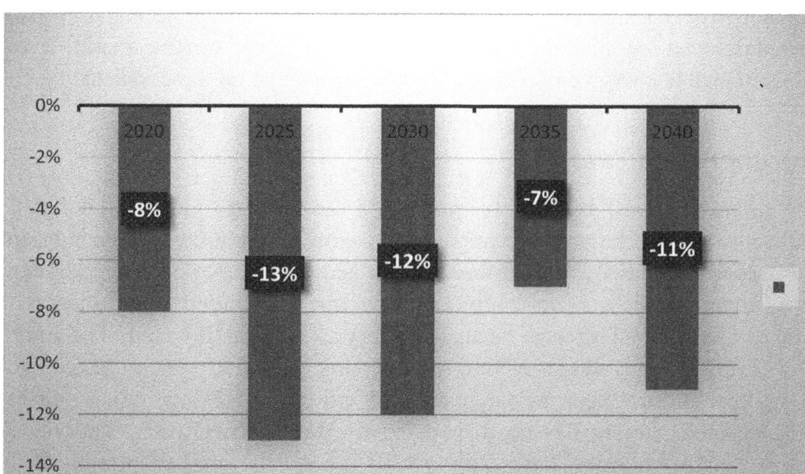

Figure 13.4 Jordan's greenhouse-gas avoided emissions, 2020–2040.
Source: From A National Green Growth Plan for Jordan (2017).

Economic indicator

Renewable energy sources are well-suited to Jordan's climate, as mentioned above. While hydropower, biofuel, and geothermal sources are acknowledged to a lesser degree, several government assessments (Updated Submission of Jordan's 1st Nationally Determined Contribution [NDC], 2021) highlight wind and solar power as the nation's most accessible renewable resources. The creation of a clean, sustainable, and economically viable energy supply has been a priority for Jordan due to various considerations, including the increasing prices of imported fuel, the inflow of refugees, the growth of the domestic sector, and the requirement to reduce greenhouse gas emissions. In addition to improving social fairness and welfare, investing in renewable energy can boost the Jordanian economy, create green jobs, and increase investment.

With the National Energy Strategy Plan covering the years 2007–20, Jordan launched its investment program in response to the uncertainty of imported energy sources and the need to guarantee energy independence. The plan included a wide range of energy sources, including renewables, nuclear power, and oil shale extraction, with the goal of attracting investments totaling $20 billion by 2020. After 3 years of recognition from Bloomberg's Climate (R. E. A, n.d.) Scope 2019 Index, this strategic endeavor established Jordan as a leader in the MENA region for creating the most favorable environment for renewable energy investment (Abu-Rumman et al., 2020). In addition, Ernst and Young (Khawaja et al., 2021) ranked Jordan as the 38th best market in the world for renewable energy investments. Accolades like this help with growth, change, and efficiency in operations by giving investors useful information and boosting confidence in the financial markets. The Green Growth National Action Plan for 2021–25 (GG-NAP) was introduced by Jordan, marking a decisive step toward encouraging a sustainable economy. An estimated $85.3 million will be needed to undertake the 12 priority activities outlined in this comprehensive strategy, which aim to improve green growth, especially in the energy sector. An industrial renewable roadmap and investment plan are being developed. The JREEEF is being accredited under the Green Climate Fund. The adoption of the United Nations Energy Sector Monitoring, Reporting, and Verification System is also a key action (ENERGY SECTOR Green Growth National Action Plan 2021–2025, 2020). With the overall goal of reaching a 7.5% economic growth rate by 2025, these initiatives are expected to mainly receive funding from the public budget, donors, partners, and private sector investors. Jordan has quickly diversified its energy mix, thanks to its impressive progress in developing renewable energy projects, which have attracted investments totaling over US$5 billion (MoEnv, 2020).

Eligible investors interested in constructing renewable energy projects under a build–own–operate framework have been invited to submit proposals in three consecutive phases by the Jordanian government. In order to attract international investment in these projects and provide investors with certainty about future earnings while reducing risk, the MEMR has also set feed tariffs. Feed tariffs, originally implemented in the first round, outline the recommended rates for electricity generation from different renewable energy sources (MoEnv, 2020; Stoyanov et al., 2021). A total of 200 MW of solar capacity and 117 MW of wind capacity were reserved for projects in the southern region of Ma'an during the first round (2012–15). The government of Jordan then instituted a bidding process with subsequent rounds of competition. Phase two, which ran from 2013 to 2019, distributed 200 MW of power to the country's eastern and northern regions. Midway through the 2018 third round of direct proposal submissions, bids dropped to $0.026 kWh^{-1}. The NEPCO paid around $0.11 kWh^{-1} for electricity that came from renewable sources, as opposed to $0.12 kWh^{-1} for electricity that came from traditional sources. Two hundred megawatts (MW) of solar power and 100 MW of wind power are included in this most recent round.

With storage prices continuing to fall, renewable energy sources like solar PV are becoming more attractive as a power sector alternative to solutions based on imported fuels. With massive investments in wind and solar power, the Jordanian government reached a major milestone in 2020, producing 2063.3 MW of electricity (R. E. A, n.d.). Also, in 2021, there were 863 applications for renewable energy permits from a variety of industries, including oil and its derivatives, natural resources, radiological and nuclear, and a significant 23% rise over the previous year.

A number of programs have been launched by the Jordanian government to encourage more investment in renewable energy. Some of these measures include improving tax incentive programs and creating an energy fund to support the construction of renewable energy installations. Investors in certain areas slated for socioeconomic development may be eligible for a tax break of up to 100% for a maximum of 10 years. To further expedite the process and avoid drawn-out competitive bidding procedures, investors can also choose to negotiate directly with the MEMR. Furthermore, by agreeing to buy power from renewable energy projects, NEPCO gives guarantees. By developing investment ideas, promoting projects to prospective international investors, and easing foreign investment within the country, the Jordan Investment Commission is crucial in attracting funding for green growth initiatives (Olawuyi, 2021).

Energy indicator

Load demand profiles are shown in Fig. 13.5 in this section. By examining this monthly demand over 5-year periods, it illustrates the trends in increasing demand in summer months (June, July, and August), which are sunny months, during which solar panels can take serious part of this demand.

Between March and May of 2020, while the global COVID-19 epidemic was in full swing, there was a sharp drop in the demand for monthly load. The widespread effects of the epidemic on many parts of people's everyday lives are mirrored in this pattern. There is a robust relationship between the monthly averages during the 5-year period and the load profiles shown in the figure. In addition, when compared to prior years, 2020 saw a decrease in both the minimum and average demand. Take 2020 as an example; in 2018, the minimum demand was 1219 MWh, and in 2019, it was 1029 MWh. In a similar vein, average demand fell 37% in 2020 and 40% in 2019 compared to the previous 2 years. Table 13.1 shows that early

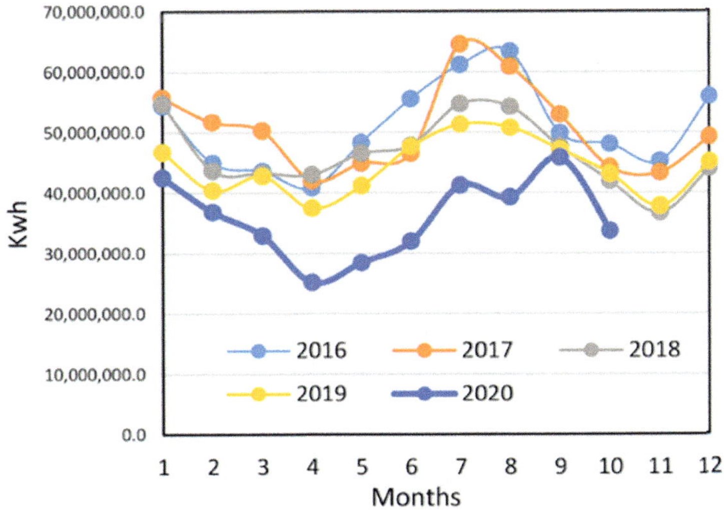

Figure 13.5 Monthly electricity demand (KWh) in Jordan's city center over the last five years, including during the current pandemic. Jordan's Reference Price of Renewable Energy Sources. *Source*: From MoEnv. (2020). Waste sector green growth national action plan. Ministry of Environment Amman, The Hashemite Kingdom of Jordan; Alasali, F. (n.d.). Impact of the COVID-19 pandemic on electricity demand and load forecasting.

Table 13.1 Summary of electrical demand from 2016 to 2020 during the time of movement restrictions.

	Electricity demand (MWh)				
	2016	**2017**	**2018**	**2019**	**2020**
Maximum demand	2098	1839	1735	1768	1296
Minimum demand	1128	1189	1219	1029	765
Average demand	1443	1492	1444	1319	941

Source: Data from Alasali, F. (n.d.). Impact of the COVID-19 pandemic on electricity demand and load forecasting.

measurement period (March 2020–May 2020), peak demand and demand activities had a direct impact on power consumption (March, April, and May) (2020).

A feed-in-tariff program for renewable power sources and an energy efficiency fund were both created by the Renewable Energy and Energy Efficiency Law. By 2020, the government hopes to have met 10% of its energy demands using renewable sources; hence, it is increasing its power generating capacity from 1.13 to 1.8 GW. The goal of current initiatives is to strengthen the country's energy sector by increasing the use of renewable energy sources. These sources include solar thermal, geothermal, bioenergy, and wind energy. The Jordanian Energy Sector Updated Strategy (2007–20) expanded upon the 2004 National Strategy by resolving issues that had previously impeded project execution. Notably, terms like "resilience" and "sustainability" were absent in both the 2007–20 strategy and its predecessor (Shahateet et al., 2021).

The plan outlines Jordan's primary energy challenges, categorized into resilience and sustainability. While not explicitly labeled as such, energy security has long been acknowledged by the government as a persistent hindrance to the country's development (MOPIC, 2017). Moreover, reports highlight the critical need for resilience in Jordan's energy sector, given its vulnerability to external shocks like price fluctuations, with significant economic implications (A National Green Growth Plan for Jordan, 2017). Rapid population growth in certain governorates over the past 3 years has significantly increased energy demand, straining national infrastructure, even impacting humanitarian efforts in refugee camps due to unsustainable energy costs (The Royal Institute of International Affairs, 2016).

Case studies on the integration of renewable energy projects in Jordan

Global energy demand is steadily increasing, particularly in developing nations, driven by economic growth. Projections outlined in the International Energy Outlook anticipate a 50% surge in global energy consumption from 2020 to 2050 (World Energy Transitions Outlook 2022, 2022). Recent fluctuations in oil and natural gas prices, influenced by events like the COVID-19 pandemic and geopolitical tensions, have strained the global economy (Le et al., 2021; World Energy Transitions Outlook 2022, 2022). Heightened concerns about climate change have underscored the urgency for cleaner energy alternatives. According to a recent report by the Renewable Energy Policy Network for the 21st Century, renewables contributed around 28.3% to global electricity generation in the first quarter of 2021, up from 26% in the same period of 2019. This uptick in renewables is attributed to higher costs associated with coal and natural gas, coupled with the increasing affordability of solar and wind energy. However, coal and natural gas still play substantial roles in global electricity supply, comprising approximately 60%. There is an expectation of increased demand for renewables due to their lower operational costs. Yet, transitioning to renewable energy for electricity generation necessitates significant investments in new transmission technology (Ministry of Energy and Mineral Resources MEMR, 2020).

In terms of renewable energy capability to meet domestic demands, Jordan ranks third among Arab nations, according to the Friedrich–Ebert–Stiftung (Le et al., 2021). As an added bonus, in 2021, excluding hydropower, Jordan had the highest proportion of renewable energy contribution among Arab countries, according to the Arab Future Energy Index Report (AFEX) (Muheisen, 2023) from the Regional Center for Renewable Energy and Energy Efficiency, which assessed the renewable energy enabling environment in the region (Dadashi et al., 2022). In addition, AFEX recognized Jordan as a frontrunner among regional nations for creating a favorable climate for investments in energy efficiency and renewable sources. The Regulatory Indicators for Sustainable Energy also found that in 2017, the MENA region's leading sustainable energy countries were Tunisia, Jordan, and the United Arab Emirates. These countries' policies were particularly strong in promoting energy efficiency, renewable sources, and energy access.

Profile of renewable energy in Jordan, a geographic overview

There are 11.2 million people living in Jordan, yet around 75% of the country's territory is desert. Because of its proximity to the Tropic of Cancer, it receives a lot of sunshine. With an average temperature of 17.2°C, the weather in Amman, the capital, is a combination of desert and Mediterranean climates (Burdon, 2019). Approximately 316 days a year, Jordan enjoys abundant sunshine. The sun is stronger in the summer because it is higher in the sky (Weather, 2020). The average amount of solar energy per square meter is approximately 5.6 kWh daily. The total annual sunshine intensity is between 1800 and 2700 kWh/m^2. Because of its favorable climate, Jordan is an excellent location for solar power installations (Alrwashdeh, 2022). Fig. 13.6 presents three important maps regarding Jordan's potential for renewable energy: (a) a map showing the amount of sunlight received annually, (b) a map showing the average wind speeds, and (c) there are 11.2 million people living in Jordan, yet around 75% of the country's territory is desert.

The wind speeds in Jordan range from 7 to 11 m/s, and the country's pleasant climate is perfect for wind turbine installations due to its consistently clear skies and mild temperatures. About 16% of Jordan's land area, or 3.6 GW, might be used to generate wind power, according to the country's MEMR (Alrwashdeh, 2022). Investments in wind and solar energy have great potential in southern regions like Ma'an, Tafilah, and Aqaba, as shown in Fig. 13.6 (Ministry of Energy and Mineral Resources MEMR, 2020; Shatnawi et al., 2021). With an average of 7300 Wh/m^2/day, Ma'an and Aqaba stand out among other areas when it comes to high levels of direct normal radiation (Alrwashdeh, 2018; Shatnawi et al., 2021).

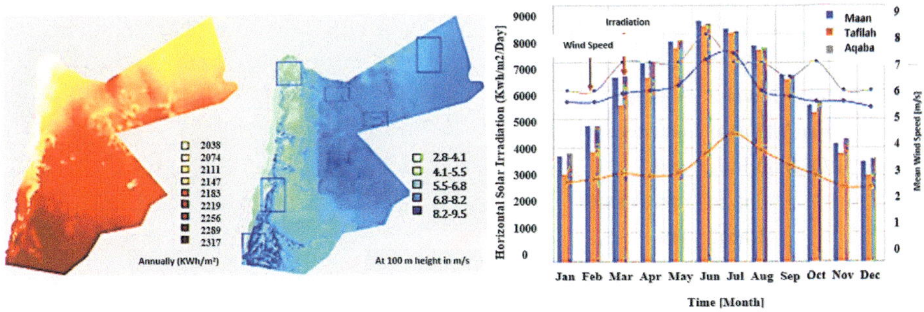

Figure 13.6 Renewable energy potential maps of Jordan.
Source: From Hamed, T. A., & Bressler, L. (2019). Energy security in Israel and Jordan: The role of renewable energy sources. *Renewable Energy, 135*, 378–389. https://doi.org/10.1016/j.renene.2018.12.036.

Investigating renewable energy projects in Jordan

In this section, we will look at how renewable energy sources can help Jordan fulfill its energy needs. It delves into Jordan's integration of renewable energy across different sectors, assessing the progress made and identifying both challenges and opportunities for further expansion. It also takes a look at the national strategies and policies that have helped Jordan generate and use renewable energy in the last 10 years. The purpose of conducting capacity assessments of renewable energy projects is to provide light on the renewable energy industry's present status in the nation. Since 2015, Jordan has achieved substantial advancements in developing many solar PV projects. The majority of these activities entail the signing of power purchase agreement proposals between 2015 and 2018. These agreements involve stakeholders from different development areas, including Ma'an, Mafraq's King Hussein Bin Talal Development Area, as well as additional investors in Aqaba and Amman. By 2021, Jordan's total installed solar energy capacity had reached 1831 MW, accounting for 75% of the whole renewable energy capacity. This increase was mostly due to the launch of a solar energy project with a capacity of 200 MW in 2020 (Ministry of Energy and Mineral Resources MEMR, 2020; MoEnv, 2020). Fig. 13.7 outlines the main areas for executing solar power projects.

In Jordan, solar and wind energy are the most easily accessible renewable options, while bioenergy, hydropower, and geothermal energy offer fewer

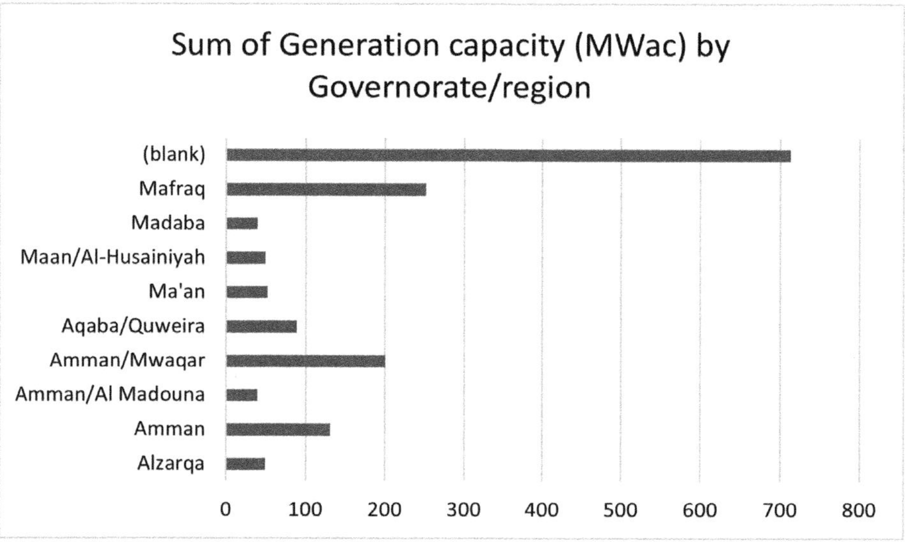

Figure 13.7 Summary of solar energy projects in Jordan.
Source: From Salah, A. A., Shalby, M. M., & Basim Ismail, F. (2023). The status and potential of renewable energy development in Jordan: Exploring challenges and opportunities. *Sustainability: Science, Practice and Policy, 19*(1). https://doi.org/10.1080/15487733.2023.2212517.

opportunities. Biomass, however, holds significance among renewables, yet ensuring sustainability is paramount for its broader adoption across various industries. Exploring innovative methods to electrify specific sectors is crucial. Jordan has committed to reducing its greenhouse gas emissions by 14% by the year 2030 as part of its NDC. This pledge emphasizes the dedication to the development and utilization of renewable energy sources, while also promoting financial investments in industry. Jordan's progress highlights the feasibility of shifting toward a decarbonized energy blend while guaranteeing reliable electricity supply. In 2018, the implementation of solar PV and wind power technologies resulted in the reduction of around 1.5 million metric tons of carbon emissions (Muheisen, 2023). In addition, the JREEEF promoted the use of solar water heaters, which were previously the main renewable energy source for heating and cooling. In 2021, Jordan had a total renewable energy capacity of 2445.7 MW. This included 1498 MW from projects that supplied electrical businesses, and 947.6 MW from noncommercial systems owned by people and institutions. This signifies a 26% surge in energy production in relation to the previous year, 2020 (The Jordan Times, n.d.).

According to the reference price list shown in Fig. 13.8 used to determine the highest price for buying electricity produced from renewable sources, different tariff limits have been set for various types of renewable energy. These include rates of \$0.12 for wind energy, \$0.19 for solar thermal, \$0.17 for photovoltaic, \$0.13 for biomass, and \$0.08 for biogas. These rates will be binding for the successful bidders participating in direct procurement processes for renewable energy projects. Moreover, if a facility is entirely of Jordanian origin, there is a possibility of a 15% tariff increase (Hosseini, 2018). Given the significant decrease in expenses associated with renewable energy technologies in the last 10 years, it would be wise for the Jordanian government to consider enhancing the prominence of renewables in the future energy mix.

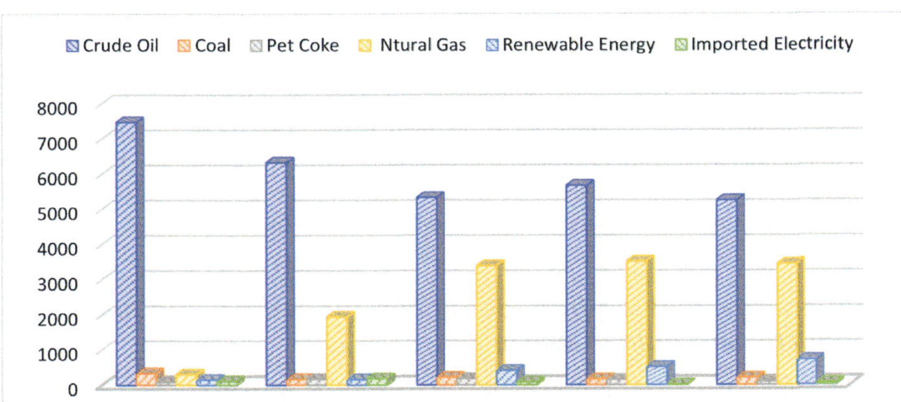

Figure 13.8 Illustrates the imports of crude oil, oil products, and coal from 2014 to 2018, measured in 1000 tons imports of crude oil.
Source: Data from worlddata (n.d.). http://www.worlddata.info.

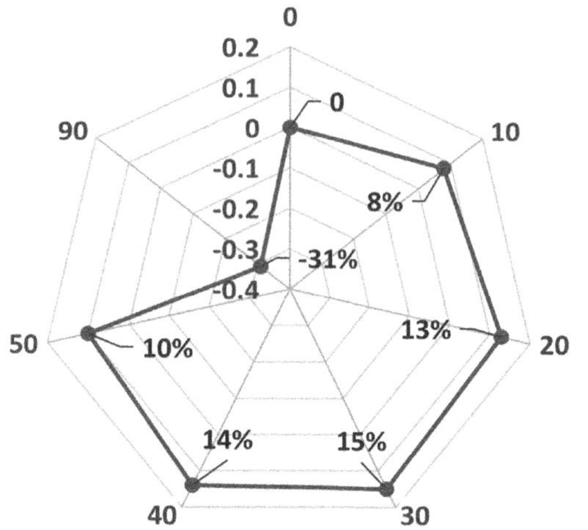

Figure 13.9 Solar array tilt angle and energy output.
Source: Data from worlddata (n.d.). http://www.worlddata.info.

Solar energy in Jordan

Solar energy is an auspicious and enduring substitute for conventional fossil fuels that has garnered escalating scrutiny in recent years. Solar panels and solar thermal collectors harness the sun's plentiful energy to generate electricity and heat, respectively. Solar energy has diverse applications, encompassing residential and commercial buildings, transportation, and industrial processes (Dehghan et al., 2021). The effectiveness of solar panels is influenced by several factors, one of which is the degree of inclination of the panels in relation to the sun. Research has indicated that modifying the inclination angle of solar panels according to the latitude of the site can substantially enhance their energy production. Furthermore, the alignment of the panels in relation to the sun's azimuth angle can also affect their energy production (Trotter et al., 2018).

Fig. 13.9 depicts the correlation between the tilt angle of a solar array and its energy production efficiency. The energy production efficiency of a solar array demonstrates an upward trend as the tilt angle increases until it reaches a peak, beyond which it begins to decline. This optimal tilt angle is determined by the latitude and local climate conditions of the installation site. The figure highlights the importance of proper tilt angle selection in maximizing the energy output efficiency of a solar array.

Biogas energy, derived from decay and rot, animal and manure byproducts, and human waste

The geopolitical landscape, driven by Russia's invasion of Ukraine, has sparked significant transformations in energy markets and policies, expected to endure for decades. Despite current policy frameworks, the energy sector is swiftly evolving toward cleaner, more cost-effective, and safer energy systems. Projections suggest that there will be a decrease in the demand for coal, a stability of the demand for natural gas by the end of the 2020s, and a decrease in the demand for oil due to the rising sales of electric vehicles by the mid-1930s, with further decreases expected by the middle of the 21st century (World Energy Outlook 2022 shows the global energy crisis can be a historic turning point towards a cleaner and more secure future, n.d.). White-rot and brown-rot fungi are essential in speeding up the decomposition of wood by releasing oxidative enzymes, especially in unregulated conditions. Enzymatic delignification is an efficient and low-energy process that exhibits significant substrate specificity when conditions are suitable (Floudas et al., 2012). To achieve future energy scenarios and meet climate change mitigation goals, widespread and sustainable utilization of bioenergy is imperative. However, this poses several challenges, including strategic land use management to mitigate negative impacts such as deforestation, greenhouse gas emissions, threats to human health, loss of biodiversity, and impacts on food production. Globally, utilizing byproducts from biofuel production as animal feed can enhance land use efficiency and reduce greenhouse gas emissions (Anderson et al., 2018; Park et al., 2022). Substituting grains with these byproducts in animal feed composition can improve land usage by reducing the area needed for grain production (Corré et al., 2016; de Souza et al., 2019).

In both the United States and Europe, the process of producing ethanol from different sources results in the creation of valuable byproducts that are acceptable for animal feed (Mumm et al., 2014). These byproducts contribute to the high nutritional content of the feed. Sugarcane yeasts, bagasse, and molasses can enhance milk and beef production in cows. However, the inefficient utilization of bagasse in outdated boilers calls for upgrading combined heat and power systems or exploring alternative options like importing eucalyptus chips to repurpose surplus bagasse effectively (de Souza et al., 2022). Food waste proves to be a valuable asset suitable for utilization as input material for anodes in Microbial Fuel Cell systems due to its accessibility and the ease with which it can be converted into a carbon source (Hung et al., 2019).

The Anammox process, when initiated, can involve inoculation with enhanced species or a combination of cultures. Various studies have demonstrated distinct growth traits and dominance of anammox bacteria under specific operating conditions using different sludge inoculants. While bioreactors have been successful in producing anammox bacteria, traditional microbiological techniques still pose challenges in identifying and enriching these bacteria. Precise taxonomic classification of microbial populations in both nascent sludge and mature Anammox sludge is essential to fully comprehend the microbial environment (MoonilKima, 2023).

Bioenergy status in Jordan

The bioenergy sector in Jordan demonstrates progress in renewable energy, contributing an estimated 11% to the energy mix in 2020 (Komendantova et al., 2021). However, the specific portion attributed to biomass energy production remains uncertain, likely constituting less than 0.1% due to limited bioenergy output (Al-Kurdi, 2021). Despite this, Jordan holds considerable potential for biomass generation sourced from agricultural residues, agroindustrial waste, and municipal waste. While the biomass energy sector in Jordan is slowly expanding, the country currently generates only 3.5 MW of electricity from biomass resources, For the time being, home energy production is the most common application of biomass. Biomass energy mostly consists of tree trunks, chips, bark, and sawdust, with wood accounting for around 44% of its composition. However, the production of biomass energy can be achieved with any type of organic material. Corn cobs and fruit pits are among the agricultural byproducts that can be utilized as additional sources of biomass. One frequent way to generate electricity is by using wood waste and timber. Instead of depending only on utility connections, industries that generate trash often use a process called cogeneration to generate electricity internally. Much of the steam and energy that paper and sawmills produce is actually used internally. However, these businesses usually have to buy extra power from utilities because of their high energy demands (The National Energy Education Development, NEED, 2019).

Six individual samples were analyzed, including olive leaves, green palm leaves, dry palm leaves, and olive pomace obtained from three different olive presses. The objective was to assess their gas production potential (GB21) to verify their capacity for biogas generation. Within the sample set, olive pomace variant three exhibited the most substantial methane output, registering at 103 NLKgFM-1, as depicted in below in Fig. 13.10 (Al-Addous et al., 2017).

Information pertaining to the manufacturing process of olive oil and its secondary products:

1. Olive oil peat serves as a raw material for the cement industry in clinker production.
2. A considerable proportion of the olive peat that is generated is dried and directly employed as a source of heat in residential settings.

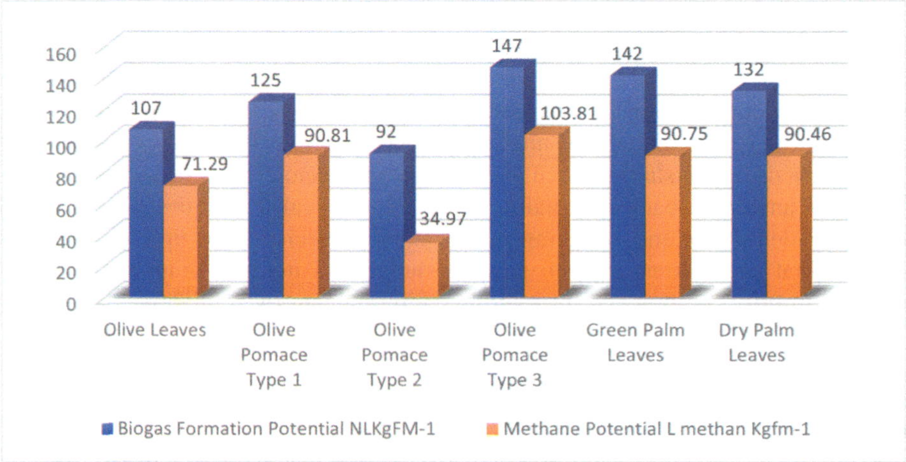

Figure 13.10 Olive leaves and pomace for their potential.
Source: Data from Al-Addous, M., Alnaief, M., Class, C., Nsair, A., Kuchta, K., & Alkasrawi, M. (2017). Technical possibilities of biogas production from olive and date waste in Jordan. *BioResources, 12*(4), 9383–9395. https://doi.org/10.15376/biores.12.4.9383-9395.

3. A standard quantity of olive pellets, weighing 1 ton, is normally composed of 35% peat, 17% oil, and 48% pomace (Al-Fayez, n.d.).
4. Each olive oil milling line generated approximately 200 tons of peat in 2018 (worlddata, n.d.).
5. Many olive oil production facilities and companies operate multiple milling lines.

Challenges and opportunities

While Jordan has made significant strides in expanding its renewable energy capacity, several challenges lie ahead in terms of infrastructure, investments, affordability, and conflicts of interest with the fossil fuel economy and electricity utility. One of the primary challenges is the need for a reliable infrastructure to support the energy transition. Integrating significant amounts of renewable energy into the grid requires a large and flexible grid infrastructure. The freeze on large-scale renewable energy projects from 2019 to 2022 reflects the challenges of integrating renewable energy into the grid without sufficient storage capacity. The identified challenges and barriers can be summarized in the following points:

- Policy and regulatory frameworks
- Grid infrastructure limitations
- Financial constraints
- Technological challenges
- Social and cultural acceptance
- Interconnection with neighboring countries
- Strengthening regulatory and policy frameworks
- Enhancing grid infrastructure and energy storage
- Promoting public–private partnerships
- Fostering innovation and research collaboration
- Implementing financial incentives and support mechanisms
- Enhancing energy efficiency and demand-side management
- Building institutional capacity and human resources
- Strengthening international cooperation and collaboration
- Raising public awareness and stakeholder engagement

Conclusion

A substantial decline in electricity consumption—estimated at around 10%—was caused by the economic downturn brought on by the COVID-19 epidemic, which had a profound effect on Jordan's power market. However, NEPCO's long-term payment obligations to gas and power suppliers under existing contracts did not reflect this drop. Also, from the middle of March 2020 until the middle of May 2020, payments from the distribution companies to NEPCO were suspended since their ability to recover energy purchasing costs decreased (National Electricity Power Energy Annual Reports, 2019). Due to severe working capital problems, NEPCO was forced to defer payments after subsequent payment delays had a significant impact on the organization's operational cash flows. As a result, as of April, NEPCO could only pay its green and traditional private independent power producers (IPPs) 75% of what they were due. These IPPs provide approximately half of the country's electricity. This issue jeopardized Jordan's power supply and the industry's foundational contractual integrity. That kind of unpredictability might make Jordan less appealing as an FDI destination and put its position as a renewable energy industry role model in the SEMED area at risk (ENERGY SECTOR Green Growth National Action Plan 2021–2025, 2020). In order to resolve these issues, the bank worked closely with NEPCO, the Jordanian government, investors, and other IFIs. As a result of these initiatives, NEPCO has agreed to pay off its debts and restore strict payment discipline with all renewable energy IPPs by the middle of 2021. The planned loan will not be signed until NEPCO has negotiated a comparable agreement with traditional power suppliers. Given the enormous investment needed for Jordan's and the region's continuing shift to renewable energy, this coordinated effort is a critical step toward regaining investor trust (VISP, n.d.).

References

A National Green Growth Plan for Jordan. (2017).

Abu-Rumman, G., Khdair, A. I., & Khdair, S. I. (2020). Current status and future investment potential in renewable energy in Jordan: An overview. *Heliyon*, *6*(2), e03346. Available from https://doi.org/10.1016/j.heliyon.2020.e03346.

Al-Addous, M., Al Hmidan, S., Jaradat, M., Alasis, E., & Barbana, N. (2020). Potential and feasibility study of hybrid wind–hydroelectric power system with water-pumping storage: Jordan as a case study. *Applied Sciences*, *10*(9), 3332. Available from https://doi.org/10.3390/app10093332.

Al-Addous, M., Alnaief, M., Class, C., Nsair, A., Kuchta, K., & Alkasrawi, M. (2017). Technical possibilities of biogas production from olive and date waste in Jordan. *BioResources*, *12*(4). Available from https://doi.org/10.15376/biores.12.4.9383-9395.

Alasali, F. (n.d.). Impact of the COVID-19 pandemic on electricity demand and load forecasting.

Al-Fayez, E. (n.d.). General syndicate of contemporaries and Jordanian Olive producers. http://www.civilsociety-jo.net.

Al-Kurdi, L.M. (2021).*Biofuels in Jordan: Perspectives*. AMMAN: Echoing Sustainability in MENA.

Al-Masri, H. M. K., Al-Quraan, A., AbuElrub, A., & Ehsani, M. (2019). Optimal coordination of wind power and pumped hydro energy storage. *Energies*, *12*(22), 4387. Available from https://doi.org/10.3390/en12224387.

Alrwashdeh, S. (2018). Map of Jordan Governorates wind distribution and mean power density. *International Journal of Engineering & Technology*, *7*.

Alrwashdeh, S. S. (2022). Energy sources assessment in Jordan. *Results in Engineering*, *13*100329. Available from https://doi.org/10.1016/j.rineng.2021.100329.

Anderson, R., Keshwani, D., Guru, A., Yang, H., Irmak, S., & Subbiah, J. (2018). An integrated modeling framework for crop and biofuel systems using the DSSAT and GREET models. *Environmental Modelling and Software*, *108*. Available from https://doi.org/10.1016/j.envsoft.2018.07.004, http://www.elsevier.com/inca/publications/store/4/2/2/9/2/1.

Burdon, D. (2019). *Handbook of the Geology of Jordan: To accompany and explain the three sheets of the 1:250,000 Geological Map of Jordan East of the Rift by Albert M. Quennell*. Government of the Hashemite Kingdom of Jordan

Corré, W. J., Conijn, J. G., Meesters, K. P. H., & Bos, H. L. (2016). Accounting for co-products in energy use, greenhouse gas emission savings and land use of biodiesel production from vegetable oils. *Industrial Crops and Products, 80*. Available from https://doi.org/10.1016/j.indcrop.2015.11.062, http://www.elsevier.com/inca/publications/store/5/2/2/8/2/5.

Dadashi, Z., Mahmoudi, A., & Rashidi, S. (2022). Capacity and strategies of energy production from renewable sources in Arab countries until 2030: A review from renewable energy potentials to environmental issues. *Environmental Science and Pollution Research, 29*(32), 47837–47866. Available from https://doi.org/10.1007/s11356-022-20544-z, https://link.springer.com/journal/11356.

de Souza, N. R. D., Cavalett, O., & Junqueira, T. L. (2022). Techno-economic and environmental assessment of bioenergy and livestock integrated systems in Brazil. *Sustainable Production and Consumption, 32*, 580–592. Available from https://doi.org/10.1016/j.spc.2022.05.013.

de Souza, N. R. D., Fracarolli, J. A., Junqueira, T. L., Chagas, M. F., Cardoso, T. F., Watanabe, M. D. B., Cavalett, O., Venzke Filho, S. P., Dale, B. E., Bonomi, A., & Cortez, L. A. B. (2019). Sugarcane ethanol and beef cattle integration in Brazil. *Biomass and Bioenergy, 120*. Available from https://doi.org/10.1016/j.biombioe.2018.12.012, http://www.journals.elsevier.com/biomass-and-bioenergy/.

Dehghan, M., Pfeiffer, C. F., Rakhshani, E., & Bakhshi-Jafarabadi, R. (2021). A review on techno-economic assessment of solar water heating systems in the middle east. *Energies, 14*(16), 4944. Available from https://doi.org/10.3390/en14164944.

EDAMA (n.d.). *EDAMA clean energy sector survey Covid-19 impacts—Executive summary of outcomes*. Amman: EDAMA

ENERGY SECTOR Green Growth National Action Plan 2021–2025. (2020). Ministry of Environment website. https://moenv.gov.jo/ebv4.0/root_storage/en/eb_list_page/2022_jordan_energy_v03.pdf.

Floudas, D., Binder, M., Riley, R., Barry, K., Blanchette, R. A., Henrissat, B., Martínez, A. T., Otillar, R., Spatafora, J. W., Yadav, J. S., Aerts, A., Benoit, I., Boyd, A., Carlson, A., Copeland, A., Coutinho, P. M., De Vries, R. P., Ferreira, P., Findley, K., … Hibbett, D. S. (2012). The paleozoic origin of enzymatic lignin decomposition reconstructed from 31 fungal genomes. *Science (New York, N.Y.), 336*(6089), 1715–1719. Available from https://doi.org/10.1126/science.1221748, http://www.sciencemag.org/content/336/6089/1715.full.pdf.

Generated energy surges by 8% last year, consumed energy by 6.6%. *J. News*. (2023).

Global Petrol Prices. (2022). https://www.globalpetrolprices.com/ 2022.

Hamed, T. A., & Bressler, L. (2019). Energy security in Israel and Jordan: The role of renewable energy sources. *Renewable Energy, 135*. Available from https://doi.org/10.1016/j.renene.2018.12.036, http://www.journals.elsevier.com/renewable-and-sustainable-energy-reviews/.

Hosseini, F. (2018). *Case study on policy reforms to promote renewable energy in Jordan*. Economic and Social Commission for Western Asia.

Hung, Y. H., Liu, T. Y., & Chen, H. Y. (2019). Renewable coffee waste-derived porous carbons as anode materials for high-performance sustainable microbial fuel cells. *ACS Sustainable Chemistry and Engineering, 7*(20). Available from https://doi.org/10.1021/acssuschemeng.9b02405, http://pubs.acs.org/journal/ascecg.

IEA. (n.d.). *World Energy Outlook 2022 shows the global energy crisis can be a historic turning point towards a cleaner and more secure future*. IEA.

IGA. (n.d.). *Geothermal power database*. Bonn: IGA.

International Energy Agency (n.d.). http://www.iea.org.

Jordan News (2022). https://www.jordannews.jo.

Khawaja, M. K., Ghaith, M., & Alkhalidi, A. (2021). Public-private partnership versus extended producer responsibility for end-of-life of photovoltaic modules management policy. *Solar Energy, 222*, 193–201. Available from https://doi.org/10.1016/j.solener.2021.05.022, http://www.elsevier.com/inca/publications/store/3/2/9/index.htt.

Kiwan, S., & Al-Gharibeh, E. (2020). Jordan toward a 100% renewable electricity system. *Renewable Energy, 147*. Available from https://doi.org/10.1016/j.renene.2019.09.004, http://www.journals.elsevier.com/renewable-and-sustainable-energy-reviews/.

Komendantova, L. M. N., Al-Salaymeh, A., Al Twassi, S., AlBeek, R., & Hassouneh, K. (2021). On the crossroad—renewable energy sources or shale oil? Understanding patterns of social attitudes in Jordan. *OPEC Energy Review*.

Latlong (n.d.). http://www.latlong.net.

Le, T.-H., Le, A. T., & Le, H.-C. (2021). The historic oil price fluctuation during the Covid-19 pandemic: What are the causes. *Research in International Business and Finance, 58*, 101489. Available from https://doi.org/10.1016/j.ribaf.2021.101489.

MEMR, (2013). The Hashemite Kingdom of Jordan , Annual Report 2013. https://www.memr.gov.jo/ebv4.0/root_storage/en/eb_list_page/annual_report_2013.pdf.

MEMR. (2020). *Summary of Jordan energy strategy 2020–2030*. MEMR

Ministry of Energy and Mineral Resources (MEMR). (2020). *Annual report 2019*. Amman: MEMR.

MoEnv. (2020). *Waste sector green growth national action plan*. Ministry of Environment Amman, The Hashemite Kingdom of Jordan.

MoonilKima F., Identification of bacterial communities in conventional wastewater treatment sludge to inform inoculation of the anammox process. Chemosphere (2023),

MOPIC, (2017). *Jordan's way to sustainable development first national voluntary review on the implementation of the 2030 agenda*.

Muheisen, R.A. (2023). Fuel sector investments, renewable energy transition key for Jordan to cushion possible crisis impacts. *Jordan Times*.

Mumm, R. H., Goldsmith, P. D., Rausch, K. D., & Stein, H. H. (2014). Land usage attributed to corn ethanol production in the United States: Sensitivity to technological advances in corn grain yield, ethanol conversion, and co-product utilization. *Biotechnology for Biofuels, 7*(1). Available from https://doi.org/10.1186/1754-6834-7-61.

Myyas, R. N., & Almajali, M. R. (2022). Effect of COVID-19 on energy sector in Jordan. *International Journal of Energy Economics and Policy, 12*.

National elecrecity power energy annual reports. (2019).

Olawuyi, D. S. (2021). *From energy consumers to energy citizens*. Brill, 10.1163/9789004465442_006.

Prabheesh, K., Padhan, R., & Garg, B. (2020). COVID-19 and the oil price-stock market Nexus: Evidence from net oil-importing countries. *Energy Research Letters p., 1*(2).

Park, G.W., Gong, G., Joo, J., Song, J., Lee, J., Lee, J.P., Kim, H.T., Ryu, M.H., Sirohi, R., Zhuang, X., & Min, K. (2022). Recent progress and challenges in biological degradation and biotechnological valorization of lignin as an emerging source of bioenergy: A state-of-the-art review. *Renewable and Sustainable Energy Reviews, 157*, 112025.

R.E. A. (n.d.). *Renewable readiness assessment: The Hashemite Kingdom of Jordan*.

Salah, A. A., Shalby, M. M., & Basim Ismail, F. (2023). The status and potential of renewable energy development in Jordan: Exploring challenges and opportunities. *Sustainability: Science, Practice and Policy*, *19*(1). Available from https://doi.org/10.1080/15487733.2023.2212517.

Shahateet, M. I., Khazali, M., Albaali, G., Sweis, N., & Saidi, A. G. (2021). Barriers to improving energy efficiency: insights from energy services companies in Jordan. *International Journal of Energy Economics and Policy*, *11*(2). Available from https://doi.org/10.32479/ijeep.10811.

Shatnawi, N., Abu-Qdais, H., & Qdais, F. A. (2021). Selecting renewable energy options: An application of multi-criteria decision making for Jordan. *Sustainability: Science, Practice and Policy*, *17*(1). Available from https://doi.org/10.1080/15487733.2021.1930715.

Stoyanov, L., Bachev, I., Zarkov, Z., Lazarov, V., & Notton, G. (2021). Multivariate analysis of a wind–PV-based water pumping hybrid system for irrigation purposes. *Energies*, *14*, 3231.

The Jordan Times. (n.d.). http://www.TheJordanTimes.com.

The National Energy Education Development, NEED (2019). https://www.need.org/wp-content/uploads/2019/10/BiomassAtAGlance_11x17.pdf.

The Royal Institute of International Affairs. (2016). *Refugees and energy resilience in Jordan*. Chatham House, the Royal Institute of International Affairs

Trotter, P. A., Maconachie, R., & McManus, M. C. (2018). Solar energy's potential to mitigate political risks: The case of an optimised Africa-wide network. *Energy Policy*, *117*, 108–126. Available from https://doi.org/10.1016/j.enpol.2018.02.013.

Updated submission of Jordan's 1st Nationally Determined Contribution (NDC) (2021). http://extwprlegs1.fao.org/docs/pdf/jor205905E.pdf.

VISP N.L.F., (n.d.). Document of the European Bank for reconstruction and development.

Weather. (2020). Weather Online Ltd.

World Energy Transitions Outlook 2022 (2022). International REnewable Energy Agency.

Worlddata (n.d.). http://www.worlddata.info.

Zittis, G., Almazroui, M., Alpert, P., Ciais, P., Cramer, W., Dahdal, Y., Fnais, M., Francis, D., Hadjinicolaou, P., & Howari, F. (n.d.). Climate change and weather extremes in the Eastern Mediterranean and Middle East. *Reviews of Geophysics*, *60*.

Part 4

Optimization and control for power systems with renewables

Energy management in energy communities

14

Morsy Nour[1,2], Matteo Troncia[1] and José Pablo Chaves-Ávila[1]
[1]Institute for Research in Technology (IIT), ICAI School of Engineering, Comillas Pontifical University, Madrid, Spain, [2]Department of Electrical Engineering, Faculty of Energy Engineering, Aswan University, Aswan, Egypt

Introduction

For decades, the power systems have been operating in a supplier-centric structure involving a small number of active participants (i.e., mainly dispatchable fossil fuel-based power plants). Environmental challenges drive many countries to adopt renewable energy sources (RESs) in the form of large power plants or small-generation near-end customers besides other sustainable technologies (International Renewable Energy Agency (IRENA), 2020). Therefore, the supplier-centric structure is changing to a customer-centric structure due to the technological advancements that enable electricity generation and storage by small devices that can be installed near end customers with viable costs. Future power systems are expected to encounter economic and technical challenges because of the high penetration of distributed energy resources (DERs) like small RESs, small energy storage (ES), electric vehicles (EVs) with unidirectional or bidirectional chargers, flexible demand, etc. (Cuenca et al., 2021). These DERs are connected to the main power system and governed by regulations that are designed for the suitability of large participants. They may not be suitable for capturing the aimed benefits of DERs.

As a result, many management approaches and business models are proposed in the literature and studied in pilot projects to manage DERs to maximize the benefits of involved stakeholders and provide numerous services, while keeping the power system operation within technical limits (Guerrero et al., 2020). The end customers can manage their DERs individually in an uncoordinated manner to maximize their revenues, minimize costs, maximize self-consumption, and so forth under the applied tariff design at the distribution level using building management systems or home energy management systems (HEMS). On the other hand, a large number of customers can aggregate their DERs in the form of virtual power plants to obtain common objectives, participate in different electricity markets, and provide different ancillary services to the power system (Naval & Yusta, 2021; Nour, Magdy, Chaves-Ávila, et al., 2023). Moreover, they can be coordinated in the form of a microgrid that could operate in grid-connected or isolated mode (Bastawy et al., 2022; Nour, Magdy, Bakeer, et al., 2023).

Towards Future Smart Power Systems with High Penetration of Renewables. DOI: https://doi.org/10.1016/B978-0-443-29871-4.00014-2

Furthermore, the end customers can form an energy community (EC) based on the European Union (EU) concepts of a sharing economy to coordinate their DERs, share energy among themselves, decrease the energy exchange with the main grid, and operate as a single entity enabling access to different electricity markets and provision of different ancillary services using flexible devices (i.e., ES, EVs, controllable loads, etc.) (Barnes et al., 2022; Scarabaggio et al., 2021). By doing so, the ECs could provide social, environmental, and economic benefits (Gianaroli et al., 2024). ECs have the potential to increase independence from the main grid by decreasing energy purchases from retailers, increasing the consumption of locally generated energy within EC, and decreasing the energy costs for individuals and the whole EC (Saif et al., 2022, 2023). This results in reducing the energy required from large power plants and transmitting it through transmission lines and can decrease investment requirements and energy losses.

On the other hand, the EC implementation faces many obstacles and challenges, such as insufficient regulations developments or nonexistence of regulations in some countries, a lack of end customers' interest and awareness, a lack of investors' confidence, and complexity of administrative procedures (López et al., 2024).

Many articles in the literature studied ECs' legal frameworks, types, business models, structures, energy sharing, and services they can provide (López et al., 2024). Queiroz et al. (2023) analyzed the performance of different types of energy-sharing coefficients which allocate the local energy among members, in EC formed, by public buildings based on the coefficients proposed in the Portuguese legislation for EC. The findings show that time variable (i.e., dynamic) coefficients provide better performance than fixed (i.e., static) coefficients and that EC results in benefits for the entire EC and EC members. Another article studies the performance of EC under 144 scenarios for a Spanish case study for 1 year (Manso-Burgos et al., 2022). The findings show that the EC can fulfill self-consumption, environmental, and economic goals in most pessimistic scenarios.

In Lazzari et al. (2023), the genetic algorithms is used to optimize the planning and operation of EC to minimize economic and environmental benefits. The optimization algorithm finds the optimal combination of EC members and energy allocation of locally generated energy. In Berg et al. (2024), the benefits of ECs and the associated impacts on distribution networks are assessed considering different member configurations and the presence of different types of customers (residential, commercial, and both). The study duration is 1 year and considered Norwegian and Spanish case studies. The findings show that the types of customers have a significant effect on EC performance in costs, self-consumptions, and the associated impacts on the distribution network.

Moreover, datasets are shared and open-source tools are developed to facilitate the research of ECs considering different settings (Vecchi et al., 2024). In Berg and Löschenbrand (2022), a data set that represents a Norwegian EC is shared. The data includes PV generation data, EV charging patterns, electricity prices, and house consumption. Gomes et al. (2023) developed a Python-based open-source tool that enables the creation and simulation of ECs to operate the different sources that are part of the ECs and using different optimization algorithms. Considering the limitation in available datasets for ECs, Velosa et al. (2023) developed an open-source tool that enables generating generation and demand datasets for ECs to simulate many scenarios and test different algorithms. The tool enables the creation of ECs with different sizes, members, and appliances.

The organization of this chapter is as follows. Section II provides an overview of EC concepts, the common elements of ECs, the legal figures of ECs defined by EU regulations, and modeling approaches for energy sharing in ECs. Section III presents the optimization models of HEMS and EC. Section IV presents the study results and analyses the findings. The conclusion of the chapter is presented in section V.

Energy communities' legal figures and management approaches

Many countries worldwide have developed regulations for incentivizing RESs and other sustainable technologies (Chen et al., 2014; Lu et al., 2020). The EU targets a decrease in emissions by 55% by 2030 and to be climate-neutral by 2050. This requires a significant development in energy policies to be achieved. The EU revised the Clean Energy Package for all Europeans proposed in 2016 with new directives (i.e., 2018/2001 and 2019/944) (European Parliament, 2018). These directives promote the transition to a sustainable energy model and introduced the concept of ECs for the first time. These directives must be transposed to countries' national regulations for ECs' successful development. Therefore, following the EU directives, several countries developed their national legislation for the deployment of ECs (Iazzolino et al., 2022).

In general, the EU realizes ECs as a method of managing collectively, fundamentally on the initiative of citizens, a variety of energy and management activities, which may involve both energy generation and consumption, storage, and other activities, like the operation of these energy systems or management of excess energy on the market (López et al., 2024). The following subsections discuss the common elements of ECs, the legal figures of ECs defined by EU regulations, and modeling approaches for energy sharing in ECs.

Common elements of energy communities

ECs are dynamic entities that adapt their structure and operations to the specific needs and characteristics of their environments, whether urban or rural. Their overarching goal is to create a localized energy ecosystem that efficiently balances production, consumption, and storage, thereby fostering sustainable and resilient local energy systems. These communities strive to minimize their reliance on external energy suppliers by managing their energy resources effectively. The following are the main elements of ECs.

Producers

Households equipped with photovoltaic (PV) panels are typical examples of producers within ECs The energy they generate can either be consumed within the community or sold to external parties, contributing to the community's sustainability and economic viability (Walker & Devine-Wright, 2008).

Consumers

EC members, including residential households and facilities with EV charging stations, utilize the energy produced. This direct consumption model strengthens the link between production and consumption, enhancing energy efficiency (Schreuer & Weismeier-Sammer, 2010).

Storage

ECs often incorporate storage systems, which may be owned by the community as a whole or by individual participants. These storage solutions enable the temporary storage of generated energy, allowing for the optimization of energy usage in response to production and demand fluctuations (Parag & Sovacool, 2016).

Energy distribution within the community

Excess energy produced by community members can be channeled to community storage facilities or directly to other consumers within the EC. This internal distribution mechanism reduces energy wastage and improves system efficiency (European Parliament, 2018).

Interaction with external energy market(s)

ECs interact with external energy markets in two main scenarios: purchasing additional energy when community demand exceeds local production and selling surplus energy when production outstrips local demand. This interaction not only provides economic benefits but also enhances the community's energy resilience (COM/2016/0860., 2016).

Legal figures of energy communities defined in the EU regulation

The concept of ECs in the EU is dynamic and encompasses various forms under EU legislation, reflecting the diverse ways in which individuals and entities can collectively engage in energy activities. The first explicit reference to ECs in the EU policy appears in the European Green Deal, highlighting their importance in achieving the EU's energy and climate goals. Subsequent directives and regulations provide definitions for five terms that encapsulate different aspects of ECs:

Jointly-Acting Active Customer (JAAC) describes individuals or groups that collectively engage in electricity generation, consumption, or storage (European Parliament,

2019; European Parliament, 2018). This term emphasizes active participation in the energy market, focusing on collaboration among customers.

Jointly-Acting Renewable Self-Consumers (JARSC) refers to entities that jointly produce renewable energy for their own consumption (European Parliament, 2018). This model promotes the shared use of renewable energy resources among participants, enhancing sustainability and self-sufficiency.

Citizen Energy Community (CEC) highlights the role of citizens and local entities in forming ECs (European Parliament, 2019). It focuses on social and environmental benefits, encouraging local engagement and collective actions in the energy sector.

Renewable Energy Community (REC) specifically focuses on communities engaged in renewable energy activities (European Parliament, 2018; Regulation (EU)). This term underscores the collective efforts to produce, consume, store, and sell renewable energy within a community, contributing to the transition toward a cleaner energy system.

Closed distribution networks (CDNs) refer to energy networks designed for a specific, limited group of users, often within a confined area or serving specific types of operations (European Parliament, 2019). They are distinct from public networks and cater to specific energy needs of their users.

These definitions are provided and contextualized within the following EU legislative texts:

- Directive 2019/944 (common rules for the internal electricity market) (European Parliament, 2019): This directive outlines rules for the internal market in electricity, aiming to ensure the supply of electricity, consumer protection, and higher levels of competitiveness. It acknowledges the role of CECs and introduces frameworks to facilitate their establishment and operation within the internal electricity market.
- Directive 2018/2001 (promotion of the use of energy from renewable sources) (European Parliament, 2018): This directive sets ambitious goals for the EU to increase the share of energy from renewable sources in its overall energy mix. It introduces the concept of RECs, encouraging member states to support these communities as a means to achieve broader renewable energy targets.
- Regulation 2018/1999 (governance of the Energy Union and Climate Action) (*Regulation (EU) 2018/1999 of the European Parliament and of the Council of 11 December 2018 on the Governance of the Energy Union and Climate Action, amending Regulations (EC) No 663/2009 and (EC) No 715/2009 of the European Parliament and of the Council*, 2018): This regulation establishes the framework for the governance of the Energy Union, integrating climate and energy policies across the EU. It aims to ensure the achievement of the EU's energy and climate objectives and commitments under the Paris Agreement, acknowledging the importance of community and citizen engagement in these efforts.

Table 14.1 resumes the legal acts that define the framework for those legal figures (Troncia, 2024).

Comparison of the legal figures for energy communities

The EU's approach to ECs and consumer participation in the energy market encompasses various configurations, each with distinct characteristics and regulatory

Table 14.1 Legal framework of the five figures that can be considered as energy communities in the European legislation.

Act	Jointly-acting renewable self-consumers (JARSC)	Renewable Energy Community (REC)	Jointly-acting active customer (JAAC)	Citizen Energy Community (CEC)	Closed distribution system
Regulation 2018/1999 (*Regulation (EU) 2018/1999 of the European Parliament and of the Council of 11 December 2018 on the Governance of the Energy Union and Climate Action, amending Regulations (EC) No 663/2009 and (EC) No 715/2009 of the European Parliament and of the Council,* 2018)	–	Art. 2 and 20 Annexes I and IX	–	–	–
Directive 2018/2001 (European Parliament, 2018)	Art. 2 and 21	Art. 2, 7, 15, 18, and 22	Art. 18	–	–
Directive 2019/944 (European Parliament, 2019)	–	–	Art. 2,15, 16, and 20	Art. 2, 6, 16, and 59	Art. 38

Source: From Troncia, M. (2024). *Regulatory analysis, proposals for efficient flexibility mechanisms and demonstrators' framework* (Original work published 2024).

frameworks. These configurations are designed to foster greater involvement in renewable energy generation, consumption, and sharing, while also addressing social, economic, and environmental goals. Here is a detailed exploration of these different EC figures:

JAAC and JARSC are not required to be defined with an explicit purpose while CECs and RECs may have an environmental, economic, and social purpose, but not a financial one.

Considering JAAC and JARSC, they can be constituted as a REC if the corresponding purposes are satisfied.

JAAC and JARSC are not defined as a legal entity. It is an agreement between members. There is no provision mentioning in what form they can be organized.

While CECs and RECs are defined as an autonomous legal entity with these characteristics:

- EC entry has to be voluntary and open
- It is the form of membership and governance that distinguishes them.

JAAC and JARSC are characterized by geographical boundaries defined as the customer's premises and the same building for the JARSC. CECs are not characterized by geographical boundaries while RECs require that the members are located in the vicinity of the renewable project, precise criteria are defined on a country-by-country basis.

JAAC and CECs have to deal with electricity, and there is no explicit limit, but it is understood that the technology chosen is PV or RES.

JARSC and RECs have to deal with renewable electricity.

Table 14.2 compares the legal figures for ECs.

Modeling approaches for energy sharing in energy communities

Several approaches are proposed in the literature to manage ECs and to model energy sharing between community members. The EC local energy sharing could be modeled as an optimization problem, a game theoretic problem, or an auction. Moreover, energy sharing between EC members could be done using energy sharing coefficients (López et al., 2024), as shown in Fig. 14.1.

Auction

Auctions are bargaining mechanisms that are often mediated by a mediator. Distributed ledger technologies, such as blockchain, can eliminate the mediator and conduct the auction in a decentralized way by encoding the market terms in a smart contract (Nour et al., 2022). The auction can be conducted as a single auction, with just one side of the market sending bids (Capper et al., 2022). This is frequent when one side of the market is represented by a single agent. For example, one buyer receives offers from many sellers. There is also a double auction, in which buyers submit bids and sellers submit ask bids (Khorasany et al., 2018).

Table 14.2 Comparison of the legal figures for energy communities (Troncia, 2024).

	Purpose	Is it incorporated as a legal entity?	Geographical boundaries	Energy types
Jointly-acting active customer (JAAC)	No explicit purpose If it fulfils the purposes, it can be constituted as an REC.	No It is an agreement between members. There is no provision mentioning in what form they can be organized.	Yes Within its premises (other premises if Ms allows)	Electricity There is no explicit limit, but it is understood that the technology chosen is PV.
Jointly-acting rene wable self-consumers (JARSC)			Yes Same building	Electricity (renewable)
Citizen Energy Comm unity (CEC)	Environmental Economic Social Financial	Autonomous legal entity EC entry has to be voluntary and open It is the form of membership and governance that distinguishes them.	No	Electricity There is no explicit limit, but it is understood that the technology chosen is RES
Renewable Energy Comm unity (REC)			Yes: Local *	Electricity (renewable)

Source: From Troncia, M. (2024). *Regulatory analysis, proposals for efficient flexibility mechanisms and demonstrators' framework* (Original work published 2024).

Optimization

The EC can be modeled as an optimization with various objectives such as minimizing costs, maximizing revenues, maximizing consumption of local RES generation, and so forth. The optimization method can be centralized, distributed, and multilevel. The centralized method requires the presence of a central entity like an EC manager (Nguyen et al., 2018). Scalability and privacy concerns arise when

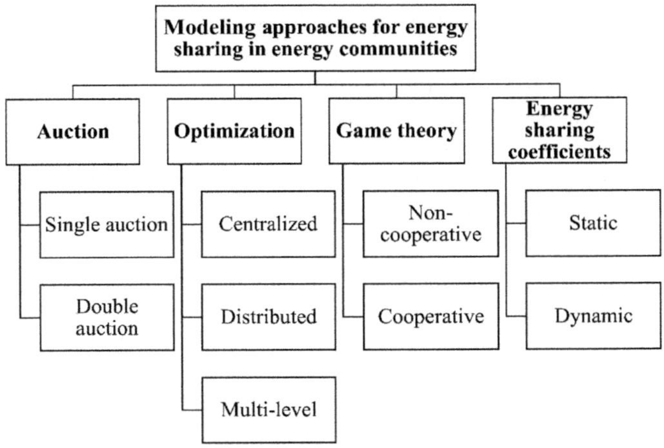

Figure 14.1 Modeling approaches for energy sharing in energy communities.

managing lots of DERs in a centralized arrangement. Distributed optimization algorithms are suitable for dealing with these challenges (Khorasany et al., 2018). In multilevel optimization, the problem is divided into several sections.

Game theory

Game theory is a method for analyzing reasonable market participants' decision-making, in which every participant's strategic behavior influences and is influenced by the behavior of other participants in a competitive scenario. Game theory was used in numerous local energy trading research to represent the players' competing interests (Fioriti et al., 2021).

Energy sharing coefficients

Previous approaches provide promising performance in energy sharing between end customers. However, they are very complex and may require high costs, considering that they are used to manage energy sharing in small quantities. Many regulations enable energy sharing between community members using the energy sharing coefficient (Gruber et al., 2023; Queiroz et al., 2023). This coefficient specifies the percentage of total energy that each community member receives. Energy sharing coefficients define how local RESs production is shared among community members. Energy-sharing coefficients can be classified into static or dynamic. In static energy sharing coefficients, the percentage of energy that each EC member receives is fixed and agreed upon by EC members ahead of time. In dynamic energy sharing coefficients, the percentage of energy that each EC member receives depends on the percentage of this demand to the total demand of EC members, and therefore, it varies over time (Gruber et al., 2023).

Problem formulation and case study

This section briefly introduces the mathematical models of HEMS and EC. The models are represented as linear optimization problems, and linear programming is used to solve them. The model represents a day ahead optimal dispatch of the houses' DERs. The grid constraints are not considered in this study (Nour, Chaves-Avila, et al., 2023; Nour, Troncia, et al., 2023).

Objective functions

For the HEMS case, the objective function is to minimize the network charges costs (i.e., contracted power costs), minimize energy purchase costs, and maximize revenues of selling energy to the retailer for each house. The optimization problem is subjected to the constraints of power balance at each house node and the house DER limits. Each house optimizes its DERs individually to minimize the objective function without interacting or coordinating with other houses. Each house can exchange (i.e., buy or sell) energy only with the retailer.

The objective function of the EC case is to minimize the network charges costs, minimize energy purchase costs, and maximize revenues of selling energy to the retailer for all the EC. The optimization problem is subjected to the constraints of local energy trading, power balance at each house node, and DER limits. The optimization is done for all the houses in the EC to minimize the objective function (i.e., all EC). Each house can exchange (i.e., buy or sell) energy with the retailer or other houses in the EC. The EC is managed by a central entity like the EC manager (i.e., aggregator). The EC manager receives data and measurements of houses and sends control signals to manage houses' DERs. More details of the mathematical formulation can be found in Nour et al. (2023, 2024).

$$min\left(\sum\nolimits_{per\in P} p_{per}^{cp} \times CP_{per} + \sum\nolimits_{t} (p_t^b \times G_t - p_t^s \times F_t)\right) \tag{14.1}$$

The optimization is subject to power balance constraints and DER constraints.

$$min\left(\sum\nolimits_{per\in P} \sum\nolimits_{h} p_{per}^{cp} \times CP_{per,h} + \sum\nolimits_{t} \sum\nolimits_{h} (p_t^b \times G_{t,h} - p_t^s \times F_{t,h})\right) \tag{14.2}$$

It is subject to power balance constraints, DER constraints, and local energy trading constraints.

where p_{per}^{cp} is the cost of contracted power at period *per*, $CP_{per,h}$ is the contracted power of house h at period *per*, p_t^b is the price of purchasing energy from the retailer at instant t, p_t^s is the price of selling energy to the retailer at instant t, $G_{t,h}$ is the energy purchased by house h at instant t from the retailer, and $F_{t,h}$ is the energy sold by house h to the retailer at instant t.

Case study

In this study, a residential community containing 55 houses supplied by the same MV/LV transformer and connected to the same low-voltage distribution network is considered as a case study. The data used is for Madrid, Spain. The demand profiles of houses' inflexible demands are anonymized profiles obtained from Iberdrola. PV generation profiles for Madrid, Spain, are obtained from the RenewableNinja website. The prices of buying and selling energy from/to the retailer are obtained from the Spanish TSO for July 2021 and can be found in REE 2024). The contracted power costs (i.e., network charges) applied to residential customers with contracted power less than 15 kW in Madrid, Spain, are given in Table 14.3. The contracted power has two prices during the day. The peak period (from 8 to 24) has a higher cost than the off-peak period (from 24 to 8). On weekends, the off-peak period cost applies to all day hours. Each house energy exchange (i.e., buy or sell) cannot exceed the contracted power in the day and at each period. A 5% tax applies to energy and contracted power costs in Madrid, Spain.

Some houses have no DERs installed, and others have different combinations of DERs (i.e., PV, ES, and EV). The DERs installed at each house can be seen in Table 14.4. Fig. 14.2 shows the aggregated demand profiles of the 55 houses, the PV generation of any house, and the energy prices. 60% (i.e., 33 houses) of the houses have a PV generation, and the installed PV generators have a 5 kWp rating. It is assumed that installed PV generation in any house houses has the same rating.

40% (i.e., 22 houses) of the houses have ES, and the ES has a 13.5 kWh energy rating and a 5 kW charger power rating. It is assumed that the ES installed in the community has the same ratings. 33% (i.e., 18 houses) of the houses have an EV, and the EV has a 24 kWh energy rating and 3.6 kW charger power rating that can charge or discharge the EV battery (i.e., bidirectional charger). It is assumed that the EVs installed in the community have the same ratings. The EVs are connected every day from 5 PM to 8 AM the next day for charging and used for mobility in other hours of the day. At departure time in the morning, the EV state of charge cannot be lower than 75%.

Table 14.3 The contracted power costs in peak and off-peak periods applied to residential customers in Madrid, Spain.

Contracted power costs	Peak	Off-peak
Transmission and distribution costs (€/kW)	23.469833	0.961130
Policy costs (€/kW)	4.970533	0.319666
Total regulated costs with a 5% tax (€/kW)	29.8623843	1.3448358

Source: From Nour, M., Chaves-Avila, J.P., Troncia, M., & Sanchez-Miralles, A. (2024). Mitigating the impacts of community energy trading on distribution networks by considering contracted power network charges. *IEEE Access, 12*, 26991–27004. https://doi.org/10.1109/ACCESS.2024.3365630.

Table 14.4 The DERs installed at each house in the EC.

House	PV	ES	EV	House	PV	ES	EV	House	PV	ES	EV	House	PV	ES	EV
1	×	×	–	15	×	×	–	29	–	–	–	43	×	–	–
2	×	×	×	16	×	–	×	30	×	×	–	44	–	–	–
3	×	×	–	17	–	–	–	31	–	–	×	45	×	×	–
4	–	–	–	18	×	×	–	32	×	–	–	46	–	–	×
5	×	×	–	19	–	–	–	33	×	×	–	47	–	–	–
6	–	–	–	20	×	×	×	34	×	–	–	48	×	×	–
7	×	–	×	21	–	–	–	35	–	–	×	49	×	–	×
8	×	–	–	22	–	–	–	36	–	–	–	50	×	×	×
9	×	×	×	23	×	×	–	37	×	×	–	51	–	–	–
10	–	–	–	24	×	–	–	38	–	–	–	52	×	×	–
11	–	–	–	25	×	–	×	39	×	–	×	53	×	×	×
12	×	×	×	26	–	–	–	40	×	×	–	54	×	×	×
13	–	–	–	27	×	×	–	41	×	–	×	55	×	×	×
14	–	–	–	28	–	–	×	42	–	–	–				

DERs, Distributed energy resources; *EC*, energy communities; *ES*, energy storage; *EV*, electric vehicles; *PV*, photovoltaic.
Source: From Nour, M., Chaves-Avila, J.P., Troncia, M., Ali, A., & Sanchez-Miralles, A. (2023). Impacts of community energy trading on low voltage distribution networks. *IEEE Access, 11*, 50412–50430. https://doi.org/10.1109/ACCESS.2023.3278090.

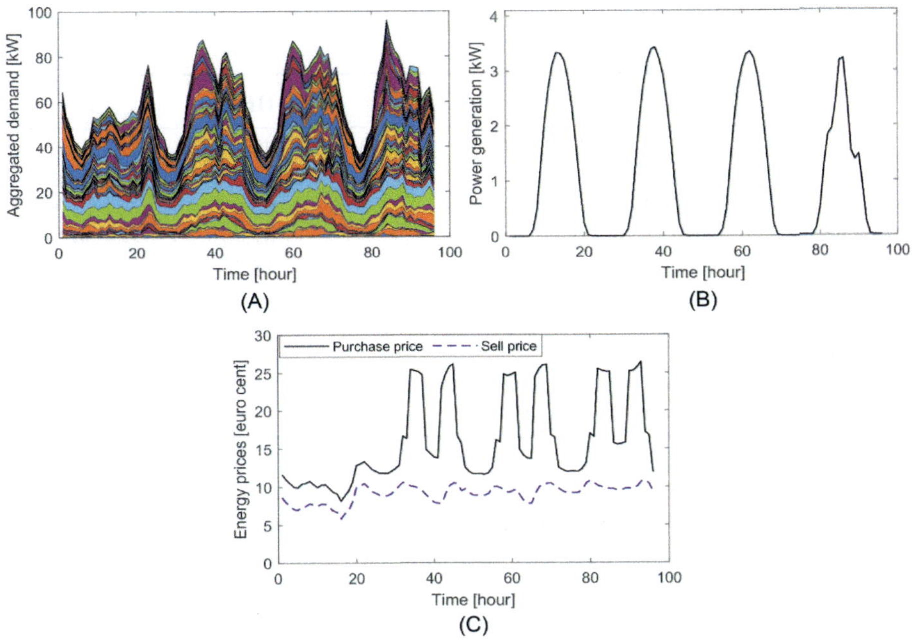

Figure 14.2 Input data for the optimization model: (A) houses aggregated demand, (B) PV generation for a single house in the EC, and (C) the retailer's prices for energy purchase and selling *ECs*, Energy communities; *PV*, photovoltaic.
Source: From Nour, M., Chaves-Avila, J.P., Troncia, M., & Sanchez-Miralles, A. (2024). Mitigating the impacts of community energy trading on distribution networks by considering contracted power network charges. *IEEE Access, 12*, 26991–27004. https://doi.org/10.1109/ACCESS.2024.3365630.

Results and discussions

In this section, the benefits of forming an EC that enables the sharing of energy between EC members and collective coordination of all EC DERs are compared with individual management of DERs using HEMS. Table 14.5 compares the performance of HEMS and EC approaches for 1 month in summer (i.e., July) for all 55 houses connected to the same distribution network.

The results show that with the formation of EC by houses, many benefits could be acquired like decreasing energy purchased from the retailer by 31.4% (i.e., higher independence from the main grid) and decreasing energy sold to the retailer by 94.4%, which indicates that a higher percentage of local PV generation is consumed in the EC instead of selling it back to the retailer. Moreover, the formation of EC significantly increases the percentage of demand covered by EC DERs. Furthermore, the total operation costs (i.e., costs of contracted power + costs of energy purchased from the retailer − revenues of energy sold to the retailer) of the 55 houses during the month decrease by 30.4% when they form an EC compared to

Table 14.5 Comparison of the performance of HEMS and EC approaches for 1 month (i.e., July).

	HEMS	EC
Energy purchased from the retailer (kWh)	38607.21	26499.84
Energy sold to the retailer (kWh)	13811	776.11
Energy shared within the EC (kWh)	0	17165.85
Demand covered by the retailer (%)	81.75	56.11
Demand covered by EC DERs (%)	18.25	43.89
Peak value of energy purchased from the retailer (kW)	142.47	221.97
Total operation costs of all houses (€)	4872.66	3392.80
EC cost reduction (%)	–	30.4
Costs of energy purchased from the retailer (€)	6141.44	3450.98
Revenues of energy sold to the retailer (€)	1268.77	58.18

DERs, Distributed energy resources; *EC*, energy communities; *HEMS*, home energy management system.

individual optimization of DERs in the case of HEMS. On the other hand, the formation of EC increased the peak of energy imported from the retailer.

The two approaches are simulated for 1 month. However, Fig. 14.3 shows only 4 days for better visibility to analyze the house behavior in each management approach. As shown in Fig. 14.3A, in the case of HEMS, the energy purchase from the retailer never reaches zero since there are houses with no DERs installed and must buy energy from the retailer. Moreover, the PV generation of houses with PV could be lower than the house demand. Furthermore, houses with PV and other flexible devices like ES or EV could cover all of their demand at some hours from their own DERs but not all day hours.

On the other hand, the formation of EC resulted in many hours with zero energy purchases from the retailer. This results from the local energy exchange between houses of EC at lower costs than the retailer prices. However, it is clear that the formation of EC results in an increase in the peak of energy purchases from the retailer compared to HEMS. A recent study shows that this increase in the peak of EC energy purchases from the retailer can be significantly decreased by efficient network tariff design (Nour et al., 2024).

In the case of HEMS, if any house has a surplus energy, its only option is to sell energy to the retailer, as illustrated in Fig. 14.3B. The figure demonstrates that HEMS results in higher energy sold to the retailer compared to the case with houses forming an EC. The houses usually sell energy at day hours when there is surplus PV generation. The EC approach resulted in a significant decrease in the energy sold to the retailer, and there are many days with zero exports to the retailer. Fig. 14.3C shows the energy shared between EC houses. The exchange mainly occurs during the daytime when PV generation is present. However, the presence of ES and EV enables extending the hours with energy sharing between houses since they store energy from PV generation or even energy purchased from the retailer at low prices hours and exchange it with other houses in the EC at hours with higher prices.

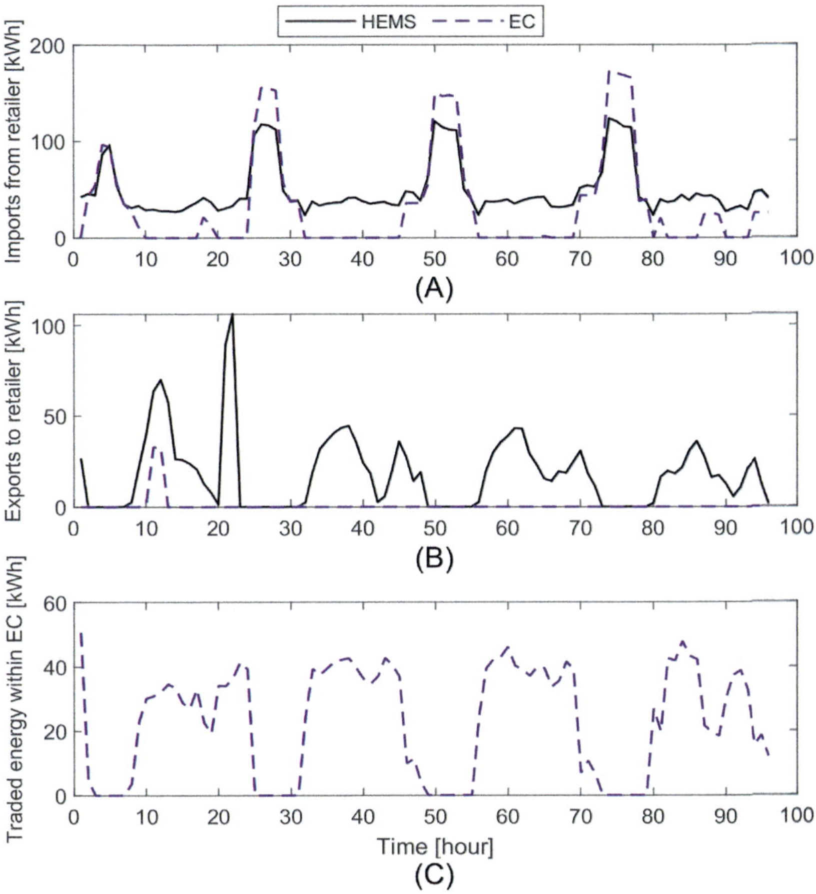

Figure 14.3 Comparison of HEMS and EC based on (A) energy imports from the retailer, (B) energy exports to the retailer, and (C) energy traded within EC. *EC*, Energy community; *HEMS*, home energy management system.

Fig. 14.4 shows the total charging power of installed ES and EV in the 55 houses for HEMS and EC. Fig. 14.3A demonstrates that there are hours with very high energy imports from the retailer for both HEMS and EC. Moreover, when houses form an EC, this results in an increase in the peak of the energy purchased from the retailer compared to HEMS, at which each house individually optimizes its DERs. To understand the reason for this, Fig. 14.4 displays the aggregated charging power of ES and EV of all houses for both management approaches. It can be seen that ES and EV have high charging power at the early hours of all presented days, which benefits from low retailer prices or satisfies the constraint of EV state of charge at departure time in the morning. In the case of EC, the charging power of ES at early day hours is higher than the charging power for HEMS and coincides

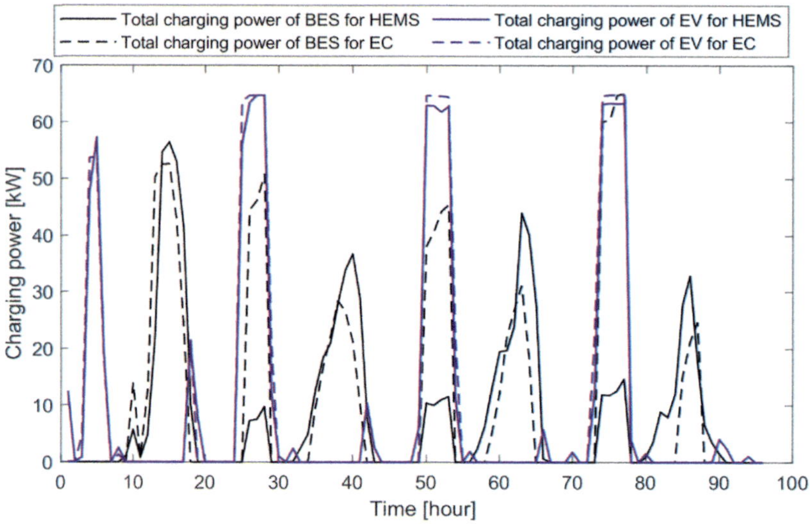

Figure 14.4 Total charging power of installed ES and EV in the 55 houses for HEMS and EC. *ES*, Energy storage; *EV*, electric vehicle.

with high charging power of EV. This is the reason for the increase in the peak of energy purchased from the retailer.

Conclusion

This chapter provides an overview of the main aspects of ECs, including the common elements of ECs, the legal figures of ECs defined by EU regulations, and modeling approaches for energy sharing in ECs. Moreover, it presented a study that evaluates the benefits obtained by residential customers if they form EC, which enables collectively coordinating all of their DERs together and enabling energy sharing between customers compared to individual optimization of customers' DERs using HEMS. The findings show that the formation of EC by residential customers increases self-consumption, increases self-sufficiency, decreases the energy purchase from the retailer, and decreases the total operation costs compared to HEMS. In contrast, the formation of EC increases the peak of energy purchase from the retailer compared to HEMS. This could result in higher impacts on the distribution network supplying houses.

Acknowledgment

This work was supported in part by the Ministerio de Ciencia, Innovación y Universidades (MCIN)/Agencia Estatal de Investigación (AEI)/10.13039/501100011033 under Project

TED2021−131365B-C41, and in part by the European Union "NextGenerationEU"/Plan de Recuperación, Transformación y Resiliencia (PRTR).

Abbreviations

CEC	Citizen Energy Community
DERs	Distributed energy resources
EC	Energy community
ES	Energy storage
EU	European Union
EV	Electric vehicle
HEMS	Home energy management system
Ms	Member states
PV	Photovoltaic
REC	Renewable Energy Community
RED	Renewable Energy Directive
RESs	Renewable energy sources

AI disclosure

During the preparation of this work, AI-assisted technologies (ChatGPT) have been used in the writing process to improve readability and language. After using this tool/service, the author(s) reviewed and edited the content as needed and take(s) full responsibility for the content of the publication.

References

Barnes, J., Hansen, P., Kamin, T., Golob, U., Musolino, M., & Nicita, A. (2022). Energy communities as demand-side innovators? Assessing the potential of European cases to reduce demand and foster flexibility. *Energy Research & Social Science*, *93*, 102848. Available from https://doi.org/10.1016/j.erss.2022.102848.

Bastawy, M., Ebeed, M., Ali, A., Shaaban, M. F., Khan, B., & Kamel, S. (2022). Optimal day-ahead scheduling in micro-grid with renewable based DGs and smart charging station of EVs using an enhanced manta-ray foraging optimisation. *IET Renewable Power Generation*, *16*(11). Available from https://doi.org/10.1049/rpg2.12531, https://ietresearch.onlinelibrary.wiley.com/journal/17521424.

Berg, K., Hernandez-Matheus, A., Aragüés-Peñalba, M., Bullich-Massagué, E., & Farahmand, H. (2024). Load configuration impact on energy community and distribution grid: Quantifying costs, emissions and grid exchange. *Applied Energy*, *363*, 123060. Available from https://doi.org/10.1016/j.apenergy.2024.123060.

Berg, K., & Löschenbrand, M. (2022). A data set of a Norwegian energy community. *Data in Brief*, *40*, 107683. Available from https://doi.org/10.1016/j.dib.2021.107683.

Capper, T., Gorbatcheva, A., Mustafa, M. A., Bahloul, M., Schwidtal, J. M., Chitchyan, R., Andoni, M., Robu, V., Montakhabi, M., Scott, I. J., Francis, C., Mbavarira, T., Espana, J. M., & Kiesling, L. (2022). Peer-to-peer, community self-consumption, and transactive energy: A systematic literature review of local energy market models. *Renewable and Sustainable Energy Reviews, 162*, 112403. Available from https://doi.org/10.1016/j.rser.2022.112403.

Chen, W. M., Kim, H., & Yamaguchi, H. (2014). Renewable energy in eastern Asia: Renewable energy policy review and comparative SWOT analysis for promoting renewable energy in Japan, South Korea, and Taiwan. *Energy Policy, 74*(C). Available from https://doi.org/10.1016/j.enpol.2014.08.019, http://www.journals.elsevier.com/energy-policy/.

COM/2016/0860. (2016). Final—Communication from the commission to the European Parliament, the Council, the European Economic and Social Committee, The committee of the regions and the European Investment Bank. Clean energy for all Europeans.

Cuenca, J. J., Jamil, E., & Hayes, B. (2021). State of the art in energy communities and sharing economy concepts in the electricity sector. *IEEE Transactions on Industry Applications, 57*(6). Available from https://doi.org/10.1109/TIA.2021.3114135, https://ieeexplore.ieee.org/servlet/opac?punumber = 28.

European Parliament. (2018). Directive (EU) 2018/2001 of the European Parliament and of the Council of 11 December 2018 on the promotion of the use of energy from renewable sources.

European Parliament (2019). *Directive (EU) 2019/944 of the European Parliament and of the Council of 5 June 2019 on common rules for the internal market for electricity and amending directive.*

Fioriti, D., Frangioni, A., & Poli, D. (2021). Optimal sizing of energy communities with fair revenue sharing and exit clauses: Value, role and business model of aggregators and users. *Applied Energy, 299*, 117328. Available from https://doi.org/10.1016/j.apenergy.2021.117328.

Gianaroli, F., Preziosi, M., Ricci, M., Sdringola, P., Ancona, M. A., & Melino, F. (2024). Exploring the academic landscape of energy communities in Europe: A systematic literature review. *Journal of Cleaner Production, 451*. Available from https://doi.org/10.1016/j.jclepro.2024.141932, https://www.sciencedirect.com/science/journal/09596526.

Gomes, E., Pereira, L., Esteves, A., & Morais, H. (2023). PyECOM: A Python tool for analyzing and simulating energy communities. *SoftwareX, 24*, 101580. Available from https://doi.org/10.1016/j.softx.2023.101580.

Gruber, L., Klatzer, T., & Wogrin, S. (2023). Developing a framework for multiple participation in energy communities. *TechRxiv*. Available from https://doi.org/10.36227/techrxiv.23294795, https://www.techrxiv.org/.

Guerrero, J., Gebbran, D., Mhanna, S., Chapman, A. C., & Verbič, G. (2020). Towards a transactive energy system for integration of distributed energy resources: Home energy management, distributed optimal power flow, and peer-to-peer energy trading. *Renewable and Sustainable Energy Reviews, 132*. Available from https://doi.org/10.1016/j.rser.2020.110000, https://www.journals.elsevier.com/renewable-and-sustainable-energy-reviews.

International Renewable Energy Agency (IRENA). (2020). *Global renewables outlook: Energy transformation 2050*. IRENA.

Iazzolino, G., Sorrentino, N., Menniti, D., Pinnarelli, A., De Carolis, M., & Mendicino, L. (2022). Energy communities and key features emerged from business models review. *Energy Policy, 165*, 112929. Available from https://doi.org/10.1016/j.enpol.2022.112929.

Khorasany, M., Mishra, Y., & Ledwich, G. (2018). Market framework for local energy trading: A review of potential designs and market clearing approaches. *IET Generation, Transmission and Distribution*, *12*(22). Available from https://doi.org/10.1049/iet-gtd.2018.5309, http://www.ietdl.org/IET-GTD.

Lazzari, F., Mor, G., Cipriano, J., Solsona, F., Chemisana, D., & Guericke, D. (2023). Optimizing planning and operation of renewable energy communities with genetic algorithms. *Applied Energy*, *338*, 120906. Available from https://doi.org/10.1016/j.apenergy.2023.120906.

López, I., Goitia-Zabaleta, N., Milo, A., Gómez-Cornejo, J., Aranzabal, I., Gaztañaga, H., & Fernandez, E. (2024). European energy communities: Characteristics, trends, business models and legal framework. *Renewable and Sustainable Energy Reviews*, *197*, 114403. Available from https://doi.org/10.1016/j.rser.2024.114403.

Lu, Y., Khan, Z. A., Alvarez-Alvarado, M. S., Zhang, Y., Huang, Z., & Imran, M. (2020). A critical review of sustainable energy policies for the promotion of renewable energy sources. *Sustainability (Switzerland)*, *12*(12). Available from https://doi.org/10.3390/su12125078, https://res.mdpi.com/d_attachment/sustainability/sustainability-12-05078/article_deploy/sustainability-12-05078-v2.pdf.

Manso-Burgos, Á., Ribó-Pérez, D., Gómez-Navarro, T., & Alcázar-Ortega, M. (2022). Local energy communities modelling and optimisation considering storage, demand configuration and sharing strategies: A case study in Valencia (Spain). *Energy Reports*, *8*. Available from https://doi.org/10.1016/j.egyr.2022.08.181.

Troncia, M. (2024). Regulatory analysis, proposals for efficient flexibility mechanisms and demonstrators' framework.

Naval, N., & Yusta, J. M. (2021). Virtual power plant models and electricity markets—A review. *Renewable and Sustainable Energy Reviews*, *149*, 111393. Available from https://doi.org/10.1016/j.rser.2021.111393.

Nguyen, S., Peng, W., Sokolowski, P., Alahakoon, D., & Yu, X. (2018). Optimizing rooftop photovoltaic distributed generation with battery storage for peer-to-peer energy trading. *Applied Energy*, *228*. Available from https://doi.org/10.1016/j.apenergy.2018.07.042, http://www.elsevier.com/inca/publications/store/4/0/5/8/9/1/index.htt.

Nour, M., Chaves-Avila, J. P., & Sanchez-Miralles, A. (2022). Review of blockchain potential applications in the electricity sector and challenges for large scale adoption. *IEEE Access*, *10*. Available from https://doi.org/10.1109/ACCESS.2022.3171227, http://ieeexplore.ieee.org/xpl/RecentIssue.jsp?punumber = 6287639.

Nour, M., Chaves-Avila, J. P., Troncia, M., Ali, A., & Sanchez-Miralles, A. (2023). Impacts of community energy trading on low voltage distribution networks. *IEEE Access*, *11*. Available from https://doi.org/10.1109/ACCESS.2023.3278090, http://ieeexplore.ieee.org/xpl/RecentIssue.jsp?punumber = 6287639.

Nour, M., Chaves-Avila, J. P., Troncia, M., & Sanchez-Miralles, A. (2024). Mitigating the impacts of community energy trading on distribution networks by considering contracted power network charges. *IEEE Access*, *12*. Available from https://doi.org/10.1109/ACCESS.2024.3365630, http://ieeexplore.ieee.org/xpl/RecentIssue.jsp?punumber = 6287639.

Nour, M., Magdy, G., Bakeer, A., Telba, A. A., Beroual, A., Khaled, U., & Ali, H. (2023). A new fractional-order virtual inertia support based on battery energy storage for enhancing microgrid frequency stability. *Fractal and Fractional*, *7*(12). Available from https://doi.org/10.3390/fractalfract7120855, http://www.mdpi.com/journal/fractalfract.

Nour, M., Magdy, G., Chaves-Ávila, J. P., Sánchez-Miralles, Á., & Jurado, F. (2023). A new two-stage controller design for frequency regulation of low-inertia power system with virtual synchronous generator. *Journal of Energy Storage*, *62*. Available from https://

doi.org/10.1016/j.est.2023.106952, http://www.journals.elsevier.com/journal-of-energy-storage/.

Nour, M., Troncia, M., Chaves-Avila, J. P., Sanchez-Miralles, A. Mitigating the impacts of community energy trading on low voltage distribution networks by considering contracted power cost. (2023), IEEE.

Parag, Y., & Sovacool, B. K. (2016). Electricity market design for the prosumer era. *Nature Energy*, *1*(4). Available from https://doi.org/10.1038/NENERGY.2016.32, http://www.nature.com/nenergy/.

Queiroz, H., Lopes, R. A., Martins, J., Silva, F. N., Fialho, L., & Bilo, N. (2023). Assessment of energy sharing coefficients under the new Portuguese renewable energy communities regulation. *Heliyon*, *9*(10). Available from https://doi.org/10.1016/j.heliyon.2023.e20599, http://www.journals.elsevier.com/heliyon/.

Regulation (EU) 2018/1999 of the European Parliament and of the Council of 11 December 2018 on the Governance of the Energy Union and Climate Action, amending Regulations (EC) No 663/2009 and (EC) No 715/2009 of the European Parliament and of the Council.

REE, Red Electrica de España (2024). System Operator Information SystemAvailable from https://www.esios.ree.es/.

Saif, A., Khadem, S. K., Conlon, M., & Norton, B. (2022). Hosting a community-based local electricity market in a residential network. *IET Energy Systems Integration*, *4*(4). Available from https://doi.org/10.1049/esi2.12062, https://ietresearch.onlinelibrary.wiley.com/journal/25168401.

Saif, A., Khadem, S. K., Conlon, M. F., & Norton, B. (2023). Impact of distributed energy resources in smart homes and community-based electricity market. *IEEE Transactions on Industry Applications*, *59*(1). Available from https://doi.org/10.1109/TIA.2022.3202756, https://ieeexplore.ieee.org/servlet/opac?punumber = 28.

Scarabaggio, P., Carli, R., Jantzen, J., & Dotoli, M. (2021). Stochastic model predictive control of community energy storage under high renewable penetration. In *29th Mediterranean conference on control and automation, MED 2021* (pp. 973–978). Italy: Institute of Electrical and Electronics Engineers Inc. http://ieeexplore.ieee.org/xpl/mostRecentIssue.jsp?punumber = 9480136, https://doi.org/10.1109/MED51440.2021.94803539781665422581.

Schreuer, A., & Weismeier-Sammer, D. (2010). *Energy cooperatives and local ownership in the field of renewable energy technologies: A literature review*. Vienna: WU Vienna University of Economics and Business.

Vecchi, F., Stasi, R., & Berardi, U. (2024). Modelling tools for the assessment of renewable energy communities. *Energy Reports*, *11*. Available from https://doi.org/10.1016/j.egyr.2024.03.048, https://www.sciencedirect.com/science/journal/23524847.

Velosa, N., Gomes, E., Morais, H., & Pereira, L. (2023). PROCSIM: An open-source simulator to generate energy community power demand and generation scenarios. *Energies*, *16* (4), 1611. Available from https://doi.org/10.3390/en16041611.

Walker, G., & Devine-Wright, P. (2008). Community renewable energy: What should it mean? *Energy Policy*, *36*(2). Available from https://doi.org/10.1016/j.enpol.2007.10.019.

Energy management and power flow in microgrids

15

Mohammad Bayat[1,2], *Ali Asghar Ghadimi*[1,2], *Abouzar Samimi*[3] *and Sahar Moslemi*[4]

[1]Department of Electrical Engineering, Institute of Advanced Technology, Arak University, Arak, Iran, [2]Department of Electrical Engineering, Faculty of Engineering, Arak University, Arak, Iran, [3]Department of Electrical Engineering, Arak University of Technology, Arak, Iran, [4]Department of Electrical Engineering, Science and Research Branch, Islamic Azad University, Tehran, Iran

Introduction

The presence of various distributed generation (DG) units and different loads with stochastic nature in microgrids (MGs) necessitates additional control algorithms, that is, energy management systems (EMS) for optimal power flow (PF) between components of the MG to achieve stable and economic operation. An EMS is essential in DG-based systems with more than one generation unit or storage device for setting the operating point of each unit in the MG to coordinate energy flow between the various units for efficient, reliable, and economic operation, resulting in demand–supply balance (He et al., 2019; Khan et al., 2019; Zhou et al., 2016). EMS manages in real time the energy flow within the MG components as well as between the MG and main grid based on the previously set strategies (Leonori et al., 2020). Energy management (EM) should first ensure that the load is met and thereafter maximize efficiency and minimize operation cost (Yu et al., 2019). The EMS of MGs in the literature consider single MGs, in islanded and grid-tied modes in addition to multi-MGs in which EM is considered for an MG community. The overall different strategies used in EMS are studied and compared in this chapter. The basic objective for comparing the different strategies for EM is to investigate the pros and cons of each method when applied to different configurations of MG. This will help to decide the strategy to be adopted for any system, depending on the system configuration and the amount of data exchange. Fig. 15.1 demonstrates a typical classification of EMS widely used in the literature. Several important issues need to be understood in EM, specifically, classification based on various aspects, classification of loads adopted, and demand response programs and comparison of optimization techniques used in EM, as described in the following subsection

Moreover, PF analysis, as the elementary tool in power systems and MGs, plays a significant role for other important assessments such as control, protection, design, planning, and specially EM among others. Unlike the conventional power systems, in MGs, particularly in islanded mode, due to the limit and nonrobust generation source

Towards Future Smart Power Systems with High Penetration of Renewables. DOI: https://doi.org/10.1016/B978-0-443-29871-4.00015-4

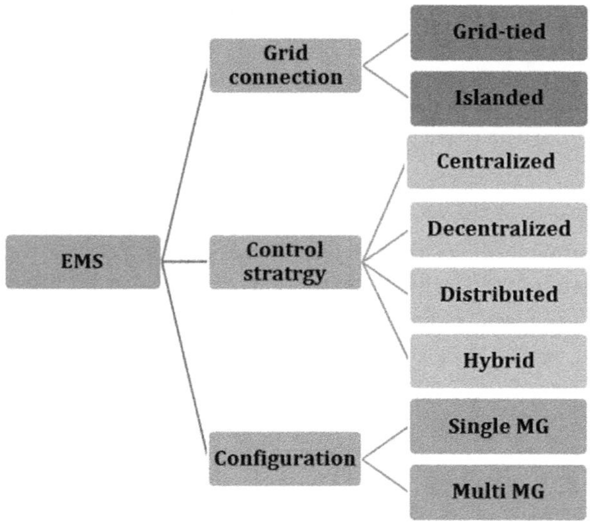

Figure 15.1 Classification of EMS. *EMS*, Energy management systems.
Source: From Bayat, M., Koushki, M. M., Ghadimi, A. A., Tostado-Véliz, M., & Jurado, F. (2022). Comprehensive enhanced Newton Raphson approach for power flow analysis in droop-controlled islanded AC microgrids. *International Journal of Electrical Power & Energy Systems, 143*, 108493.

as a slack bus (SB) to keep the voltage and frequency of the network at the rated values, the corresponding buses are not categorized into traditional ones. In the grid-connected mode, the main grid plays the role of the SB ensuring that the system frequency and its local bus voltages are constant. However, in the islanded mode, the voltage and the frequency of the MG are affected by any generation and consumption imbalances of real and reactive powers. Therefore, it is crucial to implement an appropriate scheme to adopt a proper PF calculation technique in MGs in order to achieve an optimal EM. The subsection 15.4 goes into more elaborate details about the mathematical formulation of the proposed PF algorithm.

Energy management systems and power flow in microgrids

Importance of energy management in microgrids

EM is essential in MGs for the efficient and smooth flow of energy among the various distributed energy resources (DERs) and storage devices. The importance of EM in MGs operation can be highlighted in several key aspects:

- Reliable operation
- Economic operation

- Resilient operation
- Maximum integration of renewable energy sources (RESs)
- Environmentally friendly operation

The intermittency of renewable energy resources calls for sufficient energy storage and multiple energy resources to ensure secure and reliable power supply, leading to high costs. EM is essential in MGs with combinations of renewable energy resources, dispatchable sources, storage systems, and loads to ensure optimal PF between the individual units for the system to work with maximum reliability and minimum cost.

Structures and classification of energy management systems in microgrids

The EMS in MGs can be classified from different viewpoints such as the connection with the main grid, configuration, control strategies, solving technique, and so forth. In the following, a brief explanation about some important issues of these viewpoints is provided (Allwyn et al., 2023).

Grid connection status and configuration of energy management systems in microgrids

MGs can operate in islanded mode either when there is an interruption of supply from the main grid or in locations where extension of the main grid is practically impossible due to distance or geographic constraints. Thus, based on grid connection status, EM studies can be classified as **grid-tied systems**, which can obtain energy from the grid in case of shortage of energy supplied from DERs and **off-grid or isolated systems**, which solely depend on the energy from DERs.

Control strategies of energy management systems

The control strategies used in EMS of MGs are broadly classified into three categories as **centralized**, **decentralized**, and **distributed** as well as **hierarchical** structures. In centralized control, a central control (CC) unit sends the command signals for control and protection to local control units and breakers, which are dedicated to individual units. The CC unit incorporates algorithms based on requirement to find the optimal operating point considering several factors such as cost, emission, grid availability, reliability, resilience, and so forth. A central controller with the capability of bulk data handling is responsible for decision making based on data availability attained by communication between the controller and the MG components (Olivares et al., 2014). Having said that, in interconnected power systems extending to a wide area, it is difficult to implement a centralized approach due to the requirement of vast communication and computation needs. In contrast, in decentralized control architecture, several subsystems are controlled by themselves, but they share information. However, the decentralized approach requires a high level of coordination, which becomes problematic due to the availability of only local data considering extended MGs. Hence, a hierarchical approach, instead of using a fully centralized or fully decentralized approach, is recommended in such systems. Generally, this structure consists of two or three levels: the primary, secondary, and tertiary controls. Control of power electronic converters, islanding detection, droop control, and so forth come under primary control. EMS come under secondary control to ensure reliable, secure, and economical operation in MGs, and tertiary control coordinates operation of multiple MGs (Olivares et al., 2014). Hierarchical EMS combine the advantages of centralized and decentralized control in terms of being flexible compared to the centralized approach and reduced operation cost compared to the decentralized approach (Wang et al., 2018). Distributed control requires a minimal communication network and essential information from other units, with the advantages of privacy protection, lower computational effort, and ease of extension (Shi et al., 2015; Xing et al., 2019).

The role of demand response programs in the energy management systems in microgrids

According to the U.S Department of Energy, demand response programs (DRPs) are a tariff set up to encourage pattern change in electricity consumption either responding to varied electricity prices over time or to incentives offered to reduce electricity consumption during periods of high market price or when grid reliability is at risk (QDR, 2006). It is a technique used for EM to increase benefits to both suppliers and consumers by adjusting the load curve using flexible and interruptible loads (Yuan et al., 2019). DRPs are of two types: **incentive-based** (Shahryari et al., 2019) and **price-based**. Incentive-based programs are further classified as interruptible/curtailable and capacity market programs (Mansour-lakouraj & Shahabi, 2019). The incentive-based programs encourage customers to participate in MG scheduling by offering financial benefits, and, in the interruptible/curtailable program, customers reduce consumption for incentive benefits. In the price-based scheme, the customer curtails demand in response to increasing market prices (Chaudhary & Rizwan, 2018). Time of use (TOU), critical peak pricing (CPP) and real-time pricing (RTP) are the types of price response in DRPs where the electricity price is previously determined in TOU and CPP but changes hourly in RTP (Zhao et al., 2013). In the TOU type of demand response, MGs can participate in DRPs by reducing the peak-load and flattening the load curve by shifting a part of the load from peak period when energy cost is high to off-peak period with lower energy price resulting in reduced operation cost (Mehdizadeh et al., 2018). The total load remains the same, but the operation time is shifted to reduce peak-load.

Loads in MGs are broadly classified as static (fixed) and flexible loads. Static loads are nonmanageable and the power demand should be met at any point of time, for example, hospital ICU and data centers. Flexible (responsive/adjustable/time-shiftable/nonvital) loads can assist in DRPs by responding to price signals (Karimi & Jadid, 2020) and are classified as curtailable and shiftable (deferred) loads, which can respond to economic incentives. In curtailable loads, the ability to shift the usage time is limited, but can be traded off with satisfaction of end-user based on DRP (Jin et al., 2017). In shiftable loads, the operation time can be adjusted, even though it consumes a certain amount of energy during the scheduled period (Jiang et al., 2019). Shiftable loads can be classified as interruptible (elastic) loads and noninterruptible loads. In elastic loads, operational time and power rating can be controlled by end-users depending on the required comfort level, and noninterruptible loads are flexible in terms of operational time, but cannot be interrupted during operation (Rahim et al., 2019).

Other schemes related to energy management systems in microgrids

A few other classifications are based on the signal type, solving platform, and other factors which are mentioned below. Based on the type of power, MGs are classified as AC and DC MGs. Platforms used in solving EM problems are of two types: the offline and online (Wu & Wang, 2018) approach. The majority of research work on EM is based on an offline approach due to the complexity and high cost involved in the online approach. Offline methods are mostly centralized and are incapable of addressing the real-time changes in the system, which can be overcome using online methods (Dong & Chen, 2019). Online methods can deal with real-time changes, leading to a more efficient system, but are expensive compared to offline methods. According to the method of parameter consideration, EMS are classified as deterministic and stochastic methods in which the renewable generations, load, and electricity price are considered as known parameters with forecasted parameters being considered to be perfect in the deterministic approach, whereas uncertainties in forecasting are considered in the stochastic approach (Zhang et al., 2018). Simulation techniques used in EM are of two categories: real-time simulation and non-real-time simulation. In real-time simulation, the simulator executes the event in exactly the same time as the actual event occurs. Non-real-time simulation is the traditional technique in which the behavior of the system is calculated either slower or faster than the actual occurrence of the event, depending on system complexity.

The factors to be addressed for EM studies in islanded MGs are reliability, stability, cost, and emissions. While most of the related works consider a probabilistic or prediction-based approach, more studies considering the dynamic fluctuations of load and RERs are essential, as this replicates the case of practical MGs. Storage devices play an important role in islanded MGs, and the commonly used storage device is the battery, while supercapacitors and other storage devices are considered in very few cases. Degradation of the battery and fuel cell are factors needed to be addressed in storage devices. The economic analysis of considering the degradation of storage devices has not been presented in any of the works. This is essential as consideration of degradation in storage devices may result in increased energy cost due to reduced usage of storage. Hence, an economic analysis is vital in this field. Economic feasibility of adopting diverse storage techniques such as P2H storage, supercapacitor, and pumped hydro storage is never presented in literature. Resilience is one important factor that is least considered in any of the existing literature related to EM in islanded MGs. More attention is required in this matter, as the system should be able to restore to normal in case of adverse events.

Problem models of energy management systems in microgrids

The deployment of more than one energy source in MG systems requires the use of efficient approaches for managing energy flow. This requires the development and deployment of EMS. EMS should be able to effectively coordinate energy sharing and trading among all electrical networks while supplying loads according to the operational conditions and economic constraints with secure, reliable, and efficient power system operation. In fact, optimization techniques, demand-side management, PF tools, and power quality management are needed to achieve different EMS objectives while satisfying multiple constraints. Several approaches have been proposed and have used diversified objective functions and constraints together with optimization methods for efficient EM, as depicted in Fig. 15.2.

The deployment of EM control strategies specifies the main objective functions, which could be related to the operation cost, pollution, reliability, and power quality (Cannata et al., 2019; Hannan et al., 2020; Jiménez-Fernández et al., 2019; Li et al., 2017). For instance, the main aim of using economic objective functions is to minimize the electricity price. Different formulations have been studied for cost minimization in MGs.

The deployment of EM approaches, which consider the pollution factor as an objective function, take time to validate, since the whole procedure should consider

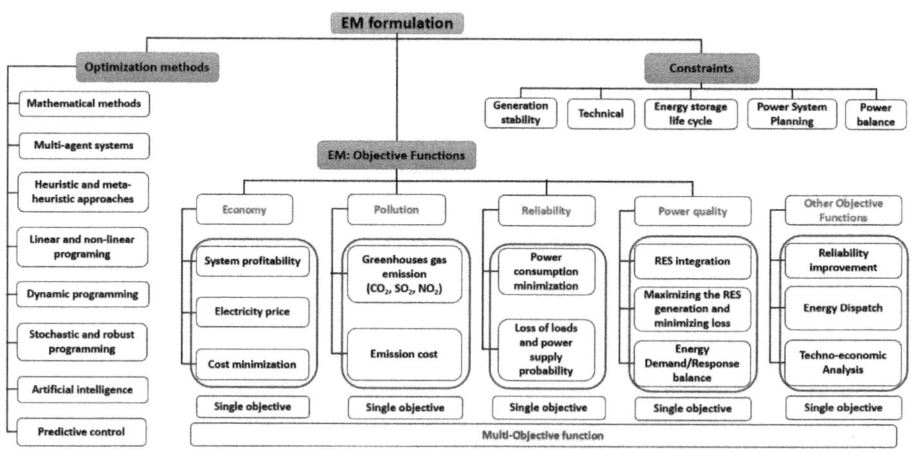

Figure 15.2 Energy management control structure. Objective functions, constraints, and optimization methods for EMS of MGs. EMS, Energy management systems; *MGs*, microgrids.

Source: From Elmouatamid, A., Ouladsine, R., Bakhouya, M., El Kamoun, N., Khaidar, M., & Zine-Dine, K. (2021). Review of control and energy management approaches in micro-grid systems. *Energies, 14*(1). https://doi.org/10.3390/en14010168.

the life cycle of the different deployed equipment. In fact, every new energy source technology which is promoted as being "renewable" or "sustainable" is subject to an energy balance analysis in order to calculate the net energy yield. The energy analysis considers not only the data for present generation systems, but also the data for the probable improvements in production and energy system technology. The equivalent CO_2, generated during the fabrication of each component, should be calculated and compared to the equivalent energy which is generated during its life cycle. We consider that this energy is generated by traditional sources in order to estimate the equivalent CO_2 emission and that, by comparing these two elements of CO_2 generation, the profitability of the system concerning the pollution objective can be defined (Elmouatamid et al., 2021).

Like economic and pollution aspects, the term "reliability" covers different aspects concerning the system operation cost, profitability, fails and maintenance, and productivity. Consequently, as mentioned above, RESs have a significant cost and consume a lot of energy in their fabrication. In order to maximize the profitability and system's reliability, the production of these sources should be maximized. Therefore, the main aim is to maximize the use of renewable energy generation, minimizing the loss of energy, keeping the storage energy system at a good SOH, and ensuring a safety and efficient supply of energy to the loads. However, the power quality, particularly the power loss, is still a main issue for the system's reliability. Therefore several works have proposed suitable EM methods and control techniques to minimize the power loss in MG systems (Elmouatamid et al., 2021).

Optimization methods for energy management systems

After defining the system's constraints and objective functions, suitable optimization methods are required to accordingly ensure the exchange of PF between the installed RES/storage and the MGs on the one hand and between MGs and the utility grid on the other hand. The rest of this subsection is dedicated to an overview of main methods from literature.

Optimization techniques used in MGs are deterministic, analytical, heuristic, stochastic, and hybrid-based models. Linear programming (LP), non-LP, dynamic programming, mixed integer LP (MILP), greedy algorithms, Lagrange relaxation, interior point method, and dynamic programming are methods used under the deterministic approach. GAs, artificial bee colony algorithm, ant lion optimizer, particle swarm optimization (PSO), gray wolf optimization, cuckoo search, mine blast algorithm, the whale optimization algorithm, the moth swarm algorithm, the harmony search, electromagnetic field optimization, and intelligent flower pollination algorithm all come under the heuristic/metaheuristic approach. Metaheuristic algorithms are based on natural processes that reach the optimum value by continuous iterations (Shaheen et al., 2021). Some other metaheuristic algorithms seen in the literature are tree–seed algorithm (El-Fergany & Hasanien, 2018), sine–cosine algorithm (Attia et al., 2018), salp-swarm algorithm (El-Fergany & Hasanien, 2020), sunflower optimization (Hussien et al., 2021), cuttlefish optimization (Hussien et al., 2020), mean variance mapping optimization (Erlich et al., 2014), and Hunger Games search algorithm iterations (Shaheen et al., 2021). Optimization problems in power in optimal PF and EM problems are solved using variations or modified versions of the above-mentioned algorithms (Marcelino et al., 2021; Shaheen et al., 2021). A hybrid system (Kim et al., 2020) uses a combination of any of the two methods, but it has the drawback of longer computation time (Mellouk et al., 2019), which can be rectified by parallelism. The pros and cons of widely used optimization techniques/algorithms are presented in Table 15.1. In problems with several mutually opposing objective functions, multiobjective optimization yields a set of optimal values instead of a single-optima while satisfying a number of equality and inequality constraints (Aghajani & Ghadimi, 2018). The rolling time horizon strategy is a combination of any optimization algorithm mentioned above along with prediction algorithms to incorporate the uncertainties involved in DERs.

Navigating power flow dynamics in islanded microgrids

In this section, our focus is on MGs in the AC Island mode, particularly in the area of medium voltage. Hence, we consider a balanced three-phase configuration. Subsequently, we will present various methods and approaches to address the navigation of PF dynamics in islanded MGs. PF dynamics in islanded MGs is a critical task that demands careful attention. In an islanded MG, where the power system operates independently from the main grid, understanding and managing PF becomes crucial for ensuring stability and optimizing resource utilization. In the following first, the modeling of components of MG including DERs, loads, and network is presented. Then PF formulation is presented and finally different solution methods are presented.

Islanded microgrid modeling for power flow analysis

The PF analysis is the foundation for studying a power grid's steady-state characteristics. This study needs an appropriate model of the system components, which will be discussed in this subsection for islanded microgrids (IMGs) (Bayat et al., 2022).

Droop-controlled distributed energy resources model

The DER refers to both single DGs and those equipped with energy storage systems (ESSs). In the grid-connected mode of MG operation, the small size DERs are usually controlled as *PV* or *PQ* buses, while in the islanded mode, the DERs are operated as voltage source to control the MG voltage and frequency and to achieve appropriate power sharing of the load demands (Bayat et al., 2015).

Generally, two DER models for dispatchable generating units, including synchronous machine-based DER (SM-DER) and electronically interfaced DER (EI-DER), are used.

Table 15.1 Comparison of widely used optimization algorithms (Allwyn et al., 2023).

Algorithm	Strength	Weakness
Linear programming (LP)	Computational reliability and speed	Most of the real-world problems are nonlinear, which can result in substantial losses when it is developed as linear model.
Mixed integer linear programming (MILP)	Guaranteed convergence of the solution with global optimum values. Fast and effective solvers available commercially.	Optimum solution cannot be obtained if the constraints are contradictory and nonlinear effects cannot be accounted.
Nonlinear programming	Results are more reliable as the model exactly replicates the nonlinear characteristics of the system under study.	Optimal results cannot be obtained if the functions used in computation are not differentiable. Requirement for feasible starting point to obtain global optima. Complicated algorithms and not feasible for postoptimal sensitivity analysis.
Dynamic programming	The sensitivity analysis can be inferred from problem solution, formulation can be done based on initial value problems.	Increase in the number of variables increases exponentially with the increasing number of functions requiring more memory space.
Heuristic approach	Optimum solution obtained with reasonable computation time and memory space while preserving the nonlinear characteristics, efficient and upgradable.	Some hidden solution may result in local optima instead of the global optima.
Greedy algorithm	Assured solution with high speed and ease of use. Results are independent of the tolerance which is the only adjustable factor.	Only suboptimal values are obtained and objective functions are not considered directly.
Lagrange relaxation	Can handle general constraints for both sequencing and routing problems. More suitable bounds evaluated in short computational period.	Optimality is not guaranteed and identical solutions can cause oscillations.
Interior point method	Better convergence characteristics besides being less sensitive to initial values, parameter tuning is not necessary.	Even after reaching the maximum number of iterations, the solution may still be far from optimal value.

(*Continued*)

Table 15.1 (Continued)

Algorithm	Strength	Weakness
Genetic algorithm (GAs)	Robust with globally optimum results and applicable to multimodal and large-scale optimization problems.	Substantial number of fitness functions are to be evaluated and can result in premature convergence.
Artificial bee colony algorithm (ABCA)	Applicable to a wide range of functions, flexible, low risk of premature convergence, and results are global optimum values. Fewer control parameters required, great convergence, robust	Higher cost of computation and requires more time and memory space. Often fails to find global optima, premature convergence.
Particle swarm optimization	Simple in coding, less sensitive to the nature of objective function and initial points, short computation period.	Not feasible for real-time applications and difficult to find the optimal design parameters.
Ant lion optimizer	Scalable, flexible, easy, good balance maintained between exploration and exploitation, superior performance.	Slow and premature convergence.

Source: From Allwyn, R.G., Al-Hinai, A., & Margaret, V. (2023). A comprehensive review on energy management strategy of microgrids. Energy Reports, 9, 5565–5591. https://doi.org/10.1016/j.egyr.2023.04.360.

SM-DER adjusts its real power by the static response of the prime mover, that is,

$$P_{Gi} = P_{Goi} + \Delta P_{Gi} \tag{15.1}$$

$$\Delta P_{Gi} = -\frac{P_{Gi\_\max}}{R_i} \tag{15.2}$$

where P_{Gi} and P_{Goi} are the injected real power and the reference set point power of the ith DG, respectively, while R_i denotes the speed droop of the corresponding DG. This is while the terminal voltage magnitude is set to the specified value V. Constant voltage operation is possible only when Q_{Gi} is between their limits. If the reactive power exceeds the limits, the generating bus is considered as a PQ bus with reactive power specified to the Q limit that has been violated.

A growing trend toward EI-DERs is evident in the power supply of MGs, in which the droop equation is distinguished based on the output impedance of the DER. If the imaginary part of the output impedance of a DER is greater than its real part, the active and reactive powers are related to frequency and voltage, respectively. This droop type is called the conventional droop or inductive droop. Some distribution networks like cable type or low voltage networks are resistive. Therefore, the reactive and active powers are related to frequency and voltage, respectively. In this case, the droop type is called inverse droop or resistive droop. If they could not get over each other, the droop type is considered a mixed type, and this condition is known as the complex output impedance of the DER. In this situation, the active and reactive powers are related to both frequency and voltage. The possible combinations are categorized by two binary variables, x and y (Kumar et al., 2019). Considering this, for the DER unit located in bus i,

$$P_{Gi} = P_{0i} + G_{pfi}(f - f_0) + G_{pvi}(V_i - V_{0i}) \tag{15.3}$$

$$Q_{Gi} = Q_{0i} + G_{qfi}(f - f_0) + G_{qvi}(V_i - V_{0i}) \tag{15.4}$$

where

$$\begin{aligned} G_{pfi} &= -\frac{1}{m_i}\left(\frac{x_i}{x_i + y_i}\right), G_{qfi} = \frac{1}{m_i}\left(\frac{y_i}{x_i + y_i}\right) \\ G_{pvi} &= -\frac{1}{n_i}\left(\frac{y_i}{x_i + y_i}\right), G_{qvi} = -\frac{1}{n_i}\left(\frac{x_i}{x_i + y_i}\right) \end{aligned} \tag{15.5}$$

In these equations, P_{0i}, Q_{0i}, f_0, and V_{0i} are set points of active power, reactive power, system frequency, and voltage magnitude of bus i, respectively. f and V_i are system frequency and voltage magnitude of bus i. m_i and n_i are the frequency droop coefficient and the voltage magnitude droop coefficient, respectively. Different combinations of x and y in Eq. (15.3) result in different types of droop formulations. Inductive (conventional) droop is achieved when $x = 1$ and $y = 0$, resistive (inverse) when $x = 0$ and $y = 1$, and complex when $x = y = 1$.

Load model

The demanded active and reactive power of loads in the ith bus (i.e., P_{Li} and Q_{Li}) can be represented as in (Bayat et al., 2021),

$$P_{Li} = P_{Lni}\left(1 + L_{pfi}(f - f_n)\right)\left(\frac{V_i}{V_{ni}}\right)^{L_{pvi}} \tag{15.6}$$

$$Q_{Li} = Q_{Lni}\left(1 + L_{qfi}(f - f_n)\right)\left(\frac{V_i}{V_{ni}}\right)^{L_{qvi}} \tag{15.7}$$

where P_{Lni} and Q_{Lni} are active and reactive power demands of loads, respectively, in the nominal operating condition of f_n and V_{ni}. The parameters L_{pvi} and L_{qvi} in the exponential model represent the voltage dependency of P_{Li} and Q_{Li}. Constant power (CP), constant current, and constant impedance (CZ) load models can be expressed by selecting the L_{pvi} and L_{qvi} as 0, 1, and 2, respectively. Moreover, L_{pfi} and L_{qfi} denote the coefficients of frequency dependence of loads.

Network model

Branches, including feeders and transformers, are expressed by the π model in PF studies. Using this model, the bus admittance matrix $\overline{\boldsymbol{Y}}_{bus}$ can be easily calculated. The elements of this matrix include real part $\boldsymbol{G}$ and imaginary part $\boldsymbol{B}$. Since the system frequency is variable in islanded MGs, these parts, which depend on the frequency, are not considered constant, unlike conventional methods. Considering that balanced networks are studied in this paper, only the positive sequence of impedance and admittance of branches are used.

The branch between bus i and bus k of the system has a series admittance element in the π model that consists of r_{ik} and x_{ik}. In addition, $b_{c,ik}$ denotes the shunt admittances resulting from the capacitance of the line between buses i and k. The $\overline{\boldsymbol{Y}}_{bus}$ matrix is a well-known representation of a network in PF studies, in which its elements are associated with these parameters as follows:

$$\begin{aligned} \overline{y}_{ik} &= \frac{1}{r_{ik} + jx_{ik}} = g_{ik} + jb_{ik} \\ g_{ik} &= \frac{r_{ik}}{r_{ik}^2 + x_{ik}^2}, b_{ik} = -\frac{x_{ik}}{r_{ik}^2 + x_{ik}^2} \end{aligned} \tag{15.8}$$

and the shunt elements like compensators are:

$$\overline{y}_{i0} = g_{i0} + jb_{i0}, \overline{y}_{k0} = g_{k0} + jb_{k0} \tag{15.9}$$

Based on these parameters, The affected elements of $\overline{\boldsymbol{Y}}_{bus}$ matrix are:

$$\begin{aligned} \overline{Y}_{ii} &= G_{ii} + jB_{ii} = \left(g_{i0} + \sum\nolimits_{k\in u} g_{ik}\right) + j\left(b_{i0} + \sum\nolimits_{k\in u}\left(b_{ik} + \frac{b_{c,ik}}{2}\right)\right) \\ \overline{Y}_{ik} &= \overline{Y}_{ki} = G_{ik} + jB_{ik} = -g_{ik} - jb_{ik} \\ \overline{Y}_{kk} &= G_{kk} + jB_{kk} = \left(g_{k0} + \sum\nolimits_{i\in w} g_{ki}\right) + j\left(b_{k0} + \sum\nolimits_{i\in w}\left(b_{ki} + \frac{b_{c,ki}}{2}\right)\right) \end{aligned} \tag{15.10}$$

where u and w are the set buses connected to bus i and bus k, respectively.

Power flow solution method for droop-controlled IMGs

In the following, first, a brief description of PF problem and its conventional solution methods (i.e., in grid-connected MGs) is provided and then different solution methods for PF analysis in DCIMGs s will be investigated.

Power flow problem and its conventional solution method

Applying Kirchhoff's current law to each bus of the system results in a set of algebraic nonlinear equations known as PF equations or power balance equations, which must be solved by iterative methods (Saadat, 1999). These PF equations for each bus i are as,

$$P_{Ci} = V_i \sum_{k=1}^{N} V_k[G_{ik}\cos(\delta_i - \delta_k) + B_{ik}\sin(\delta_i - \delta_k)] \tag{15.11}$$

$$Q_{Ci} = V_i \sum_{k=1}^{N} V_k[G_{ik}\sin(\delta_i - \delta_k) - B_{ik}\cos(\delta_i - \delta_k)] \tag{15.12}$$

where P_{Ci} and Q_{Ci} are respectively the calculated real and reactive powers supplied by bus i. Besides, the scheduled real and reactive powers at bus i are:

$$P_i = P_{Gi} - P_{Li} \tag{15.13}$$

$$Q_i = Q_{Gi} - Q_{Li} \tag{15.14}$$

The calculated powers Eqs. (15.11) and (15.12) are respectively compared to the scheduled values Eqs. (15.13) and (15.14) to obtain the mismatch matrix, which yields ΔP and ΔQ.

$$\Delta P_i(P_{Gi}, P_{Li}, V, \delta, G, B) = P_{Ci} - (P_{Gi} - P_{Li}) \tag{15.15}$$

$$\Delta Q_i(Q_{Gi}, Q_{Li}, V, \delta, G, B) = Q_{Ci} - (Q_{Gi} - Q_{Li}) \tag{15.16}$$

where ΔP_i and ΔQ_i are, respectively, the active and reactive power mismatches at bus i that must be zero at each steady-state equilibrium point. In bus i, voltage magnitude is expressed as Vi, and δi is the voltage angle. The real part of *ith* row and *jth* column element in $\overline{\boldsymbol{Y}}_{bus}$ is Gij and imaginary part of it is Bij, and the number of buses is exhibited N. The last terms in Eqs. (15.15) and (15.16), Pi and Qi, represent the net active and reactive powers injected to bus i, respectively. It is noteworthy that Eqs. (15.15) and (15.16) are expressed in the polar form, and representing the voltage of each bus i instead of Vi and δi in Cartesian form results in another form of these equations. Also, these equations can be written for the current balance instead of the power balance (Zimmerman & Murillo-Sánchez, 2016).

PF equations are comprising of six sets of variables: net injected powers P and Q, V, δ, G, and B. The first four variables are related to system buses, and the remaining are related to system branches that are normally specified and constant. A single generator bus is typically chosen as a SB. Magnitude and phase angle of the SB voltage are specified, and the injected active and reactive powers are unknown. The remaining generator buses are typically classified as PV buses, where the values of injected active power and voltage magnitude are given. Finally, all nongenerator buses, known also as load buses, are classified as *PQ* buses, in which their injected active and reactive powers are determined. So, for each bus, two variables are unknown and a 2*N*-set of equations with 2*N* unknown variables is derived. Active and reactive net injected powers of SB, as well as reactive net injected power of PV buses, are omitted, since they already appear in each one of the equations. The rest of the equations for unknown voltage magnitudes and phase angles which are obviously nonlinear is solved by employing iterative techniques such as Newton Raphson (NR) and Gauss Seidel procedures. Although the SB definition is physically impossible and it is a mathematical artifact of the solution procedure, P_{SB}, Q_{SB}, and reactive power generation level of PV buses must always be checked to ensure that they are within the physical limitations of the correspondence generators. However, the calculated values for the injected powers by SB to the network are usually positive due to technical aspects; mathematically, negativity of either of these values does not cause any problems. In this case, SB instead of generating manner treats as a load to balance generation and consumption in the network.

There are different ways to deal with the PF problem (PFP) in IMGs. In general, these methods can be categorized into two approaches: (I) The loop-based approaches that utilize the continuation PF (CPF) solution methods in an iterative manner. The idea of these approaches is proposed in Kryonidis et al. (2018) (2-loop), Kontis et al. (2019) (3-loop), and Bayat et al. (2021) and Kumar et al. (2019) (1-loop). (II) Methods that, by modifying existing methods, add frequency and voltage magnitude of the reference bus (RB) as extra variables to other variables. A valid and widely cited approach in this regard is the enhanced NR (ENR) approach (Bayat et al., 2015; Mumtaz et al., 2015) and (Bayat et al., 2015) and an efficient method based on higher order modified midpoint (HOMMP).

Utilizing convectional power flow methods for IMG

Despite the PF in grid-connected MG that can be handled using the conventional PF methods, solving the PF problem in droop-controlled IMG (DCIMG) requires a different method. In grid-connected mode, SB stands for providing/absorbing the shortage/excess of active/reactive power for balancing PFs in the system. Since the main grid provides large enough capacity for the MG requirement, the abovementioned limitations (in the previous subsection) are always satisfied, and consequently, the frequency of the system is maintained constant. Thus, the use of the SB concept in IMGs is invalid. Therefore, the unknown variables for the PF problem in IMG are bus voltages and frequency. As a bottom line, the lack of the SB leads to not being constant in the system frequency value. This is the main difference between the PF in IMG and grid-connected MG.

Since the 1-loop method is more efficient than the others (i.e., 2-loop and 3-loop methods), hereunder this method is investigated. In this method, the concept of the virtual SB (VSB) is presented. The voltage angle of the VSB is prespecified (usually is set to be zero) to provide a reference for the rest of the system buses.

However, the voltage magnitude of the VSB, V_{VSB}, and the frequency of the IMG, f_{IMG}, are unknown (Kryonidis et al., 2018). A dummy generator/load (DGL) is embedded in VSB to emulate the interchanging power of the IMG with the upstream main grid (see Fig. 15.3). Moreover, the value of f_{IMG} effects: (i) the network, especially B parameters, (ii) supplied power by droop-controlled DGs, and (iii) consumed power by frequency dependent loads. As explained in previous section (PFP and Its Conventional Solution Method), all these values must also be predefined to perform the PF analysis using CPF methods. With this regard, DERs are modeled as a component with specified active and reactive powers, like PQ buses.

Since there is no real equivalent for DGL, the PFP in the IMG is organized in the form of an optimization problem with the objective of forcing DGL active and reactive powers to zero. V_{VSB} and f_{IMG} are the variables of the problem subjecting CPF equations as an equality constraint. The feasible region for such a problem associated with CPF equations is generically nonconvex (Molzahn & Hiskens, 2019). A method consisting of two nested loops is presented in Kryonidis et al. (2018) to iteratively solve the presented optimization problem. The inner loop adjusts the f_{IMG}, and afterward, the outer loop is used to adjust V_{VSB}, while each loop performs a CPF analysis in each iteration. This method, due to the decoupled adjusting procedure, can basically only handle conventional droop characteristics of DGs. Modeling of voltage dependent droop characteristics Eq. (15.2), by using an

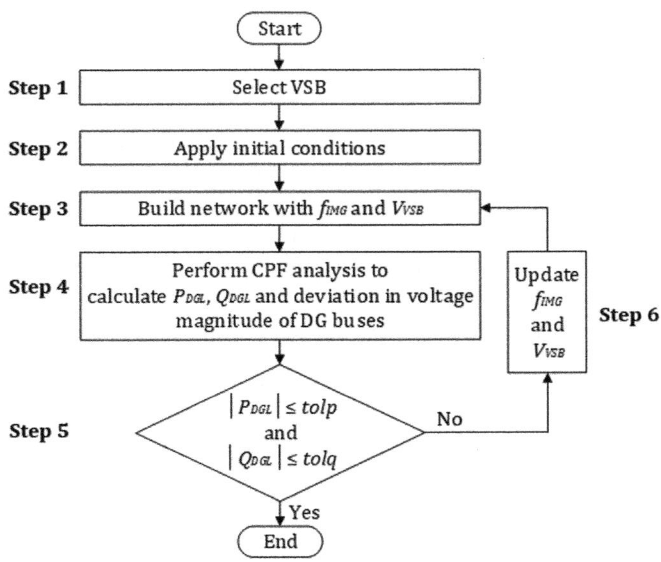

Figure 15.3 Flowchart of the iterative 1-loop method.
Source: From Bayat, M., Ghaseminezhad, K., & Ghadimi, A. A. (2021). An efficient iterative approach for power flow solution of droop-controlled islanded AC microgrids through conventional methods. *International Journal of Electrical Power & Energy Systems, 130*, 106962.

iterative procedure, especially in extended MGs with high penetration of DER units, may lead to divergence result (Kontis et al., 2019). In order to overcome this issue, Kontis et al. (2019) proposed an additional loop by introducing a predefined item called "*integral term*" that is incorporated in the outer loop, exactly after the satisfaction of the frequency control loop. The aim of this added loop, which the maximum iterate of it is set to be almost a large number, is simulating the $Q - V$ droop as much accurate as possible (Kontis et al., 2019). It is important to note that the conventional PF analysis in the outer loop of Kryonidis et al. (2018) is iteratively performed in this loop. As stated in Kontis et al. (2019), the aim of the aforementioned loop may not be met according to its maximum permissible iteration limit. In this case, the whole PF analysis procedure should be repeated using the higher value for the *integral term*.

To overcome these drawbacks, a novel method based on a single iterative loop is proposed in this paper to simultaneously adjust the f_{IMG} and V_{VSB}. The salient advantages of the 1-loop method can be summarized as:

1. All droop characteristics model of DGs can be attended in the DCIMG analysis,
2. The method is computationally efficient, owing to the fact that it calls only one CPF analysis in each loop,
3. Compared to the previous loop-based approaches, no additional parameters required for convergence guarantee in large-scale MGs.

Fig. 15.3 exhibits the flowchart of the proposed method. It is explained as the following steps:

Step 1: The selection of the VSB is a key difference in the proposed method with the methods presented in Kryonidis et al. (2018) and Kontis et al. (2019). In these methods, V_{VSB} is updated through a binary search method, hence each bus in IMG can be randomly selected as VSB, while in the present method, VSB should be stand in buses including droop-controlled DGs. Further details concerning this procedure can be found in step 6.

Step 2: Initial operating conditions are applied to the system. The DGs are adopted with their active and reactive power injections in CPF equations, and these quantities are relevant to the value of f_{IMG}. The voltage magnitude of the correspondence bus and actually droop coefficients, m and n; f_{IMG} and voltage magnitude of DG buses among V_{VSB} could possess reasonable values in the range of associated $P_{\min}/P_{\max}$ and $Q_{\min}/Q_{\max}$. Voltage magnitudes for the other buses as well as phase angle of DG buses are considered as a flat start (i.e., 1 for V and 0 for δ).

Step 3: The iterative procedure starts from this step. The network components affected by f_{IMG} (i.e., B in Y_{bus}, P_L, and Q_L and P_G and Q_G) are calculated at f_{IMG}.

Step 4: CPF analysis is performed to obtain active and reactive power injections by DGL (i.e., P_{DGL} and Q_{DGL}) and voltage magnitude of DG buses. These values are used to update f_{IMG} and V_{VSB}.

Step 5: If the absolute values of P_{DGL} and Q_{DGL} are respectively smaller than the predefined tolerance *tolp* and *tolq*, the entire procedure is terminated.

Step 6: If step 5 was not satisfied, this step would be activated and P_{DGL} and Q_{DGL} must be reallocated to all DGs through the f_{IMG} and V_{VSB} adjustment procedure, so the desired condition of leading DGL to zero for injected power must be prepared. Further details about the proposed adjustment procedure are expressed in the following.

In each iteration k, the sum of the change in injected active and reactive powers of all DG units must be respectively equal to the resulted P_{DGL} and Q_{DGL} at the end of **step 4**, as:

$$\sum_{i=1}^{N_{DG}} \Delta P_{Gi}{}^{k}\left(\Delta f^{k}, \Delta V_{i}{}^{k}\right) = P_{DGL}{}^{k} \tag{15.17}$$

$$\sum_{i=1}^{N_{DG}} \Delta Q_{Gi}{}^{k}\left(\Delta f^{k}, \Delta V_{i}{}^{k}\right) = Q_{DGL}{}^{k} \tag{15.18}$$

where $\Delta f = f_{IMG} - f_0$, and

$$\Delta P_{G}i(\Delta f, \Delta V_i) = pf_i \Delta f + pv_i \Delta V_i \tag{15.19}$$

$$\Delta Q_{G}i(\Delta f, \Delta V_i) = qf_i \Delta f + qv_i \Delta V_i \tag{15.20}$$

where N_{DG} stands for the number of droop-controlled DG units. Frequency dependent parameters of Δ_{PGi} and Δ_{QGi} are respectively expressed as follows,

$$pf_i = -1/m_i(x_i/(x_i + y_i)) \tag{15.21}$$

$$qf_i = 1/m_i(y_i/(x_i + y_i)) \tag{15.22}$$

Similarly, voltage dependent parameters of Δ_{PGi} and Δ_{QGi} are given by Eqs. (15.23) and (15.24).

$$pv_i = -1/n_i(y_i/(x_i + y_i)) \tag{15.23}$$

$$qv_i = -1/n_i(x_i/(x_i + y_i)) \tag{15.24}$$

Upon substituting Eqs. (15.19) and (15.20) in Eqs. (15.17) and (15.18), (15.25) and (15.26) are obtained.

$$\left(\sum_{i=1}^{N_{DG}} pf_i\right) \Delta f^{k} + pv_{VSB} \Delta V_{VSB}{}^{k} + \sum_{i \neq VSB}^{N_{DG}} i = 1(pv_i \Delta V_i{}^{k}) = P_{DGL}{}^{k} \tag{15.25}$$

$$\left(\sum_{i=1}^{N_{DG}} qf_i\right) \Delta f^{k} + qv_{VSB} \Delta V_{VSB}{}^{k} + \sum_{i \neq VSB}^{N_{DG}} i = 1(qv_i \Delta V_i{}^{k}) = Q_{DGL}{}^{k} \tag{15.26}$$

where ΔV_i^k is the variation in voltage magnitude of DG of bus i, in iteration k. Since DGs are modeled with their injected active and reactive powers, DG bus voltages, except for VSB, are resulted from the CPF analysis. Therefore, the last terms in the L.H.S. of Eqs. (15.25) and (15.26) will be known at the end of each iteration, which can be respectively replaced with R and S for simplicity. Hence, two algebraic equations with two unknown variables and are obtained, which can be easily solved as follows,

$$\Delta f^k = \frac{({P_{\mathrm{DGL}}}^k - R)qv_{\mathrm{VSB}} - ({Q_{\mathrm{DGL}}}^k - S)pv_{\mathrm{VSB}}}{\left(\sum_{i=1}^{N_{\mathrm{DG}}} pf_i\right)qv_{\mathrm{VSB}} - \left(\sum_{i=1}^{N_{DG}} pf_i\right)pv_{\mathrm{VSB}}} \tag{15.27}$$

$$\Delta {V_{\mathrm{VSB}}}^k = \frac{\left(\sum_{i=1}^{N_{\mathrm{DG}}} pf_i\right)qv_{\mathrm{VSB}} - \left(\sum_{i=1}^{N_{DG}} qf_i\right)({P_{\mathrm{DGL}}}^k - R)}{\left(\sum_{i=1}^{N_{\mathrm{DG}}} pf_i\right)qv_{\mathrm{VSB}} - \left(\sum_{i=1}^{N_{DG}} pf_i\right)pv_{\mathrm{VSB}}} \tag{15.28}$$

It is worth noting that at the beginning of each iteration, the voltage of DG units that are not located in VSB, resulted from the CPF analysis, is affected from the prespecified value for the voltage magnitude of VSB. Particularly, after the calculation of $\Delta {V_{\mathrm{VSB}}}^k$ using Eq. (15.28), a correction factor between 0 and 1 is required to guarantee appropriate power sharing among all DG units, especially for the DG unit, which is located in the VSB ($\mathrm{DG_{VSB}}$). The factor typically corrects the contribution of $\mathrm{DG_{VSB}}$ in power sharing Eq. (15.18), which is obtained by trial and error and Eq. (15.29) causes better convergence speed.

$$CF_{\Delta V_{\mathrm{VSB}}} = \frac{\frac{1}{n_{\mathrm{VSB}}}}{\sum_{i=1}^{N_{\mathrm{DG}}} \frac{1}{n_{\mathrm{VSB}}}} \tag{15.29}$$

Finally, the values obtained for Δf^k and $\Delta {V_{\mathrm{VSB}}}^k$ are used to calculate the relevant updated values in the next iteration as follows,

$${f_{\mathrm{IMG}}}^{k+1} = {f_{\mathrm{IMG}}}^k + \Delta f^k \tag{15.30}$$

$${V_{\mathrm{VSB}}}^{k+1} = {V_{\mathrm{VSB}}}^k + CF_{\Delta V_{\mathrm{VSB}}} \Delta {V_{\mathrm{VSB}}}^k \tag{15.31}$$

It is obvious that during the updating procedure (i.e., step 6), the injected active and reactive powers by all DG units must comply with their permissible operating limits. Hence, the DG units which are constrainedly fixed in their limits do not participate in the updating procedure. Accordingly, N_{DG} in all expressions stands for the total number of participating DG units.

Enhanced Newton Raphson method

The ENR method is a variation of the conventional NR (CNR) method used for solving PF problems in electrical systems. It is an iterative algorithm that aims to converge toward a solution by updating the voltage and current values in each iteration.

The net injected complex power at bus i is

$$\overline{S}_i = P_i + jQ_i = \vec{V}_i \vec{I}_i^{*} = \vec{V}_i \left(\sum_{j=1}^{N} \overline{Y}_{ij} \vec{V}_j \right)^{*} \tag{15.32}$$

In the CNR method, in grid-connected operation mode, the Jacobian matrix is formed. Since the voltage magnitude of the SB is prespecified and the system frequency is constant, the active and reactive mismatch powers for the SB are not considered. Moreover, in the CNR method, $\boldsymbol{\Delta P}$ for PV buses are not considered.

$$\begin{bmatrix} \boldsymbol{\Delta P} \\ \boldsymbol{\Delta Q} \end{bmatrix} = -[\boldsymbol{J}] \begin{bmatrix} \boldsymbol{\Delta \delta} \\ \boldsymbol{\Delta V} \end{bmatrix} \tag{15.33}$$

where

$$\Delta \overline{S}_i(\boldsymbol{\delta}, \boldsymbol{V}) = \overline{S}_{Ci}(\boldsymbol{\delta}, \boldsymbol{V}) - \overline{S}_i = \overline{S}_{Ci}(\boldsymbol{\delta}, \boldsymbol{V}) + \overline{S}_{Li} - \overline{S}_{Gi} \tag{15.34}$$

in which $\overline{S}_i$ denotes complex power at bus i. As stated in Eq. (15.34), in the conventional PF formulation, $\overline{S}_i$ is constant. Therefore, the elements of the Jacobian matrix $\boldsymbol{J}$ are equal to the differentiation of $\overline{S}_{Ci}$ with respect to variables $\boldsymbol{\delta}$ and $\boldsymbol{V}$; in this regard, f is constant. Once the Jacobian matrix is obtained, variables for the $(t+1)$ iteration can be calculated as

$$\begin{bmatrix} \boldsymbol{\delta} \\ \boldsymbol{V} \end{bmatrix}^{(t+1)} = \begin{bmatrix} \boldsymbol{\delta} \\ \boldsymbol{V} \end{bmatrix}^{(t)} - \left[\boldsymbol{J}^{(t)}\right]^{-1} \begin{bmatrix} \boldsymbol{\Delta P} \\ \boldsymbol{\Delta Q} \end{bmatrix}^{(t)} \tag{15.35}$$

However, in IMGs, the scheduled powers P_i and Q_i are not constant, and they change based on the system's operating point, which depends on droop and load parameters. Moreover, in islanded MGs, the concept of using a SB is invalid. Therefore, two unknown variables, f and V_{RB} (voltage magnitude of the RB), are added to the PF problem, which needs two extra equations. In Mumtaz et al. (2015), the active and reactive power equations for the total system losses are considered, while in this work, mismatch equations for the RB are considered to cope with this issue.

$$P_{C(\mathrm{RB})} = V_{\mathrm{RB}} \sum_{k=1}^{N} V_k \left[G_{(\mathrm{RB})k} \cos(\delta_{\mathrm{RB}} - \delta_k) + B_{(\mathrm{RB})k} \sin(\delta_{\mathrm{RB}} - \delta_k) \right] \tag{15.36}$$

$$Q_{C(\mathrm{RB})} = V_{RB} \sum_{k=1}^{N} V_k \left[G_{(\mathrm{RB})k} \sin(\delta_{\mathrm{RB}} - \delta_k) - B_{(\mathrm{RB})k} \cos(\delta_{\mathrm{RB}} - \delta_k) \right] \tag{15.37}$$

Therefore, Eq. (15.34) is modified as follows,

$$\Delta\overline{S}_i(f,\boldsymbol{\delta},\boldsymbol{V}) = \overline{S}_{Ci}(f,\boldsymbol{\delta},\boldsymbol{V}) - \overline{S}_i(f,\delta_i,V_i) = \overline{S}_{Ci}(f,\boldsymbol{\delta},\boldsymbol{V}) + \overline{S}_{Li}(f,V_i) - \overline{S}_{Gi}(f,V_i) \tag{15.38}$$

and the mismatch equations will be

$$\begin{bmatrix} \boldsymbol{\Delta P} \\ \boldsymbol{\Delta Q} \end{bmatrix} = -[\boldsymbol{J}_E]\begin{bmatrix} \Delta f \\ \boldsymbol{\Delta\delta} \\ \boldsymbol{\Delta V} \end{bmatrix} \tag{15.39}$$

where $\boldsymbol{\Delta P}$ and $\boldsymbol{\Delta Q}$ comprise ΔP_{RB} and ΔQ_{RB}, and $\boldsymbol{\Delta V}$ comprises V_{RB}. Besides, the voltage angle of the RB is prespecified and is not considered in $\boldsymbol{\Delta\delta}$, like the CPF method. Hence, variables for the $(t+1)$ iteration can be calculated as

$$\begin{bmatrix} f \\ \boldsymbol{\delta} \\ \boldsymbol{V} \end{bmatrix}^{t+1} = \begin{bmatrix} f \\ \boldsymbol{\delta} \\ \boldsymbol{V} \end{bmatrix}^{t} - \left[\boldsymbol{J}_E{}^{(t)}\right]^{-1}\begin{bmatrix} \boldsymbol{\Delta P} \\ \boldsymbol{\Delta Q} \end{bmatrix}^{t} \tag{15.40}$$

The enhanced Jacobian matrix $\boldsymbol{J}_E$ contains six submatrices as follows,

$$[\boldsymbol{J}_E] = \begin{bmatrix} \boldsymbol{J}_{Pf(N\times 1)} & \boldsymbol{J}_{P\delta(N\times(N-1))} & \boldsymbol{J}_{PV(N\times N)} \\ \boldsymbol{J}_{Qf(N\times 1)} & \boldsymbol{J}_{Q\delta(N\times(N-1))} & \boldsymbol{J}_{QV(N\times N)} \end{bmatrix} \tag{15.41}$$

where

$$\begin{gathered}
\boldsymbol{J}_{Pf} = \begin{bmatrix} \frac{\partial\Delta P_1}{\partial f} & \cdots & \frac{\partial\Delta P_N}{\partial f} \end{bmatrix}^T, \qquad \boldsymbol{J}_{Qf} = \begin{bmatrix} \frac{\partial\Delta Q_1}{\partial f} & \cdots & \frac{\partial\Delta Q_N}{\partial f} \end{bmatrix}^T \\
\boldsymbol{J}_{P\delta} = \begin{bmatrix} \dfrac{\partial\Delta P_1}{\partial\delta_2} & \cdots & \dfrac{\partial\Delta P_1}{\partial\delta_N} \\ \vdots & \ddots & \vdots \\ \dfrac{\partial\Delta P_N}{\partial\delta_2} & \cdots & \dfrac{\partial\Delta P_N}{\partial\delta_N} \end{bmatrix}, \quad \boldsymbol{J}_{Q\delta} = \begin{bmatrix} \dfrac{\partial\Delta Q_1}{\partial\delta_2} & \cdots & \dfrac{\partial\Delta Q_1}{\partial\delta_N} \\ \vdots & \ddots & \vdots \\ \dfrac{\partial\Delta Q_N}{\partial\delta_2} & \cdots & \dfrac{\partial\Delta Q_N}{\partial\delta_N} \end{bmatrix} \\
\boldsymbol{J}_{\mathrm{PV}} = \begin{bmatrix} \dfrac{\partial\Delta P_1}{\partial V_1} & \cdots & \dfrac{\partial\Delta P_1}{\partial V_N} \\ \vdots & \ddots & \vdots \\ \dfrac{\partial\Delta P_N}{\partial V_1} & \cdots & \dfrac{\partial\Delta P_N}{\partial V_N} \end{bmatrix}, \quad \boldsymbol{J}_{QV} = \begin{bmatrix} \dfrac{\partial\Delta Q_1}{\partial V_1} & \cdots & \dfrac{\partial\Delta Q_1}{\partial V_N} \\ \vdots & \ddots & \vdots \\ \dfrac{\partial\Delta Q_N}{\partial V_1} & \cdots & \dfrac{\partial\Delta Q_N}{\partial V_N} \end{bmatrix}
\end{gathered} \tag{15.42}$$

where T means transpose operation. Without the loss of generality, bus 1 is considered as RB. Moreover, like in the CPF method, the equations relating $\boldsymbol{\Delta P}$ and voltage magnitude variables for the buses operated in PV mode are not considered.

Detailed expressions of the partial terms in Eq. (15.42) are as follows.

$$\frac{\partial \Delta P_i}{\partial f} = \frac{\partial P_{Ci}}{\partial f} + \frac{\partial P_{Li}}{\partial f} - \frac{\partial P_{Gi}}{\partial f} = \frac{\partial P_{Ci}}{\partial f} + P_{Lni}L_{pfi}\left(\frac{V_i}{V_{ni}}\right)^{L_{pvi}} - G_{pfi} \tag{15.43}$$

$$\frac{\partial \Delta Q_i}{\partial f} = \frac{\partial Q_{Ci}}{\partial f} + \frac{\partial Q_{Li}}{\partial f} - \frac{\partial Q_{Gi}}{\partial f} = \frac{\partial Q_{Ci}}{\partial f} + Q_{Lni}L_{qfi}\left(\frac{V_i}{V_{ni}}\right)^{L_{qvi}} - G_{qfi} \tag{15.44}$$

$$\frac{\partial \Delta P_i}{\partial \delta_i} = \frac{\partial P_{Ci}}{\partial \delta_i} \tag{15.45}$$

$$\frac{\partial \Delta Q_i}{\partial \delta_i} = \frac{\partial Q_{Ci}}{\partial \delta_i} \tag{15.46}$$

$$\frac{\partial \Delta P_i}{\partial V_i} = \frac{\partial P_{Ci}}{\partial V_i} + \frac{\partial P_{Li}}{\partial V_i} - \frac{\partial P_{Gi}}{\partial V_i} = \frac{\partial P_{Ci}}{\partial V_i} + P_{Lni}\big(1 + L_{pfi}(f - f_n)\big)\left(\frac{L_{pvi}}{V_{ni}}\right)\left(\frac{V_i}{V_{ni}}\right)^{L_{pvi}-1} - G_{pvi} \tag{15.47}$$

$$\frac{\partial \Delta Q_i}{\partial V_i} = \frac{\partial Q_{Ci}}{\partial V_i} + \frac{\partial Q_{Li}}{\partial V_i} - \frac{\partial Q_{Gi}}{\partial V_i} = \frac{\partial Q_{Ci}}{\partial V_i} + Q_{Lni}\big(1 + L_{qfi}(f - f_n)\big)\left(\frac{L_{qvi}}{V_{ni}}\right)\left(\frac{V_i}{V_{ni}}\right)^{L_{qvi}-1} - G_{qvi} \tag{15.48}$$

where $\frac{\partial P_{Ci}}{\partial \delta_i}$ and $\frac{\partial Q_{Ci}}{\partial \delta_i}$ in Eqs. (15.45) and (15.46) and the first terms of Eqs. (15.47) and (15.48) (i.e., $\frac{\partial P_{Ci}}{\partial V_i}$ and $\frac{\partial Q_{Ci}}{\partial V_i}$) are similar to the Jacobian matrix of the conventional method. In other words, since $\frac{P_{Gi}}{P_{Li}}$ and $\frac{Q_{Gi}}{Q_{Li}}$ are only related to the voltage magnitude of the corresponding bus (i.e., bus i), the off-diagonal elements of Eqs. (15.45) and (15.46) are similar to the Jacobian of the conventional method. It should be noted that other terms in Eqs. (15.43) and (15.44) and Eqs. (15.47) and (15.48) for diagonal elements of the modified Jacobian matrix are not considered in Mumtaz et al. (2015). In this paper, in order to calculate $\frac{\partial P_{Ci}}{\partial f}$ and $\frac{\partial Q_{Ci}}{\partial f}$, Eqs. (15.11) and (15.12) are rewritten as,

$$P_{Ci} = V_i^2 G_{ii} + V_i \sum_{k=1, k\neq i}^{N} V_k[G_{ik}\cos(\delta_i - \delta_k) + B_{ik}\sin(\delta_i - \delta_k)] \tag{15.49}$$

$$Q_{Ci} = -V_i^2 B_{ii} + V_i \sum_{k=1, k\neq i}^{N} V_k[G_{ik}\sin(\delta_i - \delta_k) - B_{ik}\cos(\delta_i - \delta_k)] \tag{15.50}$$

Therefore, differentiating P_{Ci} and Q_{Ci} with respect to f yields,

$$\frac{\partial P_{Ci}}{\partial f} = V_i^2 \frac{\partial G_{ii}}{\partial f} + V_i \sum_{k=1, k\neq i}^{N} V_k\left[\frac{\partial G_{ik}}{\partial f}\cos(\delta_i - \delta_k) + \frac{\partial B_{ik}}{\partial f}\sin(\delta_i - \delta_k)\right] \tag{15.51}$$

$$\frac{\partial Q_{Ci}}{\partial f} = -V_i^2\frac{\partial B_{ii}}{\partial f} + V_i\sum_{k=1,k\neq i}^{N} V_k\left[\frac{\partial G_{ik}}{\partial f}\sin(\delta_i-\delta_k) - \frac{\partial B_{ik}}{\partial f}\cos(\delta_i-\delta_k)\right] \tag{15.52}$$

Knowing that $x_{ik}=l_{ik}f$ and $b_{c,ik}=c_{ik}f$, where x, l, b, c, and f are all in per unit, the elements affected by the system frequency are identified. To calculate the differentiation of P_{Ci} and Q_{Ci} with respect to the frequency, the differentiation of branch parameters (i.e., r, x, and b) and shunt elements connected to buses (i.e., g_0 and b_0) need to be calculated. Note that

$$\begin{aligned}\frac{\partial x}{\partial f} &= l = \frac{x}{f}\\ \frac{\partial b}{\partial f} &= c = \frac{b}{f}\end{aligned} \tag{15.53}$$

Hence,

$$\frac{\partial G_{ii}}{\partial f} = \sum_{k\in u}\frac{\partial g_{ik}}{\partial f} = \sum_{k\in u} -\frac{2r_{ik}x_{ik}}{\left(r_{ik}^2+x_{ik}^2\right)^2}\times\frac{x_{ik}}{f} \tag{15.54}$$

$$\frac{\partial G_{ik}}{\partial f} = -\frac{\partial g_{ik}}{\partial f} = \frac{2r_{ik}x_{ik}}{\left(r_{ik}^2+x_{ik}^2\right)^2}\times\frac{x_{ik}}{f} \tag{15.55}$$

$$\frac{\partial B_{ii}}{\partial f} = \frac{\partial b_{i0}}{\partial f} + \sum_{k\in u}\left(\frac{\partial b_{ik}}{\partial f} + \frac{1}{2}\frac{\partial b_{c,ik}}{\partial f}\right) = \frac{b_{i0}}{f} + \sum_{k\in u}\left(-\frac{\left(r_{ik}^2+x_{ik}^2\right)-2x_{ik}^2}{\left(r_{ik}^2+x_{ik}^2\right)^2}\times\frac{x_{ik}}{f} + \frac{1}{2}\frac{b_{c,ik}}{f}\right) \tag{15.56}$$

$$\frac{\partial B_{ik}}{\partial f} = -\frac{\partial b_{ik}}{\partial f} = \frac{\left(r_{ik}^2+x_{ik}^2\right)-2x_{ik}^2}{\left(r_{ik}^2+x_{ik}^2\right)^2}\times\frac{x_{ik}}{f} \tag{15.57}$$

The higher order modified midpoint method

Unlike the NR, the high order Newton-like (HONL) solvers present an order of convergence higher than two, so that they are expected to converge employing a fewer number of iterations. If together with this feature the computational complexity of the method is light, the resulting algorithm should be more competitive than the conventional NR (Tostado-Véliz et al., 2023). The applicability of one of HONL methods called HOMMP method to PF in IMGs has been investigated in

Tostado-Véliz et al. (2023). This novel technique presents a multistep structure, by which its order of convergence is equal to $N + 1$ (where $\boldsymbol{N}$ is the number of steps); however, its computational effort is comparable to the conventional NR. A visual explanation of this method alongside the ENR method and the 1-loop-based method is explained in Fig. 15.4. As it can be seen from the flowcharts in Fig. 15.4, the overall procedure of the HOMMP is quite similar to that implemented for other standard solvers. Actually, only the so-called modified midpoint loop has to be added. This routine, however, does not entail extra calculations, since it only needs the Jacobian matrix and the values of $\alpha's$, β, and N as entries. For these steps, the computational processes already implemented to calculate the Jacobian and PF equations can be reused, thus saving codification efforts. This structure allows to easily reuse available codes for the NR. In this way, the implementation of the HOMMP in commercial software tools is straightforward, requiring only some minor modifications. The reader can refer to Tostado-Véliz et al. (2023) for more details. It is worth noting that $\boldsymbol{x} \in \mathbb{R}^n$ is the state vector defined as $[f, \boldsymbol{\delta}, \boldsymbol{V}]^\top$ **and** $\mathbf{g}: \mathbb{R}^n \longmapsto \mathbb{R}^n$ are the PF equations.

Case studies and simulation results

This section is mainly presented in two parts. The first part provides PF results for two test systems, one small and one large, each with unique topologies and control strategies, different topologies (i.e., radial or meshed), different droop control strategies, and different load models are presented. In the second part, an MV MG is considered for performing EM strategies in both grid-connected and islanded mode of operation (Bayat et al., 2021).

Simulations are conducted in the MATLAB® environment, in which the MATPOWER package (Zimmerman & Murillo-Sánchez, 2016) with some modification is used for applicability in islanded mode. All simulations are performed using an Intel Core i7–4510U, 2.0–2.6 GHz, RAM 8 GB, personal computer. Note that all the reported parameters in the tables and the figures are in per unit, unless otherwise stated.

Power flow studies in IMGs

To assess the performance of the proposed PF formulation and solution method, this section utilizes two test systems. The first is a radial system with 7 buses, commonly used for its simplicity. The second is a larger-scale system with 118 buses operating in a meshed configuration, which better reflects real-world power grids. In all test systems and cases, the amounts of P_0 and Q_0 are set to zero. Also, the convergence tolerance is set to 10^{-5}. The reported computational times have been obtained as the mean value of 1000 simulations to minimize the influence of other computational activities.

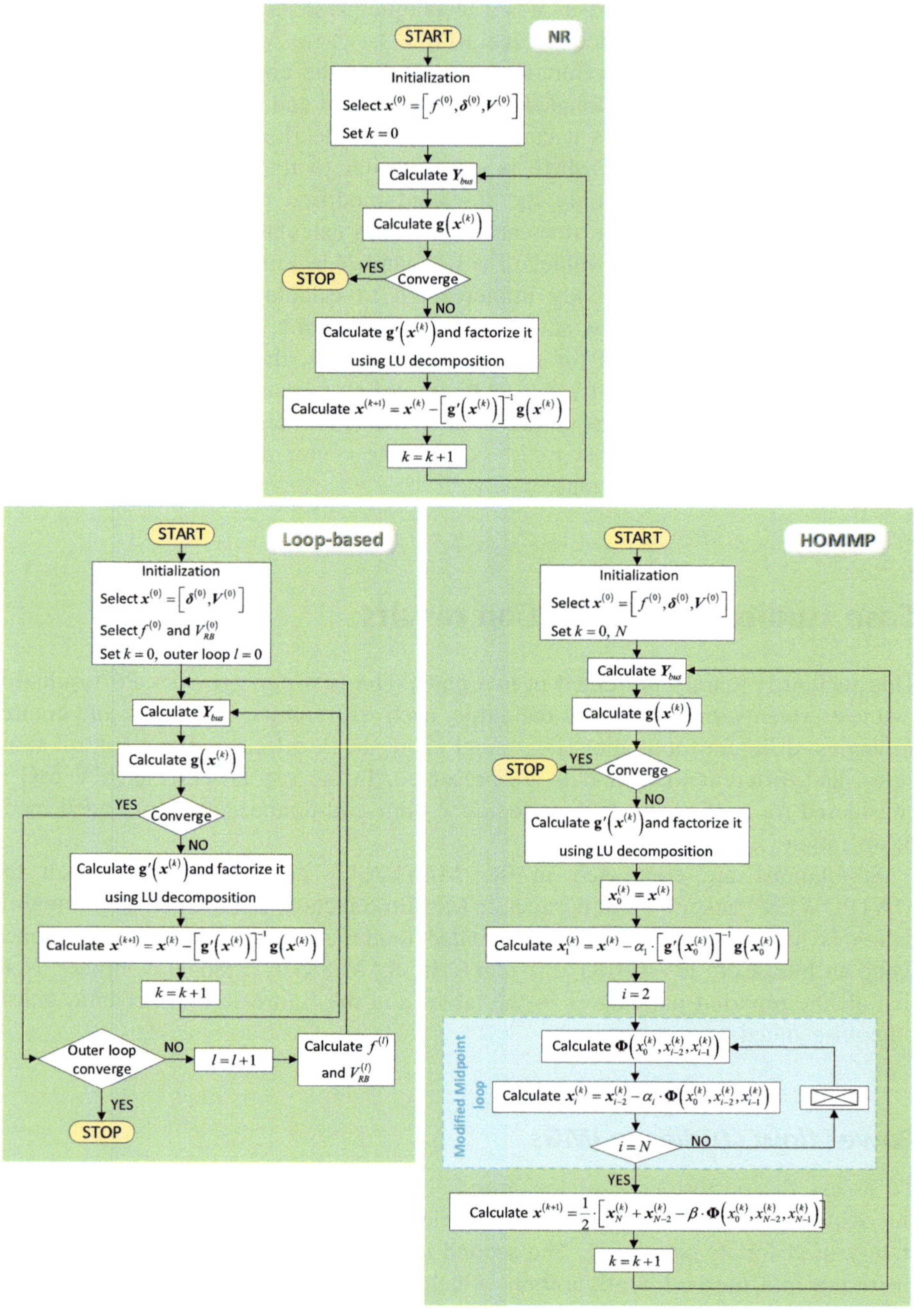

Figure 15.4 Flowcharts of the enhanced Newton Raphson. Flowcharts of the ENR, 1-loop-based solver and HOMMP for PF in IMGs. *HOMMP*, Higher order modified midpoint; *ENR*, enhanced NR. *Source:* From Tostado-Véliz, M., Bayat, M., Ghadimi, A. A., & Jurado, F. (2023). A novel Newton-like method with high convergence rate for efficient power-flow solution in isolated microgrids. *IET Generation, Transmission & Distribution, 17*(6), 1368–1382.

The 7-bus test system

A small-scale IMG, which is introduced in Bayat et al. (2021), operating in 0.4 kV line to line level and 50 Hz frequency, is shown in Fig. 15.5 This system includes three DGs, one CZ load, and a CP load. Lines and loads information, as well as information of DGs with a base power of 10 kVA, are depicted in this figure. In this figure, DGL is embedded to bus 4 (which is needed for loop-based methods). The obtained results are the same, regardless of the VSB location. Different droop characteristics and coefficients are investigated in Bayat et al. (2022, 2021). For brevity, only the results for the case that the droop is inductive with the characteristics of $f_0 = 1.02$, $V_{01} = V_{02} = V_{03} = 1.00$, $m_1 = m_3 = 0.08$, $n_1 = n_3 = 0.3227$, $m_2 = 0.04$ and $n_2 = 0.1614$ (Bayat et al., 2021) is reported here. In addition to the steady-state values of time domain (TD) simulations, the obtained results of 1-loop method (Bayat et al., 2021), 2-loop method (Bayat et al., 2021), 3-loop method (Bayat et al., 2021), ENR (Bayat et al., 2022), and (Tostado-Véliz et al., 2023) approaches are reported in Table 15.2. The results for all PF solution methods are identical. However, the computational times vary significantly. TD simulations conducted by PSIM software take the longest at 120 seconds, followed by the 3-loop (0.212 seconds), 2-loop (0.082 seconds), and 1-loop (0.018 seconds) methods. ENR takes 0.030 seconds, and HOMMP emerges as the fastest method, requiring only 0.012 seconds.

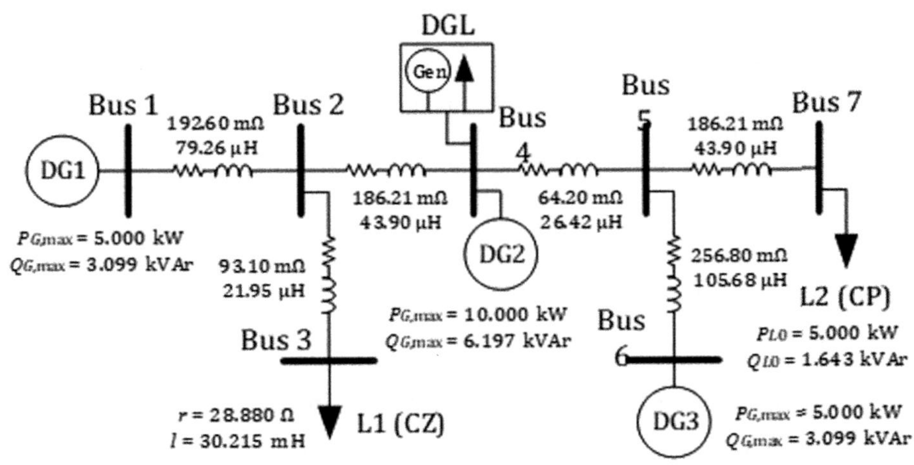

Figure 15.5 The 7-bus test system (Bayat et al., 2021).
Source: From Bayat, M., Koushki, M. M., Ghadimi, A. A., Tostado-Véliz, M., & Jurado, F. (2022). Comprehensive enhanced Newton Raphson approach for power flow analysis in droop-controlled islanded AC microgrids. *International Journal of Electrical Power & Energy Systems, 143*, 108493.

Table 15.2 Power flow results for the 7-bus test system.

Solution method			**PSIM**		**PF solution**	
Frequency			1.00044		1.00048	
Voltage magnitude of	1	0.97346		0.97378		
buses	2	0.97027		0.97063		
	3	0.96738		0.96774		
	4	0.97304		0.97341		
	5	0.97188		0.97229		
	6	0.97612		0.97645		
	7	0.96564		0.96611		
Generation	DG1	0.2445	0.0811	0.2440	0.0813	
	DG2	0.4890	0.1639	0.4881	0.1648	
		DG3	0.2445	0.0743	0.2440	0.0730
Load		L1	0.4678	0.1538	0.4682	0.1540
		L2	0.5000	0.1640	0.5000	0.1643

Source: From Bayat, M., Koushki, M. M., Ghadimi, A. A., Tostado-Véliz, M., & Jurado, F. (2022). Comprehensive enhanced Newton Raphson approach for power flow analysis in droop-controlled islanded AC microgrids. *International Journal of Electrical Power & Energy Systems, 143*, 108493.

The 118-bus test system

In the second test case, six DG units have been added to an 11 kV distribution system presented in Zimmerman and Murillo-Sánchez (2016) and Zhang et al. (2007) to provide the islanded operation of the network depicted in Fig. 15.6. The primary system, which is operated radially, has 15 tie switches. In this study, five of them (drawn in blue in Fig. 15.6) are considered closed to provide a weakly meshed network. Data about the line impedances and system loads can be found in Zimmerman and Murillo-Sánchez (2016) and Zhang et al. (2007). Moreover, DG information is provided in Bayat et al. (2022). Three different cases are considered, in which the droop control modes are inductive, resistive, and complex, respectively. The base power is considered 10 MVA, and the system frequency and voltage set points for all DGs are set to 1 and 1.03 *p.u*, respectively. The results of the PF analysis using the ENR and HOMMP method are presented in Table 15.3.

Centralized energy management strategy in microgrids

As mentioned before, different types of EM strategies are utilized for MGs. In this section, a centralized framework proposed in Bayat et al. (2015) is presented. In the presented EMS, DERs, reactive power compensators (RPCs), and loads are supposed to be equipped with local controllers (LCs) with the capability to communicate with the Central Energy Management System (CEMS). The communication basement uses an incident command system (ICS) structure as proposed in

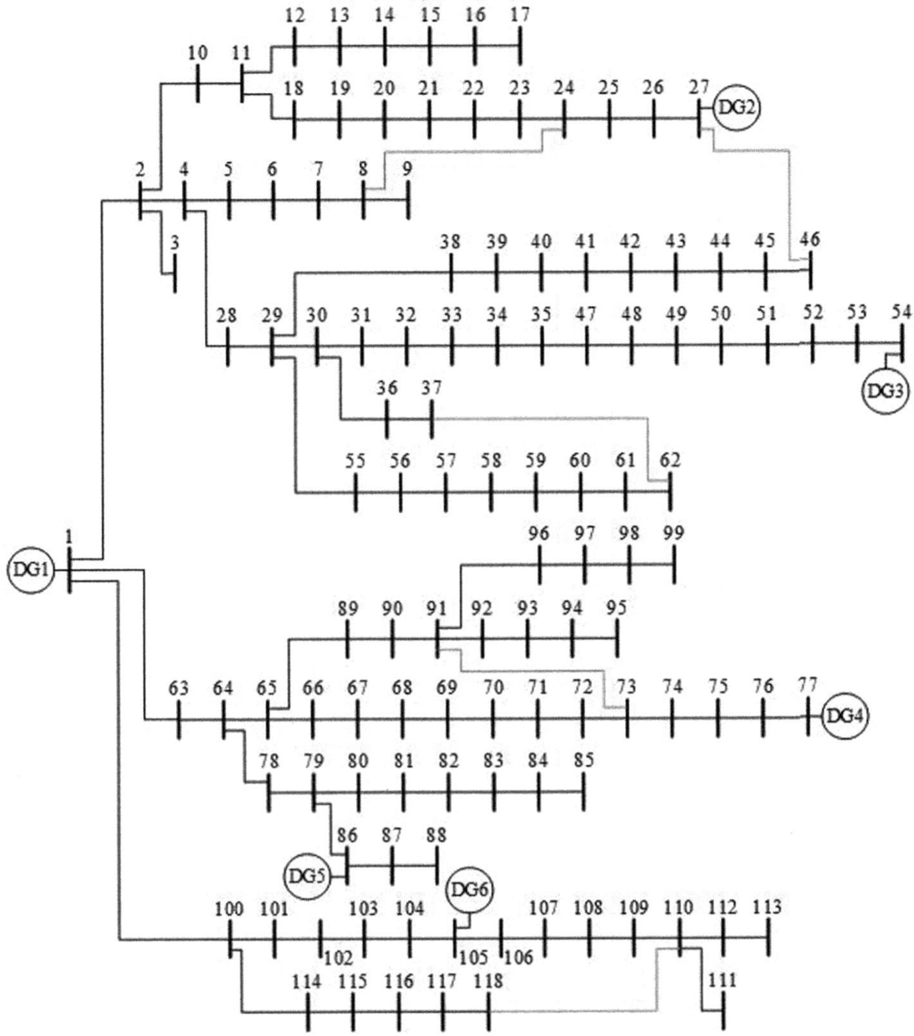

Figure 15.6 The modified 118-bus distribution system operating as an IMG.
Source: From Bayat, M., Koushki, M. M., Ghadimi, A. A., Tostado-Véliz, M., & Jurado, F. (2022). Comprehensive enhanced Newton Raphson approach for power flow analysis in droop-controlled islanded AC microgrids. *International Journal of Electrical Power & Energy Systems, 143*, 108493.

Rogers et al. (2010) and Aquino-Lugo et al. (2011). Having access to such a communication platform, the control strategy can be summarized as follows;

1. Depending on the MG operation mode, the CEMS computes a set of set points and sends them through a hierarchy where they are received by the LCs. In the ICS framework, the

Table 15.3 PF results for the 118-bus test system- Active and reactive power of DGs, system frequency and voltage magnitude of DG buses and computational time for ENR method.

Test cases	**Case 1**			**Case 2**			**Case 3**		
Frequency	**0.98944**			**1.01270**			**0.99731**		
#DG	P_G	Q_G	V	P_G	Q_G	V	P_G	Q_G	V
DG1	0.6326	0.5706	0.9737	0.7700	0.4720	0.9446	0.7070	0.5459	0.9085
DG2	0.3173	0.2263	0.9961	0.2969	0.2360	0.9803	0.3070	0.2262	1.0243
DG3	0.3173	0.1881	0.9914	0.2587	0.2360	0.9642	0.2860	0.2051	0.9296
DG4	0.3173	0.1936	0.9921	0.2704	0.2360	0.9673	0.2905	0.2097	0.9506
DG5	0.3173	0.2324	0.9904	0.3033	0.2360	0.9666	0.3114	0.2306	0.9865
DG6	0.4226	0.3213	0.9927	0.4214	0.3151	0.9678	0.4232	0.3155	0.9415
Comp. Time (s)	0.0164			0.0142			0.0112		

DG, Distributed generation; *ENR*, enhanced NR.
Source: From Bayat, M., Koushki, M. M., Ghadimi, A. A., Tostado-Véliz, M., & Jurado, F. (2022). Comprehensive enhanced Newton Raphson approach for power flow analysis in droop-controlled islanded AC microgrids. *International Journal of Electrical Power & Energy Systems, 143*, 108493.

intermediate agents (see Aquino-Lugo et al., 2011) take over the responsibility of communication between the CEMS and the LCs.

2. Each LC then controls the DER, the RPC or the controllable loads under its supervision to meet the requests.

Proposed control strategy

In the grid-connected mode, the CEMS aims at minimizing the total operation cost of the MG preserving the system voltage as close as possible to the nominal value. In the islanded mode, however, maintaining a reliable power supply is more preferable than other economic profits. In this mode, keeping the frequency and voltages of the MG in their predefined acceptable range is the main purpose. It is worth noting that in this section, it is supposed that the DERs have the conventional droop (inductive) characteristics. In this mode, two cases can occur:

Case 1: If the generation capacity of DGs is higher than the total consumption of the MG (i.e., sum of loads and losses), the MG continues to supply the load demand, while the DGs work within their operational limits. In this case which is referred to as "light loading," the voltage and the frequency deviation can be simply minimized by making use of the droop characteristics of DERs. To better understand this supply-side control, a simple MG containing two EI-DERs supplying a load is shown in Fig. 15.7A. P_{Gob-a} and P_{Goa-b} denote the real power reference set points of the DER i at frequency f_o, while "b" and "a" denote before and after control, respectively. It is known that by shifting up the $P-f$ droop characteristics of the DER (determining the new reference set points for P_{Gi} (i.e., P_{Goi-a} in reference frequency f_o), the frequency deviation after the control Δf_a, is less than the frequency deviation before the control, Δf_b. It is noted that by defining new set points for real and reactive powers of DERs, the load demanded power and hence the system losses are changed. Owing to this fact, as it is seen from Fig. 15.7B $P_{G1}+P_{G2}$ which are equal to the total consumption of the MG takes different values before and after control. Similarly, to minimize the voltage deviation of bus i to which DER_i is connected, a new reactive power set point for Q_{Gi} (i.e., Q_{Goi-a}) can be determined, which is shown in Fig. 15.7C.

Case 2: This is the case in which the generation capacity of DGs is less than the sum the total consumption of the MG. This case is referred to as "heavy loading" in which all DGs with any reference setpoint are likely to exceed their real power limits so that a severe frequency and/ or voltage deviations is occurred which cannot be determined from droop characteristics. In this case, those ESS with adequate energy to discharge will quickly respond to match supply and demand in order to counteract frequency excursion. It is noted that CEMS decides on charging or discharging of ESSs to cope with such a situation. If the total load consumption is compensated for by ESSs and DGs, a portion of DGs' capacity is assumed to be released. Hence, the DERs' set points are recalculated similar to Case1 to mitigate the voltage and frequency deviation. However, if the DERs cannot supply the total load, a portion of loads must be disrupted to provide production and consumption balance.

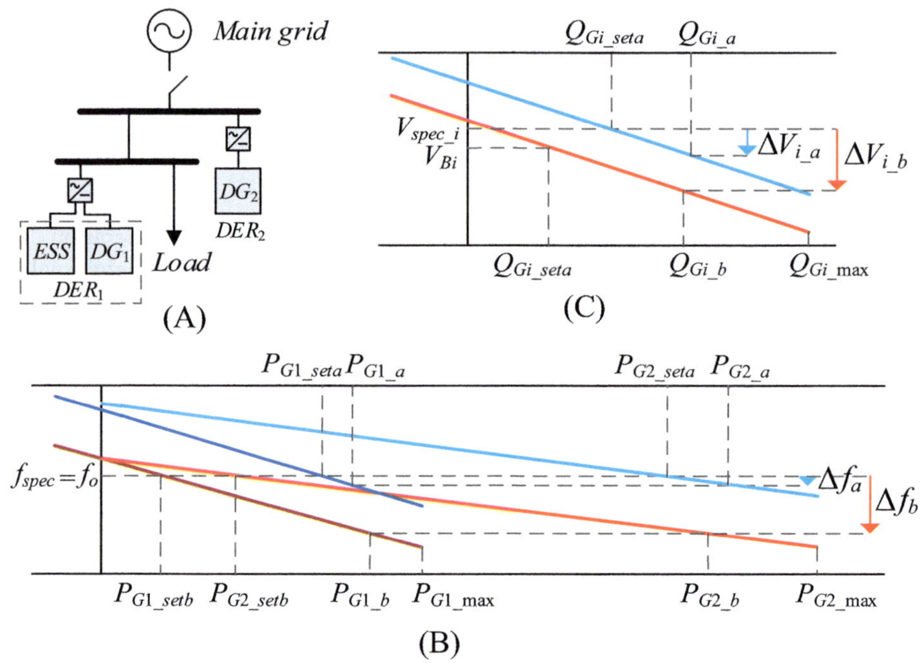

Figure 15.7 Typical example of supply-side secondary control: (A) system architecture, (B) frequency control, and (C) voltage control.
Source: From Bayat, M., Sheshyekani, K., Hamzeh, M., & Rezazadeh, A. (2015). Coordination of distributed energy resources and demand response for voltage and frequency support of MV microgrids. *IEEE Transactions on Power Systems, 31*(2), 1506–1516.

Optimization problem

Islanded mode

In the islanded mode of operation, the highest priority is to keep reliable and high-quality power supply to customers, rather than economic interests. With this regard, the goal is to minimize the frequency deviation index (FDI) and the voltage deviation indices (VDI) represented by Eqs. (15.58) and (15.59) respectively:

$$\mathrm{FDI} = (f - f_s)^2 = (\Delta f)^2 \tag{15.58}$$

$$\mathrm{VDI} = \sum_{i=1}^{U} (V_i - V_{si})^2 = \sum_{i=1}^{U} (\Delta V_i)^2 \tag{15.59}$$

where U is the number of MG buses whose voltage magnitudes violate the predefined acceptable range, and f_s and V_{si} are specified values for frequency and voltage, respectively. As discussed earlier, depending on the generation and consumption mismatch, the MG will operate either under light loading or heavy loading.

- **Light loading**: In this case, the control variables include new references set points P_{Goi} and Q_{Goi} and the operating set points of RPCs.
- **Heavy loading:** In this case, following the fast response of ESSs and if all DGs still operate at their maximum limits, a certain amount of loads is usually disrupted to fulfill Eqs. (15.58) and (15.59), so that the voltage and frequency of the MG are maintained close to their nominal values. The proposed control algorithm aims at minimizing the total amount of the manipulated load. Hence, an additional objective function is defined as manipulated load index (MLI) as:

$$\mathrm{MLI} = \sum_{i=1}^{N_s} \left[(\Delta P_{Li})^2 + (\Delta Q_{Li})^2\right] \tag{15.60}$$

where N_s is the number of the most effective buses whose load changes have the highest impact on the MG FDI and VDI. Moreover, working with the most effective buses helps favorably with minimizing the communication burden between different MG agents.

Grid-connected mode

In the grid-connected mode, the objective is to simultaneously minimize the MG total operating cost which is represented by operating cost index (OCI) and the VDI.

$$\mathrm{OCI} = \sum_{i=1}^{N_G} F_i(P_{Gi}) + \rho P_M \tag{15.61}$$

where $F_i(P_{Gi})$ is the generation cost of the MG dispatchable DER i whose output real power is P_{Gi}. The generation cost is traditionally expressed by a quadratic function; however, as a common practice, it is approximated by a piecewise linear model (Khodaei, 2013). The real-time market price at point of common coupling for the considered time interval is presented by ρ, while P_M is the power exchanged between the MG and the main grid. The second term in Eq. (15.61) represents the cost of power transfer from the grid. P_M would be negative when the MG excess power is sold back to the grid (Khodaei, 2013). The aforementioned objectives are subject to conventional PF constraints, namely, (i) the equality constraints and (ii) the inequality constraints (i.e., branch flow limits, limitations of DERs, and the voltage magnitude limitation of system buses). So, the control variables in the grid-connected mode are real and reactive power set points of DERs (i.e. P_{Gi} and Q_{Gi}) and the operating set points of RPCs.

It is worth nothing that as an additional constraint, the CEMS decides on the charging or discharging operation mode and the amount of exchanged power of any ESS. For the grid-connected MG, the operation mode of each was determined based on the market price, the operation cost, and voltage limitations. For example, if the MG total operation cost exceeds the limit specified by the CEMS, Dr can participate in control strategy to fulfill the operator requirements. In this case to simultaneously minimize OCI and VDI by use of minimum manipulated load demand MLI, the most effective candidate buses with the highest impact on VDI can be used. Hence, the grid-connected mode is also considered for three objective functions as well as islanded mode of operation.

Solution method

There are different optimization methods to solve the proposed problem. In Bayat et al. (2015), the multiobjective PSO (MOPSO) (Coello et al., 2004) is used due to its generality and simplicity (AlRashidi & El-Hawary, 2008). This provides a set of nondominated pareto optimal solutions and thus a set of alternative solutions to manage the system. Finally, one of the solutions is selected based on the operator choice (Bayat et al., 2015).

Selection of the most effective buses

The way for selecting the most effective buses can be better understood from the Jacobian matrix. The negative inverse of extended Jacobian matrix (i.e. $[J_E]^{-1}$ in Eq. (15.40) describes how the state variables (f, δ, V) change due to real and reactive power mismatches at all buses. These changes are linearized relationships which are called sensitivities. The sensitivities of Δf to ΔP and ΔQ are given by the submatrix J_{Pf} and J_{Qf}, respectively. Similarly, the sensitivities of ΔV to ΔP and ΔQ are given by the submatrix J_{PV} and J_{QV}, respectively (Bayat et al., 2015).

Classification of loads

In the proposed algorithm, load classification is an essential task for the participation of demand response in the management of the MG. In Bayat et al. (2015), the MG loads are simply classified as controllable and noncontrollable loads. The loads belonging to the former category can be manipulated to regulate the voltage and/or the frequency of the MG. The loads can be classified according to their controllability degree. This information for load controllability is available through data provided by the system analyzer or the manufacturers (Rogers et al., 2010). The CEMS is constantly notified about the load changes and the integration of new devices by receiving data from the LCs communicating with loads fitted with smart meters over a Δt time horizon. The time horizon is defined by the system operator (Bayat et al., 2015). Once the MG switches to the islanded mode, depending on the surplus or shortage of local real and reactive power generation, the MG is likely to face with unacceptable frequency and/or voltage deviations. Hence, the CEMS acts to correct these violations using a-priori knowledge about the recent system model. In case of shortage of local power generation, the loads participate in the overall control task. To this aim, two types of loads in terms of their controllability are considered; *Cat*1 specifies a set of loads which have maximum controllability so that they can be controlled at any time, while *Cat*2 defines the loads whose demands can be controlled only over a specified time.

$$P_{Loi} = P_{Cat1i} + P_{Cat2i} + P_{NCi} \tag{15.62}$$

$$Q_{Loi} = Q_{Cat1i} + Q_{Cat2i} + Q_{NCi} \tag{15.63}$$

where P_{Cat1i} (Q_{Cat1i}) and P_{Cat2i} (Q_{Cat2i}) denote the aggregated real (reactive) power associated to *Cat*1 and *Cat*2, respectively, while P_{NCi} (Q_{NCi}) is the noncontrollable part of the real (reactive) load power of bus *i*. Note that each single load can reveal floating controllability over different timescales. This kind of classification can help with modification and selecting the targeted load buses for voltage and frequency control.

Case study

The single line diagram of a typical 20 kV MG involving five radial feeders is shown in Fig. 15.8. The MG is connected to the upstream utility grid via a static switch. This system consists of five dispatchable DG units, one nondispatchable DG unit, and three ESSs. Units 1 and 6 are SM-DERs, while units 2–4 refer to EI-DERs equipped with their individual parallel ESSs. It is assumed that each ESS has a $P-f$ droop characteristics similar to its associated DER. *DG5* is a nondispatchable PV unit, which is treated as a negative load. The output generation power of the PV unit is assumed to be intermittent over the 24-hours interval, as shown in Fig. 15.9A. It is assumed that the PV operates at unity power factor. The technical specifications of different DERs (DGs and ESSs) can be found in Bayat et al. (2015). Without losing any generality, the maximum charging or discharging power of each ESS is assumed to be 30% of the generation capacity of its respective DG. The characteristics of ESSs are given in Bayat et al. (2015). The MG feeds seven three-phase balanced aggregated loads, four of them (i.e., *Load* 2, *Load* 4, *Load* 1–2, and *Load* 3–2) are frequency- and voltage-dependent loads. All loads are assumed with a lagging power factor of 0.85. Two RPCs are connected to bus 14 and 18, improving the loud power factor of these load buses to maximum 0.95.

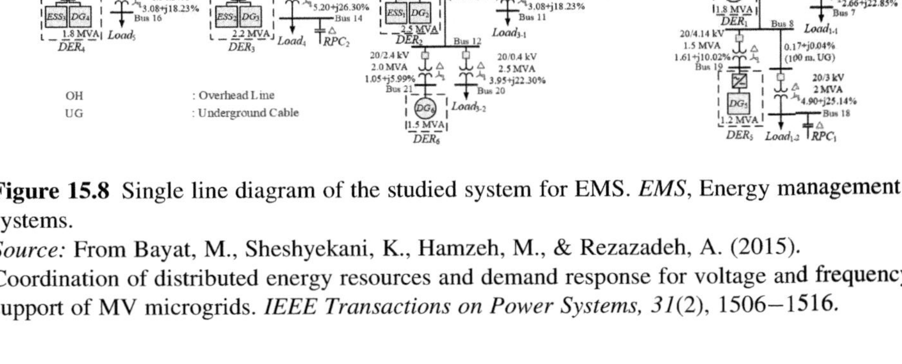

Figure 15.8 Single line diagram of the studied system for EMS. *EMS*, Energy management systems.

Source: From Bayat, M., Sheshyekani, K., Hamzeh, M., & Rezazadeh, A. (2015). Coordination of distributed energy resources and demand response for voltage and frequency support of MV microgrids. *IEEE Transactions on Power Systems, 31*(2), 1506–1516.

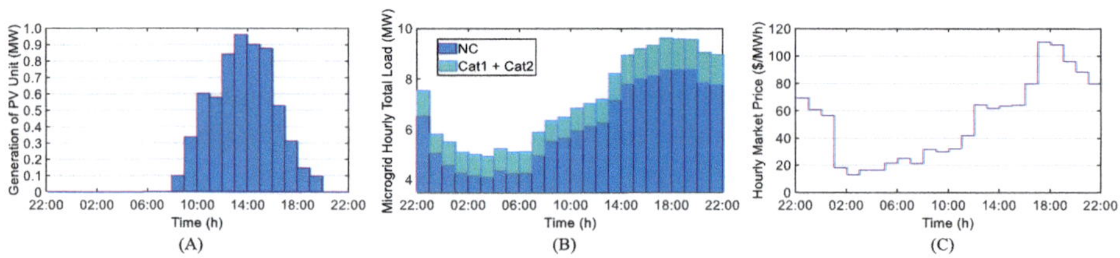

Figure 15.9 Forecasted PV unit power generation, load, and market price variation over 24 h. (A) Forecasted power generation by a nondispatchable PV unit during 24 h, (B) microgrid hourly total specified load during 24 h, and (C) the hourly market price during 24 h.
Source: From Bayat, M., Sheshyekani, K., Hamzeh, M., & Rezazadeh, A. (2015). Coordination of distributed energy resources and demand response for voltage and frequency support of MV microgrids. *IEEE Transactions on Power Systems, 31*(2), 1506–1516.

Note that the RPCs are supposed to inject any amount of reactive power within their limits to the system. The hourly total demanded power of the MG over a 24-hours interval is shown in Fig. 15.9B. The controllable part of the total load (i.e., $Cat1 + Cat2$) is discriminated in this figure. The classification of the loads as well as the relevant data for controllability of the loads of $Cat2$ over daily hours is presented in Bayat et al. (2015). The hourly market price over the same time interval is also shown in Fig. 15.9C. The speed droop R for SM-DERs and the m for EI-DERs are set to 1%, and the n coefficient for EI-DERs is considered to be 5%. While L_{pf} and L_{qf} are considered 1.5 and 1.0, respectively, the voltage dependency of loads is characterized by ZIP model with the share of 0.05, 0.1, and 0.85 for impedance, current, and power relation parts, respectively.

DG1 is modeled as SM-DER equipped with an AVR with specified voltage magnitude of $1p.u.$ It is noted that the nominal value for MG bus voltages, V_S, is $1p.u.$ while the specified MG frequency is considered to be $f_s = 50\text{Hz}$. All buses whose voltage magnitude exceeds $\pm 2.5\%$ from nominal value are taken into account in the objective function VDI. The performance of the proposed control strategy is evaluated for both grid-connected and islanded modes of operation assuming two possible generation and demand imbalances (i.e., light loading and heavy loading cases). It is worth nothing that, in the simulation, the maximum acceptable cost of $430\$/h$ is considered for the MG.

The MG operation mode over the considered 24-hour interval is listed in Table 15.4. The MG total operation cost, the VDI, the frequency, and the charging/discharging schedule of ESSs during 24-hours interval before and after applying the proposed control procedure are shown in Fig. 15.10. As it can be seen in Fig. 15.10, the proposed control strategy generally succeeds to mitigate the frequency and voltage deviations while minimizing the total operation cost. To further study the performance of the proposed control strategy under different load patterns, the case of light loading (during time interval 22:00–14:00 hours) and heavy loading (during time interval 14:00–22:00 hours) is elaborately discussed in Bayat et al. (2015).

Table 15.4 Microgrid operation mode over 24 h.

Time interval	22:00–9:30	9:30–11:30	11:30–16:30	16:30–18:30	18:30–22:00
Operation mode	Grid-connected	Islanded	Grid-connected	Islanded	Grid-connected

Source: From Bayat, M., Sheshyekani, K., Hamzeh, M., & Rezazadeh, A. (2015). Coordination of distributed energy resources and demand response for voltage and frequency support of MV microgrids. *IEEE Transactions on Power Systems, 31*(2), 1506–1516.

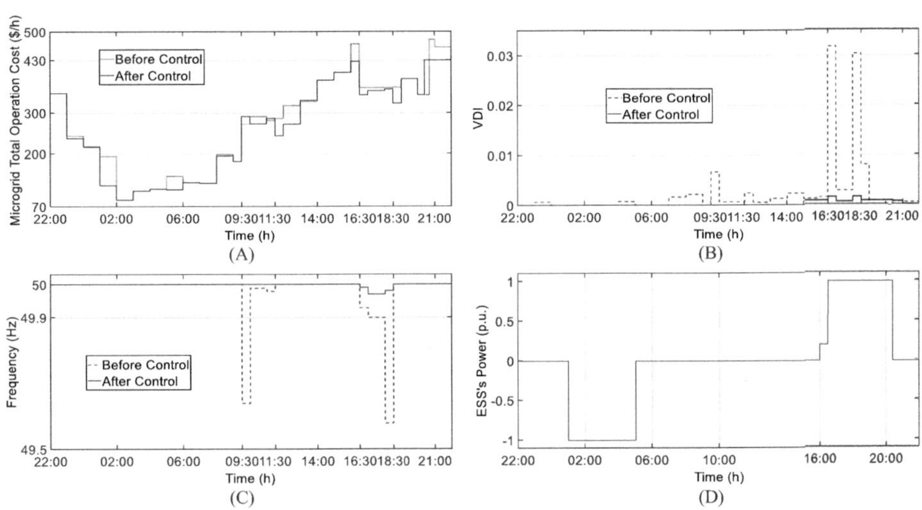

Figure 15.10 Describing microgrid operation and control. (A) Microgrid total operation cost during 24 h before and after control, (B) VDI of the system during 24 h before and after control, (C) system frequency during 24 h before and after control, and (D) charging/discharging schedule of ESSs during 24 h. *VDI*, Voltage deviation indices.
Source: From Bayat, M., Sheshyekani, K., Hamzeh, M., & Rezazadeh, A. (2015). Coordination of distributed energy resources and demand response for voltage and frequency support of MV microgrids. *IEEE Transactions on Power Systems, 31*(2), 1506–1516.

Conclusion

A brief survey of literature pertaining to different issues under the topic of EMS in MGs has been investigated in this chapter. The initial part of this chapter covers the general topics related to EM, followed by a critical review of the various aspects of EM, in particular the configuration of the EMS and the structures and classification of EMS in MGs as well as the optimization techniques. EM is a comprehensive topic requiring extended research, as it is required for hybrid renewable energy units to work in an economic, reliable, and competent manner. Apart from that, problem models of the EMS and the role of DRPs in EM have been explored. The pros and cons of different strategies for EM have been compared broadly in this chapter from different viewpoints. Also, a PF analysis method has been presented for islanded MGs. The method is based on the well-known NR scheme and tries to enhance it to overcome the limitation of using the scheme in droop-controlled isolated MGs with variable frequency and without an SB. In the presented PF analysis method, any type of droop control scheme can be utilized. Finally, the PF has been utilized for two test systems, one small and one large, each with unique topologies and control strategies, different topologies (i.e., radial or meshed), different droop control strategies, and different load models. Then, an MV MG is considered for performing EM strategies in both grid-connected and islanded mode of operation.

References

Aghajani, G., & Ghadimi, N. (2018). Multi-objective energy management in a micro-grid. *Energy Reports, 4*. Available from https://doi.org/10.1016/j.egyr.2017.10.002, http://www.journals.elsevier.com/energy-reports/.

AlRashidi, M. R., & El-Hawary, M. E. (2008). A survey of particle swarm optimization applications in electric power systems. *IEEE Transactions on Evolutionary Computation, 13*(4).

Allwyn, R. G., Al-Hinai, A., & Margaret, V. (2023). A comprehensive review on energy management strategy of microgrids. *Energy Reports, 9*. Available from https://doi.org/10.1016/j.egyr.2023.04.360, http://www.journals.elsevier.com/energy-reports/.

Aquino-Lugo, A. A., Klump, R., & Overbye, T. J. (2011). A control framework for the smart grid for voltage support using agent-based technologies. *IEEE Transactions on Smart Grid, 2*(1).

Attia, A. F., El Sehiemy, R. A., & Hasanien, H. M. (2018). Optimal power flow solution in power systems using a novel Sine-Cosine algorithm. *International Journal of Electrical Power and Energy Systems, 99*. Available from https://doi.org/10.1016/j.ijepes.2018.01.024.

Bayat, M., Sheshyekani, K., & Rezazadeh, A. (2015). A unified framework for participation of responsive end-user devices in voltage and frequency control of the smart grid. *IEEE Transactions on Power Systems, 30*(3). Available from https://doi.org/10.1109/TPWRS.2014.2344133.

Bayat, M., Sheshyekani, K., Hamzeh, M., & Rezazadeh, A. (2015). Coordination of distributed energy resources and demand response for voltage and frequency support of MV microgrids. *IEEE Transactions on Power Systems, 31*(2).

Bayat, M., Ghaseminezhad, K., & Ghadimi, A. A. (2021). An efficient iterative approach for power flow solution of droop-controlled islanded AC microgrids through conventional methods. *International Journal of Electrical Power & Energy Systems, 130*, 106962.

Bayat, M., Koushki, M. M., Ghadimi, A. A., Tostado-Véliz, M., & Jurado, F. (2022). Comprehensive enhanced Newton Raphson approach for power flow analysis in droop-controlled islanded AC microgrids. *International Journal of Electrical Power & Energy Systems, 143*, 108493.

Cannata, N., Cellura, M., Longo, S., Montana, F., Sanseverino, E. R., Luu, Q.L., & Nguyen, N. Q. (2019). Multi-objective optimization of urban microgrid energy supply according to economic and environmental criteria. In *IEEE Milan PowerTech, PowerTech 2019*. Italy: Institute of Electrical and Electronics Engineers Inc. https://doi.org/10.1109/

PTC.2019.8810776. 9781538647226 http://ieeexplore.ieee.org/xpl/mostRecentIssue.jsp?punumber = 8792346.

Chaudhary, P., & Rizwan, M. (2018). Energy management supporting high penetration of solar photovoltaic generation for smart grid using solar forecasts and pumped hydro storage system. *Renewable Energy*, *118*. Available from https://doi.org/10.1016/j.renene.2017.10.113.

Coello, C. A. C., Pulido, G. T., & Lechuga, M. S. (2004). Handling multiple objectives with particle swarm optimization. *IEEE Transactions on Evolutionary Computation*, *8*(3).

Dong, G., & Chen, Z. (2019). Data-driven energy management in a home microgrid based on bayesian optimal algorithm. *IEEE Transactions on Industrial Informatics*, *15*(2). Available from https://doi.org/10.1109/TII.2018.2820421, http://ieeexplore.ieee.org/xpl/RecentIssue.jsp?punumber = 9424.

El-Fergany, A. A., & Hasanien, H. M. (2020). Salp swarm optimizer to solve optimal power flow comprising voltage stability analysis. *Neural Computing and Applications*, *32*(9). Available from https://doi.org/10.1007/s00521-019-04029-8, http://link.springer.com/journal/521.

El-Fergany, A. A., & Hasanien, H. M. (2018). Tree-seed algorithm for solving optimal power flow problem in large-scale power systems incorporating validations and comparisons. *Applied Soft Computing*, *64*. Available from https://doi.org/10.1016/j.asoc.2017.12.026.

Elmouatamid, A., Ouladsine, R., Bakhouya, M., El Kamoun, N., Khaidar, M., & Zine-Dine, K. (2021). Review of control and energy management approaches in micro-grid systems. *Energies*, *14*(1), 168. Available from https://doi.org/10.3390/en14010168.

Erlich, I., Rueda, J.L., Wildenhues, S., & Shewarega, F. (2014). Solving the IEEE-CEC 2014 expensive optimization test problems by using single-particle MVMO. In *Proceedings of the 2014 IEEE congress on evolutionary computation, CEC 2014* (pp. 1084–1091). Germany: Institute of Electrical and Electronics Engineers Inc. https://doi.org/10.1109/CEC.2014.6900517, 9781479914883.

Hannan, M. A., Tan, S. Y., Al-Shetwi, A. Q., Jern, K. P., & Begum, R. A. (2020). Optimized controller for renewable energy sources integration into microgrid: Functions, constraints and suggestions. *Journal of Cleaner Production*, *256*, 120419. Available from https://doi.org/10.1016/j.jclepro.2020.120419.

He, X., Fang, X., & Yu, J. (2019). Distributed energy management strategy for reaching cost-driven optimal operation integrated with wind forecasting in multimicrogrids system. *IEEE Transactions on Systems, Man, and Cybernetics: Systems*, *49*(8). Available from https://doi.org/10.1109/TSMC.2019.2919941, http://ieeexplore.ieee.org/xpl/mostRecentIssue.jsp?punumber = 6221021.

Hussien, A. M. Mekhamer, S. F. & Hasanien, H. M. (2020). Cuttlefish optimization algorithm based optimal PI controller for performance enhancement of an autonomous operation of a DG System. In *2nd International conference on smart power and internet energy systems, SPIES 2020* (pp. 293–298). Egypt: Institute of Electrical and Electronics Engineers Inc. http://ieeexplore.ieee.org/xpl/mostRecentIssue.jsp?punumber = 9242535, https://doi.org/10.1109/SPIES48661.2020.9243093. 9781728166117.

Hussien, A. M., Hasanien, H. M., & Mekhamer, S. F. (2021). Sunflower optimization algorithm-based optimal PI control for enhancing the performance of an autonomous operation of a microgrid. *Ain Shams Engineering Journal*, *12*(2). Available from https://doi.org/10.1016/j.asej.2020.10.020, http://www.elsevier.com/wps/find/journaldescription.cws_home/724208/description#description.

Jiang, W., Yang, K., Yang, J., Mao, R., Xue, N., & Zhuo, Z. (2019). A multiagent-based hierarchical energy management strategy for maximization of renewable energy consumption in interconnected multi-microgrids. *IEEE Access*, *7*. Available from https://doi.org/10.1109/access.2019.2955552.

Jiménez-Fernández, S., Camacho-Gómez, C., Mallol-Poyato, R., Fernández, J., Del Ser, J., Portilla-Figueras, A., & Salcedo-Sanz, S. (2019). Optimal microgrid topology design and siting of distributed generation sources using a multi-objective substrate layer coral reefs optimization algorithm. *Sustainability*, *11*(1), 169. Available from https://doi.org/10.3390/su11010169.

Jin, M., Feng, W., Liu, P., Marnay, C., & Spanos, C. (2017). MOD-DR: Microgrid optimal dispatch with demand response. *Applied Energy*, *187*. Available from https://doi.org/10.1016/j.apenergy.2016.11.093.

Karimi, H., & Jadid, S. (2020). Optimal energy management for multi-microgrid considering demand response programs: A stochastic multi-objective framework. *Energy*, *195*, 116992. Available from https://doi.org/10.1016/j.energy.2020.116992.

Khan, M. W., Wang, J., Ma, M., Xiong, L., Li, P., & Wu, F. (2019). Optimal energy management and control aspects of distributed microgrid using multi-agent systems. *Sustainable Cities and Society*, *44*. Available from https://doi.org/10.1016/j.scs.2018.11.009, http://www.elsevier.com/wps/find/journaldescription.cws_home/724360/description#description.

Khodaei, A. (2013). Microgrid optimal scheduling with multi-period islanding constraints. *IEEE Transactions on Power Systems*, *29*(3).

Kim, R. K., Glick, M. B., Olson, K. R., & Kim, Y. S. (2020). MILP-PSO combined optimization algorithm for an islanded microgrid scheduling with detailed battery ESS efficiency model and policy considerations. *Energies*, *13*(8). Available from https://doi.org/10.3390/en13081898, https://www.mdpi.com/1996-1073/13/8/1898.

Kontis, E. O., Kryonidis, G. C., Nousdilis, A. I., Malamaki, K. N. D., & Papagiannis, G. K. (2019). Power flow analysis of islanded AC microgrids. In *2019 IEEE Milan PowerTech*. Institute of Electrical and Electronics Engineers Inc. http://ieeexplore.ieee.org/xpl/mostRecentIssue.jsp?punumber = 8792346, https://doi.org/10.1109/PTC.2019.8810612. 9781538647226.

Kryonidis, G. C., Kontis, E. O., Chrysochos, A. I., Oureilidis, K. O., Demoulias, C. S., & Papagiannis, G. K. (2018). Power flow of Islanded AC microgrids: Revisited. *IEEE Transactions on Smart Grid*, *9*(4). Available from https://doi.org/10.1109/TSG.2018.2799480.

Kumar, A., Jha, B. K., Singh, D., & Misra, R. K. (2019). Current injection-based Newton–Raphson power-flow algorithm for droop-based islanded microgrids. *IET Generation, Transmission & Distribution*, *13*(23).

Leonori, S., Paschero, M., Frattale Mascioli, F. M., & Rizzi, A. (2020). Optimization strategies for microgrid energy management systems by genetic algorithms. *Applied Soft Computing*, *86*, 105903. Available from https://doi.org/10.1016/j.asoc.2019.105903.

Li, B., Roche, R., & Miraoui, A. (2017). Microgrid sizing with combined evolutionary algorithm and MILP unit commitment. *Applied Energy*, *188*. Available from https://doi.org/10.1016/j.apenergy.2016.12.038, http://www.elsevier.com/inca/publications/store/4/0/5/8/9/1/index.htt.

Mansour-lakouraj, M., & Shahabi, M. (2019). Comprehensive analysis of risk-based energy management for dependent micro-grid under normal and emergency operations. *Energy*, *171*. Available from https://doi.org/10.1016/j.energy.2019.01.055, http://www.elsevier.com/inca/publications/store/4/8/3/.

Marcelino, C. G., Avancini, J. V. C., Delgado, C. A. D. M., Wanner, E. F., Jiménez-Fernández, S., & Salcedo-Sanz, S. (2021). Dynamic electric dispatch for wind power plants: A new automatic controller system using evolutionary algorithms. *Sustainability*, *13*(21), 11924. Available from https://doi.org/10.3390/su132111924.

Mehdizadeh, A., Taghizadegan, N., & Salehi, J. (2018). Risk-based energy management of renewable-based microgrid using information gap decision theory in the presence of peak load management. *Applied Energy, 211*. Available from https://doi.org/10.1016/j.apenergy.2017.11.084, http://www.elsevier.com/inca/publications/store/4/0/5/8/9/1/index.htt.

Mellouk, L., Ghazi, M., Aaroud, A., Boulmalf, M., Benhaddou, D., & Zine-Dine, K. (2019). Design and energy management optimization for hybrid renewable energy system- case study: Laayoune region. *Renewable Energy, 139*. Available from https://doi.org/10.1016/j.renene.2019.02.066.

Molzahn, D. K., & Hiskens, I. A. (2019). A survey of relaxations and approximations of the power flow equations. *Foundations and Trends® in Electric Energy Systems, 4*(1–2). Available from https://doi.org/10.1561/3100000012.

Mumtaz, F., Syed, M. H., Hosani, M. A., & Zeineldin, H. H. (2015). A novel approach to solve power flow for islanded microgrids using modified Newton Raphson with droop control of DG. *IEEE Transactions on Sustainable Energy, 7*(2).

Olivares, D. E., Mehrizi-Sani, A., Etemadi, A. H., Cañizares, C. A., Iravani, R., Kazerani, M., Hajimiragha, A. H., Gomis-Bellmunt, O., Saeedifard, M., Palma-Behnke, R., Jiménez-Estévez, G. A., & Hatziargyriou, N. D. (2014). Trends in microgrid control. *IEEE Transactions on Smart Grid, 5*(4). Available from https://doi.org/10.1109/TSG.2013.2295514.

Rahim, S., Javaid, N., Khan, R. D., Nawaz, N., & Iqbal, M. (2019). A convex optimization based decentralized real-time energy management model with the optimal integration of microgrid in smart grid. *Journal of Cleaner Production, 236*, 117688. Available from https://doi.org/10.1016/j.jclepro.2019.117688.

Rogers, K. M., Klump, R., Khurana, H., Aquino-Lugo, A. A., & Overbye, T. J. (2010). An authenticated control framework for distributed voltage support on the smart grid. *IEEE Transactions on Smart Grid, 1*(1).

Saadat, H. (1999). Power flow analysis. *Power System Analysis*, 1999.

Shaheen, M. A. M., Hasanien, H. M., Turky, R. A., Ćalasan, M., Zobaa, A. F., & Aleem, S. H. E. A. (2021). OPF of modern power systems comprising renewable energy sources using improved chgs optimization algorithm. *Energies, 14*(21). Available from https://doi.org/10.3390/en14216962, https://www.mdpi.com/1996-1073/14/21/6962/pdf.

Shahryari, E., Shayeghi, H., Mohammadi-ivatloo, B., & Moradzadeh, M. (2019). A copula-based method to consider uncertainties for multi-objective energy management of microgrid in presence of demand response. *Energy, 175*. Available from https://doi.org/10.1016/j.energy.2019.03.129.

Shi, W., Xie, X., Chu, C.-C., & Gadh, R. (2015). Distributed optimal energy management in microgrids. *IEEE Transactions on Smart Grid, 6*(3). Available from https://doi.org/10.1109/TSG.2014.2373150.

Tostado-Véliz, M., Bayat, M., Ghadimi, A. A., & Jurado, F. (2023). A novel Newton-like method with high convergence rate for efficient power-flow solution in isolated microgrids. *IET Generation, Transmission & Distribution, 17*(6).

Wang, D., Qiu, J., Reedman, L., Meng, K., & Lai, L. L. (2018). Two-stage energy management for networked microgrids with high renewable penetration. *Applied Energy, 226*. Available from https://doi.org/10.1016/j.apenergy.2018.05.112.

Wu, N., & Wang, H. (2018). Deep learning adaptive dynamic programming for real time energy management and control strategy of micro-grid. *Journal of Cleaner Production, 204*. Available from https://doi.org/10.1016/j.jclepro.2018.09.052.

Xing, X., Xie, L., & Meng, H. (2019). Cooperative energy management optimization based on distributed MPC in grid-connected microgrids community. *International Journal of*

Electrical Power and Energy Systems, *107*. Available from https://doi.org/10.1016/j.ijepes.2018.11.027.

Yu, D., Zhu, H., Han, W., & Holburn, D. (2019). Dynamic multi agent-based management and load frequency control of PV/Fuel cell/ wind turbine/ CHP in autonomous microgrid system. *Energy*, *173*. Available from https://doi.org/10.1016/j.energy.2019.02.094.

Yuan, D., Lu, Z., Zhang, J., & Li, X. (2019). A hybrid prediction-based microgrid energy management strategy considering demand-side response and data interruption. *International Journal of Electrical Power & Energy Systems*, *113*. Available from https://doi.org/10.1016/j.ijepes.2019.05.045.

Zhang, D., Fu, Z., & Zhang, L. (2007). An improved TS algorithm for loss-minimum reconfiguration in large-scale distribution systems. *Electric Power Systems Research*, *77*(5-6).

Zhang, Y., Meng, F., Wang, R., Zhu, W., & Zeng, X.-J. (2018). A stochastic MPC based approach to integrated energy management in microgrids. *Sustainable Cities and Society*, *41*. Available from https://doi.org/10.1016/j.scs.2018.05.044.

Zhao, Z., Lee, W. C., Shin, Y., & Song, K.-B. (2013). An optimal power scheduling method for demand response in home energy management system. *IEEE Transactions on Smart Grid*, *4*(3). Available from https://doi.org/10.1109/TSG.2013.2251018.

Zhou, B., Li, W., Chan, K. W., Cao, Y., Kuang, Y., Liu, X., & Wang, X. (2016). Smart home energy management systems: Concept, configurations, and scheduling strategies. *Renewable and Sustainable Energy Reviews*, *61*. Available from https://doi.org/10.1016/j.rser.2016.03.047.

Zimmerman, R. D., & Murillo-Sánchez, C. E. (2016). Matpower optimal scheduling tool MOST 1.0. *User's Manual*, 2016.

Forecasting techniques for power systems with renewables

16

Paul Arévalo[1,2], Darío Benavides[1,3] and Danny Ochoa-Correa[2]
[1]Department of Electrical Engineering, University of Jaén, Jaén, Spain, [2]Faculty of Engineering, Department of Electrical, Electronics and Telecommunications Engineering (DEET), Universidad de Cuenca, Balzay Campus, Cuenca, Azuay, Ecuador, [3]Department of Electrical Engineering, University of Málaga, Malaga, Spain

Introduction

The increasing capacity of renewable energy offers a key opportunity to address the growing global energy crisis. Despite reaching 27% for renewable energy and 24% for natural gas in 2019 (Zhu et al., 2022), challenges persist due to the variable nature of renewable energy production. Governments and policymakers advocate for increased penetration of renewables, with projections suggesting that their contribution could reach 85% by 2050, highlighting their crucial role in the future energy landscape (Alkabbani et al., 2021). Global electricity demand rises with population and economic growth, primarily fueled by conventional fossil fuel-based power plants emitting harmful greenhouse gases (Islam & Roy, 2023). To address this, the integration of renewable energy sources (RES) like wind, solar, hydropower, and biogas is essential for sustainable energy production. Advances in wind and photovoltaic (PV) technologies have significantly increased renewable energy generation (Li et al., 2020). However, the inherent uncertainty of RES requires innovative solutions. Technologies such as smart inverters and machine learning enhance integration, while research in frequency control aims to manage high RE penetrations (Wen et al., 2019). The volatility of renewable energy production underscores the need for advanced forecasting techniques to ensure system balance. Accurate predictions are crucial for informed decision-making in energy system management. This chapter explores energy forecasting concepts, highlighting the importance of precision and introducing advanced techniques to address specific challenges (Alkabbani et al., 2021).

The evolution of RES necessitates robust forecasting methods for efficient grid integration and sustainable energy planning. In a Sicily wind energy study (Cellura et al., 2008), traditional approaches, including statistical analyses and geostatistical interpolation, were employed to estimate wind probability distribution functions, emphasizing the need for reliable tools, such as artificial neural networks (ANNs), for spatial wind speed estimation. In Germany, autoregressive integrated moving average (ARIMA)-based forecasting models have optimized renewable energy generation (Fattah et al., 2023). A hybrid methodology, derived from four years of

Towards Future Smart Power Systems with High Penetration of Renewables. DOI: https://doi.org/10.1016/B978-0-443-29871-4.00016-6

data, improved yearly accuracy by approximately 89.5%. The REN4KAST software platform was introduced to address the stochastic nature of renewable energy generation. Examining wind speed forecasts in Karnataka, a weather model with a 3 km resolution identified regions suitable for wind power plants (Gangopadhyay et al., 2019), recommending new locations based on factors like transport facilities and diurnal patterns. The economic viability of a compressed air energy storage (CAES)-PV system was addressed through a novel model-free deep reinforcement learning (DRL) method (Dolatabadi et al., 2023). This approach optimized CAES energy arbitrage using a hybrid forecasting model based on 2-D convolutional neural networks (CNNs) and bidirectional long short-term memory (BLSTM) units.

On the other hand, various techniques and procedures for forecasting generation from random REs exist. A comprehensive review of the study highlighted the importance of tailored forecast products for end-users (Sweeney et al., 2020). Emphasis was placed on integrating blockchain technology for decentralized yet trusted data exchange, addressing uncertainties, and enhancing predictive dispatch tools for system stability with higher renewable energy integration. The dynamic nature of solar and wind energy, subject to fluctuations on various timescales, presents challenges in managing stable and reliable energy supplies. Accurate forecasting becomes imperative for optimal energy production management in intermittent and stochastic RES (Notton et al., 2018). The review emphasizes the cost implications of variability and stochastic variations in REs, highlighting the economic benefits of efficient forecasting. It underscores the role of forecasting in mitigating integration costs, becoming an essential tool for electrical grid managers.

Integrating renewables and the increasing demand for electricity necessitate efficient energy management strategies. Smart grid technology emerges as a solution to enhance energy efficiency, particularly in the context of renewable energy integration (Sharma et al., 2023). The comprehensive overview provided in this survey paper explores various techniques for demand forecasting across residential, industrial, and agricultural sectors. Load forecasting, demand response management, and implementing smart meters are crucial for achieving energy efficiency and green energy goals. Electricity price forecasting is fundamental for grid management and renewable energy integration. Complex generation mix data and heteroskedasticity pose challenges for accurate forecasting, addressed in this study using advanced analytics methods like principal component analysis (PCA) (Nyangon & Akintunde, 2024). The research emphasizes the importance of PCA in improving forecasting accuracy, particularly in highly renewable energy markets like the California independent system operator, contributing to grid stability and optimal resource utilization. Furthermore, the pivotal role of renewable energy in a low-carbon society requires understanding the impact of weather and climate on energy demand and supply. A systematic review of seasonal forecasting for wind and wind power in the Iberian Peninsula and the Canary Islands sheds light on the importance of teleconnection patterns and the stratosphere in predicting wind variability (Bayo-Besteiro et al., 2022). The limited existing literature underscores the need for increased research efforts to enhance the operation and planning of renewables. Integrating variable and nondispatchable renewable power generation requires assessing variability and accurate forecasting. This review paper surveys previous research on solar, wind, wave, and tidal energy resources, emphasizing the need for coherent studies that analyze variability and forecasting methods consistently (Widén et al., 2015). The conclusion stresses the importance of comparable studies, metrics, and evaluations across different renewable energy resources.

Although the reviewed studies have significantly contributed to the understanding of renewable energies and their forecasts, there is a lack of integration that addresses traditional methods, meteorological models, machine learning, and ANNs in a holistic manner. Most existing approaches focus on specific aspects, without providing a comprehensive and comparative view. For example, the study presented in Sweeney et al. (2020) emphasizes the importance of forecast products tailored to end-users, highlighting the need to integrate blockchain technology for decentralized yet reliable data exchange. However, a greater integration of diverse methods is required to more comprehensively address the challenges associated with variability and stochasticity in REs, as evidenced in previous studies (Bayo-Besteiro et al., 2022; Widén et al., 2015). This lack of global integration underscores the need for future research aimed at developing more unified and comprehensive approaches, leveraging the strengths of different methodologies to improve the accuracy and applicability of forecasts in the context of renewable energies. Zhu et al. (2022) highlighted that the continual increase in renewable energy capacity presents a crucial opportunity to address the growing energy crisis and the ever-rising global demand. Islam and Roy (2023) pointed out that, despite the significant electricity generation by conventional fossil fuel-based power plants, they also emit environmentally harmful greenhouse gases. Alkabbani et al. (2021) mentioned that accuracy in predicting renewable energy generation becomes a crucial factor in addressing challenges associated with the variability and intermittency of these sources.

These findings underline the importance of considering the specific needs of end-users and exploring decentralized solutions for data exchange in forecast management within the realm of renewable energies. The lack of uniformity in renewable energy generation, in contrast to conventional resources, emphasizes the need for broader and more effective strategies to address these challenges and ensure a successful transition to a more sustainable energy matrix. Alkabbani et al. (2021) highlighted the need for improved accuracy in predicting renewable energy generation to address challenges associated with the variability and intermittency of these sources. The projected increase in the contribution of REs to the future energy matrix, as mentioned in the citation from Alkabbani et al. (2021), emphasizes the urgency of developing more advanced and accurate forecasting methods to ensure a successful and sustainable integration of these sources into global energy systems. The lack of uniformity in renewable energy generation, as evidenced in Islam and Roy (2023), poses specific challenges in planning and managing optimal electricity generation, highlighting the need for more comprehensive approaches that address the inherent variability of these sources. Collectively, these studies underscore existing gaps in current research on renewable energy forecasts, emphasizing the need for more integrative reviews and advanced methods that effectively address the variability and stochastic nature of these energy sources. In summary, to address identified gaps in the literature, this chapter makes a significant contribution by bridging this divide, presenting a unified and in-depth perspective on forecasting techniques for renewable energy integration. The following key contributions are proposed:

- Conducting a comparative and integrated analysis of the uniqueness of the load forecasting problem in integrated energy systems, highlighting its crucial differences compared to conventional electrical systems.
- Thoroughly investigating traditional methods, meteorological models, machine learning techniques, and ANNs, providing a critical understanding of their specific applicability to renewable energy generation.
- Emphasizing the unique contributions of this chapter by offering a critical analysis of case studies and practical applications. The aim is to provide valuable insights for effective implementations in real-world environments.

Fundamentals of forecasting

Basic concepts

In light of the expansive growth within the realm of RES, acquiring an in-depth comprehension of their inherently stochastic nature has become imperative. This understanding is pivotal not only for the seamless dispatch of energy but also for preemptively addressing potential stability challenges within the electrical grid. Consequently, there arises a pressing need for predictive modeling, grounded in historical data and a meticulous analysis of the power outputs from RES (Li et al., 2020). The temporal horizon, denoting the timeframe during which such prognostications are formulated, is contingent upon the specific application at hand. Emphasizing the pivotal role of precision in forecasting becomes paramount, serving as a foundational cornerstone for informed decision-making in the realm of renewable energy systems. The dependability of these prognostications holds direct implications for the efficacy of planning and operation strategies within the energy infrastructure. Armed with a profound understanding of these foundational concepts, we position ourselves to delve into the intricate realm of forecasting techniques, precisely tailored to accommodate the nuances introduced by the variability inherent in RES, thereby addressing the intricate challenge of accurately predicting the dynamics of energy supply and demand (Koltsaklis et al., 2021).

The efficacy of forecasting techniques within renewable energy systems is inherently linked to a meticulous consideration of temporal scales. Categorized into four primary groups, these temporal scales act as a dynamic framework, facilitating the nuanced adaptation of forecasting strategies to the diverse planning requirements dictated by different timescales. In the ultrashort term (0–4 h), the focus of forecasts sharpens to manage near-real-time events, thereby enabling immediate and responsive adjustments to abrupt changes. Extending this temporal window, the short-term scale (0–72 h) emerges as a crucial domain for anticipating variations in both energy generation and consumption. This extension proves pivotal in optimizing the daily operation of the energy system (Wen et al., 2019).

Stepping into the arena of medium to long-term planning (1 month–1 year) necessitates a comprehensive consideration of seasonal factors and trends extending over more extended durations. This temporal domain is indispensable for shaping strategic decisions governing the development of energy infrastructures. Finally, the long-term forecasting horizon (>1 year) transcends immediate concerns, projecting overarching trends that provide invaluable insights for policymaking and formulating transformative strategies for the evolution of the energy matrix. A nuanced and comprehensive understanding of these diverse temporal scales forms the bedrock for the effective implementation of forecasting techniques, thereby ensuring a precise, adaptive, and sustainable management approach to the intricacies of energy systems (Moret et al., 2016). To further illustrate these concepts, Fig. 16.1 provides a concrete example of a demand prediction system over the course of a day. The graph captures the dynamic interplay between actual and predicted renewable energy generation, offering a visual insight into the accuracy and responsiveness of the forecasting model within this relatively brief temporal window. This visualization serves to underscore the importance of precise short-term predictions in ensuring optimal energy management and grid stability, particularly in the context of RES with inherent variability.

Traditional forecasting methods

The traditional forecasting methods refer to established and conventional approaches used to anticipate future events or trends. These methods are generally based on historical data and observed patterns from the past. Some of the most common traditional methods include MA, ARIMA method, linear regression, time series decomposition, exponential smoothing models, relative index method, and Delphi method.

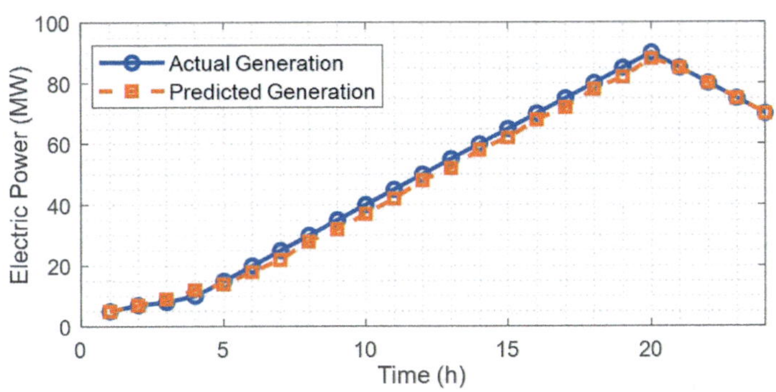

Figure 16.1 Example of renewable energy prediction for a short-term interval (24 hours). Anticipating tomorrow's energy: A 24-hour snapshot of renewable energy prediction.

Moving average and autoregressive integrated moving average methods

The MA method is a fundamental technique in time series analysis and forecasting. It involves calculating the average of a set of past observations to predict future values. In the context of solar irradiance forecasting, the MA model generates a set of random variables using past values of solar irradiance, resembling a random walk with zero mean and constant variance of white noise (Mishra et al., 2017). This method is expressed as MA (q), where "q" represents the order of the process, determining the number of past steps considered for calculating the average along the time series. In the evaluated models, MA contributes to point forecasts within the broader framework of solar irradiance forecasting. The main idea is to smooth temporal fluctuations and highlight general trends.

In recent years, the ARIMA method has emerged as an evolution of MA. Given the inherent uncertainty in predictions, models have been developed to forecast not only the mean but also the associated uncertainty, emphasizing the need for rigorous comparisons to assess their ability and accuracy in the dynamic field of solar energy. David et al. (2018) focused on intraday forecasts with time horizons of 1 to 6 hours, using models that employ only endogenous inputs, specifically past solar irradiance data. The selected models combine three point forecast models (autoregressive and moving average recursive [ARMArls] and coupled autoregressive and dynamic system) with seven probabilistic methods (linear quantile regression model, weighted quantile regression, quantile regression neural network, recursive generalized conditional heteroskedasticity, sieve bootstrap, quantile regression forest, and gradual boosting decision trees), resulting in a comprehensive comparison of 20 model combinations (David et al., 2018).

The ARIMA model is a versatile approach widely used in renewable energy forecasting, known for its simplicity and efficiency. It combines autoregressive, integrated (I), and MA components to capture temporal dependencies and patterns in the data (David et al., 2016; Malhan & Mittal, 2022). The ARIMA model is particularly effective in capturing the persistence of solar irradiance datasets, making it a robust competitor in forecasting scenarios (David et al., 2016). Specifically, ARMArls model, a variant of the ARIMA model, has proven to be robust in predicting renewable energies, demonstrating its effectiveness in providing accurate short-term forecasts of solar irradiance (David et al., 2016). In summary, the comparison of these forecasting methods reveals the complexities of solar irradiance prediction. The MA method, with its simplicity, contributes to the set of models, while the ARIMA model, and its ARMArls variant, demonstrate their effectiveness in capturing the dynamics of solar irradiance in short-term forecasts. The exhaustive evaluation involving various combinations of these methods highlights the nuanced nature of forecasts in the renewable energy domain, shedding light on the strengths and limitations of each approach (David et al., 2016, 2018; Mishra et al., 2017; Zhang et al., 2022). As the renewable energy landscape evolves, these methods serve as essential tools to optimize energy management and network stability, promoting a sustainable energy future.

However, there are some drawbacks, as the MA technique can be sensitive to noisy data or random fluctuations in the time series. Sudden oscillations can affect forecast accuracy, especially in high-variability environments. Additionally, the performance of the MA method depends on the appropriate choice of the moving window size (the number of observations used for the average). In situations where the time series exhibits complex patterns, determining the optimal window length can be a challenge. The MA assumes that the time series is stationary, and the trend is linear. In the presence of nonlinear trends or structural changes in the data, the model may provide inaccurate forecasts. Sensitivity to parameter choice is a key feature of the ARIMA method, similar to the MA. The performance of ARIMA is inherently linked to the appropriate choice of parameters, including autoregression (p), differencing (d), and MA (q) orders. Incorrect determination of these parameters can lead to inaccurate forecasts.

Another significant limitation of ARIMA is its lack of suitability for nonlinear data. The model assumes that the relationship between observations is linear, meaning that it may struggle to capture the complexity of nonlinear patterns present in the time series. The inability to effectively handle irregular seasonality is another challenge associated with ARIMA. If the time series exhibits irregular seasonal patterns or seasonality that changes over time, the model may not fit optimally, affecting forecast accuracy. Additionally, ARIMA works best with stationary time series, where the mean and variance are constant over time. In cases where the series is nonstationary, additional preprocessing, such as differencing, is required, adding complexity to the analysis. Another crucial limitation of ARIMA is its lack of consideration for external effects or exogenous variables in the model. This deficiency can limit its ability to capture external influences that may affect the time series, resulting in less accurate forecasts. Finally, ARIMA's sensitivity to abrupt changes in the time series can impact the reliability of long-term forecasts. Significant changes in series behavior over time can affect the model's effectiveness in accurately predicting future events (Malhan & Mittal, 2022). Fig. 16.2 illustrates an example prediction using an MA for a random signal with a 100-hour interval and a window size of 5. The signal was generated in MATLAB® 2021a using the "movmean" function.

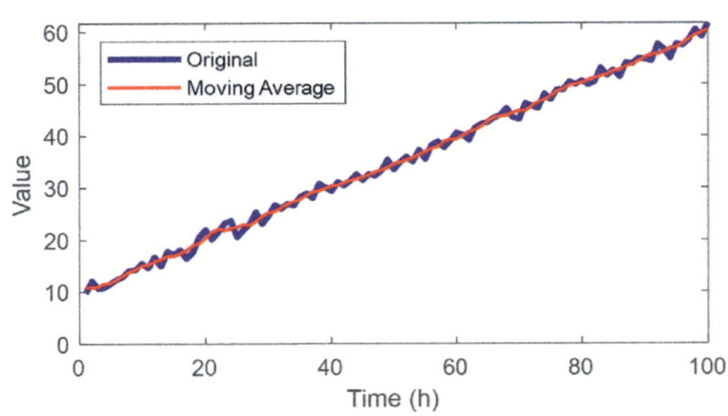

Figure 16.2 Original time series and MA forecast. Analyzing trends: Original time series vs. moving average forecast.

Linear regression in renewable energy prediction

Linear regression analyzes the relationship between two variables, such as time and the magnitude of an event. Through this relationship, the future behavior of the dependent variable can be predicted. Linear regression emerges as a fundamental technique in predicting renewable energy production, particularly in solar energy, where the variability of solar radiation poses challenges for grid management. Various studies have addressed this issue, emphasizing the importance of accurate prediction to ensure system stability and optimal resource allocation. In a study (Hassan et al., 2017), the focus was on predicting solar radiation using linear regression techniques, specifically the least squares regression. The simple linear regression equation is employed to model the relationship between solar radiation and a time-related variable, showing a 32% improvement in prediction accuracy compared to other reference models. Another study (Sharma et al., 2011) addressed solar power generation prediction in distributed systems. Both linear regression and support vector machines (SVM) with different kernel functions were utilized. SVM-based models proved to be 27% more accurate than the existing models, highlighting the effectiveness of these techniques in distributed generation prediction.

In the context of wind energy, Sanusi and Corne (2016) concentrated on wind speed prediction to optimize the integration of this RES. Multivariate linear regression is applied, emphasizing the importance of derived features, such as the gradient in wind speed, to enhance model accuracy. Uncertainty in electricity prices in renewable energy environments is tackled in a fourth study (Minhas et al., 2017). Here, linear regression is applied to forecast electricity prices, leveraging the intermittency of RES. A load control algorithm based on logistic regression is proposed to activate consumer load demands during low-price periods. Subsequently, in a study (Wahbah et al., 2020), a nonparametric model of solar irradiance probability density using local linear regression was proposed. This method surpasses both parametric and nonparametric models in terms of fit and precision. In summary, these studies demonstrate the versatility and effectiveness of linear regression in predicting key variables for renewable energy generation. The combination of linear regression with advanced machine learning techniques offers a comprehensive approach to address the challenges of variability in renewable energy production, contributing to the transition to more sustainable energy systems.

Despite its utility in predicting related variables, linear regression has some disadvantages that must be considered. It assumes a linear relationship between variables, which may not accurately reflect the complexity of certain phenomena, leading to inaccurate models in nonlinear cases. Additionally, linear regression is sensitive to outliers, where extreme data points can disproportionately impact results, affecting model robustness. It also lacks automatic handling of categorical variables, necessitating additional transformations for inclusion. Moreover, linear regression may suffer from multicollinearity, where predictor variables are highly correlated, making it challenging to precisely identify individual contributions. It may not be suitable for describing nonconstant relationships over time or in different data segments, as it assumes a constant relationship. In environments with large

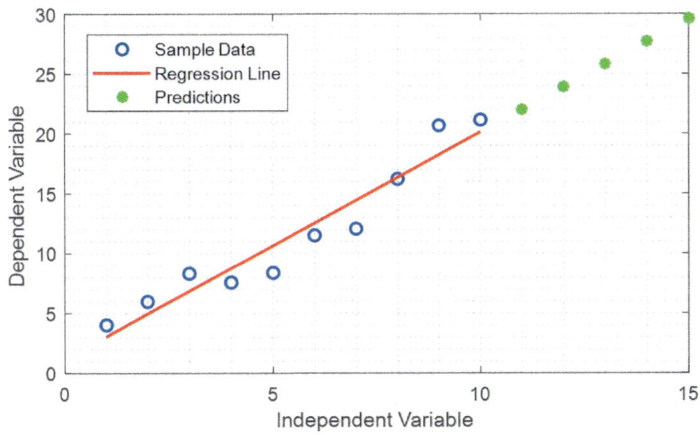

Figure 16.3 Regression prediction illustration, capturing trends and forecasting new data. Unlocking insights: Regression prediction revealing trends and forecasting future data.

datasets, linear regression can become computationally expensive and less efficient compared to more advanced machine learning approaches. In summary, while linear regression offers simplicity and ease of interpretation, its limitations in terms of linear assumptions, sensitivity to outliers, and challenges with nonlinear relationships suggest the need to carefully consider its applicability in specific contexts. Exploring more advanced approaches becomes essential when dealing with complex data. Fig. 16.3 illustrates an example of data prediction using linear regression.

Analysis and decomposition of time series in renewable energy prediction

The analysis of time series and its decomposition are crucial components in the prediction of renewable energy, contributing to understanding and forecasting complex patterns. A widely used approach involves decomposing time series data into components such as trend, seasonality, and residuals. This method enhances the understanding of underlying patterns, enabling more accurate predictions (Pujari et al., 2023). The study (Pujari et al., 2023) on wind energy prediction utilizes the Brock–Dechert–Scheinkman test to detect nonlinearity and the augmented dickey-fuller test to assess stationarity. Long-term dependencies are analyzed using the Hurst exponent. Subsequently, the time series is decomposed into trend, cycle, and seasonal patterns using the seasonal and trend decomposition using loess (STL) method. Furthermore, the study introduces an evolutionary algorithm for neural architecture search (NAS) to efficiently optimize models, comparing nonlinear autoregressive (NAR) models, LSTM neural networks, and wavelet neural networks (WNN) (Pujari et al., 2023). In another study (Malhan & Mittal, 2022), a novel prediction model was proposed for combined wind and hydroelectric power generation. The model consists of three phases: phase-I utilizes ARIMA and BiLSTM and phase-II integrates forecasts of seasonal and off-season periods obtained through a diligent search algorithm (DSA). Finally, phase-III facilitates the amalgamation of results from phases I and II to construct the proposed prediction model. Seasonal effects are crucial, and DSA identifies hidden seasonality, improving overall performance. This model offers advantages in terms of accuracy, diversity, and uniformity, demonstrating its effectiveness for long-term prediction scenarios.

On the other hand, Damaševičius et al. (2024) presented a novel recurrent neural network (RNN) model based on attention for long-term prediction of renewable energy generation, focusing on wind and solar energy datasets. Decomposition techniques are applied, and a modified metaheuristic is introduced to optimize hyperparameters. The proposed approach demonstrates an improvement in prediction accuracy, especially in addressing challenges related to the storage and management of renewable energy generation. The introduced HHO algorithm proves effective in handling the large number of hyperparameters in RNN-type networks. These studies highlight the importance of time series analysis and decomposition in improving the accuracy of renewable energy prediction models. The proposed methods, such as evolutionary NAS and ensemble prediction, contribute to overcoming challenges in nonlinear and dynamic patterns of renewable energy. The attention-based RNN approach, along with metaheuristic optimization, emerges as a promising strategy for long-term prediction. Like in any research, there are limitations, and future directions include refining optimization algorithms, exploring additional decomposition methods, and examining alternative metaheuristics for further improvements. The continuous evolution of methodologies and models remains essential for advancing the capabilities of renewable energy prediction systems.

Fig. 16.4 illustrates an example of decomposing a time series into its fundamental components using the STL method in MATLAB® 2021a. In the first subplot, the original time series is shown, which in this hypothetical example is a combination of a sinusoidal function with random noise. The second subplot presents the trend component of the series, obtained through a smoothing process that highlights long-term variations and eliminates short-term fluctuations. This trend provides information about the overall direction of the time series. The third subplot shows two additional components. The green line represents the seasonal component, capturing repetitive or cyclical patterns in the original series. Obtaining this component involves removing the trend from the original series. The blue line in the third subplot represents the residuals, which are the variations not explained by the trend and seasonality. Residuals are essential for understanding irregularities and unexpected changes in the time series. Together, this STL decomposition allows for a more detailed analysis of the time series, facilitating the identification of seasonal patterns, long-term trends, and unusual events. This approach is crucial for making more accurate forecasts and understanding the underlying structure of temporal data.

Exponential smoothing models

Exponential smoothing models play a crucial role in predicting electricity production, especially from RES. These models assign exponential weights to past values,

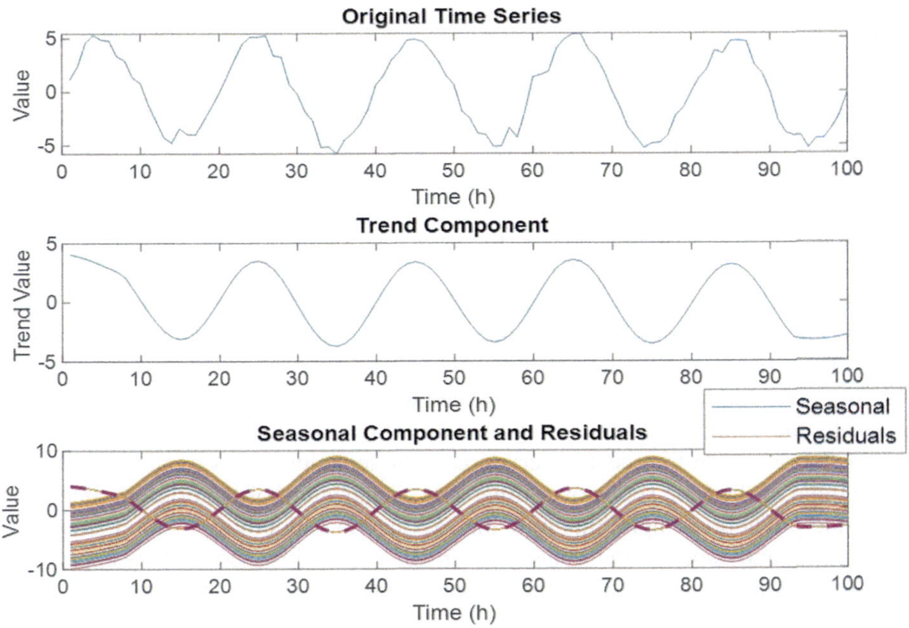

Figure 16.4 Time series decomposition: original, trend, seasonal, and residual components. Decoding time series: Original, trend, seasonal, and residual components unveiled.

giving more importance to recent observations. By adapting to changes in trends over time, they provide valuable insights into future electricity production trends, particularly in the context of renewable energy. The application of exponential smoothing models has been demonstrated in predicting electricity production in various countries, such as China, Japan, and the United States. These models have shown superior prediction accuracy compared to classical versions like error, trend, seasonal models. The ability to estimate and predict electricity production from both coal and RES is essential for informed decision-making and sustainable energy management (Tulkinov, 2023). In a study (Boland & Farah, 2021), the focus was on wind and solar systems connected to the grid in Australia. Accurate prediction of the production of these systems is crucial for increasing renewable energy penetration. Exponential smoothing is explored as an efficient alternative for predicting variance, considering the changing nature of error variance over time. The study emphasizes the importance of constructing prediction intervals, especially in situations where variance is not constant, such as in wind farms with conditional changes or due to diurnal effects in solar parks.

The discussion on predicting the production of wind or solar parks is conducted in the study (Boland & Farah, 2021), introducing two approaches: predicting the resource (wind speed or solar radiation) and using a power conversion model, or predicting directly from historical performance. The latter approach is preferred due to lower model error levels and considerations about the complexity of power conversion models. The study demonstrates the effectiveness of building prediction interval forecasts for situations where common assumptions for time series prediction may not hold. Despite their advantages, some drawbacks of exponential smoothing models include sensitivity to sudden and extreme changes in data, as well as the need to properly adjust parameters, which can be challenging in dynamic environments (Boland & Farah, 2021). Additionally, these models may not capture highly complex data patterns and are not ideal for datasets with irregular seasonality or pronounced nonlinear trends. The choice of exponential smoothing method may also depend on the specific nature of the data and require adjustments on a case-by-case basis.

In Fig. 16.5, an example of exponential smoothing applied to a time series is presented. The original time series (in blue) exhibits a sinusoidal pattern with random noise. The exponential smoothing process is carried out manually, using a smoothing parameter (alpha) set to 0.8. The dashed red line represents the predictions generated by the exponential smoothing method. This approach weighs past values and assigns more weight to recent observations. The technique aims to adapt to changes in the trend of the time series over time. The visualization of the original time series alongside predictions highlights how the exponential smoothing method smoothens fluctuations and captures the underlying trends of the time series. This approach is useful for generating forecasts that more accurately reflect the evolution of the series as new data is incorporated.

Delphi method in time series forecasting

In the context of renewable energy, this approach involves consulting experts anonymously and in a structured manner to gain insights into the future of sustainable energy. Initially, relevant experts in the field of renewable energy are identified. These may be professionals with expertise in specific technologies, energy policies, or any other crucial aspect for understanding the future landscape of RES. Once a representative group of experts is selected, key questions related to the subject are

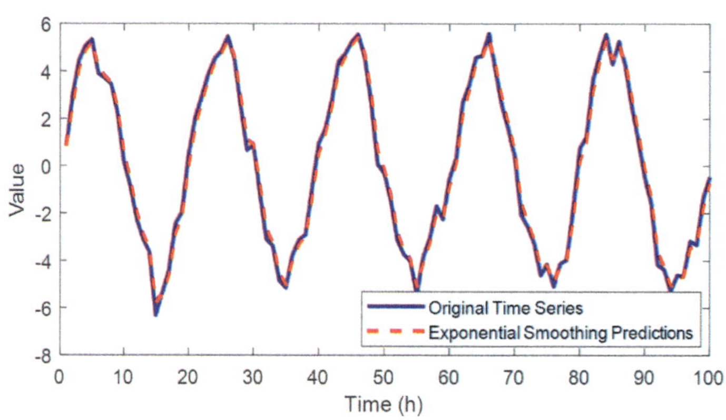

Figure 16.5 Exponential smoothing example. Illustrating smoothing power: A practical example of exponential smoothing.

formulated. These questions often address various aspects such as the expected adoption of renewable technologies, potential challenges, the influence of government policies, or even possible technological advancements in the sector. It is essential that the responses are collected anonymously to encourage honesty and openness from the experts.

The Delphi method has proven to be a valuable tool in predicting and analyzing various aspects related to energy, as evidenced in recent studies. In Panda et al. (2023), a significant transformation of the conventional energy system is highlighted due to the integration of emerging technologies, RES, and storage devices. This analysis proposes a shift toward a decentralized system to address the limited adaptability of traditional energy systems. The effects of privatization and restructuring of the electrical system are explored, along with the integration of distributed generations and consumer-side demand management with information and communication technology protocols. Challenges related to deregulated electricity markets are addressed, and demand management is suggested to handle supply and demand variability in distribution sectors. In Obrecht and Denac (2016), the focus was on the correlation between energy policy and the development of the energy industry with technological prediction. Based on a Delphi survey in Slovenia, the study addressed the importance of energy efficiency and reducing final energy consumption compared to keeping energy prices low. Findings on the future prices of various energy sources, including wood biomass, were discussed, and the feasibility of achieving the 20% renewable energy target in the European Union by 2020 was questioned. Experts emphasize the need for a radical transformation in energy policy toward efficient energy use and local renewable energy production. In another study (Ribeiro & da Silva, 2015), the Delphi method was used to provide support in strategic decision-making regarding microalgae-derived biofuels. The goal was to determine the prospects of employing microalgae in biofuel production by 2030. The method proved successful in qualitative research for long-term forecasting situations, where expert opinions are the main source of information. Results provide a clear insight into current and future issues in the microalgae-derived biofuel market. These studies illustrate the applicability of the Delphi Method in different energy research contexts, providing valuable insights that influence decision-making, policies, and future development in the field of renewable energy.

While the Delphi Method is valuable in prediction and decision-making, it is not without disadvantages and challenges. Firstly, the quality of predictions heavily depends on the selection of experts, which can introduce biases if the group is not representative. Additionally, there is a risk of conformity, where participants may be influenced by majority opinions or prominent experts, limiting the diversity of perspectives. The process can also be costly and time-consuming, involving anonymous response collection, iterations, and controlled feedback. This temporal and financial limitation can be a barrier in situations with budget or time constraints. Furthermore, the lack of direct interaction among participants may limit the ability to discuss, exchange information, and refine opinions in real time. The dependence on the quality of questions is another critical consideration. Poorly formulated questions can lead to inaccurate or unhelpful responses, compromising the validity of the results. In rapidly changing areas like renewable energy, long-term predictions can quickly become obsolete due to unexpected technological advancements.

Effective management of feedback also poses challenges, as deciding what feedback to provide to experts and how to balance divergence of opinions requires specific skills. Additionally, in contexts where a deep understanding of market dynamics is sought, the lack of direct interaction and reliance on anonymous responses can limit the thoroughness of the analysis. Despite these limitations, the Delphi method remains a valuable tool for addressing uncertainty and leveraging expert experience in long-term planning. It is crucial to use it consciously, recognizing its challenges and complementing it with other methodologies when necessary to obtain a more comprehensive and accurate perspective.

Fig. 16.6 illustrates the iterative process of the Delphi method applied to the prediction of solar energy growth over five iterations. Each subplot represents the estimates provided by a different expert (expert 1, expert 2, and expert 3) across the iterative consultations. The x-axis denotes the iteration number, while the y-axis represents the evolving estimates of solar energy growth. The markers ("o") depict the estimated values at each iteration. In the initial iteration (iteration 1), each expert offers their respective random estimates. As the Delphi method progresses, controlled feedback is applied, leading to a convergence of estimates toward a consensus in subsequent iterations (iteration 2 to iteration 5). The legends provide a clear distinction between the estimates contributed by each expert throughout the iterative process. This visual representation showcases how the Delphi method facilitates collaboration among experts, allowing their insights to inform and adjust predictions over successive iterations, ultimately aiming for a refined consensus in long-term forecasting.

Meteorological models

In the realm of renewable energy generation, various types of meteorological models are employed to forecast atmospheric conditions and optimize energy production. Some of the common types include numerical weather prediction, solar irradiance models, wind dispersion models, and data assimilation models.

Numerical weather prediction

Numerical weather prediction (NWP) models play a pivotal role in forecasting wind and solar power output, offering insights into atmospheric conditions critical for renewable energy generation. To grasp the foundations of successful NWP forecasts, a comprehensive understanding of key principles is essential (Haupt et al., 2017). These principles encompass atmospheric motion, observation sources, data assimilation, postprocessing, probabilistic predictions, and validation methods. The NWP model, exemplified by the widely used weather research and forecasting model, discretizes atmospheric equations, enabling numerical approximation crucial

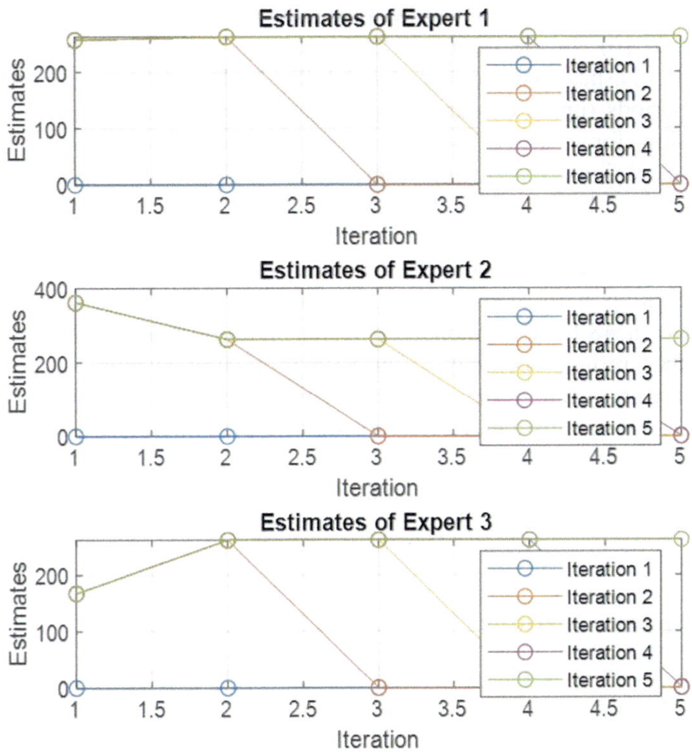

Figure 16.6 Delphi iterative convergence in solar energy growth projections. Navigating Delphi iterative convergence: Solar energy growth projections.

for accurate forecasts (Haupt et al., 2017). At the inception of the 20th century, the conceptualization of NWP was rooted in predicting weather by applying fundamental conservation laws (Haupt et al., 2017). Early attempts involved human "calculators" and grid-based calculations, laying the groundwork for modern NWP models. With the advent of digital computers, pioneering works by visionaries like Richardson and Charney led to the development of successful NWP models (Haupt et al., 2017). In solar energy forecasting, a focus on direct normal irradiance (DNI) for concentrating solar thermal (CST) systems is paramount (Alkabbani et al., 2021). This involves considering various forecasting methods, including numerical weather prediction models, statistical forecasting, and nowcasting. Challenges in immediate forecasting (up to 6 hours) necessitate harmonizing signals for high-resolution spatial-temporal predictions at the plant level (Ramírez & Vindel, 2017).

For PV power systems, accurate forecasting methods are crucial for power supply and demand balancing (Kato, 2016). Forecasting tools, including NWP models, satellite images, and measured PV power output data, contribute to various time horizons, from minutes to days. The evolution of irradiance forecasting demands a nuanced understanding of fluctuation characteristics and the application of forecasting methods in both power systems and individual energy management systems (Kato, 2016). In the realm of CST systems, forecasting and nowcasting of DNI are critical aspects (Ramírez & Vindel, 2017). The challenges lie in immediate prediction periods (up to 6 hours), requiring harmonization of signals for reliable, high-resolution forecasts. Considerations are presented regarding single versus aggregated or regional forecasting (Ramírez & Vindel, 2017). Wind power generation, characterized by its intermittent nature, demands advanced forecasting techniques (Shirzadi et al., 2023). Traditional NWP models face limitations in long-term forecasting due to computational constraints. A hybrid AI-based approach, integrating deep learning with a probability distribution, enhances wind speed forecast accuracy over a 48-hour horizon. The fusion of this improved model with NWP models further refines accuracy, offering a promising avenue for efficient wind speed forecasting in renewable energy systems (Shirzadi et al., 2023). In summary, integrating NWP models and various forecasting methodologies is crucial for optimizing the performance of renewable energy systems. These approaches provide valuable insights into atmospheric conditions, contributing to more reliable and efficient renewable energy generation.

Solar irradiance models

Solar irradiance models play a crucial role in the assessment and optimization of PV energy systems, providing essential information about the amount of incident solar radiation on the Earth's surface. In this context, five prominent models stand out: HOMER (hybrid optimization model for multiple energy resources), PVWatts, SOLPOS (solar position and intensity), SMARTS (simple model of the atmospheric radiative transfer of sunshine), and PVGIS (PV geographical information system). Each of these models addresses specific aspects related to solar irradiance prediction and has undergone detailed studies revealing their effectiveness and applicability under various conditions. The following is a brief description of each model followed by citations highlighting their utility and performance in particular contexts.

Hybrid optimization model for multiple energy resources

HOMER is a versatile tool that not only serves as a simulator and optimizer but also incorporates models for predicting solar irradiance. Its use extends to microgrid systems, and a detailed study (Shaikh et al., 2021) on the configuration, operation, and integration of multiple energy sources in microgrid systems emphasized the prominence of HOMER Pro and the particle swarm optimization algorithm. This study provides a comprehensive review of microgrid perspectives, considering adopted technologies, control strategies, existing applications, software tools, and artificial intelligence techniques.

PVWatts

PVWatts, developed by the National Renewable Energy Laboratory (NREL), is an online tool that models the performance of PV solar energy systems. This model uses meteorological data and PV system specifications to estimate solar energy production. In a comparative study (Milosavljević et al., 2022) evaluating 14 PV simulation tools, PVWatts showed lower deviation in simulated results compared to experimental measurements of a 2 kWp PV system in Niš.

Solar position and intensity

SOLPOS, developed by the NREL, focuses on calculating the sun's position and solar intensity at specific locations and times. In a study (Suh & Choi, 2017), methods for converting monthly accumulated irradiation data into hourly data were compared. The research concluded that the SOLPOS algorithm was the most suitable method, as it showed highly similar monthly irradiation patterns to measured values at seven sites in the United States and Korea.

Simple model of the atmospheric radiative transfer of sunshine

SMARTS, an acronym for simple model of the atmospheric radiative transfer of sunshine, developed by NREL, is used to simulate spectral solar irradiance. In a study (Ali & Zhao, 2022), spectral solar irradiance measured by a spectrometer in the black hills area was analyzed, and a data-based model was developed to convert measurements into absolute irradiance. This case study demonstrated that using SMARTS to determine the energy production of PV systems improves accuracy by avoiding absolute calibration of the light spectrometer.

Photovoltaic geographical information system

PVGIS, developed by the Joint Research Centre of the European Commission, provides data and models to evaluate the performance of PV systems. In a study (Lehneis et al., 2020), PVGIS was used to simulate the electricity generation of 1.612 million PV systems in Germany. This model relies on publicly accessible data and offers an alternative for calculating highly resolved electricity generation data from PV systems, demonstrating its applicability in different regions and climatic conditions.

Machine learning and neural networks for renewable energy generation prediction

Machine learning and ANNs emerge as innovative and efficient tools to precisely anticipate energy production from RES. This research field not only addresses the challenges associated with the inherent variability of these energy sources but also provides advanced solutions that can significantly enhance the planning and management of renewable energy generation. In this section, we will explore the most powerful techniques currently available and present some challenges and future trends.

Recurrent neural networks

The application of RNN in renewable energy generation forecasting signifies a significant advancement in optimizing sustainable energy prediction. RNNs, with their ability to grasp intricate temporal patterns, emerge as ideal tools for addressing the inherent variability in RES. This technique enables the modeling of dynamic relationships among climatic variables, past generation patterns, and other relevant factors, thereby achieving more accurate forecasts. In this section, we will explore how RNN, particularly the well-recognized techniques of LSTM and gated recurrent unit (GRU), offer a promising perspective to enhance prediction accuracy in the context of renewable energy generation.

Long short-term memory

LSTM stands out as a robust model for solar forecasting, adept at addressing challenges such as hyperparameter tuning and the consideration of meteorological parameters. The optimized stacked Bidirectional LSTM/LSTM model, introduced in the referenced study, integrates stacked LSTM layers, dropout architecture, and Bayesian optimization to augment performance. The model exhibits exceptional accuracy in forecasting global horizontal irradiance (GHI) and observed plane of array (POA) irradiance, achieving R^2 values of 0.99 for univariate and multivariate models using GHI data. Michael et al. (2022) underscored the model's flexibility, achieved through Bayesian optimization and the choice between BiLSTM and LSTM architectures. The conclusions highlight the efficacy of the LSTM model in delivering superior solutions for univariate and multivariate solar forecasting, surpassing benchmark models in terms of error metrics. The incorporation of POA data and the model's successful application to PV power forecasting further emphasize its versatility. The study acknowledged the LSTM model's prowess in handling extensive time series data and proposed future efforts to mitigate errors in solar irradiance forecasting under nonlinear weather conditions.

On the other hand, some studies (Shirzadi et al., 2023; Rangelov et al., 2023) discussed the hurdles posed by the unpredictable nature of RES, particularly wind and solar power. They proposed innovative AI-based solutions to enhance the precision of energy generation forecasts. In Shirzadi et al. (2023), the emphasis was on wind power, revealing the limitations of traditional numerical weather prediction models due to computational constraints. The study introduced a hybrid AI approach that integrates wind speed probability distribution with a deep learning model, showcasing improved accuracy in a 48-hour forecast for Montreal. Results indicate a substantial reduction in the root mean squared error, highlighting the efficacy of combining AI techniques for enhanced wind power prediction. Similarly, Rangelov et al. (2023) delved into challenges related to the integration of PV systems into the grid, emphasizing the volatile and intermittent nature of RES. The paper evaluates three AI methods—random forest, deep neural network, and long short-term memory network—for short-term PV output power forecasting in Berlin, Germany. The study underscores the significance of accurate PV power output forecasts in reducing disruptions, refining capacity planning, and cutting operational costs. The successful performance of the random forest and deep neural network models underscores their ability to generate precise solar power forecasts and navigate abrupt changes in PV power output. Collectively, these studies underscore the pivotal role of AI in overcoming challenges tied to renewable energy integration and refining the dependability of forecasting models.

Gated recurrent unit

GRU, as another variant of RNN, has demonstrated effectiveness in renewable energy forecasting and energy management. In a study exploring a demand-supply matching approach for renewable energy forecasting, DL models, including GRU, were employed to enhance energy management for futuristic cities with 100% renewables (Almalaq et al., 2021; Kim et al., 2021). The use of GRU in this context aligns with its ability to capture dependencies in time series data, which is crucial for forecasting intermittent RES. Furthermore, the study proposing a data-driven bidding strategy for a DER aggregator in the day-ahead power market emphasizes the role of GRU in the precise modeling of renewables and local load, showcasing its superiority over LSTM, RNN, and SVR (Kim et al., 2021). The GRU model, optimized using the ELPSO algorithm and IMI feature selection technique, outperformed other models in terms of forecasting accuracy, contributing to reduced penalty costs and improved bidding strategies in the face of uncertainties. In summary, LSTM and GRU, as sophisticated RNN variants, play crucial roles in addressing the complexities of renewable energy forecasting. The cited studies provide compelling evidence of their efficacy in capturing temporal dependencies, handling uncertainties, and outperforming traditional models. The versatility of these models, demonstrated across various scenarios and datasets, positions them as valuable tools for advancing the accuracy and efficiency of renewable energy prediction, thereby contributing to the sustainable evolution of energy systems.

Despite their notable advantages, RNN models, such as LSTM and GRU, are not without challenges and limitations in predicting renewable energy generation. One key challenge is the need for high-quality data for training, as the accuracy of these models heavily depends on the quality and representativeness of historical data. Additionally, the computational complexity associated with deeper architectures, such as deep neural networks, may require significant resources. The interpretation of results can also be challenging, especially for models based on deep learning, which may lack transparency. Moreover, extreme climatic variability and unpredictable weather conditions pose additional challenges, as models must adapt to sudden changes in environmental variables. These challenges underscore the importance of continuously addressing data quality, computational efficiency, and model interpretability to optimize accuracy in renewable energy generation prediction.

Convolutional neural networks

CNNs, commonly associated with computer vision, have found innovative applications in predicting renewable energy generation, particularly in scenarios such as wind speed maps. This discourse delves into the advancements achieved by various studies in employing CNNs for renewable energy prediction, emphasizing their potential to overcome challenges linked to the intermittent nature of renewables. In Schreiber et al. (2021), the ComPonentNet (CPNet) family of architectures utilizes CNNs for wind speed prediction. The study introduces three architectures, illustrating that processing wind components separately, as in the bottom-fused CPNet (BF-CPNet), significantly enhances forecasting accuracy. The results underscore the promising nature of CNNs in addressing the stochastic elements introduced by renewable energy, highlighting potential cost reductions for power systems. Similarly, Schreiber et al. (2021) introduced a temporal convolutional network (TCN) for day-ahead power forecasts, improving forecast accuracy by considering the diurnal cycle. The study demonstrated a notable improvement (up to 25%) compared to traditional multilayer perceptrons in renewable power forecasts. The proposed TCN architecture exhibits adaptability for multitask learning, inductive transfer learning, and even zero-shot learning, indicating its potential to overcome challenges associated with renewable power prediction. In Bastos et al. (2021), the U-Convolutional model employs CNNs for hourly wind speed prediction, displaying competitive performance against advanced CNN architectures, fully connected neural networks, and univariate models. The proposed architecture, combining U-Net and Convnet, proves promising for spatiotemporal wind speed prediction. The results underscore the importance of advanced models in supporting power systems grappling with the increasing penetration of intermittent renewables.

Dolatabadi et al. (2023) addressed the economic optimization of a CAES-PV system using DRL and a 2D-CNN-BLSTM hybrid forecasting model. By tackling the intermittent nature of solar power production, the study demonstrates the effectiveness of CNN-based forecasting in enhancing the profitability of the CAES-PV system in energy markets. Lastly, Wu et al. (2023) introduced the unrolled spatiotemporal graph convolutional network (USGCN) for distribution system state estimation and forecasting. The model, incorporating complex spatiotemporal correlations of RES, outperforms traditional methods and achieves at least a 16.42% improvement in state estimation accuracy. The USGCN's ability to efficiently capture spatiotemporal correlations highlights its potential to address challenges posed by high renewables.

The discussed studies illustrate the diverse applications of CNNs in predicting renewable energy generation. From wind speed maps to day-ahead power forecasts and distribution system state estimation, CNNs have demonstrated their effectiveness in enhancing accuracy and overcoming challenges associated with the intermittent nature of renewables. These advancements not only contribute to more reliable renewable energy predictions but also hold the potential to reduce operational costs and support the integration of renewables into modern power systems. Future work may further refine these CNN-based approaches, exploring additional features, analyzing errors in-depth, and incorporating confidence intervals to enhance spatiotemporal forecasts even further. Despite the notable progress in the application of CNNs in renewable generation prediction, challenges and limitations exist that must be addressed to optimize their utility in practical settings. Firstly, the computational intensity of CNNs, especially in handling large datasets and complex models, can lead to high processing power requirements and prolonged training times. Additionally, the need for extensive training datasets can be a drawback in areas with limited availability of high-quality data.

Generative adversarial networks

Generative adversarial networks have become pivotal tools in the domain of renewable energy prediction, particularly when confronted with limited historical data. Their primary role lies in generating synthetic data to train models effectively, and the following summaries encapsulate insights from five relevant studies. In a collaborative initiative, the Korea electrotechnology research institute and Korea Institute of energy research introduced a renewables management system for large-scale grids. Leveraging a 12-hour-ahead solar irradiance forecast model and a generative adversarial network (GAN) for generating future RES penetration scenarios, the study underscored the potential of GANs in comprehensive grid management (Song et al., 2023). To address the uncertainties of renewable energy, another study introduced normalizing flows, a deep learning technique, for scenario-based probabilistic forecasts in power systems. This novel approach competes with other generative models, including GANs, showcasing its efficacy in generating accurate weather-based scenarios (Dumas et al., 2022). The Fed-LSGAN, an innovative federated deep generative learning framework, offers a privacy-preserving solution for renewable scenario generation in power systems. By combining federated learning and least square GANs, this model outperforms centralized methods, emphasizing the versatility of GANs in overcoming privacy concerns (Li et al., 2022).

Shifting the focus to short-term cloud coverage forecasting through sky image prediction using GANs, a study employs a deep CNN topology with adversarial loss. This method aids in estimating available solar resources by providing realistic images, evaluated using a cloud coverage metric (Andrianakos et al., 2019). In the realm of wind park power modeling, various methods to estimate probability distributions for individual turbine powers are explored. Comparisons between variational Bayesian inference models, GANs, and others reveal that Bayesian models generally outperform GANs in terms of mean absolute errors (MAE) (Bentsen et al., 2022). While GANs offer significant advantages in renewable energy prediction, it is crucial to acknowledge certain disadvantages and future challenges. Computational intensity poses challenges, particularly in handling large datasets and complex models, leading to high processing power requirements and prolonged training times. Additionally, interpreting decisions made by GANs can be complex, hindering their adoption in critical environments where explainability is crucial. Sensitivity to hyperparameter configuration underscores the importance of careful selection to avoid suboptimal performance or overfitting.

Looking ahead, improving the interpretability of GANs and developing techniques for transparent decision-making stand out as key priorities. The adaptability to scenarios with limited data presents a crucial challenge, where strategies like transfer learning may play a pivotal role. Enhancing computational efficiency through exploring more efficient architectures and techniques will be essential for real-time applications. Integrating measures of uncertainty into GAN predictions is vital for informed decision-making, especially in renewable energy generation characterized by variability. Lastly, addressing the generalization of GANs to diverse geographical contexts and climatic conditions remains a significant challenge, requiring considerations of data variability and location-specific conditions for effective applicability in different regions. Overcoming these challenges will unlock the full potential of GANs in renewable energy prediction, supporting a more efficient and sustainable transition to clean energy sources.

Deep learning

Deep learning, a transformative paradigm in predicting renewable energy generation, leverages intricate architectures like deep neural networks to discern complex and nonlinear patterns within data. The following summaries encapsulate insights from five significant studies applying deep learning to renewable energy prediction. In the realm of solar energy, the traditional encoder single deep learning method proves effective for short-term forecasts. The integration of this method into the power grid addresses challenges posed by the irregular and uncontrollable nature of power generation from renewables, showcasing a 27% improvement in accuracy over conventional methods (Chang et al., 2021). For accurate short-term PV power forecasts, a novel model, transformer neural network (TransNN), integrated with CNN and utilizing variational mode decomposition, outperforms benchmark models with MAE values under 1 kW (Nguyen Trong et al., 2023).

Addressing the intermittent and irregular nature of wave energy, an optimized deep learning neural network, guided by the moth-flame optimization algorithm, excels in forecasting wave energy flux. This approach, surpassing statistical and physics-based methods, exhibits robust performance, especially over short-term horizons (Bento et al., 2021). Climate change mitigation necessitates a transition to wind energy, but the nonlinear nature of wind poses forecasting challenges. LSTM networks, NAR models, and WNN emerge as efficient models for accurate wind forecasting. An evolutionary multiobjective strategy for NAS optimally designs these models, balancing computational cost and prediction accuracy (Pujari et al., 2023). In the context of wind farm prediction, a deep, multilayered neural network demonstrates efficacy in estimating wind farm generation 24 hours in advance. Trained and validated with data from a wind farm, this model outperforms existing predictors, highlighting the prowess of deep learning in enhancing predictive generation models for wind farms (Torres et al., 2018). While deep learning offers remarkable advancements, challenges and future endeavors persist. Computational intensity, model interpretability, and hyperparameter sensitivity emerge as notable disadvantages. Future efforts should prioritize enhancing model interpretability, adapting to scenarios with limited data, improving computational efficiency, and addressing model generalization across diverse geographical and climatic conditions. Overcoming these challenges will usher in a new era of accurate and sustainable renewable energy predictions.

Reinforcement learning

RL techniques offer promising avenues for real-time optimization in the renewable energy sector. Let us delve into the application of RL in various scenarios. One innovative application involves CAES combined with PV systems. A novel DRL method is employed to optimize CAES energy arbitrage, considering the intermittent nature of solar power production. A hybrid forecasting model, integrating 2-D CNNs and BLSTM units, enhances solar irradiance predictions. The proposed framework demonstrates effectiveness in real-time scheduling and profitability, showcasing its superiority over state-of-the-art methods (Dolatabadi et al., 2023). In the domain of building energy management, a decision tree-based RL approach is introduced for accurate electricity consumption forecasting. The decision tree identifies different contexts, and a multiarmed bandit algorithm guides the selection of the most suitable forecasting algorithm. The study, validated in building and industrial contexts, highlights the methodology's adaptability and efficacy in enhancing prediction accuracy (Ramos et al., 2022). Addressing challenges in electrification and decarbonization, RL techniques are applied to demand response scenarios. Transfer learning, leveraging existing domain knowledge, is explored to mitigate sample complexity and enhance the performance of machine learning and optimization algorithms.

Despite progress, challenges such as the lack of benchmarks and consensus on techniques for avoiding negative transfer persist in the application of RL to demand response (Peirelinck et al., 2022). Virtual power plants (VPPs) offer the potential to increase power system reliability and efficiency. RL is utilized to formulate profit-maximizing strategies for a VPP aggregating solar systems and electric vehicle (EV) chargers with vehicle-to-grid support. Two-stage stochastic optimization, combining linear programs and RL techniques, optimizes energy trading decisions in day-ahead and imbalance electricity markets. The study evaluates the proposed strategies using real market data and emphasizes the potential profitability of VPPs with EV integration (Rahman et al., 2022). While RL exhibits promising applications in renewable energy prediction, challenges persist. Model interpretability, high sample complexity, and potential negative transfer in demand response scenarios are notable disadvantages. Future endeavors should focus on improving interpretability, addressing data limitations, and establishing benchmarks to overcome the identified challenges. Additionally, research must extend into real-world experiments for a more comprehensive understanding of RL's effectiveness in renewable energy applications.

Transformer neural networks

These networks, initially designed for natural language processing, have proven adaptable to various domains, including the prediction of renewable energy generation. A groundbreaking study in wind power forecasting explores the application of a Transformer coupled with a CNN using attention mechanisms. This novel approach, evaluated across different periods and time intervals, outperforms existing models for ultrashort-term and short-term forecasting. By effectively capturing complex patterns and dependencies in wind power data, the transformer algorithm, in conjunction with CNN and attention mechanisms, demonstrates superior accuracy, enhancing the integration of wind power into the grid (Mulewa et al., 2023). Similarly, the spatial-temporal graph transformer network (STGTN) is proposed for short-term wind speed forecasting. This model incorporates a transformer with an external attention mechanism and a graph convolutional layer, enabling the extraction of both temporal and spatial features. The STGTN outperforms benchmark methods in terms of MAE, root mean square error, and mean absolute percentage error, showcasing its effectiveness in capturing wind speed fluctuations and spatial dependencies (Pan et al., 2022).

Extending the application of transformers to solar energy, a study utilizes the transformer deep neural network model for solar irradiance prediction. By combining features based on spatiotemporal properties, the transformer model, typically applied in NLP or vision problems, demonstrates remarkable accuracy with a maximum worst-case mean absolute percentage error of 3.45% for one-day-ahead predictions. The model's ability to utilize space–time bonds between features and regional data proves advantageous for solar irradiance forecasting (Pospíchal et al., 2022). Expanding the scope to include VPPs, a study proposes a hybrid system for renewable energy forecasting, integrating the traditional seasonal ARIMA with a transformer neural network. Named H-transformer, this hybrid model outperforms other single models and hybrid approaches across various datasets of wind speed and solar energy. The study emphasizes the importance of accurate forecasting for the profitability and stability of renewable energy systems (Padilha et al., 2022). The integration of CNN, LSTM networks, and transformers is explored for solar energy production forecasting. This hybrid model, outperforming existing models, proves effective in providing accurate forecasts, facilitating the integration of solar energy into grids. The combination of spatial and temporal features extracted by CNN, LSTM, and transformer enhances forecasting accuracy (Al-Ali et al., 2023).

For wind speed and wind energy forecasting, a transformer-based deep neural network-integrated with wavelet transform is proposed (Liu et al., 2023). This model, evaluated in various regions in Bahia, Brazil, outperforms the LSTM baseline, demonstrating effectiveness in forecasting wind speed and power generation. The integration of the transformer model with wavelet decomposition further improves forecast accuracy (Nascimento et al., 2023). While transformer ANNs exhibit promising capabilities, challenges and areas for future research persist. Some of the identified limitations include the complex training process, the need for extensive computational power, and the scarcity of open multivariate renewable energy datasets. Future endeavors should focus on mitigating these challenges, exploring long-term forecasting, incorporating seasonal factors, and enhancing the models' robustness across different weather conditions. Additionally, efforts should be directed toward developing interpretable models, addressing optimization complexities, and fostering broader applicability in various forecasting tasks.

Challenges and future research lines in renewable generation prediction

Table 16.1 provides a concise overview of the challenges and future research directions associated with various methods employed in renewable energy prediction. The methods encompass traditional time series forecasting approaches, meteorological models, and cutting-edge machine learning techniques. The challenges highlighted include issues related to adaptability, linearity assumptions, nonlinear patterns, model sensitivity, and the need for substantial data and computational resources. Additionally, future research

Table 16.1 Challenges and future research lines in renewable energy prediction methods.

Method	Challenges	Future Research Lines
Moving average and autoregressive integrated	Sensitivity to outliers and abrupt changes.	Improvements in adaptability to sudden changes in energy generation. Integration of more advanced methods to handle variability.
Linear regression	Linearity assumptions may not capture the complexity of variability.	Development of more sophisticated nonlinear models. Integration of additional variables for improved accuracy.
Analysis and decomposition of time series	Difficulty in handling nonlinear patterns and nonstationary trends.	Research on more advanced decomposition techniques. Adaptation to complex seasonal patterns.
Exponential smoothing models	Sensitivity to parameter and model structure choices.	Development of optimization methods for automatic parameter selection. Improvements in predictive capability.
Delphi method	Dependence on experts and subjectivity in data collection.	Incorporation of methods to manage uncertainty and enhance objectivity. Evaluation of method robustness.
Meteorological models	Specific challenges for each type of meteorological model.	Research on improving the accuracy of specific models. Integration of multiple models for more robust forecasts.

(Continued)

Table 16.1 (Continued)

Method	Challenges	Future Research Lines
Machine learning and neural networks	Need for large datasets and computational resources.	Development of learning techniques with limited data. Research on more computationally efficient models.
Recurrent Neural Networks	Vanishing/exploding gradient problems in long sequences.	Research on more advanced and efficient architectures. Handling longer temporal sequences.
Convolutional Neural Networks (CNNs)	Challenges in capturing temporal relationships in sequential data.	Exploration of hybrid architectures to combine temporal and spatial features. Improvements in interpretability.
Generative Adversarial Networks (GANs)	Generation of realistic data in low generation scenarios.	Research on enhancement techniques for generation in low data availability situations. More robust validation.
Deep learning	Interpretability and explainability of complex models.	Development of methods to increase the interpretability of models. Research on explainability techniques.
Reinforcement Learning (RL)	Need for efficient algorithms for continuous adaptation.	Development of more efficient and adaptable RL algorithms. Integration with traditional approaches to improve robustness.
Transformer Neural Networks (TNNs)	Need for large amounts of data for training.	Research on training techniques with small datasets. Exploration of more computationally efficient architectures.
Federated learning (FL)	Challenges in collaboration between distributed models.	Improvements in security and efficiency of distributed collaboration. Research on more robust federated methods.
Transfer learning (TL)	Effective transfer of knowledge between domains.	Development of transfer learning techniques specific to renewable energy forecasts. Evaluation of generalization.
Ensemble methods	Effective integration of heterogeneous models for improved accuracy.	Research on more sophisticated ensemble methods. Evaluation of ensemble robustness and reliability.

lines are outlined, suggesting potential improvements, advanced techniques, and specific areas for exploration within each method to enhance the precision and reliability of renewable energy predictions.

Conclusions

In this comprehensive analysis of renewable energy generation prediction methods, a wide spectrum of approaches has been explored, ranging from classical methods to advanced techniques in machine learning and ANN. The review of fundamental concepts and traditional methods highlights their historical significance but also reveals their limitations in the face of the increasing complexity of current energy landscapes. The inclusion of meteorological models demonstrates the critical relevance of understanding and predicting atmospheric conditions to optimize renewable energy production. However, specific challenges are evident in terms of accuracy and adaptability to changing conditions. The section dedicated to machine learning techniques and ANNs reveals the innovative potential of these approaches to address the inherent variability of RES. Despite their significant contributions, challenges such as the need for large datasets and substantial computational resources are noted.

As a result of this analysis, future research directions are highlighted to advance renewable energy generation prediction. Adapting traditional methods to changing energy landscapes, improving accuracy in meteorological models, and exploring more resource-efficient approaches emerge as key areas for future investigations. In conclusion, this study provides a comprehensive overview of the current state of renewable energy generation prediction, identifying challenges and pointing towards promising directions to advance this crucial field in the transition toward a sustainable future.

References

Al-Ali, E. M., Hajji, Y., Said, Y., Hleili, M., Alanzi, A. M., Laatar, A. H., & Atrí, M. (2023). solar energy production forecasting based on a hybrid CNN-LSTM-transformer model. *Mathematics*, *11*(3). Available from https://doi.org/10.3390/MATH11030676, https://www.mdpi.com/2227-7390/11/3/676.

Ali, A. J., & Zhao, L. (2022). Solar spectral irradiance analysis and modeling: A case study in The Black Hills Area. In *North American power symposium, NAPS 2022*, https://doi.org/10.1109/NAPS56150.2022.10012211.

Alkabbani, H., Ahmadian, A., Zhu, Q., & Elkamel, A. (2021). Machine learning and metaheuristic methods for renewable power forecasting: A recent review. *Frontiers in Chemical Engineering*, *3*. Available from https://doi.org/10.3389/FCENG.2021.665415/BIBTEX.

Almalaq, A., Alshammarry, A., Alanzi Bader Alharbi, F., & Alshudukhi, M. (2021). Deep learning applied on renewable energy forecasting towards supply-demand matching. In *20th IEEE international conference on machine learning and applications, ICMLA 2021* (pp. 1345–1349). Institute of Electrical and Electronics Engineers Inc. Available from https://doi.org/10.1109/ICMLA52953.2021.00218.

Andrianakos, G., Tsourounis, D., Oikonomou, S., Kastaniotis, D., Economou, G., & Kazantzidis, A. (2019). Sky image forecasting with generative adversarial networks for cloud coverage prediction. In *10th International conference on information, intelligence, systems and applications, IISA 2019*. Institute of Electrical and Electronics Engineers Inc. https://doi.org/10.1109/IISA.2019.8900774.

Bastos, B. Q., Cyrino Oliveira, F. L., & Milidiú, R. L. (2021). U-Convolutional model for spatio-temporal wind speed forecasting. *International Journal of Forecasting*, *37*(2). Available from https://doi.org/10.1016/J.IJFORECAST.2020.10.007.

Bayo-Besteiro, S., García-Rodríguez, M., Labandeira, X., & Añel, J. A. (2022). Seasonal and subseasonal wind power characterization and forecasting for the Iberian Peninsula and the Canary Islands: A systematic review. *International Journal of Climatology*, *42*(5). Available from https://doi.org/10.1002/JOC.7359, https://rmets.onlinelibrary.wiley.com/doi/10.1002/joc.7359.

Bento, P. M. R., Pombo, J. A. N., Mendes, R. P. G., Calado, M. R. A., & Mariano, S. J. P. S. (2021). Ocean wave energy forecasting using optimised deep learning neural networks. *Ocean Engineering*, *219*. Available from https://doi.org/10.1016/J.OCEANENG.2020.108372.

Bentsen, L. Ø., Warakagoda, N. D., Stenbro, R., & Engelstad, P. (2022). Probabilistic wind park power prediction using bayesian deep learning and generative adversarial networks. *Journal of Physics: Conference Series*, *2362*(1). Available from https://doi.org/10.1088/1742-6596/2362/1/012005, https://iopscience.iop.org/article/10.1088/1742-6596/2362/1/012005/meta.

Boland, J., & Farah, S. (2021). Probabilistic forecasting of wind and solar farm output. *Energies*, *14*(16). Available from https://doi.org/10.3390/EN14165154, https://www.mdpi.com/1996-1073/14/16/5154.

Cellura, M., Cirrincione, G., Marvuglia, A., & Miraoui, A. (2008). Wind speed spatial estimation for energy planning in Sicily: Introduction and statistical analysis. *Renewable Energy*, *33*(6). Available from https://doi.org/10.1016/J.RENENE.2007.08.012.

Chang, R., Bai, L., & Hsu, C. H. (2021). Solar power generation prediction based on deep Learning. *Sustainable Energy Technologies and Assessments*, *47*. Available from https://doi.org/10.1016/J.SETA.2021.101354.

Damaševičius, R., Jovanovic, L., Petrovic, A., Zivkovic, M., Bacanin, N., Jovanovic, D., & Antonijevic, M. (2024). Decomposition aided attention-based recurrent neural networks for multistep ahead time-series forecasting of renewable power generation. *PeerJ Computer Science*, *10*, e1795. Available from https://doi.org/10.7717/PEERJ-CS.1795, https://peerj.com/articles/cs-1795.

David, M., Luis, M. A., & Lauret, P. (2018). Comparison of intraday probabilistic forecasting of solar irradiance using only endogenous data. *International Journal of Forecasting*, *34* (3). Available from https://doi.org/10.1016/J.IJFORECAST.2018.02.003.

David, M., Ramahatana, F., Trombe, P. J., & Lauret, P. (2016). Probabilistic forecasting of the solar irradiance with recursive ARMA and GARCH models. *Solar Energy*, *133*. Available from https://doi.org/10.1016/J.SOLENER.2016.03.064.

Dolatabadi, A., Abdeltawab, H., & Mohamed, Y. A.-R. I. (2023). Deep reinforcement learning-based self-scheduling strategy for a CAES-PV system using accurate sky images-based forecasting. *IEEE Transactions on Power Systems*, *38*(2). Available from https://doi.org/10.1109/TPWRS.2022.3177704.

Dumas, J., Wehenkel, A., Lanaspeze, D., Cornélusse, B., & Sutera, A. (2022). A deep generative model for probabilistic energy forecasting in power systems: Normalizing flows. *Applied Energy*, *305*. Available from https://doi.org/10.1016/J.APENERGY.2021.117871.

Fattah, R., Malik, H., Basmadjian, R., & Shaafieyoun, A. (2023). Assessing ARIMA-based forecasts for the percentage of renewables in Germany: Insights and lessons for the future. *Energies*, *16*(16). Available from https://doi.org/10.3390/EN16166005, https://www.mdpi.com/1996-1073/16/16/6005.

Gangopadhyay, A., Kanase, R., Mohan, G., Halder, M., Deshpande, M., Mukhopadhyay, P., & Srinivasan, J. (2019). Use of a weather forecast model to identify suitable sites for new wind power plants in Karnataka. *Current Science*, *117*(8). Available from https://doi.org/10.18520/CS/V117/I8/1347-1353.

Hassan, M. Z., Ali, M. E. K., Ali, A. B. M. S., Kumar, J. (2017). Forecasting day-ahead solar radiation using machine learning approach. In *4th Asia-Pacific world congress on computer science and engineering, APWC on CSE 2017* (pp. 252–258). https://doi.org/10.1109/APWCONCSE.2017.00050.

Haupt, S. E., Jiménez, P. A., Lee, J. A., & Kosović, B. (2017). *Principles of meteorology and numerical weather prediction. Renewable energy forecasting: From models to applications.* Woodhead Publishing. Available from https://doi.org/10.1016/B978-0-08-100504-0.00001-9.

Islam, S., & Roy, N. K. (2023). Renewables integration into power systems through intelligent techniques: Implementation procedures, key features, and performance evaluation. *Energy Reports*, *9*. Available from https://doi.org/10.1016/J.EGYR.2023.05.063.

Kato, T. (2016). Prediction of photovoltaic power generation output and network operation. In *Integration of distributed energy resources in power systems: Implementation, operation and control*. Available from https://doi.org/10.1016/B978-0-12-803212-1.00004-0.

Kim, H. J., Kang, H. J., & Kim, M. K. (2021). Data-driven bidding strategy for der aggregator based on gated recurrent unit-enhanced learning particle swarm optimization. *IEEE Access*, *9*. Available from https://doi.org/10.1109/ACCESS.2021.3076679.

Koltsaklis, N., Panapakidis, I. P., Pozo, D., & Christoforidis, G. C. (2021). A prosumer model based on smart home energy management and forecasting techniques. *Energies*, *14*(6). Available from https://doi.org/10.3390/EN14061724.

Lehneis, R., Manske, D., & Thrän, D. (2020). Generation of spatiotemporally resolved power production data of PV systems in Germany. *ISPRS International Journal of Geo-Information*, *9*(11). Available from https://doi.org/10.3390/IJGI9110621, https://www.mdpi.com/2220-9964/9/11/621.

Li, Y., Li, J., & Wang, Y. (2022). Privacy-preserving spatiotemporal scenario generation of renewable energies: A federated deep generative learning approach. *IEEE Transactions on Industrial Informatics*, *18*(4). Available from https://doi.org/10.1109/TII.2021.3098259.

Li, Y., Wang, C., & Li, G. (2020). A mini-review on high-penetration renewable integration into a smarter grid. *Frontiers in Energy Research*, *8*. Available from https://doi.org/10.3389/FENRG.2020.00084/BIBTEX.

Liu, J., Zang, H., Cheng, L., Ding, T., Wei, Z., & Sun, G. (2023). A Transformer-based multimodal-learning framework using sky images for ultra-short-term solar irradiance forecasting. *Applied Energy*, *342*. Available from https://doi.org/10.1016/J.APENERGY.2023.121160.

Malhan, P., & Mittal, M. (2022). A novel ensemble model for long-term forecasting of wind and hydro power generation. *Energy Conversion and Management*, *251*. Available from https://doi.org/10.1016/J.ENCONMAN.2021.114983.

Michael, N. E., Hasan, S., Al-Durra, A., & Mishra, M. (2022). Short-term solar irradiance forecasting based on a novel Bayesian optimized deep long short-term memory neural network. *Applied Energy*, *324*. Available from https://doi.org/10.1016/J.APENERGY.2022.119727.

Milosavljević, D. D., Kevkić, T. S., & Jovanović, S. J. (2022). Review and validation of photovoltaic solar simulation tools/software based on case study. *Open Physics*, *20*(1). Available from https://doi.org/10.1515/PHYS-2022-0042/MACHINEREADABLECITATION/RIS, https://www.degruyter.com/document/doi/10.1515/phys-2022-0042/html.

Minhas, D. M., Khalid, R. R., & Frey, G. (2017). Demand response load management price forecasting regression models. In *5th IEEE international conference on smart energy grid engineering, SEGE 2017* (pp. 373–378). Institute of Electrical and Electronics Engineers Inc. Available from https://doi.org/10.1109/SEGE.2017.8052828.

Mishra, S., Leinakse, M., & Palu, I. (2017). Wind power variation identification using ramping behavior analysis. *Energy Procedia, 141*. Available from https://doi.org/10.1016/J.EGYPRO.2017.11.075.

Moret, S., Bierlaire, M., & Maréchal, F. (2016). Robust optimization for strategic energy planning. *Informatica (Netherlands), 27*(3). Available from https://doi.org/10.15388/INFORMATICA.2016.103.

Mulewa, S., Parmar, A. M., & De, A. (2023). Attention based Transformer coupled with convoluted neural network for ultra-short- and short-term forecasting of multiple wind farms. *International Journal of Green Energy*. Available from https://doi.org/10.1080/15435075.2023.2244589, https://www.tandfonline.com/doi/abs/10.1080/15435075.2023.2244589.

Nascimento, E. G. S., de Melo, T. A. C., & Moreira, D. M. (2023). A transformer-based deep neural network with wavelet transform for forecasting wind speed and wind energy. *Energy, 278*. Available from https://doi.org/10.1016/J.ENERGY.2023.127678.

Nguyen Trong, T., Vu Xuan Son, H., Do Dinh, H., Takano, H., & Nguyen Duc, T. (2023). Short-term PV power forecast using hybrid deep learning model and variational mode decomposition. *Energy Reports, 9*. Available from https://doi.org/10.1016/J.EGYR.2023.05.154.

Notton, G., Nivet, M. L., Voyant, C., Paoli, C., Darras, C., Motte, F., & Fouilloy, A. (2018). Intermittent and stochastic character of renewable energy sources: Consequences, cost of intermittence and benefit of forecasting. *Renewable and Sustainable Energy Reviews, 87*. Available from https://doi.org/10.1016/J.RSER.2018.02.007.

Nyangon, J., & Akintunde, R. (2024). Principal component analysis of day-ahead electricity price forecasting in CAISO and its implications for highly integrated renewable energy markets. *Wiley Interdisciplinary Reviews: Energy and Environment, 13*(1), e504. Available from https://doi.org/10.1002/WENE.504, https://wires.onlinelibrary.wiley.com/doi/10.1002/wene.504.

Obrecht, M., & Denac, M. (2016). Technology forecast of sustainable energy development prospects. *Futures, 84*. Available from https://doi.org/10.1016/J.FUTURES.2016.09.002.

Padilha, G. A. G., Ko, J., Jung, J. J., De Mattos Neto, P. S. G., Foresta, F. L., Gao, J., & Afonso, G. (2022). Transformer-based hybrid forecasting model for multivariate renewable energy. *Applied Sciences, 12*(21). Available from https://doi.org/10.3390/APP122110985, https://www.mdpi.com/2076-3417/12/21/10985.

Pan, X., Wang, L., Wang, Z., & Huang, C. (2022). Short-term wind speed forecasting based on spatial-temporal graph transformer networks. *Energy, 253*. Available from https://doi.org/10.1016/J.ENERGY.2022.124095.

Panda, S., Mohanty, S., Rout, P. K., Sahu, B. K., Parida, S. M., Samanta, I. S., Bajaj, M., Piecha, M., Blazek, V., & Prokop, L. (2023). A comprehensive review on demand side management and market design for renewable energy support and integration. *Energy Reports, 10*. Available from https://doi.org/10.1016/J.EGYR.2023.09.049.

Peirelinck, T., Kazmi, H., Mbuwir, B. V., Hermans, C., Spiessens, F., Suykens, J., & Deconinck, G. (2022). Transfer learning in demand response: A review of algorithms for data-efficient modelling and control. *Energy and AI, 7*. Available from https://doi.org/10.1016/J.EGYAI.2021.100126.

Pospíchal, J., Kubovčík, M., & Dirgová Luptáková, I. (2022). Solar irradiance forecasting with transformer model. *Applied Sciences*, *12*(17). Available from https://doi.org/10.3390/APP12178852, https://www.mdpi.com/2076-3417/12/17/8852.

Pujari, K. N., Miriyala, S. S., Mittal, P., & Mitra, K. (2023). Better wind forecasting using evolutionary neural architecture search driven green deep learning. *Expert Systems with Applications*, *214*. Available from https://doi.org/10.1016/J.ESWA.2022.119063.

Rahman, S., Punt, L., Ardakanian, O., Ghiassi, Y., & Tan, X. (2022). On efficient operation of a V2G-enabled virtual power plant. In *9th ACM international conference on systems for energy-efficient buildings, cities, and transportation* (pp. 119–128). Association for Computing Machinery, Inc. https://dl.acm.org/doi/10.1145/3563357.3564067.

Ramírez, L., & Vindel, J. M. (2017). *Forecasting and nowcasting of DNI for concentrating solar thermal systems. Advances in concentrating solar thermal research and technology*. Woodhead Publishing. Available from https://doi.org/10.1016/B978-0-08-100516-3.00013-7.

Ramos, D., Faria, P., Gomes, L., & Vale, Z. (2022). A contextual reinforcement learning approach for electricity consumption forecasting in buildings. *IEEE Access*, *10*. Available from https://doi.org/10.1109/ACCESS.2022.3180754.

Rangelov, D., Boerger, M., Tcholtchev, N., Lämmel, P., & Hauswirth, M. (2023). Design and development of a short-term photovoltaic power output forecasting method based on random forest, deep neural network and LSTM using readily available weather features. *IEEE Access*, *11*. Available from https://doi.org/10.1109/ACCESS.2023.3270714.

Ribeiro, L. A., & da Silva, P. P. (2015). Qualitative Delphi approach of advanced algae biofuels. *Management of Environmental Quality: An International Journal*, *26*(6). Available from https://doi.org/10.1108/MEQ-03-2014-0046/FULL/PDF.

Sanusi, U., & Corne, D. (2016). Feature selection for accurate short-term forecasting of local windspeed. In *IEEE 8th international workshop on computational intelligence and applications, IWCIA-proceedings 2015* (pp. 121–126). https://doi.org/10.1109/IWCIA.2015.7449474.

Schreiber, J., Vogt, S., Sick, B., & Temporal, T.E. (2021). Convolution networks for transfer learning problems in renewable power time series forecast. In *Lecture notes in computer science (including subseries lecture notes in artificial intelligence and lecture notes in bioinformatics)*. **12978 LNAI** https://link.springer.com/chapter/10.1007/978-3-030-86514-6_8, https://doi.org/10.1007/978-3-030-86514-6_8/TABLES/4.

Shaikh, P. H., Shaikh, A., Memon, Z. A., Lashari, A. A., & Leghari, Z. H. (2021). Microgrids: A review on optimal hybrid technologies, configurations, and applications. *International Journal of Energy Research*, *45*(9). Available from https://doi.org/10.1002/ER.6666, https://onlinelibrary.wiley.com/doi/10.1002/er.6666.

Sharma, M., Mittal, N., Mishra, A., & Gupta, A. (2023). Survey of electricity demand forecasting and demand side management techniques in different sectors to identify scope for improvement. *Smart Grids and Sustainable Energy*, *8*(2). Available from https://doi.org/10.1007/S40866-023-00168-Z/TABLES/7, https://link.springer.com/article/10.1007/s40866-023-00168-z.

Sharma, N., Sharma, P., Irwin, D., & Shenoy, P. (2011). Predicting solar generation from weather forecasts using machine learning. In *IEEE international conference on smart grid communications, SmartGridComm 2011* (pp. 528–533). https://doi.org/10.1109/SMARTGRIDCOMM.2011.6102379.

Shirzadi, N., Nasiri, F., Menon, R. P., Monsalvete, P., Kaifel, A., & Eicker, U. (2023). Smart urban wind power forecasting: Integrating weibull distribution, recurrent neural networks, and numerical weather prediction. *Energies*, *16*(17). Available from https://doi.org/10.3390/EN16176208, https://www.mdpi.com/1996-1073/16/17/6208.

Song, S., Kang, S., Lee, S., & Kim, C. K. (2023). Korean renewables management system: Copulas model-based adaptive droop control strategy for energy storage systems. *IEEE Access*, *11*. Available from https://doi.org/10.1109/ACCESS.2023.3337373.

Suh, J., & Choi, Y. (2017). Methods for converting monthly total irradiance data into hourly data to estimate electric power production from photovoltaic systems: A comparative study. *Sustainability*, *9*(7). Available from https://doi.org/10.3390/SU9071234, https://www.mdpi.com/2071-1050/9/7/1234.

Sweeney, C., Bessa, R. J., Browell, J., & Pinson, P. (2020). The future of forecasting for renewable energy. *Wiley Interdisciplinary Reviews: Energy and Environment*, *9*(2), e365. Available from https://doi.org/10.1002/WENE.365, https://wires.onlinelibrary.wiley.com/doi/10.1002/wene.365.

Torres, J. M., Aguilar, R. M., & Zuñiga-Meneses, K. V. (2018). Deep learning to predict the generation of a wind farm. *Journal of Renewable and Sustainable Energy*, *10*(1). Available from https://doi.org/10.1063/1.4995334/285796, http://aip/jrse/article/10/1/013305/285796/Deep-learning-to-predict-the-generation-of-a-wind.

Tulkinov, S. (2023). Grey forecast of electricity production from coal and renewable sources in the USA, Japan and China. *Grey Systems*, *13*(3). Available from https://doi.org/10.1108/GS-10-2022-0107/FULL/PDF.

Wahbah, M., Feng, S., El-Fouly, T. H. M., & Zahawi, B. (2020). Root-transformed local linear regression for solar irradiance probability density estimation. *IEEE Transactions on Power Systems*, *35*(1). Available from https://doi.org/10.1109/TPWRS.2019.2930699.

Wen, S., Wang, Y., Tang, Y., Xu, Y., & Li, P. (2019). Proactive frequency control based on ultra-short-term power fluctuation forecasting for high renewables penetrated power systems. *IET Renewable Power Generation*, *13*(12). Available from https://doi.org/10.1049/IET-RPG.2019.0234, https://ietresearch.onlinelibrary.wiley.com/doi/10.1049/iet-rpg.2019.0234.

Widén, J., Carpman, N., Castellucci, V., Lingfors, D., Olauson, J., Remouit, F., Bergkvist, M., Grabbe, M., & Waters, R. (2015). Variability assessment and forecasting of renewables: A review for solar, wind, wave and tidal resources. *Renewable and Sustainable Energy Reviews*, *44*. Available from https://doi.org/10.1016/J.RSER.2014.12.019.

Wu, H., Xu, Z., & Wang, M. (2023). Unrolled spatiotemporal graph convolutional network for distribution system state estimation and forecasting. *IEEE Transactions on Sustainable Energy*, *14*(1). Available from https://doi.org/10.1109/TSTE.2022.3211706.

Zhang, W., Kleiber, W., Hodge, B. M., & Mather, B. (2022). A nonstationary and non-Gaussian moving average model for solar irradiance. *Environmetrics (London, Ont.)*, *33*(3), e2712. Available from https://doi.org/10.1002/ENV.2712, https://onlinelibrary.wiley.com/doi/10.1002/env.2712.

Zhu, J., Dong, H., Zheng, W., Li, S., Huang, Y., & Xi, L. (2022). Review and prospect of data-driven techniques for load forecasting in integrated energy systems. *Applied Energy*, *321*. Available from https://doi.org/10.1016/J.APENERGY.2022.119269.

Optimal energy management of energy storage and electric vehicles in power systems

17

Darío Benavides[1,2], Edisson Villa-Ávila[1,3], Adrián Criollo[3] and José A. Aguado[2]

[1]Department of Electrical Engineering, University of Jaén, Jaén, Spain, [2]Department of Electrical Engineering, University of Málaga, Malaga, Spain, [3]Faculty of Engineering, Department of Electrical, Electronics and Telecommunications Engineering (DEET), Universidad de Cuenca, Balzay Campus, Cuenca, Azuay, Ecuador

Introduction

The infrastructure for electric vehicles (EVs) in urban areas in recent years tends to increase with the adoption of renewable energy sources (RES) in the energy and transportation sector (Almatar, 2023). Several countries have joined together in the sustainable and smart mobility strategy for 2030 planned by the European Union. Similarly, the European Green Deal has planned to reduce CO_2 emissions by 90% before the year 2050 (Consilium, 2024; Marotta et al., 2023). Based on global sales of EV batteries (EVB) from 2005 to 2018, it is estimated to grow from 5.3 million in 2019 to 40 million units in 2030 (Dhakal & Min, 2021). This leads to a growth of around 13% in the demand of the energy system that must be considered in the transportation sector.

Consequently, large-scale integration of EVs implies new challenges in electrical distribution systems (EDS). Therefore, vehicle-to-everything (V2X) support has been widely studied under different modes of operation (Alnasser et al., 2020; Islam et al., 2022; Saxena et al., 2023). Some recent studies are based on mode vehicle-to-grid (V2G) (Amertet & Gebresenbet, 2024; Jiao et al., 2021; Li & Gu, Zeng, et al., 2021; Shipman et al., 2020), vehicle-to-vehicle (Alghawi & Mounsef, 2024; İnci et al., 2022; Liu, Ni, et al., 2023; Wu et al., 2024; Yassine & Hossain, 2023), vehicle-to-home (V2H) (Borge-Diez et al., 2021; He et al., 2022; Nagel et al., 2023; Shipman et al., 2020; Zafar & Slama, 2022), and vehicle-to-load (V2L) (Haroon et al., 2023; Licea, 2020; Rajamand, 2020). Its application will depend exclusively on the service that the EV offers to the electrical grid. This integration addresses the EV as a potential element in energy storage systems (ESS) linked in modern electrical systems. The advantage that the V2G mode highlights is its contribution as an energy source, allowing it to supply energy to EDS during peak

Towards Future Smart Power Systems with High Penetration of Renewables. DOI: https://doi.org/10.1016/B978-0-443-29871-4.00017-8

hours and load leveling (Jain & Jain, 2018). Similarly, it would allow for voltage regulation at the point of common coupling (PCC) through reactive power compensation (Hu et al., 2022).

On the other hand, in modern electrical systems, microgrids (MGs) and smart grids have made it possible to improve flexibility in energy management of supply and demand in EDS. In this context, energy management modeling and solution techniques have been widely studied. In Liu, Zhao, et al. (2023), the classification and energy management system (EMS) of MG with ESS are exhaustively summarized. ESSs promote open exchange and flexible trading of energy production and consumption (Li & Wang, 2021). In addition, the energy efficiency programs allow increasing the total energy productivity index of the smart MG (Dorahaki et al., 2020). In addition, Tang (2023) determined the optimal control of energy management with cooperative dispatch of EVs with positive results that allow minimizing the cost of domestic electricity and maximizing photovoltaic PV consumption. This process was carried out under the conditions of optimal control of energy management.

This research analyzes the active support of the EV in the EDS in the specific V2G&G2V operating modes with the integration of distributed RES and ESS. The proposed methodology highlights load leveling during the hours of greatest demand, allowing the EV's energy to be optimized and generating energy savings for consumers. Reference energy price values and selective schedules are used. In the case study, vanadium redox flow batteries (VRFB) are analyzed as high-capacity stationary ESS. Finally, a scenario analysis for the integration of EVs is presented. This configuration offers a more efficient and flexible electrical grid for the integration of various EV charging station (EVCS) with renewable PV and wind turbine (WT) resources.

Policies and future trends

The integration of the EV into EDS has increased in recent years. In Europe, it is expected that 100% EV integration will be covered by 2035. The growing substitution of internal combustion vehicles for EVs supported by future policies and trends has been raised (Gupta et al., 2021; Yang & Fulton, 2023). Some of the important issues include the development of charging infrastructure, battery technology, and policy frameworks (Eldjalil et al., 2017; Gupta et al., 2021; Haghani et al., 2023). As an example, a study has determined that the cost of charging an EV compared to the cost of fuel is lower in the Portuguese community (Rodrigues et al., 2021). It generates great expectations in this transition process. Some countries in Asia, South Africa, and India have also addressed positive studies on this energy transition (Huda et al., 2019; Hull et al., 2024; Mittal et al., 2024). Furthermore, a promising study has determined the changes involved in this global transition process using the generalized method of moments (Cao et al., 2024). This allows for an analysis of the results through specific policies. Likewise, energy diplomacy is considered key during this transition process (Huda, 2024).

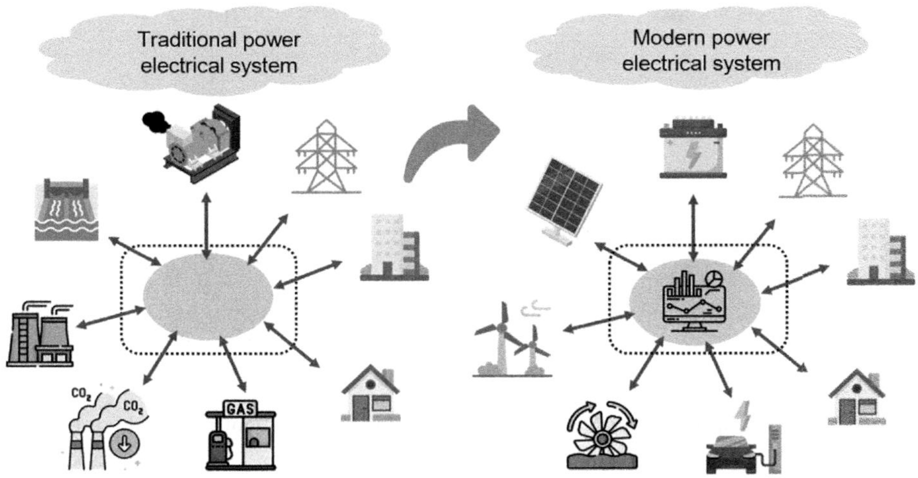

Figure 17.1 Transition from traditional to modern electrical power system.

This leads to the implementation of strategies for sustainable development in the energy and transport sectors. These strategies are based on the integration of energy generation from renewable sources and the implementation of ESS (see Fig. 17.1). As a strategic plan, it also involves the progressive replacement of gas stations with EVCSs. Integrating several sources with distributed generation implies maintaining optimal control and management of energy. In addition, considering the problems that can be generated by the intermittency of renewable resources and the demand required for EVs in a short period. Through optimal charging scheduling techniques, the flexibility of EVs can be exploited and the impact of solar and wind systems on the grid can be reduced (Mastoi et al., 2022).

Problems and constraints

The main energy limitation for the integration of EVCS lies in the need for large amounts of energy to achieve full charging of the EVs. Consequently, the daily energy demand curve is affected by the variable use of EVCS throughout the day (Gupta et al., 2021; Huda et al., 2019). During different hours, power demand experiences fluctuations due to charging and discharging of EVs. Therefore, the integration of various EVs clearly shows the direct impact of large-scale integration of EVs on the power grid (Boudina et al., 2020). As shown in Fig. 17.2A, this leads to an energy dispatch different from the traditional one, where load leveling can be carried out in which the peak load demand can be transferred to off-peak hours (Aziz et al., 2015). However, the optimal configuration of these systems requires programmed control for the charging stations. In addition, it must allow for an auto-scaling connection that does not generate forced energy requirements in current electrical networks.

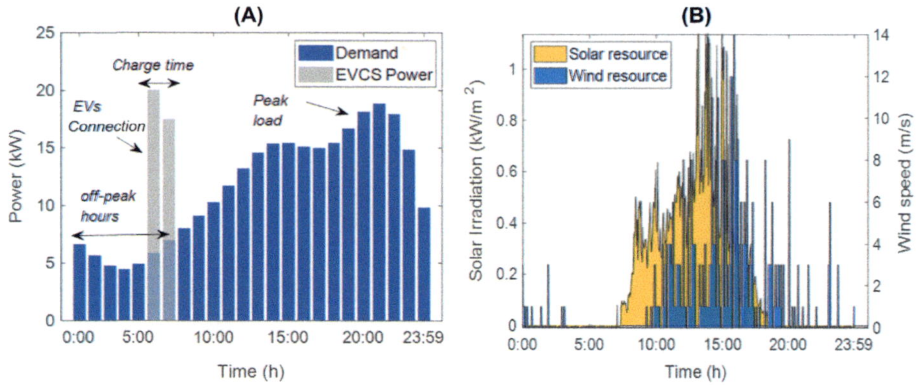

Figure 17.2 (A) Impact of EVCS on demand and (B) intermittency of renewable resources.

High capacity power is generally required to fully recharge the EV battery for a few hours a day. (for example: 0600–0800 hours). The peak power of the EVCS depends on some parameters such as the state of charge of the EV's battery and the percentage of charge required, which implies providing adequate energy management during this process (See Fig. 17.2A). On the other hand, the integration of renewable resources can be a solution, but it implies the availability of the solar or wind resource and its variability can cause intermittency in the electrical grid. Fig. 17.2B presents the typical profiles of solar and wind resources. Consequently, the use of ESS is required to optimize energy management of demand profiles and load leveling over established time periods.

Methodologies

Overview

Optimal energy management depends on the application scenarios, which could be distinguished mainly as power supply side, power grid side, and power distribution side. Some of the scenario's studies related to uncertainty of renewable resources, intermittent sources, impact of power fluctuations, voltage fluctuations, power peaks, frequency variations, standby power supply, virtual power plant, V2G, power supply reliability, failure analysis, among others. Fig. 17.3 shows an example of the workflow that can be considered for the implementation of optimal energy management models in power systems. After determining the scenario, the process begins with the proposal of new methods and strategies that will be validated using computer modeling and system simulation tools. Finally, an experimentation phase is determined as the results of the analysis of the proposals. The favorable determination of the study can be categorized into improvement in management, optimal sizing, technical viability, cost, and others of interest.

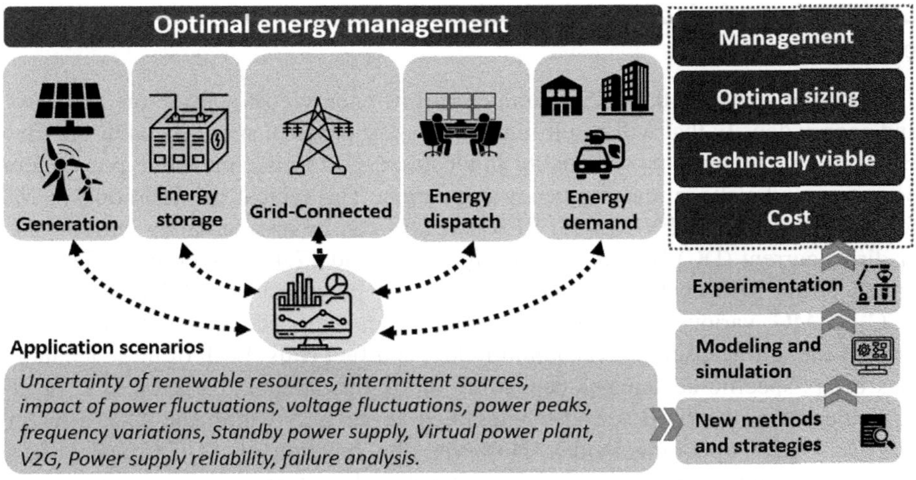

Figure 17.3 Workflow for implementation of optimal energy management models.

Operation modes: vehicle to grid and grid to vehicle

ESSs can also be linked to the charging and discharging of EVs. The V2G context implies that EVs not only consume power from the grid (G2V), but can also return power to the grid when needed, especially at specific times when energy demand is required or the cost of electricity is high. The application of V2G and G2V has been proposed as a strategy for load leveling (Eldjalil et al., 2017; Jain & Jain, 2018; Zhang et al., 2014). This could generate great support in the electrical grid. Consequently, sophisticated bidirectional converters are needed that allow for energy exchange between the EV battery and the power grid in both directions. However, the integration of these devices into the electrical grid can generate some disadvantages such as generating electromagnetic noise, harmonic distortion, and complexity in the control design. In addition, the bidirectional exchange of energy must consider the efficiency of the equipment. During the charging and discharging process of the EV battery, energy losses may occur.

Adhikary et al. (2023) reported an innovative bidirectional converter based on G2V&V2G specific operation. This system integrates a tariff analysis based on the time of use and monitoring of EV charging parameters. EV time-of-use tariffs allow for excessive coordination of charging, but the adoption of battery storage is necessary (Schwarz et al., 2020). Similarly, a decentralized V2G/G2VEVCS programming has been established based on the electricity rate and demand response programs (Liao et al., 2021). The authors offer a solution to the multivariate routing problem of EVs in the logistics sector, allowing for optimal communication between the EV, the client, the EVCS, and the dispatching entities. EV charging scheduling and electricity pricing algorithms based on the IoT environment have also been studied (Corchado, n.d.). This encourages the integration of V2G and G2V systems into the large-scale electrical grid.

Integration of electric vehicle charging station with renewable energy sources

The optimal sizing of the ESS guarantees a low energy cost considering the power and energy density that will result in the necessary physical space. In addition, these ESS can also improve power quality in electrical grid through reactive power compensation and voltage and frequency regulation. The typical configuration in MG for the integration of EVCS with RES and ESS can be an alternating current (AC) or direct current (DC) bus, as seen in Fig. 17.4A and 17.4B, respectively. The integrated power grid enables bidirectional flow of power to the ESS and EVCS in PCC. At MG, various energy storage technologies are used, such as lead batteries, lithium-ion, ultracapacitors, vanadium redox, and fuel cells. Each of these technologies offers specific advantages depending on its operational characteristics. In this particular case, the EVB are also integrated as an ESS within the MG, allowing V2G and G2V operating modes. However, the function of AC/DC and DC/DC power converters is essential for the control and performance of these systems. Consequently, it must be considered that the energy transfer is within the maximum limit of the converter equipment (Adhikary et al., 2023).

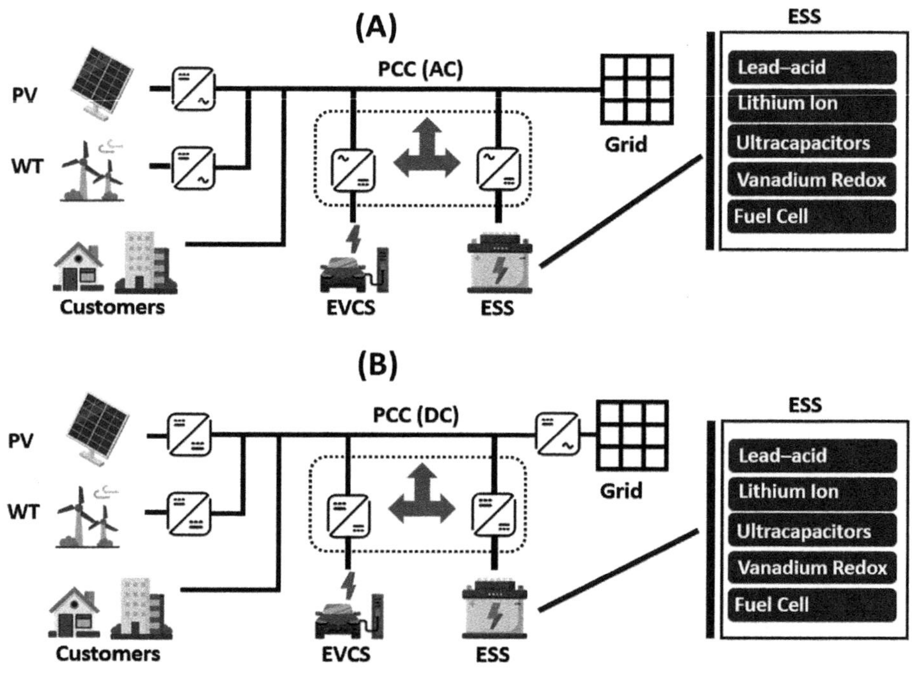

Figure 17.4 Integration of ESS in MG with active support of EVs: (A) AC bus configuration and (B) DC bus configuration.

Systems modeling

Electric vehicle battery modeling

EVB modeling can be modeled as a series voltage source with a variable resistor. The dynamic state of charge of the battery can be expressed as:

$$\mathrm{SOC}_n^{\mathrm{EV}} = \mathrm{SOC}_{n-1}^{\mathrm{EV}} - \frac{i \times t}{Q_0}$$

$\mathrm{SOC}_n^{\mathrm{EV}}$ and $\mathrm{SOC}_{n-1}^{\mathrm{EV}}$ are the SoC at two moments of time, i is the current, t is the time step, and Q_0 is the initial capacity of the battery.

The voltage of the cell is given by Ohm's law as shown in the equation below.

$$V = \mathrm{OCV}\left(\mathrm{SOC}^{\mathrm{EV}}\right) - i \times R\left(\mathrm{SOC}^{\mathrm{EV}}, T\right)$$

where V is the voltage at the terminals of the battery, T is the temperature of the battery, OCV is the open circuit voltage of the battery as a function of the $\mathrm{SOC}^{\mathrm{EV}}$, i is the current through the battery, and R is the resistance of the battery as a function of SoC and temperature (Bhoir et al., 2021).

On the other hand, the battery degradation model can be calculated using the stress factor SF with the following equation:

$$SF = \prod_{1}^{n} (I_{pn}) = I_{p1} \times I_{p2}. \times I_{pn}$$

where I_{pn} are the impacts of the parameters such as average SoC, DoD, C-rate and temperature during cycling, and so forth. Therefore, the actual degradation rate DR_{actual} can be calculated as follows:

$$DR_{\mathrm{actual}} = DR_{\mathrm{ref}} \times SF$$

where DR_{ref} stands for the reference degradation rate. SF is the stress factor calculated above.

Bidirectional alternating current/direct current converters

The bidirectional grid-connected AC/DC converter is essential for the V2G&G2V mode of operation. It is responsible for allowing for the directional flow of energy between the EV and the grid. The modeling of this system proposed in Han et al. (2020) with a grid voltage early decoupling approach. The topology of the proposed system is presented in Fig. 17.5. The three-phase voltage and current of the grid source are denoted as e_a, e_b, e_c and i_a, i_b, i_c, respectively.

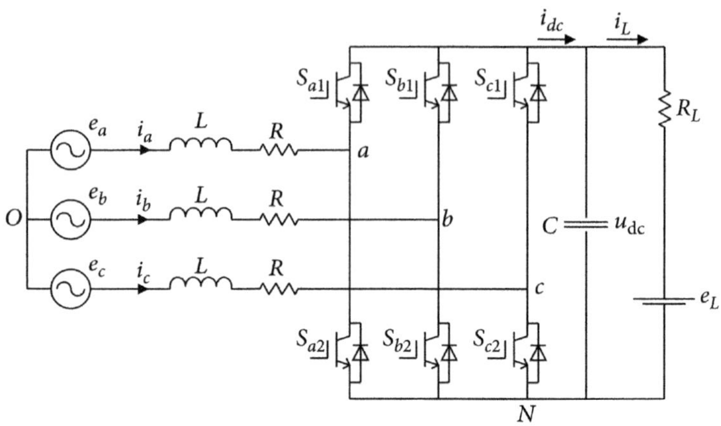

Figure 17.5 Representation of the bidirectional AC/DC converter circuit (Han et al., 2020).

$$\begin{cases} e_a = L\dfrac{di_a}{dt} + Ri_a + u_{aN} + u_{NO}, \\ e_b = L\dfrac{di_b}{dt} + Ri_b + u_{bN} + u_{NO}, \\ e_c = L\dfrac{di_c}{dt} + Ri_c + u_{cN} + u_{NO}, \\ C\dfrac{du_{dc}}{dt} = i_{dc} - i_L \end{cases}$$

where L and R are the inductance and resistance of the grid model. C is the capacitance in the DC circuit. u_{dc} is the voltage and i_{dc} is current in DC. R_L and e_L represent the dc load model. i_L is the load current.

Besides, u_{aN}, u_{bN}, and u_{cN} are the voltage between the midpoint of each phase and point N and u_{NO} is the voltage between point N and O of the grid.

On the other hand, the activation of the IGBT switches works in an inverted manner in $S_{a1}S_{b1}$, S_{c1} with respect to S_{a2}, S_{b2}, S_{c2}. Consequently, $S_k = 1$ is considered active and $S_k = 0$ inactive (See Fig. 17.5). Therefore, u_k is defined as follows:

$$u_k = u_{kN} + u_{NO} = S_k u_{dc} + u_{NO}$$

where $k = \{a, b, c\}$.

The transformation of the three-phase system (a, b, c) to two components of a stationary orthogonal $d - q$ frame can be described as follows (Han et al., 2020):

$$\begin{cases} e_d = L\dfrac{di_d}{dt} + Ri_d - \omega_0 L i_q + S_d u_{dc}, \\ e_q = L\dfrac{di_q}{dt} + Ri_q + \omega_0 L i_d + S_q u_{dc}, \\ C\dfrac{du_{dc}}{dt} = \dfrac{3}{2}\left(i_d S_d + i_q S_q\right) - i_L \end{cases}$$

where e_d and e_q are frame components of voltage $d-q$. Similarly, i_d and i_q are frame currents. ω_0 is rotation angular velocity of frame. The equations can be used for the design of controllers in control strategies applied in V2G.

Vanadium redox flow battery

The VRFB generation process is based on liquid energy sources that are stored in two tanks and pumped through electrochemical cells, in which energy sources are electrochemically charged or discharged depending on the applied voltage. Unlike traditional batteries, VRFB store energy through a reversible chemical reaction in electrochemical cells. Fig. 17.6 details this procedure where the electrodes maintain positive and negative polarity during the electrochemical process in two separate tanks. Circulation pumps move the electrolyte through the cell stack. When discharged, these reactions are reversed and electrical energy can be supplied to the grid. Although the efficiency during charging may be low, it is possible to control the exchange of energy between devices through hybrid ESS (Qiu et al., 2014). It is also proven to be a promising solution for isolated MGs as a frequency control strategy and power grid stabilizer (Mohiti et al., 2021).

The state of charge of the VRFB SOC_{VRFB} can be calculated by the voltage generated through the operation of the ions of the positive and negative cathodes, where their exchange of states allows the charging or discharging of the battery, in a similar way to the electrochemical formulation using the vanadium V^{2+} and V^{3+} to determine capacity of VRFB (Gu & Chen, 2021; Li & Hikihara, 2008).

$$SOC_{VRFB} = \frac{\left[VO_2^+\right]_T}{\left[VO^{2+}\right]_T + \left[VO_2^+\right]_T} = \frac{\left[V^{2+}\right]_T}{\left[V^{2+}\right]_T + \left[V^{3+}\right]_T}$$

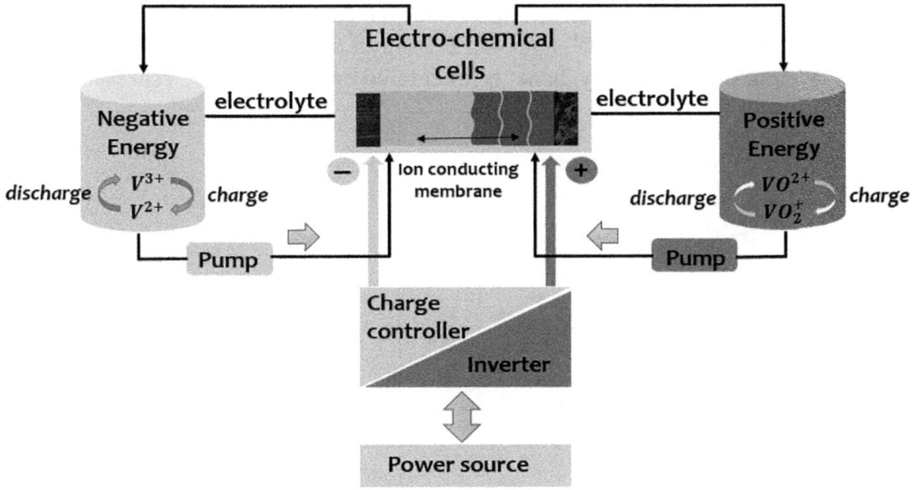

Figure 17.6 Vanadium redox flow battery principle.

The parameters VO^{2+} and VO_2^+ are in fact vanadium oxide ions from the positive cathode in a charge and discharge sequence. V^{2+}, V^{3+} represents oxidation states of vanadium from negative cathode in the tank.

Power of the vanadium redox flow (CellCube) unit is calculated by the following equations (Turker et al., 2013):

$$P_{\mathrm{OCV}} = N_c \times V_{oc} \times I_s$$

P_{OCV} represents the charging power, and the discharge power considering losses. N_c is the number of cells per string, and V_{oc} is the open circuit voltage. Similarly, the string power P_s can be calculated as follows:

$$P_S = N_c \times V_s \times I_s$$

Where V_s and I_s is the voltage and current the string, respectively.

Case study

In this section, the proposed management system that integrates PV and WT renewable sources in a MG with VRFB support is presented. In particular, an EV approach is included as energy support during peak hours. The control strategy involves double energy flow from the battery to the grid V2G and vice versa G2V. This strategy would reduce consumers' energy costs during peak hours of highest consumption. Fig. 17.7 shows the schematic diagram of the MG and the active support of the EV. The bidirectional converters of the ESS and EVCS would manage the energy to and from the EDS objectively. To improve load leveling, the G2V mode allows you to optimize energy costs in the hours of least consumption (0:00–6:00). V2G mode allows power grid support during peak consumption hours (7:00 p.m.–11:00 p.m.). On the other hand, the VRFBs would work objectively with PV and WT when there is excess renewable generation and contribution to the grid when required.

The data of the MG equipment is presented in Table 17.1. It integrates a 34 kW PV system, 15 kW WT as RES, and VRFBs with a capacity of 100 kWh for energy storage. A lithium-ion battery bank is integrated into the study, representing the

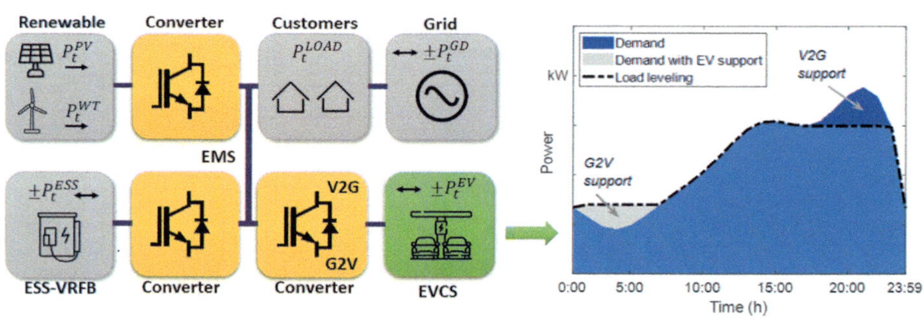

Figure 17.7 Strategy for energy management with EV support in the electrical grid.

Table 17.1 Technical specifications of the equipment employed in the experiment.

Equipment description	Model	Cells / Modules	Operating voltage	Energy
PV system 1 (15 kWp)	Monocrystalline	60 (15 × 4)	150 Vdc-450 Vdc/230 Vac	up to 50 kWh
PV system 2 (15 kWp)	Polycrystalline	60 (15 × 4)	150 Vdc-450 Vdc/230 Vac	up to 50 kWh
PV system 3 (4 kWp)	Polycrystalline	16(16 × 1)	150 Vdc-450 Vdc/230 Vac	up to 15 kWh
WT system (15 kW)	Serie E70Pro	N/A	230Vac	up to 40 kWh
Vanadium redox flow batteries	Cell Cube FB 10−100	2 Cluster	48 Vdc/230Vac	100 kWh
Lithium-ion battery (EV battery simulator)	Samsung ELPT392−0002	11 Cells	642 V/230Vac	44 kWh
Bidirectional AC/DC converter	GPTech	1	150 Vdc-450 Vdc/230 Vac	up to 50 kWh

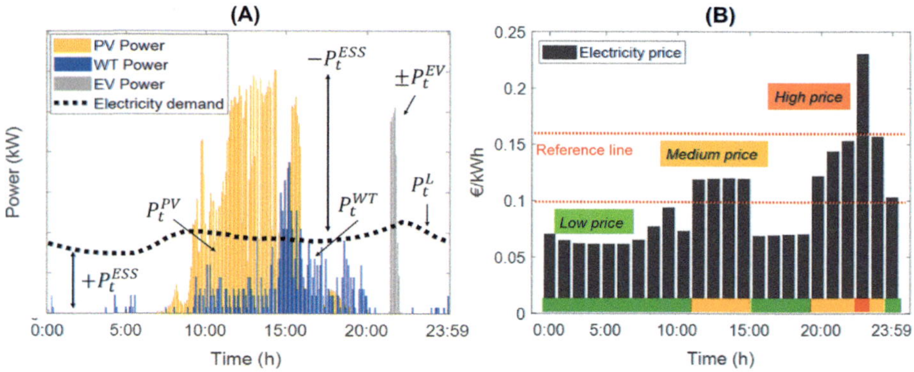

Figure 17.8 (A) Electrical demand with integration of RES and EVs. (B) Referential electricity price per hour.

power that EV batteries can deliver. It also includes the bidirectional converter that allows for the flow of energy to the public electrical grid (Espinoza et al., 2017).

The following equation defines the flow of power in the electrical grid P_t^{Gd}.

$$P_t^{\mathrm{Gd}} = P_t^{\mathrm{PV}} + P_t^{\mathrm{WT}} \pm P_t^{\mathrm{ESS}} \pm P_t^{\mathrm{EV}} - P_t^{L}$$

Where P_t^{PV} and P_t^{WT} are the powers of the PV and WT systems, respectively. $\pm P_t^{\mathrm{ESS}}$ is the output power of the ESS considering the positive flow for power delivery to the grid and negative flow for battery charging. $\pm P_t^{\mathrm{EV}}$ is the power of the EV. The positive and negative signs refer to the V2G and G2V modes. P_t^{L} is the load of the electrical demand at a time t. Fig. 17.8A details the power profiles based on the equations expressed above.

The electricity demand curve and the reference energy price play a fundamental role in the energy management model. In this context, Fig. 17.8B shows the referential electricity price per hour. The average price of the day (reference to 04/24/2024) is €0.0983 kWh^{-1}, with the lowest price of the day being 03–04 h €0.06177 kWh^{-1} and the highest price of the day being 21–22 h €0.23063 kWh^{-1} due to the constant change in the tariff value of the energy cost. The cost to the consumer can be calculated using the equation (Adhikary et al., 2023).

$$C_t^{\mathrm{Consumer}} = C_t^{G2V} \pm C_t^{V2G}$$

Where C_t^{G2V} is based on the discharging rate with real-time tariff and C_t^{V2G} is based on the charging rate with real-time tariff.

Results

Optimal schedules for vehicle-to-grid and grid to vehicle operation modes

Users can charge their EVs during hours with the lowest electricity prices (Li & Wang, 2021). It presents the optimal energy management schedule based on energy

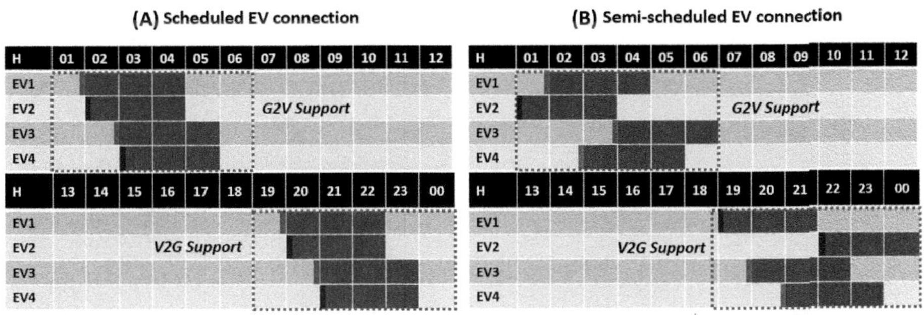

Figure 17.9 G2V and V2G energy management based on schedule of: (A) scheduled connection to the electrical grid. (B) Semischeduled connection to the electrical grid.

cost with G2V and V2G support for connection to the electrical grid, in which a scheduled sequential or semischeduled schedule planning is established in an established time range. As indicated in Fig. 17.9 with the energy reference cost, the vehicle charging schedule can be from 00:00 AM to 6:00 AM, that is, G2V. On the other hand, vehicle support to the grid (V2G) can be carried out during peak demand hours (8:00 PM to 11:00 AM). It is important to mention that scheduled EV support to the grid would significantly improve demand leveling.

Energy management model

The energy that EVs can deliver or discharge will depend on the associated types and models. Power electronic converters and energy storage devices are an important evaluation of the reliability of EVs (Singh et al., 2023). Fig. 17.10A presents two specific EV models KIA SOUL EV and IONIQ EV (Benavides et al., 2024). The required powers of the EVCS are around 50 kW and charging times of 2.3 and 2.5 h, respectively, in G2V mode. Based on the efficiency of the converters, the energy in V2G mode can be estimated in a similar way. Therefore, the charging and discharging efficiency of the EV is considered the same in the simplification of the model.

This configuration can be seen in Fig. 17.10B with the G2V and V2G energy management model in the electrical grid, in which a sequential load of four vehicles has been estimated in the G2V support schedule. On the contrary, a sequential discharge of the same EVs was scheduled during the V2G support schedule. The significant contribution of the EV can be reflected in the leveling of load in the projected demand curve. Similarly, the contribution of EV in MG is analyzed below in the case study MG applications in Fig. 17.10C. As can be seen, the excessive power of PV and WT $\left(P_t^{\mathrm{PV}} + P_t^{\mathrm{WT}}\right)$ is sent to the VRFB charging with $-P_t^{\mathrm{ESS}}$. On the contrary, when the demand P_t^L is supplied, it can deliver energy to the grid $+P_t^{\mathrm{ESS}}$. The EV contribution can be combined in G2V $-P_t^{\mathrm{EV}}$ and V2G $+P_t^{\mathrm{EV}}$ at established times of lower energy consumption. The established configuration would allow for a significant reduction in consumption from the electrical grid This analysis is presented in the following section.

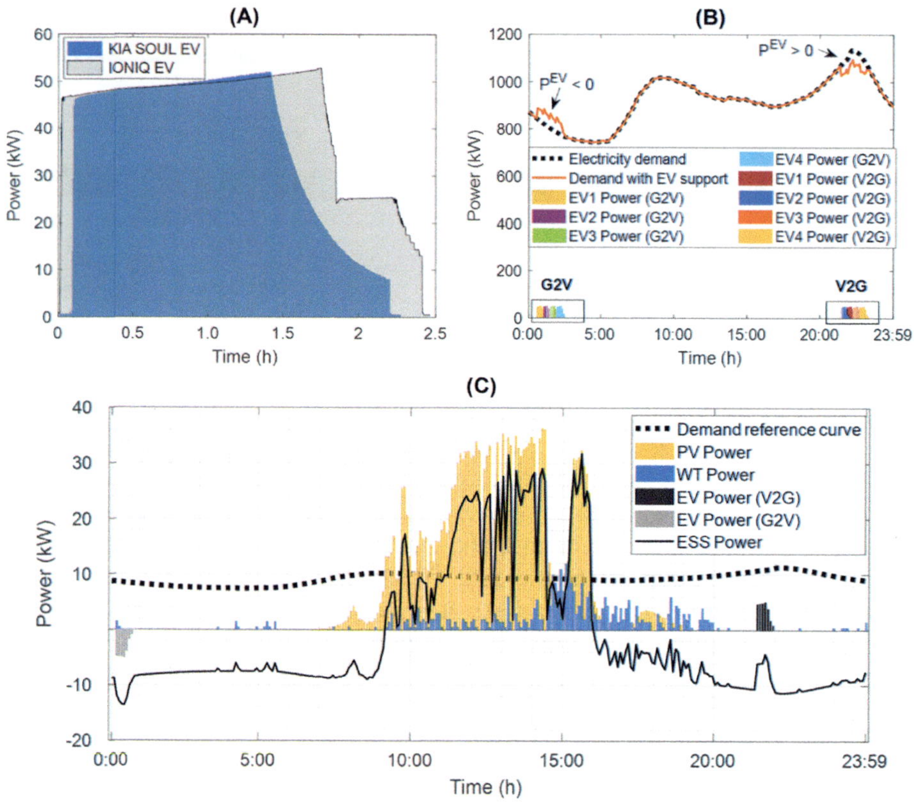

Figure 17.10 (A) Results of the EVCS experiment for two models (0 to 100%). (B) Energy management G2V and V2G in the electrical grid. (C) Energy management G2V and V2G with RES/ESS support in the electrical grid.

Optimal energy flow

According to the study presented, this section details the optimal modes of energy flow in different cases. Initially, the power generated by PV and WT systems is injected directly into the power grid. The management of the ESS is based on the following operation criteria. VRFB operation: If the total power generated by the PV and WT systems is greater than the demand, charging power is allocated for the batteries P_t^{ESS}. On the contrary, if the demand is lower, the energy is discharged to the electrical grid, as long as the SOC of the battery is within the established maximum and minimum values. In addition, the contribution of EVs P_t^{EV} is considered if they are programmed to operate at specific times in Fig. 17.9.

Table 17.2 indicates the power profiles for the case study. The cost of electricity is variable throughout the day expressed in €/kWh. Also, the maximum demand of 70 kW is considered in peak hours with a profile assigned in P_t^L. The power of the PV and WT systems adds up to 49 kW in installed power $\left(P_t^{\mathrm{PV}} + P_t^{\mathrm{WT}}\right)$.

Table 17.2 Energy dispatch results with V2G&G2V support in MG.

Hour	€/kWh	P_t^L (70kW)	P_t^{EV} (50kW)	P_t^{PV} (34kW)	P_t^{WT} (15kW)	P_t^{ESS} (30kW)	P_t^{Gd}
00:00–01:00	0.07059	21.18	0.00	0.00	0.00	6.18	15.00
01:00–02:00	0.06483	19.45	50.00	0.00	0.00	24.45	45.00
02:00–03:00	0.06247	18.74	50.00	0.00	3.00	20.74	45.00
03:00–04:00	0.06177	18.53	45.00	0.00	0.00	18.53	45.00
04:00–05:00	0.06189	18.57	0.00	0.00	0.00	3.57	15.00
05:00–06:00	0.06188	18.56	0.00	0.00	5.00	3.56	10.00
06:00–07:00	0.06561	19.68	0.00	5.00	0.00	4.68	10.00
07:00–08:00	0.07759	23.28	0.00	7.00	0.00	6.28	10.00
08:00–09:00	0.09397	28.19	0.00	15.00	0.00	8.19	5.00
09:00–10:00	0.07294	21.88	0.00	26.00	0.00	−4.12	0.00
10:00–11:00	0.11906	35.72	0.00	30.00	10.00	−4.28	0.00
11:00–12:00	0.11929	35.79	0.00	20.00	12.00	3.79	0.00
12:00–13:00	0.11989	35.97	0.00	34.00	5.00	−3.03	0.00
13:00–14:00	0.11932	35.80	0.00	33.00	10.00	−7.20	0.00
14:00–15:00	0.069	20.70	0.00	25.00	0.00	−4.30	0.00
15:00–16:00	0.06975	20.93	0.00	15.00	0.00	2.93	3.00
16:00–17:00	0.06981	20.94	0.00	10.00	15.00	−4.06	0.00
17:00–18:00	0.07046	21.14	0.00	5.00	10.00	3.14	3.00
18:00–19:00	0.12208	36.62	0.00	0.00	0.00	26.62	10.00
19:00–20:00	0.14362	43.09	−50.00	0.00	5.00	−11.91	0.00
20:00–21:00	0.15328	45.98	−50.00	0.00	0.00	−4.02	0.00
21:00–22:00	0.23063	69.19	−45.00	0.00	0.00	9.19	15.00
22:00–23:00	0.15663	46.99	0.00	0.00	0.00	26.99	20.00
23:00–24:00	0.10352	31.06	0.00	0.00	0.00	11.06	20.00

Consequently, the contribution generated by VRFB is a function of the storage capacity of 100 kWh $\left(\pm P_t^{\mathrm{ESS}}\right)$, which is why the contribution from the grid $\left(P_t^{\mathrm{Gd}}\right)$ is reduced during the hours of highest consumption. On the other hand, this configuration allows the demand curve to be leveled, as presented in Fig. 17.10B.

Cost of energy dispatch

The cost of variable energy in some countries is subject to daily and hourly fluctuations. The daily market hourly price is assigned by the operator. These variations reflect the dynamics of supply and demand, as well as the integration of renewable sources and storage systems in the electrical grid.

Table 17.3 presents a summary of electricity costs with the proposed systems. The electricity costs of the assigned power profiles referring to the previous table are considered. These values integrate the support of the EV in the grid independent of the MG and considering the support combined with the MG. The

Table 17.3 Cost of energy dispatch with V2G&G2V support in MG.

Hour	€/kWh	C_t^{consumer}	$C_t^{G2V} \pm C_t^{V2G}$	C_t^{consumer} EV support	C_t^{consumer} EV and MG support
00:00–01:00	0.07059	1.49	0.00	1.49	1.06
01:00–02:00	0.06483	1.26	3.24	4.50	2.92
02:00–03:00	0.06247	1.17	3.12	4.29	2.81
03:00–04:00	0.06177	1.14	2.78	3.92	2.78
04:00–05:00	0.06189	1.15	0.00	1.15	0.93
05:00–06:00	0.06188	1.15	0.00	1.15	0.62
06:00–07:00	0.06561	1.29	0.00	1.29	0.66
07:00–08:00	0.07759	1.81	0.00	1.81	0.78
08:00–09:00	0.09397	2.65	0.00	2.65	0.47
09:00–10:00	0.07294	1.60	0.00	1.60	0.00
10:00–11:00	0.11906	4.25	0.00	4.25	0.00
11:00–12:00	0.11929	4.27	0.00	4.27	0.00
12:00–13:00	0.11989	4.31	0.00	4.31	0.00
13:00–14:00	0.11932	4.27	0.00	4.27	0.00
14:00–15:00	0.069	1.43	0.00	1.43	0.00
15:00–16:00	0.06975	1.46	0.00	1.46	0.21
16:00–17:00	0.06981	1.46	0.00	1.46	0.00
17:00–18:00	0.07046	1.49	0.00	1.49	0.21
18:00–19:00	0.12208	4.47	0.00	4.47	1.22
19:00–20:00	0.14362	6.19	−3.24	−0.99	0.00
20:00–21:00	0.15328	7.05	−3.12	−0.62	0.00
21:00–22:00	0.23063	15.96	−2.78	5.58	3.46
22:00–23:00	0.15663	7.36	0.00	7.36	3.13
23:00–24:00	0.10352	3.21	0.00	3.21	2.07
TOTAL:		81.90	0.00	65.82	23.32

analysis shows savings of around 19.63% and 71.53% of the total consumption value. Consequently, it allows for a positive change to improve the integration of EV and MG systems with grid support.

Conclusions

ESS are essential to achieve a transition toward a cleaner and more sustainable energy matrix. Their flexibility and ability to adapt to changing needs make them essential in modern energy management. This chapter addressed the perspective of optimal energy management based on the use of ESS and EVs.

It has been observed that the incentive on the government side allows for an adoption of ESS and EV. The proposal to generate a more sustainable energy matrix has promoted research and implementation of innovative solutions in the transportation sector with the use of VG2 and G2V.

The methodology for energy management based on ESS and EV can be applied under different criteria and scenarios. Therefore, the application of optimization algorithms can determine when to charge or discharge the battery to maximize efficiency and minimize losses. Likewise, factors such as electrical demand, the availability of renewable energy, and operating schedules must be considered to support the EV to the grid.

The cost–benefit analysis allows for determining financial viability. The results show that optimal V2G management during scheduled hours can reduce the impact of peak hours and obtain economic benefits of up to 19.63%. Likewise, with the support of renewable energies and EV, it can be up to 71.53%. It also offers notable advantages in leveling demand.

An approach to the integration of ESS and EVs into power systems in a progressive manner has been presented. For large-scale applications, it contemplates important challenges that have been limited in this chapter, such as the analysis of V2G charging cycles, infrastructure of EVCSs, and efficiency of bidirectional converters, which can be considered for future lines of development investigation.

Acknowledgments

The authors wish to thank to the Spanish Ministry of Science and Innovation and European Union NextGenerationEU/PRTR, under the research Project "Desarrollo de herramientas computacionales para microrredes multienergía con vectores de electricidad, hidrógeno y gas" (TED2021–129631B-C31).

The authors thank Universidad de Cuenca for easing access to the facilities of the Microgrid Laboratory of the Centro Científico Tecnológico y de Investigación Balzay (CCTI-B), for allowing the use of its equipment, and for authorizing its staff the provision of technical support necessary to carry out the experiments described in this article.

The icons used in this document were developed by Freepik, monkik, Smashicons and Pixel perfect, from http://www.flaticon.com.

References

Adhikary, S., Biswas, P. K., Sain, C., Thanikanti, S. B., & Nwulu, N. I. (2023). Bidirectional converter based on G2V and V2G operation with time of usage-based tariff analysis and monitoring of charging parameters using IoT. *Energy Reports*, *9*. Available from https://doi.org/10.1016/j.egyr.2023.04.358, http://www.journals.elsevier.com/energy-reports/.

Alghawi, M., & Mounsef, J. (2024). Overview of vehicle-to-vehicle energy sharing infrastructure. *IEEE Access*, *12*. Available from https://doi.org/10.1109/ACCESS.2024.3388088, http://ieeexplore.ieee.org/xpl/RecentIssue.jsp?punumber = 6287639.

Almatar, K. M. (2023). Increasing electric vehicles infrastructure in urban areas for efficiently employing renewable energy. *Environment, Development and Sustainability*. Available from https://doi.org/10.1007/s10668-023-03723-x, https://www.springer.com/journal/10668.

Alnasser, A., Sun, H., & Jiang, J. (2020). Recommendation-based trust model for vehicle-to-everything (V2X). *IEEE Internet of Things Journal*, *7*(1). Available from https://doi.org/10.1109/JIOT.2019.2950083, http://ieeexplore.ieee.org/servlet/opac?punumber = 6488907.

Amertet, S., & Gebresenbet, G. (2024). Hybridized renewable energy for smart vehicle-to-grid (V2G) systems. *Energy Exploration and Exploitation*, *42*(1). Available from https://doi.org/10.1177/01445987231209770, https://journals.sagepub.com/home/EEA.

Aziz, M., Oda, T., & Kashiwagi, T. (2015). Extended utilization of electric vehicles and their re-used batteries to support the building energy management system. *Energy Procedia*, *75*, 1938–1943. Available from https://doi.org/10.1016/j.egypro.2015.07.226, http://www.sciencedirect.com/science/journal/18766102.

Benavides, D., Arévalo, P., Villa-Ávila, E., Aguado, J. A., & Jurado, F. (2024). Predictive power fluctuation mitigation in grid-connected PV systems with rapid response to EV charging stations. *Journal of Energy Storage*, *86*. Available from https://doi.org/10.1016/j.est.2024.111230, https://www.sciencedirect.com/science/journal/2352152X.

Bhoir, S., Caliandro, P., & Brivio, C. (2021). Impact of V2G service provision on battery life. *Journal of Energy Storage*, *44*. Available from https://doi.org/10.1016/j.est.2021.103178, http://www.journals.elsevier.com/journal-of-energy-storage/.

Borge-Diez, D., Icaza, D., Açıkkalp, E., & Amaris, H. (2021). Combined vehicle to building (V2B) and vehicle to home (V2H) strategy to increase electric vehicle market share. *Energy*, *237*. Available from https://doi.org/10.1016/j.energy.2021.121608, https://www.journals.elsevier.com/energy.

Boudina, R., Wang, J., Benbouzid, M., Khoucha, F., & Boudour, M. (2020). Impact evaluation of large scale integration of electric vehicles on power grid. *Frontiers in Energy*, *14*(2). Available from https://doi.org/10.1007/s11708-018-0550-6, https://www.springer.com/journal/11708.

Cao, J., Wu, H., & Zou, Y. (2024). Advancing sustainable development: the impact of energy transition on accelerating technological progress. *Economic Change and Restructuring*, *57* (2). Available from https://doi.org/10.1007/s10644-024-09664-3, https://www.springer.com/journal/10644.

Consilium (2024). https://www.consilium.europa.eu/es/policies/clean-and-sustainable-mobility/.

Corchado, J.M. (n.d.). Engineering IoT-based open MAS for large-scale V2G/G2V. *Systems*, *11*. Available from https://doi.org/10.3390/SYSTEMS11030157.

Dhakal, T., & Min, K. S. (2021). Macro study of global electric vehicle expansion. *Foresight and STI Governance*, *15*(1). Available from https://doi.org/10.17323/2500-2597.2021.1.67.73, https://foresight-journal.hse.ru/en/archive.html.

Dorahaki, S., Dashti, R., & Shaker, H. R. (2020). Optimal energy management in the smart microgrid considering the electrical energy storage system and the demand-side energy efficiency program. *Journal of Energy Storage, 28*. Available from https://doi.org/10.1016/j.est.2020.101229, http://www.journals.elsevier.com/journal-of-energy-storage/.

Eldjalil, C. D. A., Said, D., Khoukhi, L., & Mouftah, H. T. (2017). A novel pricing policy for G2V and V2G services. *14th IEEE annual consumer communications and networking conference, CCNC 2017*, 634–635. Available from https://doi.org/10.1109/CCNC.2017.7983201.

Espinoza, J. L., Gonzalez, L. G., & Sempertegui, R. (2017). Micro grid laboratory as a tool for research on non-conventional energy sources in Ecuador. *IEEE international autumn meeting on power, electronics and computing, ROPEC 2017, 2018*, 1–7. Available from https://doi.org/10.1109/ROPEC.2017.8261615.

Gu, F. C., & Chen, H. C. (2021). Modelling and control of vanadium redox flow battery for smoothing wind power fluctuation. *IET Renewable Power Generation, 15*(15). Available from https://doi.org/10.1049/rpg2.12244, https://ietresearch.onlinelibrary.wiley.com/journal/17521424.

Gupta, R. S., Tyagi, A., & Anand, S. (2021). Optimal allocation of electric vehicles charging infrastructure, policies and future trends. *Journal of Energy Storage, 43*. Available from https://doi.org/10.1016/j.est.2021.103291, http://www.journals.elsevier.com/journal-of-energy-storage/.

Haghani, M., Sprei, F., Kazemzadeh, K., Shahhoseini, Z., & Aghaei, J. (2023). Trends in electric vehicles research. *Transportation Research Part D: Transport and Environment, 123*. Available from https://doi.org/10.1016/j.trd.2023.103881, http://www.elsevier.com/inca/publications/store/3/1/1/5/3/.

Han, J., Zhou, X., Lu, S., & Zhao, P. (2020). A three-phase bidirectional grid-connected AC/DC converter for V2G applications. *Journal of Control Science and Engineering, 2020*. Available from https://doi.org/10.1155/2020/8844073, http://www.hindawi.com/journals/jcse.

Haroon, F., Aamir, M., Waqar, A., Mian Qaisar, S., Ali, S. U., & Almaktoom, A. T. (2023). a composite exponential reaching law based SMC with rotating sliding surface selection mechanism for two level three phase VSI in vehicle to load applications. *Energies, 16*(1). Available from https://doi.org/10.3390/en16010346, http://www.mdpi.com/journal/energies/.

He, Z., Khazaei, J., & Freihaut, J. D. (2022). Optimal integration of vehicle to building (V2B) and building to vehicle (B2V) technologies for commercial buildings. *Sustainable Energy, Grids and Networks, 32*. Available from https://doi.org/10.1016/j.segan.2022.100921, http://www.journals.elsevier.com/sustainable-energy-grids-and-networks/.

Hu, J., Ye, C., Ding, Y., Tang, J., & Liu, S. (2022). A distributed MPC to exploit reactive power V2G for real-time voltage regulation in distribution networks. *IEEE Transactions on Smart Grid, 13*(1). Available from https://doi.org/10.1109/TSG.2021.3109453, https://ieeexplore.ieee.org/servlet/opac?punumber = 5165411.

Huda, M., Aziz, M., & Tokimatsu, K. (2019). The future of electric vehicles to grid integration in Indonesia. *Energy Procedia, 158*, 4592–4597. Available from https://doi.org/10.1016/j.egypro.2019.01.749, http://www.sciencedirect.com/science/journal/18766102.

Huda, M. S. (2024). Renewable energy diplomacy and transitions: An environmental peacebuilding approach. *Environmental Innovation and Societal Transitions, 50*. Available from https://doi.org/10.1016/j.eist.2024.100815, https://www.sciencedirect.com/science/journal/22104224.

Hull, C., Giliomee, J. H., Visser, M., & Booysen, M. J. (2024). Electric vehicle adoption intention among paratransit owners and drivers in South Africa. *Transport Policy, 146*. Available from https://doi.org/10.1016/j.tranpol.2023.11.015, https://www.sciencedirect.com/science/journal/0967070X.

İnci, M., Büyük, M., & Özbek, N. S. (2022). Sliding mode control for fuel cell supported battery charger in vehicle-to-vehicle interaction. *Fuel Cells, 22*(5). Available from https://doi.org/10.1002/fuce.202200105, http://onlinelibrary.wiley.com/journal/10.1002/(ISSN)1615-6854.

Islam, S., Iqbal, A., Marzband, M., Khan, I., & Al-Wahedi, A. M. A. B. (2022). State-of-the-art vehicle-to-everything mode of operation of electric vehicles and its future perspectives. *Renewable and Sustainable Energy Reviews, 166*. Available from https://doi.org/10.1016/j.rser.2022.112574, https://www.journals.elsevier.com/renewable-and-sustainable-energy-reviews.

Jain, P., & Jain, T. (2018). Application of V2G and G2V coordination of aggregated electric vehicle resource in load levelling. *International Journal of Emerging Electric Power Systems, 19*(2). Available from https://doi.org/10.1515/ijeeps-2017-0170, http://www.degruyter.com/view/j/ijeeps.

Jiao, Z., Ran, L., Zhang, Y., & Ren, Y. (2021). Robust vehicle-to-grid power dispatching operations amid sociotechnical complexities. *Applied Energy, 281*. Available from https://doi.org/10.1016/j.apenergy.2020.115912, https://www.journals.elsevier.com/applied-energy.

Li, M., & Hikihara, T. (2008). A coupled dynamical model of redox flow battery based on chemical reaction, fluid flow, and electrical circuit. *IEICE Transactions on Fundamentals of Electronics, Communications and Computer Sciences, E91-A*(7). Available from https://doi.org/10.1093/ietfec/e91-a.7.1741, http://www.jstage.jst.go.jp/article/transfun/E91.A/7/1741/_pdf.

Li, S., Gu, C., Zeng, X., Zhao, P., Pei, X., & Cheng, S. (2021). Vehicle-to-grid management for multi-time scale grid power balancing. *Energy, 234*. Available from https://doi.org/10.1016/j.energy.2021.121201, https://www.journals.elsevier.com/energy.

Li, X., & Wang, S. (2021). Energy management and operational control methods for grid battery energy storage systems. *CSEE Journal of Power and Energy Systems, 7*(5). Available from https://doi.org/10.17775/CSEEJPES.2019.00160, https://ieeexplore.ieee.org/xpl/RecentIssue.jsp?punumber = 7054730.

Liao, J. T., Huang, H. W., Yang, H. T., & Li, D. (2021). Decentralized V2G/G2V scheduling of ev charging stations by considering the conversion efficiency of bidirectional chargers. *Energies, 14*(4). Available from https://doi.org/10.3390/en14040962, https://www.mdpi.com/1996-1073/14/4/962/pdf.

Licea, M. A. R. (2020). Fault tolerant boost converter with multiple serial inputs and output voltage regulation for vehicle-to-aid services. *Energies, 13*(7). Available from https://doi.org/10.3390/en13071694, https://www.mdpi.com/1996-1073/13/7/1694.

Liu, S., Ni, Q., Cao, Y., Cui, J., Tian, D., & Zhuang, Y. (2023). A reservation-based vehicle-to-vehicle charging service under constraint of parking duration. *IEEE Systems Journal, 17*(1). Available from https://doi.org/10.1109/JSYST.2022.3145155, https://ieeexplore.ieee.org/xpl/mostRecentIssue.jsp?punumber = 4267003.

Liu, X., Zhao, T., Deng, H., Wang, P., Liu, J., & Blaabjerg, F. (2023). Microgrid energy management with energy storage systems: A review. *CSEE Journal of Power and Energy Systems, 9*(2). Available from https://ieeexplore.ieee.org/xpl/RecentIssue.jsp?punumber = 7054730, 10.17775/CSEEJPES.2022.04290.

Marotta, A., Lodi, C., Julea, A., & Gómez Vilchez, J. J. (2023). European governments' electromobility plans: An assessment with a focus on infrastructure targets and vehicle estimates until 2030. *Energy Efficiency*, *16*(8). Available from https://doi.org/10.1007/s12053-023-10163-z, https://www.springer.com/journal/12053.

Mastoi, M. S., Zhuang, S., Munir, H. M., Haris, M., Hassan, M., Usman, M., Bukhari, S. S. H., & Ro, J. S. (2022). An in-depth analysis of electric vehicle charging station infrastructure, policy implications, and future trends. *Energy Reports*, *8*. Available from https://doi.org/10.1016/j.egyr.2022.09.011, http://www.journals.elsevier.com/energy-reports/.

Mittal, G., Garg, A., & Pareek, K. (2024). A review of the technologies, challenges and policies implications of electric vehicles and their future development in India. *Energy Storage*, *6*(1). Available from https://doi.org/10.1002/est2.562, https://onlinelibrary.wiley.com/journal/25784862.

Mohiti, M., Mazidi, M., Rezaei, N., & Khooban, M. H. (2021). Role of vanadium redox flow batteries in the energy management system of isolated microgrids. *Journal of Energy Storage*, *40*. Available from https://doi.org/10.1016/j.est.2021.102673, http://www.journals.elsevier.com/journal-of-energy-storage/.

Nagel, R. G., Pires, V. F., Silveira, J. L., Cordeiro, A., & Foito, D. (2023). Financial analysis of household photovoltaic self-consumption in the context of the vehicle-to-home (V2H) in Portugal. *Energies*, *16*(3). Available from https://doi.org/10.3390/en16031218, http://www.mdpi.com/journal/energies/.

Qiu, X., Nguyen, T. A., Guggenberger, J. D., Crow, M. L., & Elmore, A. C. (2014). A field validated model of a vanadium redox flow battery for microgrids. *IEEE Transactions on Smart Grid*, *5*(4). Available from https://doi.org/10.1109/TSG.2014.2310212.

Rajamand, S. (2020). Vehicle-to-grid and vehicle-to-load strategies and demand response program with bender decomposition approach in electrical vehicle-based microgrid for profit profile improvement. *Journal of Energy Storage*, *32*101935. Available from https://doi.org/10.1016/j.est.2020.101935.

Rodrigues, R., Albuquerque, V., Ferreira, J. C., Dias, M. S., & Martins, A. L. (2021). Mining electric vehicle adoption of users. *World Electric Vehicle Journal*, *12*(4). Available from https://doi.org/10.3390/wevj12040233, https://www.mdpi.com/2032-6653/12/4/233/pdf.

Saxena, S., Farag, H. E. Z., Hilaire, L. S., & Brookson, A. (2023). A techno-social approach to unlocking vehicle to everything (V2X) integration: A real-world demonstration. *IEEE Access*, *11*. Available from https://doi.org/10.1109/ACCESS.2023.3244562, http://ieeexplore.ieee.org/xpl/RecentIssue.jsp?punumber = 6287639.

Schwarz, M., Auzepy, Q., & Knoeri, C. (2020). Can electricity pricing leverage electric vehicles and battery storage to integrate high shares of solar photovoltaics? *Applied Energy*, *277*. Available from https://doi.org/10.1016/j.apenergy.2020.115548, https://www.journals.elsevier.com/applied-energy.

Shipman, R., Waldron, J., Naylor, S., Pinchin, J., Rodrigues, L., & Gillott, M. (2020). Where will you park? Predicting vehicle locations for vehicle-to-grid. *Energies*, *13*(8). Available from https://doi.org/10.3390/en13081933, https://www.mdpi.com/1996-1073/13/8/1933.

Singh, S., Saket, R. K., & Khan, B. (2023). A comprehensive state-of-the-art review on reliability assessment and charging methodologies of grid-integrated electric vehicles. *IET Electrical Systems in Transportation*, *13*(1). Available from https://doi.org/10.1049/els2.12073, https://ietresearch.onlinelibrary.wiley.com/journal/20429746.

Tang, Y. (2023). Optimal control study of home energy management with cooperative dispatch of electric vehicles and energy storage devices. *Energy Reports*, *9*. Available from https://doi.org/10.1016/j.egyr.2023.05.098, http://www.journals.elsevier.com/energy-reports/.

Turker, B., Arroyo Klein, S., Hammer, E. M., Lenz, B., & Komsiyska, L. (2013). Modeling a vanadium redox flow battery system for large scale applications. *Energy Conversion and Management*, *66*. Available from https://doi.org/10.1016/j.enconman.2012.09.009.

Wu, Y., Aziz, S. M., & Haque, M. H. (2024). Vehicle-to-home operation and multi-location charging of electric vehicles for energy cost optimisation of households with photovoltaic system and battery energy storage. *Renewable Energy*, *221*119729. Available from https://doi.org/10.1016/j.renene.2023.119729.

Yang, H., & Fulton, L. (2023). Investigating investment plans for expanding battery and electric vehicle production in Europe. *World Electric Vehicle Journal*, *14*(12). Available from https://doi.org/10.3390/wevj14120347, https://www.mdpi.com/journal/wevj.

Yassine, A., & Hossain, M. S. (2023). Match maximization of vehicle-to-vehicle energy charging with double-sided auction. *IEEE Transactions on Intelligent Transportation Systems*, *24*(11). Available from https://doi.org/10.1109/TITS.2023.3265870, http://ieeexplore.ieee.org/xpl/RecentIssue.jsp?punumber = 6979.

Zafar, B., & Slama, S. A. B. (2022). PV-EV integrated home energy management using vehicle-to-home (V2H) technology and household occupant behaviors. *Energy Strategy Reviews*, *44*. Available from https://doi.org/10.1016/j.esr.2022.101001, http://www.journals.elsevier.com/energy-strategy-reviews/.

Zhang, Q. H., Shi, D. M., Cheng, G. Q., Wang, Z., & Sun, J. Q. (2014). Research on G2V AND V2G trade modes and information system development. *Applied Mechanics and Materials*, *556-562*, 5848–5851. Available from https://doi.org/10.4028/http://www.scientific.net/AMM.556-562.5848.

Index

Note: Page numbers followed by "*f*" and "*t*" refer to figures and tables, respectively.

C

I

J

9780443298714